Carbon-based Nanomaterials for Green Applications

IEEE Press
445 Hoes Lane
Piscataway, NJ 08854

Carbon-based Nanomaterials for Green Applications

Edited by

Upendra Kumar
Indian Institute of Information Technology Allahabad
Prayagraj
Uttar Pradesh
India

Piyush Kumar Sonkar
Department of Chemistry, MMV, Banaras Hindu University
Varanasi, Uttar Pradesh
India

Suman Lata Tripathi
School of Electronics and Electrical Engineering
Lovely Professional University
Phagwara, Punjab
India

IEEE*Press*

WILEY

Published by John Wiley & Sons, Inc., Hoboken, New Jersey.
Published simultaneously in Canada.

Library of Congress Cataloging-in-Publication Data Applied for:

Hardback ISBN: 9781394243396

Cover Design: Wiley
Cover Image: © Baac3nes/Getty Images

Set in 9.5/12.5pt STIXTwoText by Straive, Pondicherry, India

Dedicated
to Our
Authors, Collaborators, and
the Family Members

Contents

About the Editors

Dr. Upendra Kumar

Dr. Upendra Kumar is an assistant professor at the Department of Applied Sciences, Indian Institute of Information Technology Allahabad, Prayagraj. He received his bachelor and masters degrees in physics from Banaras Hindu University, Varanasi, and his PhD from IIT (BHU), Varanasi. Before joining IIIT Allahabad, he also worked as an assistant professor in Banasthali Vidtapith, Rajasthan, and gained teaching, research, and administrative experiences. He has two major research projects from the SERB, Govt. of India, and IIIT Allahabad. He has successfully supervised two PhD students, and four are still working in his mentorship. He has also supervised several MSc and MTech projects. He has published over 100+ research articles, including highly reputed international and national journals, conference proceedings, books, book chapters, and the ICDD database. He has also received a prestigious membership in the National Academy of Sciences India (NASI) and is a senior-grade member of IEEE Professional. Currently, he is also working as a secretary for the IEEE Photonics Society UP chapter. He is an academic editor for the journal *Advances in Condensed Matter Physics* (Wiley) and a member of many scientific societies such as IAPT, IPA, IEEE Young Professional, etc. He also served as the referee for international journals of Elsevier, Springer, Wiley, Taylor & Francis, and ACS, and a topical advisory panel member of many MDPI journals. His extensive research areas are electroceramics, solid oxide fuel cells, piezoelectric ceramics, microwave dielectric, machine learning, computational materials science, energy generation, and storage materials.

Personal Homepage: https://sites.google.com/view/drupendrakumar/home
Researchgate: https://www.researchgate.net/profile/Upendra-Kumar-7
Google Scholar: https://scholar.google.com/citations?user=rOYIerMAAAAJ&hl=en

Dr. Piyush Kumar Sonkar

Dr. Piyush Kumar Sonkar has received his BSc (Chemistry), MSc (Chemistry), and PhD (Chemistry) degrees from Banaras Hindu University, India. He is an assistant professor in the Department of Chemistry, MMV, Banaras Hindu University, Varanasi, India. He has teaching and research experience of 11 and 6 years, respectively. His research interests include nanomaterials, nanocomposites, metal complexes, fuel cells, electrochemical devices, supercapacitors, biosensors, chemical sensors, and new materials. He has published 47 international and national research papers in various reputed peer-reviewed journals. He has published six book chapters. He is the editor of two international books on nanomaterials. He has completed two research projects at UGC, New Delhi, and IoE BHU. Currently, one research project is under his supervision based on tribal medicines. He has presented his research work in various international/national seminars/workshops and conferences. Currently, two PhD students are working under his supervision.

Google Scholar Link:
https://scholar.google.co.in/citations?user=7WQwcscAAAAJ&hl=en
Research Gate Link: https://www.researchgate.net/profile/Piyush_Sonkar
Institute Profile Link: https://bhu.ac.in/Site/FacultyProfile/1_225?FA000575

Dr. Suman Lata Tripathi

Dr Suman Lata Tripathi is working as a professor at Lovely Professional University and has more than 22 years of experience in academics and research. She has completed her PhD in microelectronics and VLSI Design from MNNIT, Allahabad. She did her MTech in electronics engineering from UP Technical University, Lucknow, and her BTech in electrical engineering from Purvanchal University, Jaunpur. She completed her remote post-doc from Nottingham Trent University, London, UK, in 2022–2023. She has published more than 152 research papers in refereed Springer, Elsevier, IEEE, Wiley, and IOP science journals, conference proceedings, and e-books. She has also published 20 Indian patents and 4 copyrights. She has guided six PhD scholars, and three are in the submission stage. She has been listed in the world's top 2% of scientists published by Stanford University for 2024. She has organized workshops, summer internships, and expert lectures for students. She has worked as a session chair, conference steering committee member, editorial board member, and peer reviewer in international/national IEEE, Springer, Wiley, etc. journals and conferences. She is a visiting professor at Telkom University, Indonesia. She received the "Research Excellence Award" in 2019 and the "Research Appreciation Award" in 2020, 2021, and 2023 at Lovely Professional University, India. She received the IGEN Women's for Green Technology

"Women's Achievers Award" on International Women's Day, 8 March 2023. She received the best paper at IEEE ICICS-2018. She has also received a funded project from SERB DST under the TARE scheme in microelectronics devices. She has edited and authored more than 30 books in different areas of electronics and electrical engineering. She is associated with editing work with top publishers like Elsevier, CRC Taylor & Francis, Wiley-IEEE, SP Wiley, Nova Science, and Apple academic press. She is also working as a book series editor for the title "Smart Engineering Systems" CRC Press; "Engineering System Design for Sustainable Developments," "Decentralized Systems," and "Next Generation Internet" Wiley-Scrivener; and conference series editor for "Conference Proceedings Series on Intelligent Systems for Engineering designs" CRC Press Taylor & Francis. She is serving as academic editor of "*Journal of Electrical and Computer Engineering*" (Scopus/WoS, Q2), "*International Journal of Reconfigurable Computing*" (Scopus, Q3), "*Active and Passive Electronic Component*" (Scopus, Q4) Hindawi-Wiley and special issue guest editor for "*Advances in Nanomaterials and Nanoscale Semiconductor Applications*" Material, MDPI Journal (SCI IF=3.74, Q2). She is associated as a senior member of IEEE, a Fellow IETE, and a Life member of ISC, and she is continuously involved in different professional activities and academic works. Her area of expertise includes microelectronics device modeling and characterization, low-power VLSI circuit design, VLSI design of testing, advanced FET design for IoT, embedded system design, reconfigurable architecture with FPGAs and biomedical applications, etc.

Google Scholar profile link:
https://scholar.google.com/citations?hl=en&user=SdbROCAAAAAJ&view_op=list_works
Research gate profile link: https://www.researchgate.net/profile/Suman-Tripathi

Scopus profile link:
https://www.scopus.com/authid/detail.uri?authorId=57564799800
Institute Profile: https://vidwan.inflibnet.ac.in//profile/282457/

List of Contributors

R. Adharsh
Department of Electrical and Electronics Engineering
Sri Ramakrishna Engineering College
Coimbatore, Tamil Nadu
India

Satadal Adhikary
Post Graduate Department of Zoology
A. B. N. Seal College
Cooch Behar, West Bengal
India

Seraj Ahmad
CMP Degree College
University of Allahabad
Prayagraj, Uttar Pradesh
India

P. K. Ahluwalia
Department of Physics
Himachal Pradesh University
Shimla, Himachal Pradesh
India

Mohammad Imran Ahmad
Department of Chemistry
Integral University
Lucknow, Uttar Pradesh
India

Manoj Singh Adhikari
School of Electronics and Electrical Engineering
Lovely Professional University
Phagwara, Punjab
India

Ruchika Agarwal
Department of Animal Science
Kazi Nazrul University
Asansol, West Bengal
India

S. Allirani
Department of Electrical and Electronics Engineering
Sri Ramakrishna Engineering College
Coimbatore, Tamil Nadu
India

Akram Ali
CMP Degree College
University of Allahabad
Prayagraj, Uttar Pradesh
India

Himanshu Arora
Department of Chemistry, Faculty of Sciences
University of Allahabad
Prayagraj, Uttar Pradesh
India

Krishan Arora
School of Electronics and Electrical Engineering
Lovely Professional University
Phagwara, Punjab
India

Parvez Ahmed Alvi
Department of Physical Sciences
Banasthali Vidyapith
Banasthali, Rajasthan
India

Irtiqa Amin
Department of Computer Application
Doctoral Scholar School of Computer Application Lovely Professional University
Phagwara, Punjab
India

Quraazah Akeemu Amin
Division of Food Science and Technology
SKUAST Kashmir
Phagwara, Jammu and Kashmir
India

J. Anuradh
Nims Institute of Allied Medical Science and Technology
Nims University Rajasthan
Jaipur, Rajasthan
India

Suchandra Bhattacharya
Department of Chemistry
A. B. N. Seal College
Cooch Behar, West Bengal
India

Shikha Chander
Department of Chemistry
St. Francis College for Women
Hyderabad, Telangana
India

Manju Choudhary
Department of Bioscience and Biotechnology
Banasthali Vidyapeeth
NewaniTonk, Rajasthan
India

Prachi Diwakar
Department of Physical Sciences
Banasthali Vidyapith
Banasthali, Rajasthan
India

Gaganpreet
Department of Physics
Post Graduate Government College for Girls, Sector 11
Chandigarh
India

Shivanshu Garg
Department of Biochemistry, College of Basic Sciences & Humanities
G. B. Pant University of Agriculture and Technology
Pantnagar, Uttarakhand
India

Abhratanu Ganguly
Department of Animal Science
Kazi Nazrul University
Asansol, West Bengal
India

Vellaichamy Ganesan
Department of Chemistry
Institute of Science, Banaras
Hindu University
Varanasi, Uttar Pradesh
India

Sohini Goswami
Department of Animal Science
Kazi Nazrul University
Asansol, West Bengal
India

Mandakini Gupta
Department of Chemistry
Sunbeam Women's College Varuna
Varanasi, Uttar Pradesh
India

Kulsum Hashmi
Department of Chemistry
Isabella Thoburn College
Lucknow, Uttar Pradesh
India

G. Ilakkiya
Department of Electrical and
Electronics Engineering
Sri Ramakrishna Engineering College
Coimbatore, Tamil Nadu
India

Vikas Jangra
Department of Chemistry
Banaras Hindu University
Varanasi, Uttar Pradesh
India

Department of Chemistry
MMV, Banaras Hindu University
Varanasi, Uttar Pradesh
India

Seema Joshi
Department of Chemistry
Isabella Thoburn College
Lucknow, Uttar Pradesh
India

G. Kanthimathi
Department of Chemistry
Ramco Institute of Technology
Rajapalayam, Virudhunagar
Tamil Nadu
India

Pooja Kapoor
School of Basic and Applied Sciences
Maharaja Agrasen University
Baddi, Himachal Pradesh
India

Harpreet Kaur
DCA CGC Landran
Chandigarh, Punjab
India

Harpreet Kaur
Department of Chemistry
MMV, Banaras Hindu University
Varanasi, Uttar Pradesh
India

M. Karthik
Department of Electrical and
Electronics Engineering
SRM Madurai College for Engineering
and Technology
Pottapalayam, Tamil Nadu
India

Tahmeena Khan
Department of Chemistry
Integral University
Lucknow, Uttar Pradesh
India

Kahkashan Khatoon
CMP Degree College
University of Allahabad
Prayagraj, Uttar Pradesh
India

Manish Kumar
Department of Chemistry, L.N.T. College
B.R.A. Bihar University
Muzaffarpur, Bihar
India

Sunil Kumar
Department of Chemistry, L.N.T. College
B.R.A. Bihar University
Muzaffarpur, Bihar
India

Narvadeswar Kumar
Department of Chemistry
MMV, Banaras Hindu University
Varanasi, Uttar Pradesh
India

Manoj Kumar
CMP Degree College
University of Allahabad
Prayagraj, Uttar Pradesh
India

Celestine Lwendi
School of Creative Technologies
University of Bolton
Bolton
United Kingdom

Meenu Mangal
Department of Chemistry
Poddar International College
Jaipur, Rajasthan
India

Tola Jebssa Masho
Department of Chemistry, College of Natural and Computational Sciences
Wollega University
Nekemte
Ethiopia

Nidhi Mishra
Department of Applied Sciences
Indian Institute of Information Technology
Allahabad, Uttar Pradesh
India

Arumugam Murugan
Department of Chemistry
North Eastern Regional Institute of Science and Technology
Nirjuli, Arunachal Pradesh
India

Pooja Nain
Department of Soil Science, College of Agriculture
G. B. Pant University of Agriculture and Technology
Pantnagar, Uttarakhand
India

Sayantani Nanda
Department of Animal Science
Kazi Nazrul University
Asansol, West Bengal
India

Chandra Mohan Singh Negi
Department of Physical Sciences
Banasthali Vidyapith
Banasthali, Rajasthan
India

Raju Patel
School of Electronics Engineering
Vellore Institute of Technology
Chennai, Tamil Nadu
India

Lal Bahadur Prasad
Department of Chemistry
Banaras Hindu University
Varanasi, Uttar Pradesh
India

Y. Pathania
Department of Physics
DAV Post Graduate College
Sector 10, Chandigarh
India

Jai Prakash
Department of Chemistry
S. P. Jain College (Veer Kunwar Singh University, Ara)
Sasaram, Bihar
India

Himanshu Punetha
Department of Biochemistry
College of Basic Sciences & Humanities
G. B. Pant University of Agriculture and Technology
Pantnagar, Uttarakhand
India

Prem Rajak
Department of Animal Science
Kazi Nazrul University
Asansol, West Bengal
India

Natarajan Raman
Department of Chemistry
VHNSN College (Autonomous)
Virudhunagar, Tamil Nadu
India

Vikram Rathour
Department of Chemistry
Institute of Science, Banaras Hindu University
Varanasi, Uttar Pradesh
India

Saman Raza
Department of Chemistry
Isabella Thoburn College
Lucknow, Uttar Pradesh
India

S. Daphne Rebekal
Department of Chemistry
Sarah Tucker College
Tirunelveli, Tamil Nadu
India

Malar Retna
Department of Chemistry
Scott Christian College (Autonomous)
Nagercoil, Tamil Nadu
India

Shippu Sachdeva
School of Electronics and Electrical Engineering
Lovely Professional University
Phagwara, Punjab
India

Robin Kumar Samuel
Department of Chemistry
Good Shepherd College of Engineering and Technology
Kanyakumari, Tamil Nadu
India

R. Sanjeevi
Nims Institute of Allied Medical Science and Technology
Nims University Rajasthan
Jaipur, Rajasthan
India

Minakshi Sharma
Department of Physics
Om Sterling Global University
Hisar, Haryana
India

Smita Singh
Department of Chemistry
Institute of Science, Banaras Hindu University
Varanasi, Uttar Pradesh
India

Varsha Singh
Department of Chemistry
Institute of Science, Banaras Hindu University
Varanasi, Uttar Pradesh
India

Manoj Sindhwani
School of Electronics and Electrical Engineering
Lovely Professional University
Phagwara, Punjab
India

Sonam Soni
Department of Chemistry
Sunbeam Women's College Varuna
Varanasi, Uttar Pradesh
India

Piyush Kumar Sonkar
Department of Chemistry
MMV, Banaras Hindu University
Varanasi, Uttar Pradesh
India

Ponnusamy Thillai Arasu
Department of Chemistry, College of Natural and Computational Sciences
Wollega University
Nekemte
Ethiopia

Ravi Tejasvi
Department of Chemical Engineering
School of Energy Technology
Pandit Deendayal Energy University
Gandhinagar, Gujarat
India

Sandeep Tripathi
Nims Institute of Allied Medical Science and Technology
Nims University Rajasthan
Jaipur, Rajasthan
India

Suman Lata Tripathi
School of Electronics and Electrical Engineering
Lovely Professional University
Phagwara, Punjab
India

Veerabathuni Jaya Usha Praveena
Department of Physics
St. Francis College for Women
Hyderabad, Telangana
India

Varsha Yadav
School of Applied Sciences
Shri Venkateshwara University
Gajraula, Uttar Pradesh
India

Shyam Raj Yadav
Department of Chemistry
S. P. Jain College (Veer Kunwar Singh University, Ara)
Sasaram, Bihar
India

Preface

The last century's rapid industrialization and technological advancements have significantly impacted our environment. The escalation of carbon emissions and the exhaustion of natural resources underscore the critical need for a shift towards sustainable behaviors. In this context, carbon-based nanomaterials have surfaced as a viable solution for tackling significant environmental issues of our era. This book, *Carbon-based Nanomaterials for Green Applications*, examines the distinctive characteristics of carbon-based nanomaterials and their capacity to transform green technologies. Utilizing their outstanding thermal, electrical, and mechanical characteristics, these materials provide solutions for various applications, including renewable energy systems, environmental remediation, sustainable agriculture, and water purification. We aim to assemble this collection to connect fundamental research with practical applications. The book integrates contributions from prominent scholars, offering an interdisciplinary viewpoint on the design, production, and application of carbon-based nanomaterials in sustainable technology. This book comprehensively analyzes subjects, including carbon nanotubes, graphene, fullerenes, and other novel carbon nanostructures, as well as their functionalization and incorporation into electronics. This book is designed for scientists, engineers, and policymakers engaged in developing sustainable solutions for the future. We aspire that the insights and innovations articulated herein will stimulate additional study and collaboration in this critical domain.

We wish to convey our appreciation to the contributors whose knowledge and commitment have been essential in crafting this book. We are profoundly grateful for the support from our institutions and the broader scientific community, whose enthusiasm has consistently motivated us. In addressing the intricate issues of attaining sustainability, it is evident that no singular answer will be enough. The convergence of nanotechnology and environmental research presents a viable avenue for advancement. This book describes the synthesis, growth, development, characterization, and potential applications of carbon-based nanomaterials in the fields of chemical, biological perspective, environmental science, pharmaceuticals, drugs delivery, biomedical technology, device development, supercapacitors, solar cells, fuel cells, biofuel cells, electrocatalysis, sensors, biosensors, *etc*. Thus, this book "*Carbon-based Nanomaterials for Green Applications*" covers most of the significant fields of innovative materials and is helpful for the readers.

Acknowledgments

- First of all, we thank our contributing authors for their valuable contribution to this book.
- Dr. Upendra Kumar acknowledges the Department of Applied Sciences, IIIT Allahabad, Prayagraj, Uttar Pradesh, India, for all motivation and support.
- Prof. Suman Lata Tripathi, acknowledges the School of Electronics and Electrical Engineering, Lovely Professional University, Phagwara, India, for all motivation and support.
- We thank our mentors, Dr. Shail Upadhyay, Prof. B. N. Dwivedi (Department of Physics, IIT BHU), Prof. Devendra Kumar (Department of Ceramic Engineering, IIT BHU), and Prof. V. Ganesan (Department of Chemistry, Institute of Science, BHU), for their continuous help, support, and suggestions.
- We thank our research group members, Dr. Vedika, Dr. Narvadeshwar, Mr. Vikash Jangra, Ramsundar, Raj Kumar, and Vipin Gupta, for their help and support.
- We thank our colleagues, collaborators, friends, and well-wishers for providing moral support for this book.
- We thank our family members and friends for providing continuous moral support for this book.

1

Green Energy: An Introduction, Present, and Future Prospective

Manoj Singh Adhikari[1], Raju Patel[2], Manoj Sindhwani[1], and Shippu Sachdeva[1]

[1] *School of Electronics and Electrical Engineering, Lovely Professional University, Phagwara, Punjab, India*
[2] *School of Electronics Engineering, Vellore Institute of Technology, Chennai, Tamil Nadu, India*

1.1 Introduction

In recent years, the world has witnessed a growing global awareness of the requirement to transition from traditional fossil-fuel-based sources to more eco-friendly and sustainable alternatives [1–3]. Green energy, also known as renewable energy, is at the forefront of this movement. It refers to energy resultant from natural resources that are continually replenished and have minimal influence on the environment compared to fossil fuels. The adoption of green energy is crucial for falling greenhouse gas emissions and confirming a cleaner, better world for upcoming generations [4–6].

The advantages of green energy are numerous. Unlike fossil fuels, renewable energy sources do not produce harmful greenhouse gas emissions, which will add significant climate change and global warming. Green energy also helps reduce reliance on finite fossil fuel reserves, promoting energy security and independence for countries [7–9]. Additionally, it fosters financial evolution and job formation in the renewable energy field, leading to an additional supportable and diversified economy. However, green energy technologies also face challenges. Their initial setup costs can be higher than conventional energy sources, making investments and incentives crucial to encouraging widespread adoption. Moreover, the recurrent landscape of some renewable sources, like wind and solar, demands effective

Carbon-based Nanomaterials for Green Applications, First Edition. Edited by Upendra Kumar, Piyush Kumar Sonkar, and Suman Lata Tripathi.

Published 2025 by John Wiley & Sons, Inc.

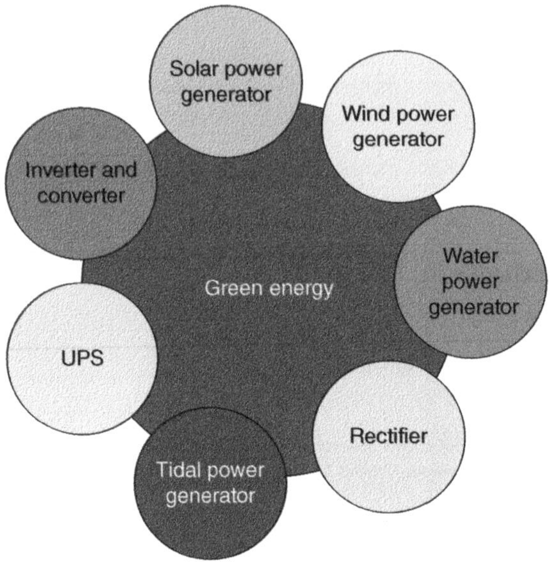

Figure 1.1 Foundation of green energy.

energy storage solutions and grid management for consistent and reliable power supply [10–12]. Figure 1.1 shows the foundation of green energy.

Despite these challenges, the global shift toward green energy continues to gain momentum. Governments, businesses, and individuals are progressively identifying the urgency of transitioning to renewable energy for a cleaner and more sustainable future. By embracing green energy solutions, we can make significant strides in justifying climate change and conserving the planet's usual properties for generations to come [13–15].

Green energy has made substantial strides in the present, proving its viability as a cleaner, sustainable alternative to conventional energy sources. With continuous innovation, supportive policies, and international partnership, the future of green energy appears promising. By overcoming existing challenges and embracing emerging technologies, green energy plays a transformative role in attaining a greener, more sustainable planet for future generations [16–18].

This chapter also discusses the role of international cooperation and collaborative efforts in advancing green energy adoption globally. As countries join forces to struggle climate change and transition approaches for low-carbon future, knowledge-sharing and technology transfer play a crucial role in expediting green energy deployment [18–24].

1.2 Present Status of Green Energy

The adoption and execution of green energy were already well underway, and it is likely that a significant progress has been made. Figure 1.2 shows the renewable energy sources and public opinions. There are some key aspects of the present status of green energy:

Increased Renewable Energy Capacity: Many nations have continued to capitalize in and expand renewable energy capacity. Solar and wind power installations have seen substantial growth globally. The rate of renewable energy technologies has also been reducing, making them more competitive with traditional fossil-fuel-based sources.

Policy Support and Commitments: Governments in the country have been establishing policies and setting targets to inspire the implementation of renewable energy. Many nations have dedicated to increase the stake of renewable energy and decrease the carbon releases for combat climate change.

Technological Advancements: There have been continuing progressions in renewable energy machineries, energy storage solutions, and grid integration. These improvements help discourse trials associated with intermittence and make renewable energy sources more reliable and efficient.

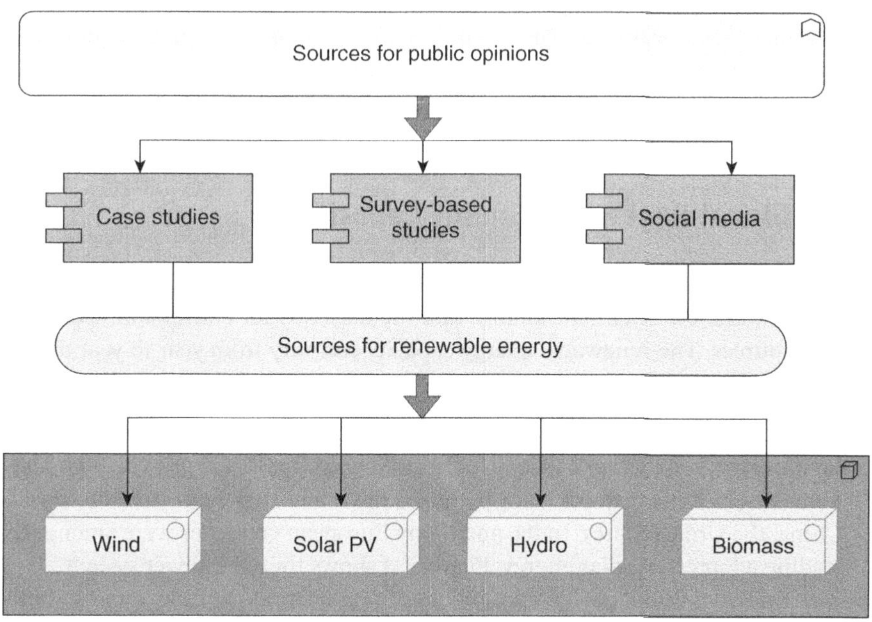

Figure 1.2 Renewable energy sources and public opinions.

Corporate Sustainability Initiatives: Many businesses and corporations have embraced renewable energy as part of their sustainability strategies. Companies are investing in renewable energy projects, purchasing renewable energy certificates, and striving to achieve carbon neutrality in their operations.

Emergence of Green Financing: The rise of green bonds and sustainable financing options has facilitated investments in renewable energy projects and initiatives. This has allowed a wider range of stakeholders to contribute to the transition toward green energy.

Community and Individual Participation: There has been increased awareness and support for green energy at the community and individual levels. More people are adopting solar panels for their homes, participating in community solar projects, and seeking ways to reduce their carbon footprint.

Challenges and Barriers: Despite the progress, challenges remain. The intermittency of solar and wind power requires better energy storage solutions and advanced grid management. Additionally, some regions may face regulatory, financial, or technical barriers to implement renewable energy projects.

Global Cooperation: International collaborations and contracts have emphasized the status of global partnership in addressing weather change to a low-carbon economy. Countries are working together to share best practices and hasten the adoption of green energy.

The status of green energy is continually evolving, and more recent developments may have occurred. The evolution to a sustainable energy future is a dynamic and ongoing process that requires continued promise from individuals, businesses, and governments worldwide.

1.3 Global Renewable Energy Capacity

Global renewable energy capacity had been steadily increasing, driven by the growing awareness of climate change and the necessity for conversion to cleaner energy sources. The renewable energy capacity can vary from year to year due to ongoing developments and policy changes. The global renewable energy capacity is discussed as:

Solar Energy: Solar power capacity had experienced significant growth, with photovoltaic (PV) installations being deployed in various regions around the world. China, the United States, India, and many European countries were among the leading adopters of solar energy. Figure 1.3 shows the solar tower system.

Wind Energy: Wind power capacity had also been expanding rapidly. Both onshore and offshore wind farms were becoming increasingly prevalent. China, Germany, the United Kingdom, and the United States were among the top countries regarding the wind energy installations.

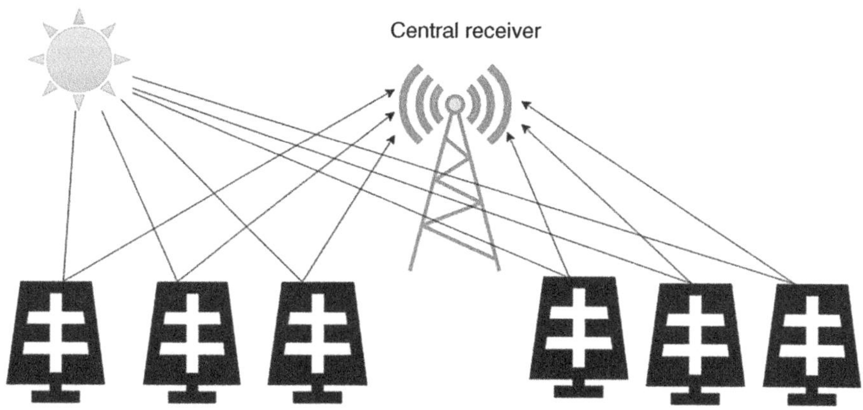

Figure 1.3 Solar tower system.

Hydropower: Hydropower had long been a significant contributor to renewable energy capacity, with large hydroelectric projects in various countries. Brazil, China, and Canada were among the top producers of hydroelectric power.

Biomass Energy: Biomass energy capacity had been growing steadily, with a focus on converting organic materials and waste into biofuels and biogas. This technology was especially relevant in agricultural regions and waste management systems.

Geothermal Energy: Geothermal power capacity, although smaller than other renewable sources, had been increasing in select regions with favorable geothermal resources, such as Iceland, the Philippines, and the United States.

Governments, international organizations, and businesses were making substantial funds in renewable energy projects and infrastructure. Various incentives, tax credits, and renewable energy targets had been established to encourage the growth of renewable energy capacity.

It is important to recognize that the renewable energy landscape is continually evolving. There might have been further advancements, policy changes, and increases in renewable energy capacity globally.

1.4 Leading Green Energy Technologies

Several green energy technologies were leading the charge in the transition toward renewable and sustainable energy sources. These technologies have been continually evolving and improving over time. There are some of the leading green energy technologies:

Wind Turbines: Wind energy is helpful to bind the kinetic energy to produce power. Such turbines, both onshore and offshore, have become more efficient and larger in size, leading to increased electricity production. Wind power is a popular choice in regions with consistent wind patterns, and it has seen substantial growth in capacity over the years.

Hydropower: Hydropower consists of the energy of falling or flowing water to produce electrical energy. Huge-scale hydroelectric power plants, as well as smaller run-of-the-river and micro-hydropower systems, contribute to the worldwide renewable energy mix. Such kind of hydropower has the advantage of being a highly reliable and controllable source of renewable energy.

Biomass Energy: Biomass energy includes converting organic materials, i.e. agricultural residues, organic waste, and wood, into energy. It is helpful for electricity generation, heating, and biofuels. Biomass is a versatile renewable resource, especially in areas with abundant agricultural or forestry residues.

Solar PV Systems: The modern solar PV converts sunlight into electricity with the help of semiconductors. It is one of the most widely deployed renewable energy technologies globally. The installation of solar panels can be on roofs, solar farms, and integrated into construction resources. Advances in solar cell efficiency and decreasing costs have made solar power more competitive with conventional energy sources. Figure 1.4 shows the custom of solar panel system.

Geothermal Energy: Geothermal power harnesses the earth's natural heat to produce electricity or deliver heating and cooling for buildings. Geothermal power plants can be found in areas with accessible geothermal reservoirs, such as geysers or hot springs.

Energy Storage Solutions: This technology plays a vital part in ensuring the reliability and stability of green energy sources. Compressed air energy storage, pumped hydro storage, batteries, and other emerging storage solutions help stock additional energy when renewable sources are abundant and issue it when needed.

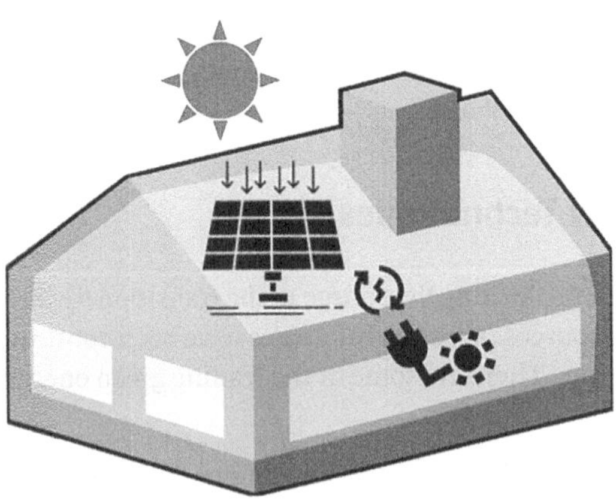

Figure 1.4 Usage of solar panel system.

Hybrid Systems: Combining multiple renewable energy sources in hybrid systems can optimize energy production and enhance grid stability. Hybrid systems are being explored to ease the intermittency challenges of individual renewable technologies.

Tidal and Wave Energy: Tidal and wave energy technologies extract energy from the natural motion of waves, oceans, and tides. While these technologies are in the initial stages of progress, they hold promise for coastal regions with strong wave or tidal energy potential.

Green Energy in Transportation: Green energy in transportation is a serious aspect of the global energies to decrease combat environment change, greenhouse gas productions, and promote sustainable mobility. Transportation is one of the contributors for carbon releases, mainly over the burning of fossil fuels in vehicles. Embracing green energy solutions for transportation can help address environmental challenges while enhancing energy efficiency and reducing dependence on finite fossil fuel resources.

Electric vehicles are the most promising green energy solutions in transportation. They run entirely on electricity, eliminating tailpipe emissions, and falling carbon footprint. The adoption of EVs has been increasing worldwide, encouraged by government incentives, advancements in battery technology, and the expansion of charging infrastructure.

Advancements in these leading green energy technologies have been driven by increasing research, investments, and government support. As the world continues to prioritize sustainability and combat climate change, these technologies are expected to show an energetic role in our energy evolution to a more renewable and low-carbon future.

1.5 Challenges in Green Energy Adoption

The adoption of green energy is crucial for a sustainable future. It faces numerous trials that are essential to be talked to accelerate its widespread implementation. The key challenges in green energy adoption contain:

Cost and Economics: The primary investment in green energy is higher related to predictable fossil-fuel-based options. While the cost of renewable energy has been reducing over time, it remains a barrier for some regions and individuals.

Intermittency and Grid Integration: Some renewable sources, such as solar and wind, are erratic in nature, because it is relied on weather situations. This inconsistency carriages challenges for grid stability and requires efficient energy storage solutions and advanced grid management.

Energy Storage: Developing effective and scalable energy storage solutions is essential to stock additional energy produced through peak times and it is used when the low renewable production. Current energy storage technologies are advancing, but there is still room for improvement and cost reduction.

Infrastructure and Grid Upgrades: Widespread adoption of green energy needs important upgrades to the present energy structure and electrical grids. Integrating renewable energy to the grid necessitates funds in smart grid technologies and transmission lines.

Land and Space Requirements: Renewable energy installations, i.e. large wind and solar farms, may require significant land or space, leading to concerns about environmental impacts and land use conflicts.

Public Awareness: Raising awareness and promoting public acceptance of green energy solutions is essential. Misconceptions or resistance from communities can slow down or prevent the implementation of renewable energy projects.

Technological Maturity: Some green energy technologies are in the initial phases of growth and look trials associated with scalability, reliability, and efficiency.

Energy Transition Workforce: The shift toward green energy requires a skilled workforce capable of installing, maintaining, and operating renewable energy systems. Developing this workforce and ensuring a just transition for workers in fossil fuel industries is critical.

Global Cooperation and Financing: International cooperation is vital to addressing climate change effectively. Developing countries may face challenges in accessing financing for renewable energy projects and need support from the international community.

Despite these challenges, progress is being made in overcoming barriers to green energy adoption. Technological advancements, policy support, and increasing public awareness are driving the change toward a further sustainable energy future. Continued collaboration among administrations, industries, and publics will be crucial in addressing these encounters and attaining a low carbon and sustainable energy system.

1.6 Prospects of Green Energy

The prospects of green energy are promising and hold significant probable to convert the global energy site. The world becomes more alert to the essential to report climate change and minimize greenhouse gas productions; green energies are predictable to show a critical part in shaping the future of energy production and consumption. There are some key prospects of green energy:

Cost Competitiveness: The sustained progressions in green energy technologies, i.e. wind power, energy storage, and solar PV, are expected to drive down costs further. As economies of scale are achieved and research and development continue, green energy sources will become progressively cost-competitive with traditional fossil fuels, creating them the more economically attractive choice.

Grid Modernization and Smart Technologies: The integration of green energy sources into smart grids will enable more efficient and reliable energy distribution. Smart technologies, such as demand response systems and advanced energy management, will optimize energy use and enhance grid stability, accommodating the inconsistency of renewable sources.

Hydrogen Economy: Hydrogen is emerging as a potential green energy carrier. Green hydrogen, produced through electrolysis powered by renewable energy, has the possible to be used in various sectors, including industry, transportation, and energy storage, contributing to a cleaner and supplementary sustainable energy system. Figure 1.5 shows the diagram of hydrogen production from sea water.

Energy Storage Breakthroughs: The development of innovative energy storage solutions, i.e. advanced batteries, pumped hydro storing, and other emerging

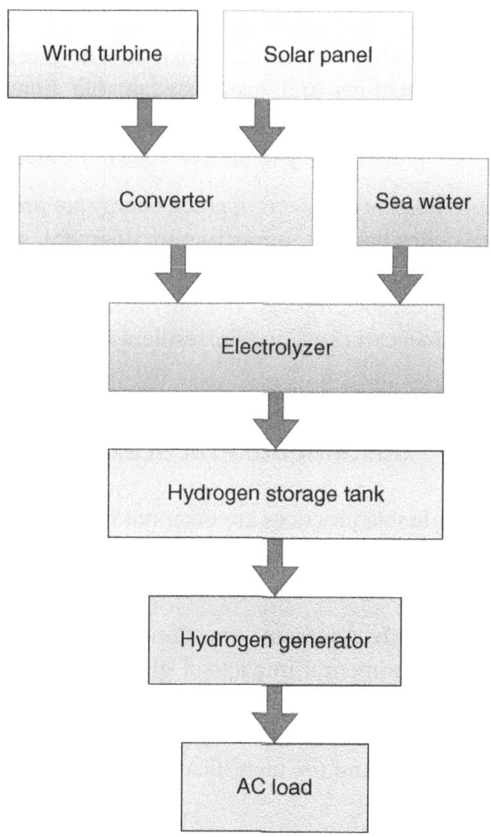

Figure 1.5 Hydrogen production from sea water.

technologies, will be a game changer for renewable energy. Efficient storage systems will report the intermittence of wind and solar power, allowing a further consistent and continuous energy supply.

Electrification of Transport: The shift toward electric vehicles and the electrification of public transportation will significantly reduce emissions from the transportation sector. This trend will be complemented by an increased adoption of renewable energy sources for charging electric vehicles, further lowering the carbon footprint of transportation.

Renewable Energy Innovation: Ongoing innovation and research in green energy technologies will drive efficiency improvements and open new opportunities for harnessing energy from untapped sources, such as marine energy, geothermal resources, and emerging PV materials.

International Collaboration and Policy Support: Global cooperation, as seen in initiatives like the Paris agreement, will show an essential part in fostering the acceptance of green energy on a larger scale. Governments worldwide are expected to implement more ambitious financial incentives, regulations, and policies to accelerate the transition to renewable energy.

Sustainable Economic Growth: The green energy area is the key to produce occupations and foster sustainable financial development. Investments in renewable energy projects and infrastructure will inspire financial activity and generate employment occasions in numerous subdivisions.

While the prospects of green energy are promising, overcoming challenges, such as policy barriers, infrastructure upgrades, and public acceptance, will be crucial for successful and widespread implementation. By embracing green energy technologies, societies can not only ease the influences of climate variation but also build an extra sustainable, resilient, and equitable future for generations to come.

1.7 Sustainable Practices in Green Energy

Sustainable practices are essential in the growth and positioning of green energy technologies to ensure their long-term environmental, social, and economic benefits. There are some key sustainable practices regarding the green energy:

Life Cycle Assessments: Conducting life cycle assessments helps evaluate the environmental impacts of green energy technologies throughout their whole life cycle, which begins with raw material extraction and finally to use and disposal. Understanding the life cycle impacts allows for informed decision-making and the identification of areas for improvement.

Resource Management: Sustainable practices in green energy involve responsible resource management. This includes using materials efficiently, recycling and reusing components when possible, and minimizing the environmental impact of resource extraction and processing.

Ecosystem and Biodiversity Protection: Ensuring that green energy projects respect and protect ecosystems and biodiversity is crucial. Site selection and planning should consider potential impacts on wildlife habitats, migratory routes, and sensitive ecological areas.

Community Engagement and Consent: Connecting local societies for the development, planning, and decision-making processes of green energy schemes fosters social acceptance and can lead to better project outcomes. Respecting indigenous rights and obtaining community consent are essential for sustainable development.

Climate and Environmental Mitigation Measures: Beyond reducing greenhouse gas emissions, sustainable practices in green energy involve implementing measures to mitigate environmental impacts. These may include noise reduction, light pollution control, and measures to prevent soil erosion and water contamination.

Energy Efficiency: Ensuring energy effectiveness in the enterprise and process of green energy organizations is critical. Optimizing energy use reduces waste and enhances the overall sustainability of the energy generation process.

Renewable Energy Integration: Effectively integrating renewable energy into existing energy systems, such as smart grid technologies and demand-side management, enhances energy efficiency and grid stability, facilitating a smoother transition to green energy sources.

Conservation and Land Use: Sustainable green energy practices involve minimizing land use impacts and prioritizing the use of non-prime or degraded land for renewable energy projects. Avoiding conversion of valuable natural habitats and agricultural land is crucial.

Monitoring and Reporting: Implementing robust monitoring and reporting systems helps track the performance of green energy projects, ensuring that they meet sustainability goals and that any issues are addressed promptly.

Social and Economic Benefits: Sustainable green energy practices seek to maximize social and economic benefits for communities where projects are located. This may include job creation, skills development, and local economic growth.

By integrating these sustainable practices into the deployment and growth of green energy technologies, we can ensure that renewable energy shows a central part in building a more supportable and resilient energy future while minimizing adverse effects to the society.

1.8 Case Studies of Successful Green Energy Projects

There have been several successful green energy projects everywhere in the biosphere that have demonstrated the viability and profits of renewable energy technologies. There are some notable case studies:

The Three Gorges Dam, China: The three Gorges Dam is the largest hydroelectric power projects in the biosphere located on the Yangtze in China. The construction of dam structure began in 1994 and was completed in 2012. It has a massive installed capacity of over 22,500 megawatts (MW) and generates a significant portion of China's electricity. The project has helped cut carbon emissions by substituting fossil-fuel-founded energy sources and has contributed to flood control and improved navigation along the Yangtze River.

Solar Power in Rajasthan, India: Rajasthan, a state in India, has made significant strides in solar energy deployment. The Bhadla Solar Park is the world's largest solar parks and is located in Rajasthan. It spans thousands of acres and has a total capacity of over 2200 MW. The state's ambitious solar energy policies, favorable climate conditions, and vast open spaces have enabled it to become a leader in solar power generation.

Hornsdale Power Reserve, Australia: It is also known as the Tesla big battery. It was constructed by Tesla in partnership with the South Australian government and Neoen. With a capacity of 150 MW and a storage capacity of 193.5 megawatt-hours (MWh), the battery has been instrumental in stabilizing the grid and providing rapid response during peak demand periods.

Offshore Wind Farms in Europe: Several offshore wind farms in Europe have been successful in harnessing wind energy from the sea. Projects like the London array offshore wind farm in United Kingdom and the Borssele wind farm in the Netherlands have demonstrated the potential of offshore wind as a reliable and clean energy source.

Geothermal Power in Iceland: Iceland is one of the key examples of effective geothermal energy operation. Around 85% of Iceland's main energy supply is from renewable sources in which geothermal power plays a significant role. The country's unique geology allows it to tap into abundant geothermal reservoirs for electricity generation and heating, making it one of the most sustainable energy systems globally.

Hydrogen Fueling Stations in Germany: Germany has been actively investing in hydrogen infrastructure, particularly hydrogen fueling stations for fuel cell vehicles. One successful example is the H_2 Mobility Germany initiative, which aims to establish a nationwide network of hydrogen refueling stations to support the growth of fuel cell vehicles.

These case studies highlight the diverse range of successful green energy projects worldwide, spanning various technologies and approaches. They demonstrate that with the right policies, investments, and commitments, renewable energy can be harnessed effectively to address energy needs sustainably and contribute to mitigating climate change.

1.9 Policy and Regulatory Framework for Green Energy

The policy and regulatory framework for green energy is a crucial aspect of fostering the conversion to renewable and sustainable energy sources. Governments nearby the world are developing and employing a range of policies and regulations to encourage the approval and growth of green energy technologies. There are some key components of the policy and regulatory framework for green energy:

Renewable Energy Targets: Many nations have set ambitious renewable energy goals to enhance the stake of renewable sources in their energy mix. These targets often include specific goals for solar, wind, hydropower, and other renewable technologies.

Power Purchase Agreements (PPAs) and Feed-in Tariffs (FiTs): PPAs and FiTs are mechanisms that provide guaranteed prices or long-term agreements for electrical energy generated from renewable sources. These incentives help attract investments in green energy projects and offer revenue stability for developers.

Renewable Portfolio Standards (RPS): RPS directives necessitate values to obtain a fixed ratio of their energy from renewable sources. Such strategy encourages utilities to invest in or purchase renewable energy to meet their portfolio obligations.

Net Metering and Feed-in Premiums: Net metering permits consumers with renewable energy systems, i.e. solar panels, to trade additional electricity back to the grid at retail rates. Feed-in premiums offer additional payments for renewable energy generation, incentivizing small-scale producers.

Tax Incentives and Subsidies: Tax credits, grants, and subsidies are offered to renewable energy projects to lower upfront costs and make green energy technologies more financially attractive.

Green Energy Certificates: Green energy certificates guarantees of origin or renewable energy certificates that signify the conservational qualities of renewable energy production. These certificates enable consumers and businesses to support renewable energy even if they are not directly connected to a renewable energy source.

Grid Connection and Interconnection: Streamlining the grid connection process and ensuring fair and nondiscriminatory access to the grid for renewable energy projects are essential for their successful integration into the existing energy infrastructure.

Energy Storage Policies: Specific regulations and incentives for energy storage technologies support grid constancy and the effective utilization of renewable energy.

Green Building Standards: Encouraging green building practices, energy-efficient design, and the usage of renewable energy technologies in construction and renovation contributes to overall energy sustainability.

Research and Development Support: Government funding and support for research, development, and novelty in green energy technologies accelerate advancements and cost reductions in the renewable energy area.

Environmental Regulations and Emission Reduction Targets: Policies targeting the decrease of greenhouse gas releases and environmental protection complement green energy initiatives and create a conducive environment for sustainable practices.

The effectiveness of the policy and regulatory framework for green energy depends on its stability, consistency, and alignment with long-term sustainability goals. By establishing clear and supportive policies, governments can encourage private sector investments, foster technological innovation, and drive the conversion toward a low-carbon and additional sustainable energy future.

1.10 Opportunities and Challenges in the Evolution to a Green Energy Future

The transition to a green energy future presents both challenges and opportunities for societies worldwide. While the shift toward renewable and sustainable energy sources is prime requirement for addressing climate change and achieving long-term environmental goals, it also requires careful planning and collaboration. Figure 1.6 shows the opportunities of renewable energy sources. There are some of the key challenges and opportunities in the evolution to a green energy future:

1.10.1 Opportunities

Climate Mitigation and Environmental Benefits: The transition to green energy significantly reduces greenhouse gas emissions, leading to a more sustainable and resilient environment. It offers an opportunity to combat climate change and protect ecosystems.

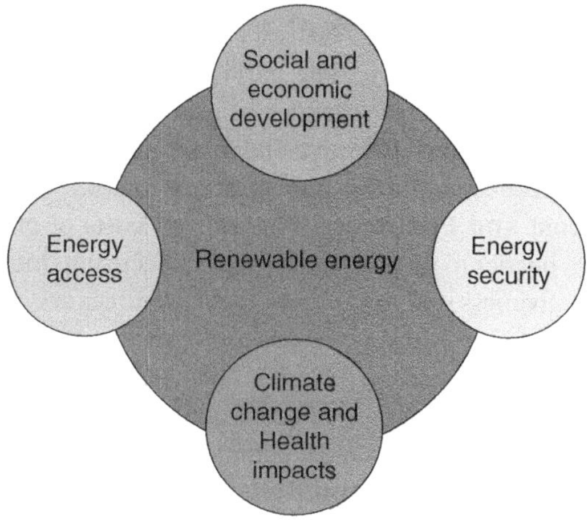

Figure 1.6 Opportunities of renewable energy sources.

Technological Advancements: Advancements in green energy skills are accelerating, leading to improved efficiency and cost reductions, making renewable energy more competitive with fossil fuels.

Job Creation: The transition to green energy can generate new-fangled job chances in the renewable energy area, contributing to monetary development and fostering innovation.

Energy Independence and Security: Investment in green energy decreases requirement on bring in fossil fuels, improving energy safety and independence for countries.

Community Empowerment: Local communities can play an active role in green energy projects, benefiting from increased energy independence and participation in decentralized energy production.

Health Benefits: Transitioning away from fossil fuels can lead to improved air quality and reduced health impacts associated with air pollution, benefiting public health.

Global Collaboration and Partnerships: The shift to green energy provides opportunities for international collaboration and partnerships to address climate change on a universal scale.

By recognizing and addressing the challenges while capitalizing on the opportunities, the transition to a green energy future can lead to a more sustainable, equitable, and prosperous world. Governments, businesses, communities, and individuals all have roles to play in driving this transformation and securing a better future for upcoming generations.

1.10.2 Challenges

Grid Integration: The variability of renewable energy sources like wind and solar creates challenges in grid management and complementary foundation and request. Effective energy storage solutions and smart grid technologies are needed to address intermittency issues.

Cost and Financing: The upfront costs of green energy technologies can be a barrier to widespread adoption. Securing financing and achieving cost competitiveness with traditional energy sources are challenges that require government incentives and private sector investment.

Infrastructure Upgrades: Transitioning to green energy requires significant upgrades to energy infrastructure, including the expansion and modernization of electricity grids and charging infrastructure for electric vehicles.

Energy Security and Reliability: Ensuring a reliable and secure energy supply during the transition is critical. Maintaining stable and consistent energy sources while integrating intermittent renewables is a challenge that requires careful planning.

Social and Economic Impacts: The transition may have economic implications for certain sectors, such as fossil fuel industries, which could result in job losses and socioeconomic challenges for affected communities. Ensuring a just and inclusive transition is essential.

Political and Regulatory Uncertainty: Changing political landscapes and shifting regulatory environments can impact the stability of green energy policies and create uncertainties for investors and developers.

Resource Constraints: Some renewable energy sources, such as rare earth metals used in certain technologies, may face supply chain challenges, creating resource constraints that need to be addressed.

1.11 Conclusion

The transition to a green energy future is a critical and urgent endeavor, determined by the requirement to combat climate change, create a sustainable world, and minimize the greenhouse gas emissions for future generations. Green energy technologies, i.e. hydro, geothermal, wind, biomass, and solar, offer immense promise in replacing fossil fuels and transforming the global energy landscape.

While challenges, such as intermittency, infrastructure upgrades, and financing obstacles, must be overcome, the opportunities presented by green energy are vast. The transition to renewables brings with it the potential for climate mitigation, environmental protection, energy security, job creation, and economic growth. It also empowers communities, fosters technological advancements, and facilitates global collaboration to tackle the defining issue of our time.

Governments, businesses, individuals, and communities must work together to create robust policy and supervisory frameworks, capitalize in research and innovation, and embrace sustainable practices in the energy sector. By doing so, we can accelerate the adoption of green energy, minimize the effect of weather change, and get the path for a brighter, cleaner, and more resilient future.

Hence, the transition to a green energy future is not merely an option; it is a necessity. Our collective efforts in embracing green energy will define the path we take toward a sustainable and thriving world, ensuring a legacy of responsible stewardship for the planet and the generations to come.

References

1 Liu, D., Chen, Y., Chai, K.K. et al. (2015). Two-dimensional optimization on user association and green energy allocation for HetNets with hybrid energy sources. *IEEE Transactions on Communications* 63 (11): 4111–4124.

2 Lin, C.C., Deng, D.J., Kuo, C.C., and Liang, Y.L. (2018). Optimal charging control of energy storage and electric vehicle of an individual on the internet of energy with energy trading. *IEEE Transactions on Industrial Informatics* 14 (6): 2570–2578.

3 Tahsin, A., Roy, P., Razzaque, M.A. et al. (2023). Energy cooperation among sustainable base stations in multi-operator cellular networks. *IEEE Access* 11: 19405–19417.

4 Jahid, A., Hossain, M.S., Monju, M.K.H. et al. (2020). Techno-economic and energy efficiency analysis of optimal power supply solutions for green cellular base stations. *IEEE Access* 8: 43776–43795.

5 Sadiq, M., Ali, S.W., Terriche, Y. et al. (2021). Future greener seaports: a review of new infrastructure, challenges, and energy efficiency measures. *IEEE Access* 9: 75568–75587.

6 Chen, C., He, B., and Tang, X. (2012). Green-aware workload scheduling in geographically distributed data centers," *IEEE International Conference on Cloud Computing Technology and Science Proceedings*, Taipei, Taiwan (3–6 December 2012), pp. 82–89.

7 Jin, Y., Wen, Y., and Chen, Q. (2012). Energy efficiency and server virtualization in data centers: an empirical investigation. *Proceedings IEEE INFOCOM Workshops*, Orlando, FL (25–30 March 2012), pp. 133–138.

8 Homsi, S., Liu, S., Chaparro-Baquero, G.A. et al. (2016). Workload consolidation for cloud data centers with guaranteed QoS using request reneging. *IEEE Transactions on Parallel and Distributed Systems* 28 (7): 2103–2116.

9 Sekhar, J., Jeba, G., and Durga, S. (2012). A survey on energy efficient server consolidation through vm live migration. *International Journal of Advances in Engineering & Technology* 5 (1): 515.

10 Fan, Q., Ansari, N., and Sun, X. (2017). Energy driven avatar migration in green cloudlet networks. *IEEE Communications Letters* 21 (7): 1601–1604.

11 Dayarathna, M., Wen, Y., and Fan, R. (2016). Data center energy consumption modeling: a survey. *IEEE Communication Surveys and Tutorials* 18 (1): 732–794.

12 Goyal, Y., Arya, M.S., and Nagpal, S. (2015). Energy efficient hybrid policy in green cloud computing, *International Conference on Green Computing and Internet of Things (ICGCIoT)*, Greater Noida, India (8–10 October 2015), pp. 1065–1069.

13 Billah, S.M.B., Kabir, K.M., Islam, M.O. et al. (2017). Hydrogen energy storage based green power plant in seashore of Bangladesh: design and optimal cost analysis. *International Conference on Innovations in Green Energy and Healthcare Technologies (IGEHT)*, Coimbatore, India (16–18 March 2017), pp. 1–5.

14 Bhattacharya, T. and Qin, X. (2020). Modeling energy efficiency of future green data centers. *International Green and Sustainable Computing Workshops (IGSC)*, Pullman, WA, USA (19–22 October 2020), pp. 1–3.

15 Delinchant, B., Wurtz, F., Ploix, S. et al. (2016). GreEn-ER living lab: a green building with energy aware occupants. *International Conference on Smart Cities and Green ICT Systems (SMARTGREENS)*, Rome, Italy (23–25 April 2016), pp. 1–8.

16 Dos Reis, T.N.F., Teixeira, M.M., and Salles Soares Neto C.D. (2023). Relation of energy consumption in green cloud computing with big data. *IEEE Green Technologies Conference*, Denver, CO, USA (19–21 April 2023), pp. 280–284.

17 Kumar, N. (2018). A study on green energy powered cognitive radio network for communication network architecture of smart grid. *International Innovative Applications of Computational Intelligence on Power, Energy and Controls with their Impact on Humanity (CIPECH)*, Ghaziabad, India (1–2 November 2018), pp. 3–7.

18 Ma, X., Huang, S., Li, Z. et al. (2020). Quantitative evaluation method of green power development level. *International Conference on Power and Energy Systems (ICPES)*, Chengdu, China (25–27 December 2020), pp. 420–425.

19 Sithan, M. and Lai, L.L. (2011). Application of green technologies in developing countries—reduced carbon emission and conservation of energy. *IEEE Power and Energy Society General Meeting*, Detroit, MI, USA (24–28 July 2011), pp. 1–7.

20 Yarlagadda, N., Verma, Y.K., Adhikari, M.S., and Mishra, V. (2023). Opto-VLSI devices and circuits for biomedical and healthcare applications. In: *A Review of Multi-material, Multi-gate MOSFET Structures* (ed. A. Kumar, S. Agarwal, V. Varshnay, et al.), 29–40. CRC Press.

21 Entezari, A., Aslani, A., Zahedi, R., and Noorollahi, Y. (2023). Artificial intelligence and machine learning in energy systems: a bibliographic perspective. *Energy Strategy Reviews* 45: 101017.

22 Legala, A. and Li, X. (2022). Hybrid data-based modeling for the prediction and diagnostics of Li-ion battery thermal behaviors. *Energy and AI* 10: 100194.

23 Machlev, R., Heistrene, L., Perl, M. et al. (2022). Explainable artificial intelligence (XAI) techniques for energy and power systems: review, challenges and opportunities. *Energy and AI* 9: 100169.

24 Kharlamova, N., Hashemi, S., and Traeholt, C. (2021). Data-driven approaches for cyber defence of battery energy storage systems. *Energy and AI* 5: 100095.

2

Properties of Carbon-Based Nanomaterials and Techniques for Characterization

Ravi Tejasvi

Department of Chemical Engineering, School of Energy Technology, Pandit Deendayal Energy University, Gandhinagar, Gujarat, India

2.1 Introduction

Carbon nanomaterials, a distinctive class of nanomaterials composed solely of carbon atoms, have emerged as a cornerstone in the realm of green energy due to their remarkable properties and versatile applications. Their unique structures, with fullerenes being 0D, carbon nanotubes (CNTs) being 1D, graphene 2D, and graphite 3D, bestow them with extraordinary electromechanical and thermochemical properties. The most common carbon nanomaterials are as listed below [1].

2.1.1 Carbon Nanotubes

CNTs are cylindrical nanostructures with diameters as small as 1 nm and lengths up to several centimeters. They exhibit extraordinary mechanical strength, with a Young's modulus of over 1 TPa and tensile strength of up to 63 GPa, which is far greater than that of steel. Electrically, they show either metallic or semiconducting behavior based on their chirality and diameter, with a current density capacity 1000 times greater than that of copper. Their thermal conductivity is also noteworthy, with values as high as 3500 $W\,mK^{-1}$ at room temperature, surpassing that of diamond.

Carbon-based Nanomaterials for Green Applications, First Edition. Edited by Upendra Kumar, Piyush Kumar Sonkar, and Suman Lata Tripathi.

Published 2025 by John Wiley & Sons, Inc.

2.1.2 Graphene

Graphene is a two-dimensional, one-atom-thick layer of carbon atoms arranged in a hexagonal lattice. It is renowned for its exceptional properties, including a tensile strength of approximately 130 GPa, which is stronger than steel. Its charge carriers exhibit extremely high mobility, up to 200,000 $cm^2\ Vs^{-1}$, and it possesses a thermal conductivity in the range of 3000–5000 $W\ mK^{-1}$. These properties make graphene an ideal material for applications in energy storage, conversion, and transmission.

2.1.3 Graphene Oxide

Graphene oxide is a single-atomic-layered material, like graphene, but highly functionalized with oxygen-containing groups. It is hydrophilic and can disperse in water, making it suitable for producing composites, coatings, and inks. While its electrical conductivity is lower than pristine graphene due to the disruption of its sp^2 bonding networks by oxygen groups, it can be partially restored through chemical reduction processes.

2.1.4 Fullerenes

Fullerenes are molecules composed entirely of carbon, forming a closed sphere, ellipsoid, or tube. Among them, C60, also known as buckminsterfullerene, is the most famous. Fullerenes have a high electron affinity and unique electronic properties, making them suitable as electron acceptors in organic photovoltaic cells.

2.2 Significance in Green Energy

In the context of green energy, these nanomaterials hold immense potential. Graphene, with its exceptional electrical conductivity (around 6000 $S\ cm^{-1}$) and large surface area (2630 $m^2\ g^{-1}$), is revolutionizing energy storage devices like batteries and supercapacitors. Its ability to facilitate fast charge transfer makes it an ideal candidate for supercapacitors, where rapid energy storage and release are crucial. CNTs, on the other hand, are not far behind, offering impressive tensile strength (up to 63 GPa) and thermal conductivity (3500 $W\ mK^{-1}$), making them suitable for enhancing the efficiency and durability of solar cells. The significance of carbon nanomaterials in green energy is further accentuated by their role in catalysis and environmental remediation. Their high surface area and tunable electronic properties make them excellent catalysts or catalyst supports, critical in energy conversion processes like fuel cells and water splitting.

Moreover, their ability to interact with various pollutants aids in the development of sensors and filters, essential for monitoring and ensuring a clean energy ecosystem.

2.2.1 Energy Storage

Carbon nanomaterials are revolutionizing the field of energy storage. Graphene and CNTs, with their high surface area ($2630\,m^2\,g^{-1}$ for graphene), provide ample space for the storage of ions in batteries and supercapacitors. For instance, supercapacitors utilizing graphene-based materials can achieve specific capacitances of over $200\,F\,g^{-1}$, significantly higher than traditional materials [2–4].

2.2.2 Solar Energy

In photovoltaic cells, carbon nanomaterials serve as excellent charge carriers, improving the efficiency of solar cells. Graphene's high transparency and conductivity make it an ideal material for transparent conducting electrodes in solar panels. The incorporation of CNTs and fullerenes has been shown to enhance the power conversion efficiency of organic solar cells, reaching values up to 15%.

2.2.3 Catalysis and Fuel Cells

Carbon nanomaterials play a crucial role as catalysts or catalyst supports in fuel cells. Their high surface area, coupled with the ability to functionalize their surface with various chemical groups, makes them highly effective in promoting electrochemical reactions. Graphene and CNTs are used in metal-air batteries and fuel cells to improve the reaction kinetics and overall efficiency.

2.2.4 Thermal Management

The exceptional thermal conductivity of carbon nanomaterials like graphene and CNTs makes them suitable for thermal management in green energy systems. Efficient heat dissipation is crucial in photovoltaic panels and energy storage systems to maintain performance and prolong lifespan.

2.2.5 Environmental Remediation

Carbon nanomaterials are also employed in environmental applications, such as in sensors and filters for detecting and removing pollutants. This contributes to the sustainability aspect of green energy technologies by ensuring minimal environmental impact.

2.3 Techniques for Characterization of Properties of Carbon Nanomaterials

It is clear as day that the carbon nanomaterials, with their remarkable mechanical, electrical, thermal, and chemical properties, are playing a pivotal role in the advancement of green energy technologies. They are at the forefront of innovation in energy storage, conversion, and conservation, pushing the boundaries of what's possible in the realm of sustainable and renewable energy sources. As research continues to unravel new properties and applications of these versatile materials, their contribution to a greener and more sustainable future appears both promising and indispensable. These properties need to be measured, and therefore the most prominent characterization techniques, to be discussed further, are listed here.

- Electrical conductivity
- Thermal conductivity
- Mechanical strength
- Surface area characterization
- Scanning electron microscopy (SEM)
  - Field effect scanning electron microscopy (FESEM)
- Energy dispersive X-ray spectroscopy (EDX)
- Transmission electron microscopy (TEM)
  - High-resolution transmission electron microscopy (HRTEM)
- Electron energy loss spectroscopy (EELS)
- Atomic force microscopy (AFM)
- Raman spectroscopy (RSA)
- Photoluminescence (PL)
  - Time-resolved photoluminescence (TRPL)
- Thermal gravimetric analysis (TGA)
  - Differential scanning calorimetry (DSC)
- Fourier transform infrared spectroscopy (FTIR)
- UV-Vis-NIR spectroscopy (UV-Vis)
- X-ray photoelectron spectroscopy (XPS)
- Small angle X-ray scattering (SAXS)
- X-ray diffraction analysis (XRD)
- Scanning electrochemical microscopy (SECM)
- Electrochemical Impedance Spectroscopy (EIS)

Table 2.1 categorizes the characterization techniques discussed in the present work, based on their applicability for physical and chemical properties' measurements as well as the breadth of the measured properties in the assessed material specimen. Table 2.2 categorizes these characterization techniques based on the reusability of the analyzed samples.

Table 2.1 An indicative categorization of the characterization techniques based on their areas of applicability and the nature of the measured properties.

Breadth of analysis	Characterization of *physical* properties	Characterization of *chemical* properties
Macro (*bulk*)	• Electrical conductivity • Thermal conductivity • Mechanical strength • Surface area • RSA • XRD	• EDX • FTIR • TGA/DSC • UV-Vis • SECM • EIS
Micro (*locale*)	• SEM/FESEM • TEM/HRTEM • AFM • SAXS • PL/TRPL	• EDX • EELS • XPS • SECM • FTIR

Table 2.2 An indicative categorization of the characterization techniques based on the reusability of the analyzed sample.

Nature of analysis	Characterization of *physical* properties	Characterization of *chemical* properties
Non-destructive *(specimen is not destroyed and can be reused for other tests)*	• Electrical conductivity • Thermal conductivity • RSA • XRD • SAXS • PL/TRPL	• FTIR • UV-Vis
Destructive (*specimen is destroyed and cannot be reused for other tests*)	• Mechanical strength • Surface area • SEM/FESEM • TEM/HRTEM • AFM	• EDX • EELS • TGA/DSC • XPS • SECM • EIS

2.3.1 Electrical Conductivity

Electrical conductivity characterization of carbon nanomaterials, particularly graphene and CNTs, is pivotal in leveraging their exceptional properties for green energy applications. The fundamental principle behind this characterization lies in understanding how these materials facilitate electron transport, a property crucial for their function in electronic devices, sensors, and energy storage systems. Figure 2.1a shows an indicative setup to measure the electrical conductivity of the carbon-based material in their powdered form. It may be noted that the variation in compaction varies the density and consequentially the electrical conductivity. Figure 2.1b shows the resultant curves for the relation between the

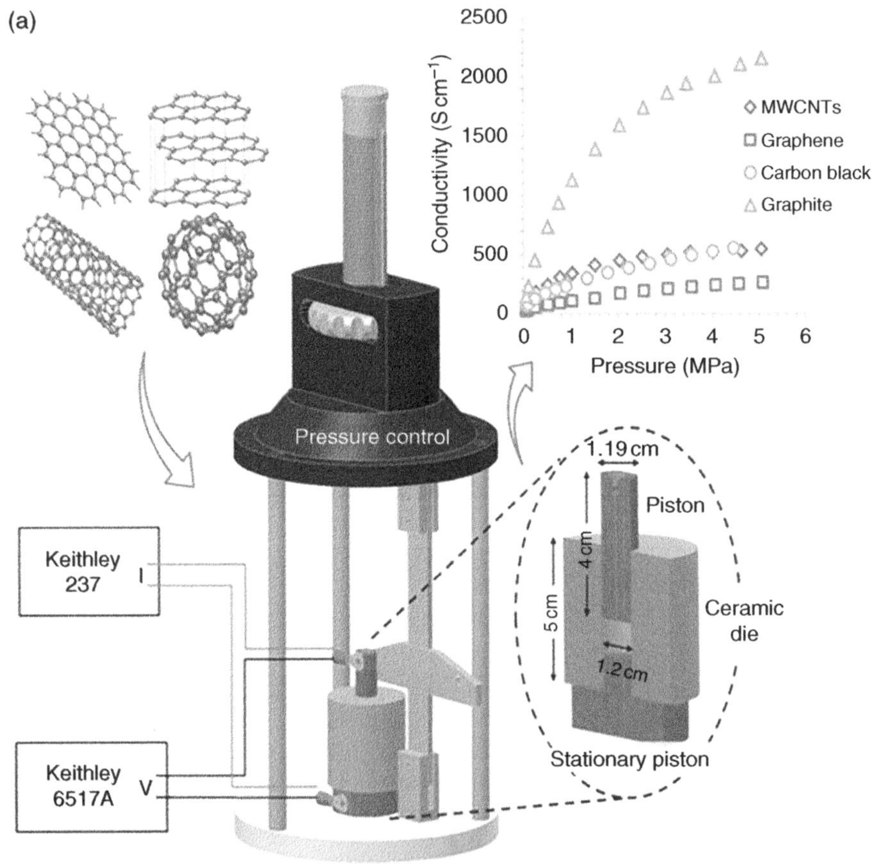

Figure 2.1 (a) Schematic representation of the experimental setup involved in the measurement of the powder density and conductivity. *Source:* Adapted from [5], (b) density as a function of pressure (ρ–P curve) for the various powder compacts. Data given are the average of three experiments with values varying less than 3%. *Source:* [5]/with permission of Elsevier, (c) electrical conductivity behavior of the different carbon powders as a function of pressure. For each material the data points represent an average of at least three identical assays differing not more than 3%. *Source:* Marinho et al. [5]/with permission of Elsevier.

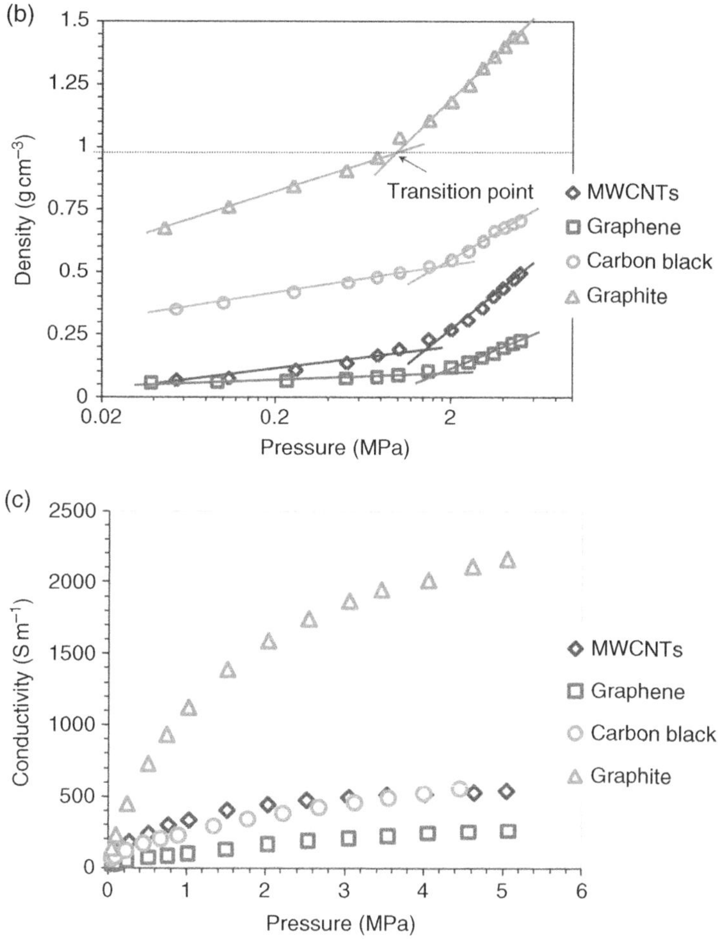

Figure 2.1 (Continued)

compacting pressures and the compacted densities, whereas Figure 2.1c shows the relation between the compacted densities and the measured electrical conductivities for various carbon-based materials.

The state-of-the-art account reveals significant advancements in this domain. A recent study delved into the characterization of metal nanoparticles/CNT/polymer composites, focusing on the uniform dispersion of metal particles on CNTs to enhance electrical conductivity. This method emphasized the role of electrostatic interactions in achieving uniform dispersion, thereby improving the conductivity of the composites. Furthermore, comprehensive research on various CNT yarns highlighted the correlation between their physical properties and structures, underscoring the fact that the electrical conductivity of CNT yarns improves with an increase in the effective length of the constituent CNTs and the density of the yarns. For instance, macroscopic CNT yarns exhibited an impressive electrical conductivity of 10,900 $S\,cm^{-1}$, a testament to the meticulous control over the physical parameters during their production.

Graphene, with its high room-temperature carrier mobility (15,000 $cm^2\,Vs^{-1}$) and large specific surface area (2675 $m^2\,g^{-1}$), also holds prominence in the field of electrical conductivity characterization. The high carrier mobility enables swift and efficient charge flow, reducing internal resistance and enhancing the power density of devices such as hybrid capacitors. Moreover, the exploration of the electrochemical properties of reduced graphene oxide nanosheets has shed light on the optimization of their charge storage capacity and kinetics, further contributing to the advancements in this field.

Looking ahead, the outlook for electrical conductivity characterization in carbon nanomaterials is quite promising. The ongoing research and continuous refinement of characterization techniques are expected to unlock new applications and improve the performance of existing technologies. The ability to fine-tune the electrical properties of carbon nanomaterials will not only bolster their application in green energy solutions but also pave the way for innovative developments across various scientific and industrial domains. As research continues to advance, the synergy between material development and characterization techniques will undoubtedly lead to a new era of materials science marked by heightened efficiency and groundbreaking applications.

2.3.2 Thermal Conductivity

Characterizing the thermal properties of carbon nanomaterials is critical in harnessing their full potential in various technological applications, especially where thermal management is crucial. The fundamental principle underlying thermal conductivity characterization in these materials is understanding how heat flows through them, which is intimately linked to their structure at the atomic level. The intrinsic thermal conductivity of these materials, often considered exceptionally high, can be significantly influenced by defects, grain boundaries, and the presence of functional groups.

Recent advancements in the field have focused on deepening our understanding of how structural, mechanical, and electronic properties of carbon nanomaterials interact under various thermal conditions. Experimental measurements of thermal properties, coupled with theoretical and computational analyses, have been instrumental in uncovering the complex relationships between thermal conductivity and other material properties. For instance, the 3ω method is a notable technique used for measuring thermal conductivity across a wide temperature range, from 30 to 750 K. This method involves heating the sample periodically and measuring the temperature oscillations at triple the frequency of heating, providing insights into the thermal properties of the material. The recent application of computational tools has also opened new avenues for predicting the performance of carbon nanomaterials in roles such as solid nano-refrigerants, particularly for cooling nanodevices.

Looking ahead, the outlook for thermal conductivity characterization in carbon nanomaterials is increasingly promising. The continual refinement of both experimental and computational methods is expected to yield even more precise insights into the thermal behavior of these materials. Moreover, with the growing interest in the thermal properties of unknown phases of carbon materials and the exploration of new thermal applications of known carbon materials, the field is poised for significant advancements. The integration of machine-learning approaches in predicting thermal properties heralds a new era of material characterization, where rapid and accurate predictions can significantly expedite the development of carbon-nanomaterial-based technologies. As the understanding of the thermal properties of carbon nanomaterials deepens, their integration into novel applications, particularly those demanding efficient thermal management, is expected to grow, marking a significant stride in the field of materials science and engineering.

2.3.3 Mechanical Strength

The nanomaterials, notably CNTs and graphene, possess extraordinary mechanical properties due to the strong sp^2 hybridization of carbon–carbon bonds. The mechanical strength of these materials is typically characterized by parameters such as tensile strength, Young's modulus, and fracture toughness. Young's modulus for CNTs, for example, can reach up to 1 TPa, and their tensile strength can be as high as 63 GPa. Graphene, on the other hand, has a breaking strength 200 times greater than steel, with a Young's modulus of approximately 1 TPa. The mechanical characterization often involves techniques like nanoindentation, AFM, and Raman spectroscopy, which can measure properties like stiffness and strength at the nanoscale. The intrinsic strength of these materials is often assessed using the formula $\sigma = E\varepsilon$, where σ is the strength, E is the Young's modulus, and ε is the strain at fracture.

State-of-the-art advancements in the mechanical characterization of carbon nanomaterials have focused on understanding the deformation and failure mechanisms at the atomic level, which is crucial for their application in high-stress environments. Recent developments include the use of in situ transmission electron microscopy (TEM) to observe the real-time mechanical deformation of CNTs and graphene under applied stress. This has provided insights into the kink band formation, buckling, and fracture mechanisms of these materials. Moreover, advancements in computational methods, like molecular dynamics simulations, have complemented experimental findings, offering detailed predictions of mechanical behavior based on atomic structure and bonding. These simulations help in predicting the mechanical response of carbon nanomaterials under different loading conditions, thereby guiding the design of materials with tailored properties for specific applications.

The outlook for the mechanical characterization of carbon nanomaterials is promising, with continuous improvements in both experimental and computational techniques expected. The integration of machine learning with computational mechanics is anticipated to offer new insights into the mechanical behavior of these materials, enabling the prediction of properties based on structural characteristics. Furthermore, as the understanding of the mechanical properties at the nanoscale improves, it will be possible to engineer carbon nanomaterials with customized mechanical properties for specific applications, particularly in the areas of aerospace, automotive, and green energy sectors.

2.3.4 Surface Area Characterization

Surface area characterization, particularly through the Brunauer–Emmett–Teller (BET) method, plays a pivotal role in the field of carbon nanomaterials, especially in applications related to green energy. The fundamental principle of the BET method revolves around physical adsorption of gas molecules on a solid surface, providing vital insights into the surface properties and pore structure of materials. The technique is predicated on the adsorption of a layer of gas molecules onto the material's surface and the subsequent formation of multiple layers, leading to a measurable pressure change. The BET equation allows for the determination of the specific surface area of materials [6, 7]. The BET equation is $\frac{1}{V(P_0 - P)} = \frac{C-1}{V_m CP} + \frac{1}{V_m C}$, where V is the volume of gas adsorbed, P_0 is the saturation pressure of the adsorbed gas, P is the equilibrium pressure at the saturation, V_m is the volume of the adsorbed gas used to form a monolayer, and C is the BET constant. Figure 2.2 shows the indicative adsorption–desorption isotherm plots generated for the estimation of the surface area through the BET adsorption.

In the context of carbon nanomaterials, the BET method has undergone significant advancements to cater to the intricate and diverse surface characteristics of these materials. Recent developments have seen enhancements in the precision and sensitivity of the BET measurements, allowing for the accurate characterization of even the most complex nanomaterials. Innovations in instrumentation and data analysis have led to the refinement of pore size distribution measurements, providing a more detailed and nuanced understanding of the porous

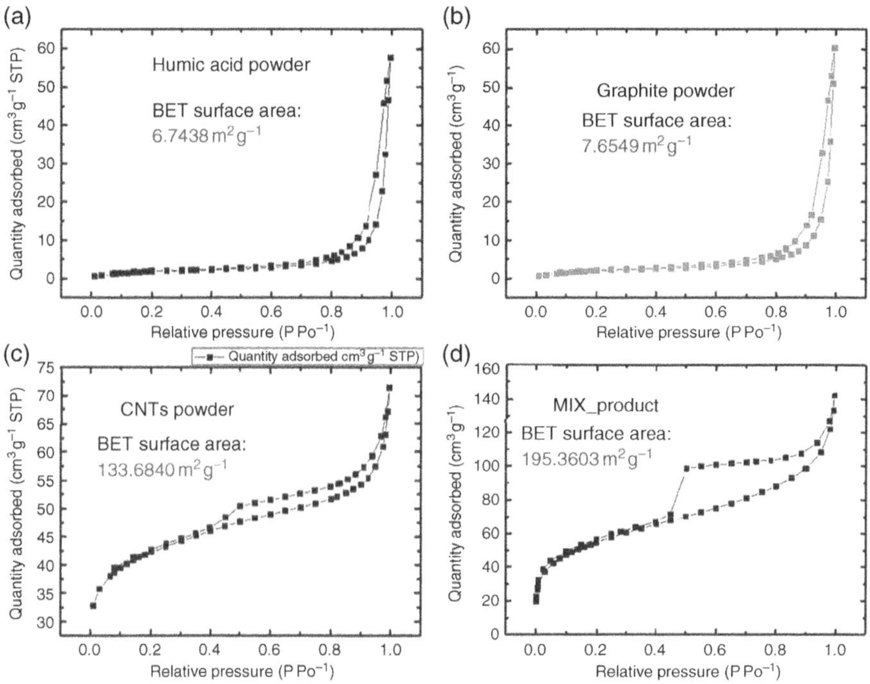

Figure 2.2 BET analysis of surface area for powders of (a) humic acid, (b) graphite, (c) CNTs, and (d) N-doped graphene/CNTs composite, respectively. *Source:* Fouda et al. [8]/ with permission of Elsevier.

structure of carbon nanomaterials. This is particularly crucial in green energy applications where the surface area and pore structure heavily affect the electrode–electrolyte interaction and consequently determine the device performance and the material's energy harvesting efficiency. Moreover, the integration of the BET method with other analytical techniques has become a state-of-the-art approach to obtain a comprehensive understanding of material properties. Techniques such as electron microscopy and spectroscopy are often employed alongside BET measurements to correlate the surface area and pore structure with the material's morphology and chemical composition.

Looking ahead, the outlook for the BET method in the characterization of carbon nanomaterials is highly promising. As the demand for more efficient and sustainable energy solutions grows, the role of accurately characterized nanomaterials becomes increasingly crucial. The continuous improvement and innovation in BET technology and its integration with other analytical methods may provide deeper insights into the structure–property relationships of carbon nanomaterials. This will not only enhance the performance of existing green energy technologies but also make way for the discovery and development of novel materials and solutions for the energy sector. The BET method, with its robust, reliable, and versatile nature, is set to remain an indispensable tool in the material scientist's arsenal, driving forward the frontier of green energy research and application.

2.3.5 Scanning Electron Microscopy

Scanning electron microscopy (SEM), including its advanced form field emission scanning electron microscopy (FESEM), plays a vital role in the characterization of carbon nanomaterials, offering insights into their surface morphology, structural integrity, and compositional details. SEM operates by scanning a focused electron beam over a sample to produce a high-resolution image. As the electrons interact with the atoms in the sample, they generate secondary electrons, backscattered electrons, and characteristic X-rays, which are detected to create an image or analyze the composition. The acceleration voltage typically ranges from 10 to 40 kV. FESEM, an advancement of SEM, uses a field emission gun as its electron source, providing a narrower electron beam with higher energy, leading to higher resolution images. This is crucial in examining the details of carbon nanomaterials, such as nanotubes and graphene layers. For instance, the integration of in situ mechanical testing in the SEM environment allows for direct observation and analysis of mechanical properties like tensile strength and Young's modulus of CNTs and graphene. These techniques employ high-resolution SEM imaging for precise strain measurements, enabling the comprehensive evaluation of mechanical characteristics. The ability to perform such detailed analysis in situ within the SEM chamber highlights the versatility of this technology in understanding the mechanical behavior of carbon nanomaterials under various conditions. Furthermore, SEM, especially FESEM, has been instrumental in exploring the surface morphology and microstructural properties of carbon nanomaterials. This includes the examination of composites reinforced with carbon CNTs or graphene, where SEM provides valuable insights into the dispersion of nanoparticles and the impact of surface modifications on the material's properties. Figure 2.3a, b shows the indicative FESEM micrographs.

The future of SEM and FESEM in the characterization of carbon nanomaterials is promising, with ongoing technological advancements continually enhancing their capabilities. The integration of SEM with other techniques, such as in situ mechanical testing, offers a more comprehensive understanding of the materials. The increasing resolution and sensitivity of FESEM will continue to reveal finer details at the nanoscale, enhancing our understanding of carbon nanomaterials and aiding in the development of new applications, especially in nanotechnology and green energy sectors. The continuous evolution of SEM and FESEM

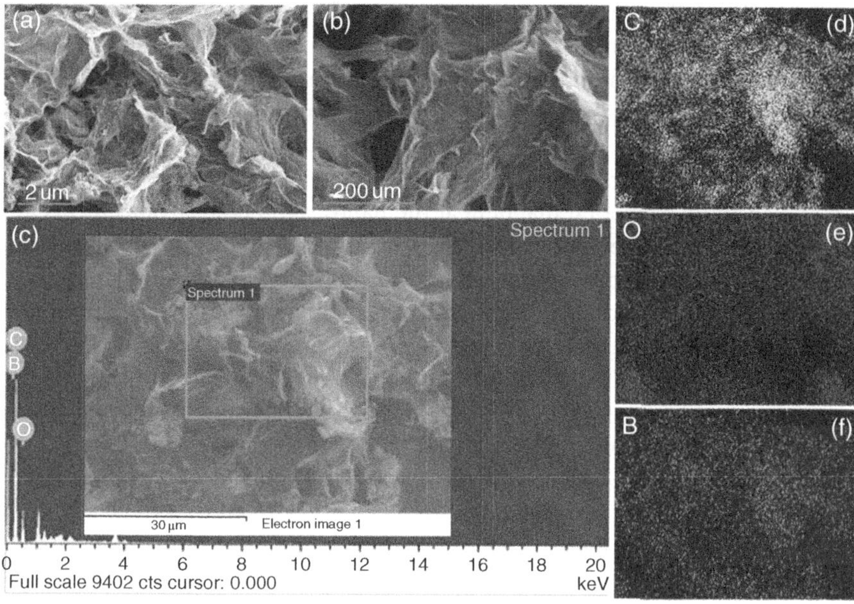

Figure 2.3 FESEM images of Boron-doped reduced graphene oxide (B-rGO) at lower (a) and higher (b) magnifications. (c) EDX spectra of B-rGO. The elemental mapping is shown in (d) carbon, (e) oxygen, and (f) boron. *Source:* Junaid et al. [9]/MDPI/CC BY 4.0.

technologies is expected to open new avenues for the exploration of carbon nanomaterials, contributing significantly to advancements in nanoscience and nanotechnology.

2.3.6 Energy Dispersive X-ray Spectroscopy

Energy dispersive X-ray spectroscopy (EDX) is a powerful analytical technique used for the elemental and chemical characterization of materials, including carbon nanomaterials such as graphene, CNTs, and fullerenes. The fundamental principle of EDX involves the interaction of X-ray beams with the sample, leading to the emission of characteristic X-rays from the sample's elements. These X-rays are then detected and analyzed to determine the elemental composition of the sample. The intensity of the X-ray signal for each element is proportional to the amount of that element in the sample, making EDX a quantitative technique as well. The interaction can be represented by the equation $\boldsymbol{I = I_0 e^{-\mu t}}$, where I is the intensity of the X-ray after passing through the sample, I_0 is the initial X-ray intensity, μ is the linear attenuation coefficient, and t is the thickness of the sample. This equation describes the attenuation of X-rays as they pass through a material, with the attenuation coefficient depending on the material's density and atomic number. As an illustration, Figure 2.3c shows an EDX spectra whereas Figure 2.3d–f show the EDX-based elemental mapping of boron-doped reduced graphene oxide (Br-rGO) [9].

Recent advancements in EDX technology have significantly improved the analysis of carbon nanomaterials. The state-of-the-art EDX systems are equipped with high-resolution detectors and advanced data processing algorithms, enabling precise and accurate elemental mapping at the nanoscale. These advancements are crucial for the characterization of carbon nanomaterials, whose properties are highly dependent on their elemental composition and structure. For instance, researchers can now use EDX to investigate the doping of CNTs with various elements to tailor their electrical and mechanical properties for specific applications. Additionally, the combination of EDX with TEM or SEM allows for the correlation of structural and elemental information, providing a more comprehensive understanding of the nanomaterials.

The outlook for EDX in the characterization of carbon nanomaterials is promising. As the technology continues to evolve, it is expected that EDX will offer even higher spatial resolution and sensitivity, enabling the analysis of individual nanostructures and defects within the carbon lattice. This will be particularly important for the development of next-generation nanomaterials with tailored properties for applications in electronics, energy storage, and catalysis. Furthermore, the integration of EDX with other analytical techniques, such as Raman spectroscopy and atomic force microscopy, is likely to provide a more holistic understanding of carbon nanomaterials, from their atomic structure to their macroscopic properties. The continuous advancements in EDX technology are expanding its capabilities, making it an indispensable tool in the field of nanotechnology.

2.3.7 Transmission Electron Microscopy

Transmission electron microscopy (TEM), including high-resolution TEM (HRTEM), serves as a cornerstone technique in characterizing carbon nanomaterials, offering unparalleled insights into their atomic structure and morphology. Fundamentally, TEM operates by transmitting a beam of electrons through a specimen. The electrons interact with the sample as they pass through, producing an image or diffraction pattern. The magnification is adjusted by changing the current through the electromagnetic lenses, allowing for observations at atomic-scale resolutions. The resolving power of TEM is theoretically limited by the wavelength of the electrons, which is defined by the equation $\lambda = \frac{h}{\sqrt{2meV}}$ where λ is the wavelength, h is Planck's constant, m is the electron rest mass, e is the electron charge, and V is the accelerating voltage. HRTEM, a variant of TEM,

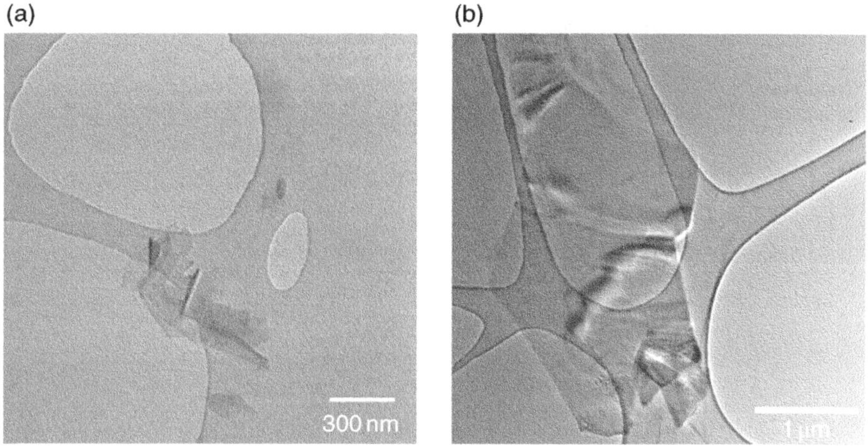

Figure 2.4 (a) TEM image of a sheet of few-layer graphene with diameter of 300–500 nm on a lacey carbon-coated grid. (b) TEM image of a larger sheet of few-layer graphene several micrometers in size. *Source:* Warner et al. [10]/Springer Nature.

furthers it by enabling the imaging of the crystallographic structure of the sample at an atomic scale by resolving individual columns of atoms. Figure 2.4 shows the atypical TEM images for graphene nanosheets [10].

Recent advancements in TEM technologies have significantly enhanced the capability to study the dynamic processes and mechanical properties of carbon nanomaterials in real-time and under varying conditions. Innovative developments such as the integration of in situ mechanical testing techniques within the TEM environment allow for direct observation of microstructural changes and deformation mechanisms at high resolutions. For instance, nanoindentation holders incorporated into TEM have facilitated the quantitative analysis of material micro-deformation and mechanical properties, revealing the intricacies of deformation mechanisms at a nanoscale. Furthermore, the implementation of piezoelectric actuators and oscillation superimposition in nanoindentation holders has broadened the spectrum of mechanical testing, enabling in situ studies of tribology, wear, friction, and material fatigue. Figure 2.5 shows a schematic of one of the 2D material (Graphene) fatigue testing setups [11].

Despite these advances, challenges persist in the field particularly related to the limitations of displacement control accuracy, stability, and the lack of force sensors in some testing setups. The development of MEMS-based stages for in situ mechanical testing in TEM is addressing these challenges, providing more refined control and feedback during experiments. These advances promise to advance our understanding of the mechanical behavior of carbon nanomaterials, crucial in their application in various domains, including green energy.

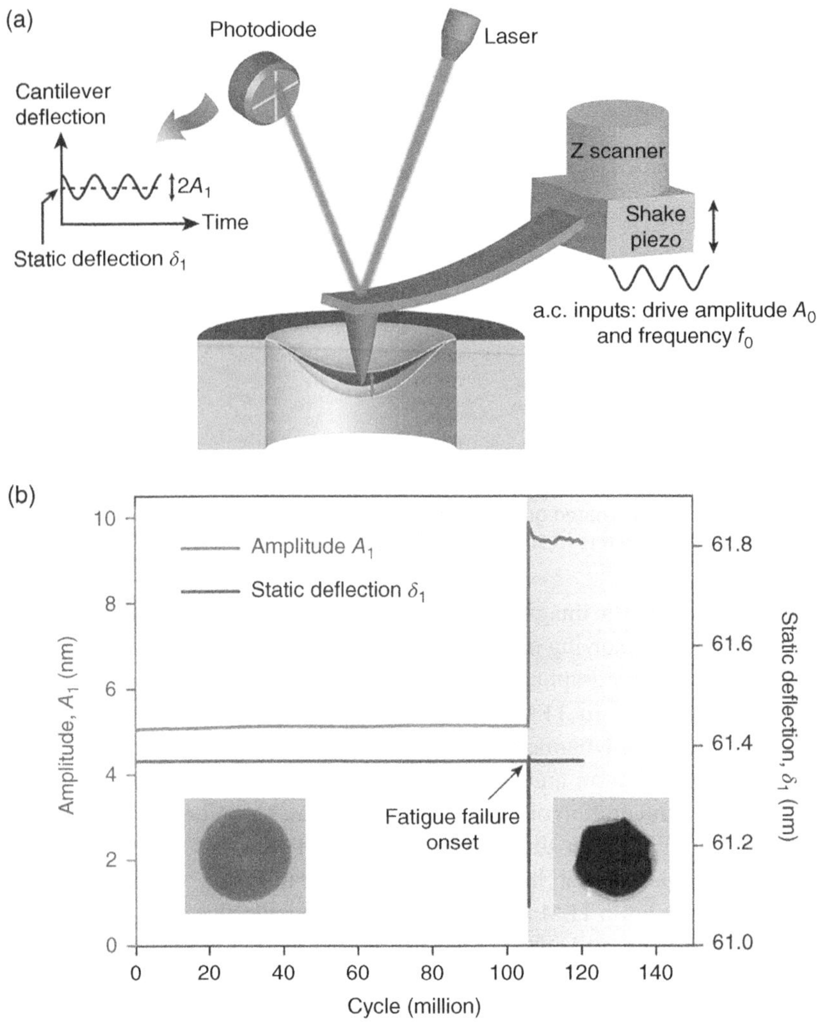

Figure 2.5 Fatigue testing of 2D materials. (a) Schematic of the fatigue testing setup. (b) Experimental data showing evolution of the amplitude and DC force signals. Abrupt jumps of the amplitude and DC deflection signals indicate the onset of fatigue failure in the film after ~106 million cycles. Insets: AFM topographic images before and after fatigue failure. Sample diameter, 2.5 μm. *Source:* Cui et al. [11]/Springer Nature.

In conclusion, TEM, and especially HRTEM, continue to play an indispensable role in the characterization of carbon nanomaterials. With ongoing technological advancements and the integration of in situ mechanical testing capabilities, our understanding of these materials is becoming ever more detailed and nuanced.

This not only enriches our fundamental knowledge of material science but also paves the way for innovative applications of carbon nanomaterials in various industries.

2.3.8 Electron Energy Loss Spectroscopy

Electron energy loss spectroscopy (EELS) is a sophisticated analytical method used in TEM to understand the electronic and chemical properties of materials, particularly carbon nanomaterials. The fundamental principle of EELS lies in inelastic scattering. When a beam of electrons with known kinetic energy interacts with a material, some electrons lose energy and undergo scattering. The energy loss of these electrons provides critical information about the electronic structure and elemental composition of the material. A key aspect of EELS is its ability to distinguish between different bonding states and measure the oxidation state and local coordination of atoms [12, 13]. This is especially useful for light elements, where the onset of ionization edges indicates ionization energy and element concentrations in thin samples [14].

Recent advancements in EELS technology have significantly enhanced its application in studying carbon nanomaterials. High-resolution electron energy loss spectroscopy (HR-EELS) now achieves meV energy resolutions, enabling detailed analysis of low-loss excitations like phonons or vibrational modes, bonding states, and surface plasmons. These improvements are crucial for carbon nanomaterials, where understanding the bonding and electronic structure at the atomic level is essential. Additionally, developments in TEM-EELS instrumentation, like magnetic prisms for electron dispersion and focusing, and energy filters for image formation and spectroscopy, have improved the quality and versatility of EELS data. Furthermore, the introduction of monochromators has enabled high-energy resolution EELS, enhancing the study of electronic structures and bonding effects. These advancements facilitate the acquisition of detailed spectral data and enable elemental mapping at high spatial resolutions.

Looking ahead, the outlook for EELS in the study of carbon nanomaterials is promising. The continuous evolution of EELS instrumentation and techniques, combined with the development of more powerful computers and data analysis methods, paves the way for more comprehensive and accurate material characterization. This includes the possibility of spectrum-imaging techniques that collect complete spectral information for each point in an image, allowing for a more holistic understanding of the material's properties. Moreover, the increasing integration of EELS with other characterization methods, such as X-ray spectroscopy and deep learning algorithms for data analysis, is expected to further expand its capabilities and applications in material science, particularly in the nanoscale characterization of carbon-based materials. EELS has established itself as a critical tool in the characterization of carbon nanomaterials. Its ability to provide detailed insights into the material's electronic structure and elemental composition at the nanoscale makes it indispensable in current and future materials research. With ongoing technological advancements and integration with other analytical techniques, EELS is set to play an even more significant role in advancing our understanding of carbon nanomaterials and their applications.

2.3.9 Atomic Force Microscopy

Atomic force microscopy (AFM) plays a pivotal role in characterizing carbon nanomaterials, leveraging its capacity to image surfaces at the atomic scale. The fundamental principle of AFM involves a cantilever with a sharp tip that scans the material's surface. The interaction between the tip and the surface causes deflections in the cantilever, which are measured to obtain topographical, mechanical, and chemical properties of the surface at the nanoscale.

Recent advancements in AFM have considerably enhanced its capabilities, particularly in time-resolved and nanomechanical applications. A novel development is the time-resolved force microscopy using the delay-time modulation method. This technique has broadened the scope of AFM, extending its applicability beyond academic research to industries and medicine, enabling the exploration of ultrafast light-induced phenomena at the nanoscale. This advancement signifies a substantial leap in capturing dynamic processes with high temporal resolution, which was previously challenging for conventional AFM techniques. Another state-of-the-art development in AFM is related to subsurface nanometrology. Current research is focused on employing quantum sensing techniques, such as using skyrmions—subatomic quasiparticles—to probe materials beyond the surface level. This approach aims to surpass the limitations of current AFM techniques in detecting subsurface features, paving the way for more profound insights into material properties at a deeper level. Additionally, there have been significant innovations in the sensor technology used in AFM. Recent developments involve integrating electronic circuits directly into the AFM sensor, enabling more efficient and accurate data processing. This advancement simplifies the AFM operation, making it more user-friendly and applicable in diverse fields, including measurements on liquid samples, which expands its versatility significantly. The AFM has also made notable strides in visualizing the nanomechanical properties of materials. Recent studies have demonstrated the ability of AFM to observe nano-deformation behavior and nanoscale stress distribution in materials in real space. This capability is particularly valuable in understanding the deformation behavior of polymeric materials, which differs significantly from that of metals and inorganic materials. The use of AFM in these applications provides critical insights into the microstructural changes and stress distribution during material deformation, which is essential for advancing material research and development.

The outlook for AFM in the characterization of carbon nanomaterials is highly promising. The advancements in time-resolved measurements, subsurface exploration, sensor technology, and visualization of nanomechanical properties have significantly broadened the applicability and effectiveness of AFM. These developments not only enhance our understanding of material properties at the nanoscale but also open new avenues for innovative applications in various scientific and industrial fields.

2.3.10 Raman Spectroscopy

Raman spectroscopy (RSA), a vibrational spectroscopic technique, has emerged as an indispensable tool for characterizing carbon nanomaterials, thanks to its nondestructive nature and high sensitivity to structural variations. Fundamental principles of RSA revolve around the Raman effect, where monochromatic light interacts with molecular vibrations, resulting in a shift in the wavelength of the scattered light. This shift is known as the Raman shift and is unique for different crystals and materials in the sample. For carbon nanomaterials, characteristic peaks in the Raman spectrum, such as the D band (indicative of defects or disorder in the carbon lattice) and the G band (representing the E2g mode of sp2-bonded carbon atoms), provide insights into their structural attributes, electronic properties, and purity. Furthermore, the ratio of the intensities of these bands (I(D)/I(G)) serves as a crucial metric for assessing the degree of disorder or defects within the carbon structure.

The G band (~1580/cm) arises from the in-plane vibration of sp^2 carbon atoms and is a primary feature in the Raman spectra of graphene-related materials. The shape, width, and position of the G band can reveal the number of layers, the degree of disorder, and the electronic structure of graphene. The 2D band or G′ band (~2700/cm) is a second-order process and is particularly sensitive to the stacking order and the number of layers in multilayer graphene. In single-layer graphene, the 2D band is sharp and single-peaked, while in bilayer graphene, it splits into multiple peaks. Like graphene, CNTs exhibit the D band (~1350/cm) related to the presence of defects and the G band. The relative intensity of the D and G bands can provide information about the level of defects in the CNT structure. The radial breathing mode (100–300/cm) is characteristic of the radial expansion and contraction of CNTs and is highly sensitive to the diameter of the nanotubes. This mode is exclusively observed in CNTs and is not present in other graphitic materials. Fullerenes exhibit a complex Raman spectrum due to their spherical geometry and unique vibrational modes. The Raman spectrum of C60, for instance, shows characteristic peaks that correspond to the pentagonal pinch mode and other vibrational modes specific to the fullerene structure.

State-of-the-art advancements in Raman spectroscopy have significantly enhanced its application in studying carbon nanomaterials, enabling deeper insights into their intricate structures and properties. Recent developments have seen the integration of Raman spectroscopy with other techniques, amplifying its sensitivity and spatial resolution. These advancements have facilitated the probing of single molecules and monitoring of real-time chemical reactions at the nanoscale. Additionally, the advent of machine learning and advanced computational methods in spectral data analysis has opened new avenues for the rapid and accurate interpretation of complex Raman spectra, further expanding the potential of Raman spectroscopy in the detailed characterization of carbon nanomaterials, including graphene, CNTs, and fullerenes.

The outlook for Raman spectroscopy in the realm of carbon nanomaterials is exceedingly promising. The continuous evolution of this technique, coupled with its integration with emerging technologies, paves the way for groundbreaking discoveries and innovations. As the understanding of carbon nanomaterials deepens and their applications proliferate across various sectors, Raman spectroscopy is expected to play a pivotal role in unveiling their full potential. Future research is likely to focus on enhancing the sensitivity and resolution of Raman techniques, exploring novel hybrid spectroscopic methods, and harnessing the power of artificial intelligence for advanced data analysis. Consequently, Raman spectroscopy is set to remain at the forefront of carbon nanomaterials research, driving advancements in nanotechnology, materials science, and related interdisciplinary fields.

2.3.11 Photoluminescence

Photoluminescence (PL) has emerged as a pivotal technique for investigating the optical properties of carbon nanomaterials, particularly single-walled carbon nanotubes (SWCNTs). This method, instrumental in the realm of nanotechnology and materials science, provides insights into the electronic structure and purity of these nanomaterials. Recent advancements in the functionalization of SWCNTs have garnered considerable attention due to the observed enhancements in their PL properties. Specifically, PL has been a key player in unveiling the intricate interaction between light and carbon nanomaterials at the quantum level.

The process of PL involves the absorption of photons by a material and the subsequent emission of photons because of electron transitions between different energy states within the material. In carbon nanomaterials like SWCNTs, the PL properties are profoundly influenced by their inherent structure and the external perturbations through functionalization or environmental interactions. The high aspect ratio, quantum confinement, and reduced dimensionality of SWCNTs make their PL properties particularly sensitive and tunable.

Recent research has shown that the PL properties of SWCNTs can be extensively adjusted by employing various functionalization techniques. This control opens up new avenues for the application of SWCNTs in optoelectronic devices, sensors, and bio-imaging, where precise tuning of the emission wavelength is crucial.

Moreover, advancements in chiral separation techniques have enriched the understanding and control of PL wavelength in SWCNTs. By selecting SWCNTs with specific chiral indices and applying targeted chemical modifications, researchers can now tailor the excitation and emission wavelengths to suit specific applications, adding a new dimension of customization to the already versatile carbon nanomaterials.

As the field progresses, the integration of computational models and advanced spectroscopic techniques continues to deepen the understanding of PL in carbon nanomaterials, paving the way for novel applications and enhancing the performance of existing technologies. The intricate interplay between the structure, functionalization, and environmental interactions of carbon nanomaterials remains a rich domain for exploration, promising exciting developments in nanotechnology, materials science, and related fields.

2.3.12 Time-Resolved Photoluminescence

Time-resolved photoluminescence (TRPL) is an advanced spectroscopic technique that plays a pivotal role in characterizing the electronic and optical properties of carbon nanomaterials. This technique is instrumental in probing the dynamics of excitons and electron–hole pairs that are crucial for understanding the optical responses of these materials.

In TRPL, the sample is excited by a short pulse of light, and the subsequent emission or PL is monitored as a function of time. The decay of this PL signal provides insights into the lifetimes of excited states, revealing intricate details about recombination processes, trapping, and energy transfer mechanisms within the material.

Recent studies have significantly advanced our understanding of TRPL in carbon nanomaterials like SWCNTs. These nanotubes exhibit unique PL properties due to their one-dimensional structure and quantum confinement effects. For instance, functionalization techniques such as oxidation and alkylation have been shown to effectively modulate the PL properties of SWCNTs. These chemical modifications can shift the intrinsic PL peak of SWCNTs, enabling PL wavelength optimization and enhancing the PL efficiency. This level of control, coupled with advancements in chiral separation techniques, has significantly expanded our capability to tailor the optical properties of SWCNTs for various applications.

Recent advancements in TRPL have introduced sophisticated methods like transient interferometric scattering microscopy, enhancing the resolution and sensitivity of PL measurements. This technique employs a high numerical aperture confocal microscope setup, allowing for high-resolution imaging and efficient suppression of out-of-focus contributions. Such advancements facilitate the sensitive detection of transient probe intensity changes due to the creation of few excitons per excitation pulse. Furthermore, the implementation of transient interferometric scattering aids in the detailed understanding of signal formation processes in carbon nanomaterials. It provides insights into the scattering and absorption cross-sections of the nano-objects under investigation, offering a more comprehensive understanding of their optical properties. These techniques have empowered researchers to probe the ultrafast dynamics of excitons in carbon nanomaterials, unraveling their complex photophysical behaviors. Understanding these dynamics is crucial for optimizing the performance of carbon-based optoelectronic devices, as it impacts their efficiency, stability, and response times.

The rich insights gained from TRPL studies of carbon nanomaterials have profound implications for various technological domains, particularly in the development of photonic and optoelectronic devices. By elucidating the mechanisms of exciton dynamics, recombination processes, and charge transfer in these materials, TRPL paves the way for designing more efficient solar cells, light-emitting diodes, and photodetectors. As research in this field progresses, we can anticipate more sophisticated TRPL methodologies, offering even finer temporal and spatial resolution. These advancements will undoubtedly deepen our understanding of the fundamental processes governing the optical behavior of carbon nanomaterials, driving innovation and enabling new applications in the realm of green technology and beyond.

2.3.13 Thermal Gravimetric Analysis and Differential Scanning Calorimetry

Thermal gravimetric analysis (TGA) and differential scanning calorimetry (DSC) are pivotal techniques in the study of carbon nanomaterials, offering insights into their thermal stability, decomposition mechanisms, and phase transitions. Fundamentally, TGA involves the measurement of mass change in a material as a function of temperature or time under a controlled atmosphere. It is characterized by three main variations: dynamic, static, and quasistatic TGA. Dynamic TGA increases temperature over time while recording mass, providing simultaneous identification of gas release and corresponding temperatures. Static TGA holds the temperature constant, aiding in understanding the decomposition at specific temperatures. Quasistatic TGA involves heating in intervals, ideal for materials with complex decomposition behavior at various temperatures. DSC, on the other hand, measures heat flows associated with material transitions as a function of temperature or time. It provides data on endothermic (heat absorption) and exothermic (heat release) reactions, crucial for studying thermal properties and behaviors of carbon nanomaterials [15–17].

Recent advancements in TGA and DSC of carbon nanomaterials focus on enhancing the understanding of their thermal properties. For example, studies have examined the thermal behavior of functionalized CNTs and graphene. The thermal decomposition of graphene oxide (GO), for instance, involves exothermic reactions releasing gases like H_2O, CO_2, and CO, crucial for understanding its energetic properties. Modern TGA and DSC instruments are often integrated with other analytical techniques like FTIR and mass spectrometry for more comprehensive analysis. These combinations allow for detailed investigation of thermal decomposition processes, including the determination of kinetic parameters and activation energies of reactions.

The outlook for TGA and DSC in the study of carbon nanomaterials is promising. With continuous technological advancements, these techniques are becoming more refined, allowing for more precise and comprehensive thermal analysis. They are essential in developing new carbon nanomaterials with optimized thermal properties for various applications, ranging from energy storage to material science. The integration of TGA and DSC with other analytical methods will likely continue to evolve, providing deeper insights into the complex thermal behaviors of carbon nanomaterials. This will not only enhance our understanding of these materials but also pave the way for innovative applications in various fields.

2.3.14 Fourier Transform Infrared Spectroscopy

Fourier transform infrared spectroscopy (FTIR) stands as a pivotal technique in the characterization of carbon nanomaterials, thanks to its high sensitivity and ability to discern functional groups and molecular structures. FTIR operates on the principle that molecules absorb specific frequencies of IR radiation, causing molecular vibrations that are unique to their structural attributes [18, 19]. The spectrum of absorbed IR radiation forms a molecular fingerprint, revealing intricate details about molecular structure, functional groups, and chemical bonds. The absorbance spectrum, a core component of FTIR analysis, is represented by the equation $\boldsymbol{A} = \mathbf{log}(\boldsymbol{I_0}/\boldsymbol{I})$ where A is the absorbance, I_0 is the intensity of the incident light, and I is the intensity after passing through the sample. Moreover, the relationship between absorbance and molecular concentration is elegantly described by Beer's Law, $\boldsymbol{A} = \boldsymbol{\varepsilon l c}$ linking absorbance (A) to the molar absorptivity (ε), path length (l), and concentration (c) of the sample.

Recent advancements have significantly augmented the capabilities of FTIR in analyzing carbon nanomaterials. Cutting-edge techniques and instruments have been developed, offering enhanced resolution, sensitivity, and speed. These improvements have expanded the applicability of FTIR, enabling real-time monitoring of chemical reactions, surface analysis of nanomaterials, and even the study of complex biological systems at the nanoscale.

Looking forward, the prospects of FTIR in the realm of carbon nanomaterials are exceptionally promising. Ongoing advancements in computational methods, data analysis techniques, and instrument design are poised to further enhance the resolution and sensitivity of FTIR. This, in turn, will deepen our understanding of nanomaterials and catalyze breakthroughs in nanotechnology, materials science, and various applied fields, marking FTIR as an indispensable tool in modern scientific research and technological development.

2.3.15 UV-Vis-NIR Spectroscopy

Ultraviolet, visible, and near-infrared region (UV-Vis-NIR) spectroscopy has emerged as a critical tool for characterizing carbon nanomaterials due to its ability to provide detailed electronic and molecular information for solid, liquid, and gas phases for both natural and synthetic materials. Fundamentally, this spectroscopy technique relies on the absorption or emission of light in the UV-Vis-NIRs of the electromagnetic spectrum. The interaction of light with the nanomaterials results in spectra that reveal intrinsic properties like band-gap energies, electronic transitions, and molecular vibrations [20, 21].

Recent advancements in this field have underscored the versatility and precision of UV-Vis-NIR spectroscopy when combined with chemometric methods. This combination has unlocked new potentials in discriminant analysis, classification, identification, and quantitative analysis across various applications. Recent papers have highlighted the use of this technique in different contexts, such as identifying wood species, detecting oil adulteration, and environmental monitoring, showcasing the method's adaptability and the depth of information it can provide. Notably, the use of machine learning and other advanced data analysis techniques has significantly enhanced the ability to interpret complex spectral data, leading to insightful and sometimes unexpected findings.

Looking forward, the outlook for UV-Vis-NIR spectroscopy in the study of carbon nanomaterials is highly promising. The technique's minimal requirement for sample preparation, capability for remote sensing, and rapid data acquisition and analysis make it an attractive choice for both research and practical applications. As machine learning and other advanced analytical techniques continue to evolve, they will further enhance the capability of UV-Vis-NIR spectroscopy to provide deeper and more nuanced insights into the properties and behaviors of carbon nanomaterials. The ongoing integration of these technologies promises to drive significant advancements in the field, offering a powerful toolkit for both scientific exploration and practical problem-solving in various industries.

2.3.16 X-ray Photoelectron Spectroscopy

X-ray photoelectron spectroscopy (XPS) has emerged as a pivotal technique in characterizing carbon nanomaterials and in studying their surface chemistry as well as bonding structures. XPS operates on the principle of the photoelectric effect, where an X-ray photon is absorbed by a material, causing the emission of core electrons. The kinetic energy of these emitted electrons is measured, providing insights into the material's elemental composition, chemical state, and electronic structure.

The process of XPS involves three main stages [22, 23]. First, an X-ray photon transfers its energy to a core electron. This causes the atom to emit the photoelectron, leaving behind a core hole. The photoelectron then travels to the surface of the material and escapes into the vacuum, where its energy is measured. The binding energy of the core electron, which is crucial to the XPS spectrum, is affected by initial state energy, final state effects (including many-body interactions and core hole relaxation), and inelastic scattering processes that reduce the kinetic energy of the photoelectrons as they move to the surface.

Understanding this process in detail allows for precise analysis of materials, especially in the characterization of doped carbon nanomaterials like graphene and CNTs. The analysis focuses on measuring binding energies [24], which are essential for compositional analysis through XPS [25]. Moreover, interactions between the photoelectron and the core hole, as well as inelastic scattering processes, contribute to the overall photoemission intensity background. This background can be mathematically modeled using a convolution of Gaussian and Lorentzian profiles, known as the Voigtian profile. In metallic systems, such as graphite or metallic CNTs, an additional asymmetry toward higher binding energies is observed, described by the Doniach–Šunjić form. This asymmetry arises from the excitation of numerous low-energy electron–hole pairs that screen the core hole. Figure 2.6 shows an atypical XPS spectra for graphene oxide and that of its synthesis precursors [26]

Recent developments in XPS characterization techniques have considerably broadened its application spectrum. Advanced XPS techniques not only provide qualitative and quantitative insights into the chemical nature of material surfaces but also offer a deeper understanding of the depth-wise variation of topology, morphology, and homogeneity. These advancements facilitate the study of new interfaces in various materials, including semiconductors and electrodes, solid synthesis pathways, catalytic and surface processes, and electronically active materials[27]. One significant development is Tougaard background analysis, a nondestructive technique that helps in obtaining depth profile and structural information from the background profiles of complex composite materials. Furthermore, XPS is now used for overlayer thickness analysis in nanocomposite

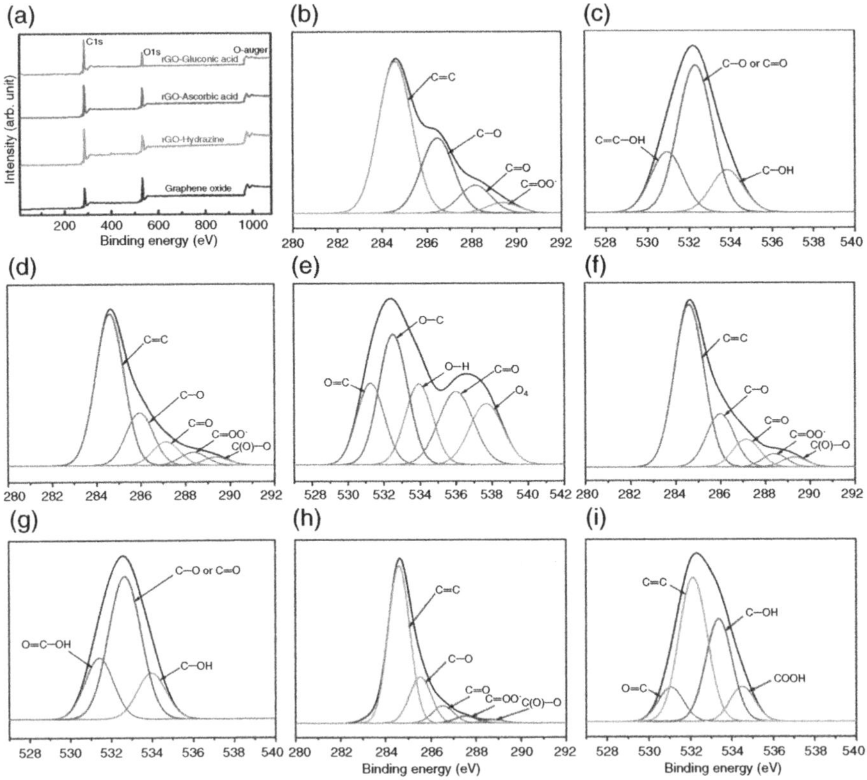

Figure 2.6 XPS spectra of (a) all data, graphene oxide (b) C one second and (c) O one second, hydrazine hydrate (d) C one second and (e) O one second, L-ascorbic acid (f) C one second and (g) O one second, and gluconic acid (h) C one second and (i) O one second. *Source:* Na et al. [26]/Springer Nature.

materials [28]. These enhancements in XPS, such as the use of angle-resolved XPS (ARXPS) for measuring film thickness or mapping layer structures, have made it an indispensable tool in materials science, especially in studying solid interfaces crucial for many electronic systems.

As it is, XPS stands as a robust and sophisticated technique in the realm of surface analysis and characterization of carbon nanomaterials. The method's ability to deduce chemical states and surface compositions with precision has been instrumental in the study and development of nanomaterials. With ongoing advancements in XPS techniques and analytical methods, its application scope is continuously expanding, promising deeper insights into the material properties and fostering advancements in nanotechnology and materials science. The future of XPS in carbon nanomaterials characterization seems poised for further innovative breakthroughs, making it an essential tool for researchers in the field.

2.3.17 Small Angle X-ray Scattering

Small angle X-ray scattering (SAXS) is a powerful technique used for analyzing the structural characteristics of carbon nanomaterials, among other substances. The core principle of SAXS lies in its ability to provide insights into the size, shape, and distribution of nanoparticles within a material [29, 30]. The fundamental measurement in SAXS is the intensity of X-rays scattered by the material as a function of the scattering angle. The scattered intensity is typically plotted against the scattering vector q, which is related to the scattering angle 2θ and the X-ray wavelength λ, by the equation $q = \frac{4\pi \sin\theta}{\lambda}$. This scattering profile is then interpreted to extract structural information about the material, such as particle size distributions, pore sizes, and surface-to-volume ratios, using various computational models and fitting techniques.

In the realm of carbon nanomaterials, recent advancements in SAXS technology and methodology have significantly enhanced the scope and precision of structural analysis. Integration of SAXS with complementary techniques like small angle neutron scattering and various high-resolution structural methods has led to a more comprehensive understanding of nanoscale phenomena. This hybrid approach allows for the study of complex biomacromolecular systems, offering detailed insights into their structure and dynamics in near-native environments. The development of dedicated synchrotron beamlines and the use of automated, high-throughput data analysis have further streamlined the SAXS process, enabling rapid structural screening and time-resolved studies of biological processes.

Furthermore, the field of SAXS has benefited from the integration of advanced computational tools and algorithms. Techniques such as iterative phase retrieval and machine-learning-based methods like deep learning have been employed to enhance the resolution and accuracy of structural models derived from SAXS data. These methods have enabled researchers to construct detailed ab initio models of biomacromolecules, offering insights into their shape and structural dynamics. The use of hybrid resolution molecular dynamic (MD) methods and advanced inferential structure determination techniques has further extended the capabilities of SAXS in providing detailed structural ensembles of flexible and heterogeneous systems.

Looking ahead, the future of SAXS in the study of carbon nanomaterials appears promising. The continuous evolution of instrumental technology, coupled with advancements in computational modeling and data analysis, is set to further refine the resolution and accuracy of SAXS-derived structural information. The integration of SAXS with complementary techniques and the development of more sophisticated analytical methods are expected to enhance its applicability across a wide range of materials and biological systems. As a result, SAXS is poised to remain a critical tool in the elucidation of complex nanostructures and in driving forward the frontiers of nanomaterial research.

2.3.18 X-ray Diffraction Analysis

X-ray diffraction (XRD) analysis is a powerful technique widely used for the characterization of carbon nanomaterials. It provides critical information about the structural properties of these materials, including their phase composition, crystallinity, and atomic arrangement. This technique has been instrumental in advancing our understanding of the unique properties of carbon nanomaterials and their applications in various fields, especially in green energy. XRD works by irradiating a material with incident X-rays and then measuring the intensities and scattering angles of the X-rays that are reflected by the material's crystal lattice[7, 31]. The principle is based on the constructive interference of monochromatic X-rays and a crystalline sample. When the conditions satisfy Bragg's Law $\boldsymbol{n\lambda = 2d\sin\theta}$, where n is an integer, λ the wavelength of incident wave, d the distance between the crystal planes, and θ the angle of incidence), the reflected X-rays from the crystal planes generate a diffraction pattern that is specific to the material's crystal structure and can be used to deduce various structural properties.

Carbon nanomaterials, including graphene, CNTs, fullerenes, and graphene oxide, exhibit distinct diffraction patterns that provide insights into their structure. For instance, in case of graphene and graphene oxide, XRD can identify the degree of oxidation in graphene oxide. The characteristic peak around 26.5° in graphene corresponds to the (002) plane of graphite[32], indicating the interlayer spacing. Similarly, for CNTs, XRD can differentiate between SWCNT and MWCNT. The diffraction patterns of CNTs show peaks corresponding to the hexagonal graphite structure, and the intensity and position of these peaks can reveal the alignment, diameter, and chirality of the nanotubes. For fullerenes, the XRD patterns of exhibit several sharp Bragg peaks. The interlayer spacing in graphite, as determined by the (002) peak in XRD, is approximately 3.35 Å. The crystallite size of carbon nanomaterials can be calculated from the broadening of XRD peaks using the Scherrer equation [31], $\boldsymbol{L = \frac{K\lambda}{\beta\cos\theta}}$ where L is the crystallite size, K is the shape factor, λ is the X-ray wavelength, β is the full width at half maximum of the peak, and θ is the Bragg angle.

Recent advancements in XRD techniques have significantly enhanced our ability to characterize carbon nanomaterials. High-resolution XRD is a recent technique that offers improved resolution and sensitivity, allowing for the detection of subtle structural changes and defects in carbon nanomaterials. The development of in situ XRD analysis techniques have enabled the real-time observation of structural changes in carbon nanomaterials during various processes, such as chemical reactions, charging/discharging in batteries, or mechanical stress. Three-dimensional (3D) XRD provides a 3D view of the internal structure of carbon nanomaterials, offering detailed insights into their crystallography and defects. Advanced algorithms and software have been developed for the quantitative analysis of XRD data, enabling precise determination of phase composition, crystallite size, strain, and other structural parameters.

In conclusion, XRD is a vital technique for the characterization of carbon nanomaterials, providing essential information about their crystal structure, phase composition, and other structural properties. The recent advancements in XRD techniques have further enhanced its capabilities, making it an essential material characterization technique. As we continue to explore and develop new carbon nanomaterials for various applications, especially in green energy, the role of XRD in their characterization and understanding is expected to become even more significant.

2.3.19 Scanning Electrochemical Microscopy

Scanning electrochemical microscopy (SECM) is a potent analytical technique that offers localized electrochemical probing of a surface. This method employs an ultramicroelectrode (UME) tip, which is meticulously maneuvered in proximity to the sample surface. The tip generates a current through electrochemical reactions, which is finely modulated by the distance between the tip and the sample surface, as well as the sample's conductivity and reactivity [33, 34]. SECM has found its way into various applications, including corrosion analysis, the study of biological systems, and the characterization of catalytic processes.

With the advent of carbon nanomaterials known for their unique properties like high surface area, electrical conductivity, and mechanical strength, SECM has emerged as a critical tool in understanding these materials' electrochemical behavior. The characterization typically involves moving the UME in a raster pattern over the sample while recording the current to generate a two-dimensional activity map. The fundamental governing equation for the feedback mode of SECM, where the tip current is influenced by the substrate's electrochemical activity, is given by $\boldsymbol{i_{\text{tip}} = 4nFC_{\text{o}}\sqrt{D/\pi t} + k_{\text{s}}A_{\text{tip}}C_{\text{o}}}$ where i_{tip} is the tip current, n is the number of electrons transferred, F the Faraday constant, D the diffusion coefficient of the mediator, C_{O} is the concentration of the mediator in bulk solution, t is the time, k_{S} is the heterogeneous rate constant of the substrate reaction, and A_{tip} is the tip area.

In the context of carbon nanomaterials, SECM's recent advancements have significantly enhanced our understanding and capabilities [35]. The evolution of UMEs into nanoelectrodes has been a game-changer, offering higher spatial resolution and sensitivity. These advancements allow the probing of individual CNTs or graphene flakes, unraveling their intrinsic electrochemical properties detached from the collective behavior of the bulk material. Furthermore, the integration of SECM with other scanning probe techniques, like AFM, has permitted the simultaneous acquisition of topographical and electrochemical data, providing a more comprehensive picture of the sample's surface characteristics.

Moreover, the development of novel SECM methodologies, such as shear force-based distance control, has improved the precision of tip–sample distance regulation, crucial for high-resolution imaging and minimizing tip–sample interactions that may lead to sample damage or tip crash. The introduction of new modes, like the redox competition mode, has expanded the applicability of SECM, enabling the study of kinetic processes and charge transfer mechanisms at the nanoscale.

The future of SECM in the realm of carbon nanomaterials is promising and is anticipated to grow for a key role in the exploration of newer morphologies and interphases in device fabrication domain. The ongoing miniaturization of the UMEs and the integration with more sophisticated imaging and analytical techniques will likely open new avenues for research. This integration is expected to lead to a more profound understanding of the electrochemical properties at the single-molecule or single-entity level, offering insights into the fundamental processes that govern the behavior of these materials. As SECM continues to evolve, its contribution to the field of carbon nanomaterials is poised to be indispensable, driving innovations and breakthroughs in energy storage, sensing, catalysis, and beyond.

2.3.20 Electrochemical Impedance Spectroscopy

Electrochemical impedance spectroscopy (EIS) has proven to be an invaluable technique in the characterization and analysis of carbon nanomaterials, especially given their complex interfacial electrochemical properties. Fundamentally, EIS measures the impedance of a system over a range of frequencies, providing insights into the behavior of charge carriers in materials, interfaces, and their composite structures [36]. A pivotal aspect of EIS in carbon nanomaterials lies in its ability to elucidate the intricate interplay between electrical and structural properties. The electrochemical processes are often represented using equivalent circuits comprising resistors and capacitors to model the resistance and the capacitance, respectively, at the electrode–electrolyte interfaces as well as in the bulk phases [37, 38]. Figure 2.7 shows atypical EIS response plots [39].

Recent advances have significantly broadened the scope and applicability of EIS in the realm of carbon nanomaterials, such as CNTs and MXene [40, 41], a novel class of two-dimensional materials. Investigations have revealed that the electrical and thermal conductivities of composite nanomaterials can be substantially enhanced by the incorporation of CNTs, a phenomenon intricately linked to the microstructural changes within the material. These changes are effectively probed by EIS, providing a window into the dynamic responses of these materials under varying external conditions like electrical fields or shear stresses in microfluidic

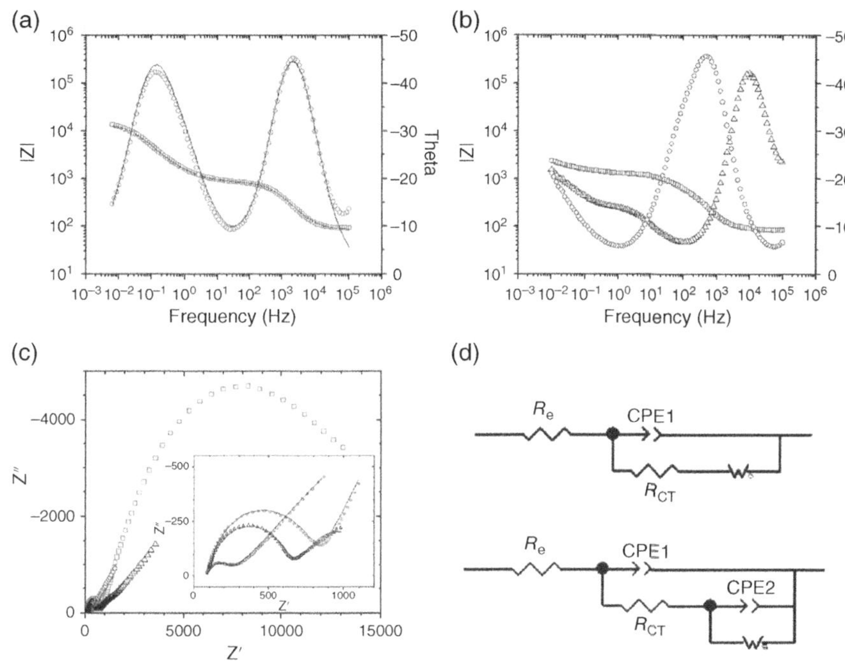

Figure 2.7 (a) Impedance modulus and phase angle diagrams versus frequency of Bode plots for GO/GCE. (b) Phase angle diagrams vesus frequency for ERG/GCE (o) and PRG/GCE (△). Impedance modulus versus frequency for ERG /GCE (□). (c) Nyquist diagram for GO/GCE (□), ERG/GCE (o), and PRG/GCE (△). Solid line (–) corresponds to the fitting of the experimental data. Inset: High-frequency interval. The solution is 0.1 M PBS (pH = 7), 0.1 M KCl containing 10 mM $K_3Fe(CN)_6$/10 mM $K_4Fe(CN)_6$. (d) Electrical equivalent circuits. *Source:* Casero et al. [39]/with permission of Elsevier.

environments. Such insights are pivotal for optimizing the performance of these materials in practical applications ranging from sensing to energy storage.

Looking ahead, the outlook for EIS in the analysis of carbon nanomaterials is exceedingly promising. With its ability to provide detailed insights into the electrochemical properties and interfacial dynamics of carbon nanomaterials, EIS stands as a cornerstone technique, poised to drive innovations across various scientific and technological domains.

2.4 Conclusion

The work provides a detailed overview of carbon-based nanomaterials, highlighting their essential role in enhancing green energy technologies. It covers the wide-ranging properties of these materials, such as electrical, thermal, and mechanical characteristics, focusing on types like graphene, CNTs, and fullerenes. It thoroughly explores various sophisticated methods used to analyze these materials, including SEM, TEM, AFM, and EIS, emphasizing how these techniques are crucial for understanding the materials' intricate structures and capabilities.

Furthermore, the work points out the significant potential of these nanomaterials in transforming key industries, especially in improving energy storage, conversion, and environmental cleanup. This exploration not only underscores the materials' complexity and promise but also suggests that continued research could lead to major scientific and technological breakthroughs. By offering an in-depth look at both the properties and the advanced methods used to study these nanomaterials, the document sets the stage for future progress in nanotechnology and sustainable energy solutions.

References

1 Sharma, S., Basu, S., Shetti, N.P. et al. (2022). Versatile graphitized carbon nanofibers in energy applications. *ACS Sustainable Chemistry & Engineering* 10 (4): 1334–1360.
2 Tripathi, S.L. and Padmanaban, S. (ed.) (2020). *Green Energy*. Wiley.
3 Kumar, U. and Upadhyay, S. (2020). Sr2SnO4 ruddlesden popper oxide: future material for renewable energy applications. In: *Green Energy* (ed. S.L. Tripathi, and S. Padmanaban), 221–250. Wiley.
4 Dhayal, V., Hashmi, S.Z., Kumar, U. et al. (2020). Spectroscopic studies, molecular structure optimization and investigation of structural and electrical properties of novel and biodegradable Chitosan-GO polymer nanocomposites. *Journal of Materials Science* 55 (30): 14829–14847.
5 Marinho, B., Ghislandi, M., Tkalya, E. et al. (2012). Electrical conductivity of compacts of graphene, multi-wall carbon nanotubes, carbon black, and graphite powder. *Powder Technology* 221: 351–358.
6 Ghane, N., Sadrnezhaad, S.K., and Hosseini, H.S.M. (2020). Combustion synthesis of g-C3N4/Fe2O3 nanocomposite for superior photoelectrochemical catalytic performance. *Applied Surface Science* 534 (May): 147563.
7 Jamalzadeh, Z., Haghighi, M., and Asgari, N. (2013). Synthesis, physicochemical characterizations and catalytic performance of Pd/carbon-zeolite and Pd/carbon-CeO2 nanocatalysts used for total oxidation of xylene at low temperatures. *Frontiers of Environmental Science & Engineering* 7 (3): 365–381.
8 Fouda, A.N., Duraia, E.S.M., and Almaqwashi, A.A. (2021). Facile and scalable green synthesis of N-doped graphene/CNTs nanocomposites via ball milling. *Ain Shams Engineering Journal* 12 (1): 1017–1024.
9 Junaid, M., Khir, M.H.M., Witjaksono, G. et al. (2020). Boron-doped reduced graphene oxide with tunable bandgap and enhanced surface plasmon resonance. *Molecules* 25 (16): 3646.

10 Warner, J.H., Rümmeli, M.H., Ge, L. et al. (2009). Structural transformations in graphene studied with high spatial and temporal resolution. *Nature Nanotechnology* 4 (8): 500–504.

11 Cui, T., Mukherjee, S., Sudeep, P.M. et al. (2020). Fatigue of graphene. *Nature Materials* 19 (4): 405–411.

12 González, P.G. and Pliego-Cuervo, Y.B. (2013). Physicochemical and microtextural characterization of activated carbons produced from water steam activation of three bamboo species. *Journal of Analytical and Applied Pyrolysis* 99: 32–39.

13 Bosman, M., Watanabe, M., Alexander, D.T.L., and Keast, V.J. (2006). Mapping chemical and bonding information using multivariate analysis of electron energy-loss spectrum images. *Ultramicroscopy* 106: 11–12. SPEC. ISS.), 1024–1032.

14 Ganesan, R., Dhinasekaran, D., Paramasivam, T. et al. (2012). Preparation and characterization of polyindole-iron oxide composite polymer electrolyte containing LiClO 4. *Polymer-Plastics Technology and Engineering* 51 (3): 225–230.

15 Aboyade, A.O., Görgens, J.F., Carrier, M. et al. (2013). Thermogravimetric study of the pyrolysis characteristics and kinetics of coal blends with corn and sugarcane residues. *Fuel Processing Technology* 106: 310–320.

16 Siddiqi, M.H., Liu, X.M., Hussain, M.A. et al. (2022). Evaluation of physiochemical, thermal and kinetic properties of wheat straw by demineralising with leaching reagents for energy applications. *Energy* 238: 122013.

17 Awal, A. and Sain, M. (2011). Spectroscopic studies and evaluation of thermorheological properties of softwood and hardwood lignin. *Journal of Applied Polymer Science* 122 (2): 956–963.

18 Ojha, D.K., Kumar, V.S.P., and Vinu, R. (2021). Analytical pyrolysis of bagasse and groundnut shell briquettes: kinetics and pyrolysate composition studies. *Bioresource Technology Reports* 15: 100784.

19 Mendham, A., Dines, T., Withnall, R. et al. (2009). IR/Raman spectroscopy and DFT calculations of cyclic di-amino acid peptides. Part III: comparison of solid state and solution structures of cyclo(L-Ser-L-Ser). *Journal of Raman Specroscopy* 40 (11): 1498–1507.

20 Rochkind, M., Pasternak, S., and Paz, Y. (2015). Using dyes for evaluating photocatalytic properties: a critical review. *Molecules* 20 (1): 88–110.

21 Singh, A., Tejasvi, R., Karmakar, S., and Basu, S. (2021). α-Fe2O3 nanorods decorated with NiMnO3 co-catalyst as photoanode for enhanced oxygen evolution reaction in photoelectrochemical water splitting. *Materials Today Communications* 27: 102231.

22 Tejasvi, R. (2022). Shape tailoring of TiO2 nanostructures during thermochemical synthesis by etchant-oxidant combinatorics. *Materials Today Proceedings* 57 (xxxx): 130–134.

23 Tejasvi, R. and Basu, S. (2020). Thermochemical method of synthesizing stemmed nanoflower TiO2/eC3N4 heterojunction structures with enhanced solar water splitting. *Journal of Physics: Energy* 2 (3): 035002.

24 Biesinger, M.C., Lau, L.W.M., Gerson, A.R., and Smart, R.S.C. (2010). Resolving surface chemical states in XPS analysis of first row transition metals, oxides and hydroxides: Sc, Ti, V, Cu and Zn. *Applied Surface Science* 257 (3): 887–898.

25 Shaikh, N.S., Padalkar, N.S., Lokhande, V.C. et al. (2022). Graphene-based aqueous magnesium ion hybrid supercapacitors with an appealing energy density advanced by a KI additive. *Energy and Fuels* 36 (13): 7186–7193.

26 Na, Y.I., Song, Y.I., Kim, S.W., and Suh, S.J. (2017). Study on properties of eco-friendly reduction agents for the reduced graphene oxide method. *Carbon Letters* 24 (1): 1–9.

27 Huang, H., Li, Y.X., Wang, H.L., and Jiang, W.F. (2021). In situ fabrication of ultrathin-g-C3N4/AgI heterojunctions with improved catalytic performance for photodegrading rhodamine B solution. *Applied Surface Science* 538: 148132.

28 Fedoseeva, Y.V., Shlyakhova, E.V., Makarova, A.A. et al. (2023). X-ray spectroscopy study of defect contribution to lithium adsorption on porous carbon. *Nanomaterials* 13 (19): 2623.

29 Yokoyama, H. (2012). Small angle X-ray scattering studies of nanocellular and nanoporous structures. *Polymer Journal* 45 (1): 3–9.

30 Narayanan, T., Chèvremont, W., Zinn, T., and Meneau, F. (2023). Small-angle X-ray scattering in the era of fourth-generation light sources. *urn:issn:1600-5767* 56 (4): 939–946.

31 Tejasvi, R., Sharma, M., and Upadhyay, K. (2015). Passive photo-catalytic destruction of air-borne VOCs in high traffic areas using TiO2-coated flexible PVC sheet. *Chemical Engineering Journal* 262 (3): 875–881.

32 Tejasvi, R. and Basu, S. (2020). Formation of C3N4 thin films through the stoichiometric transfer of the bulk synthesized gC3N4 using RFM sputtering. *Vacuum* 171: 108937.

33 Lee, Y. and Bard, A.J. (2002). Fabrication and characterization of probes for combined scanning electrochemical/optical microscopy experiments. *Analytical Chemistry* 74 (15): 3626–3633.

34 Wang, X., Han, L., Xin, H., and Mirkin, M.V. (2019). TEM-assisted fabrication of sub-10 nm scanning electrochemical microscopy tips. *Analytical Chemistry* 91 (24): 15355–15359.

35 Filice, F.P. and Ding, Z. (2019). Analysing single live cells by scanning electrochemical microscopy. *Analyst* 144 (3): 738–752.

36 Lee, H.J., Bai, S.J., and Song, Y.S. (2017). Microfluidic electrochemical impedance spectroscopy of carbon composite nanofluids. *Scientific Reports* 7 (1): 1–11.

37 Wang, S., Zhang, J., Gharbi, O. et al. (2021). Electrochemical impedance spectroscopy. *Nature Reviews Methods Primers* 1 (1): 1–21.

38 Kern, R., Sastrawan, R., Ferber, J. et al. (2002). Modeling and interpretation of electrical impedance spectra of dye solar cells operated under open-circuit conditions. *Electrochimica Acta* 47 (26): 4213–4225.

39 Casero, E., Parra-Alfambra, A.M., Petit-Domínguez, M.D. et al. (2012). Differentiation between graphene oxide and reduced graphene by electrochemical impedance spectroscopy (EIS). *Electrochemistry Communications* 20 (1): 63–66.

40 Aguedo, J., Lorencova, L., Barath, M. et al. (2020). Electrochemical impedance spectroscopy on 2D nanomaterial MXene modified interfaces: application as a characterization and transducing tool. *Chemosensors* 8 (4): 127.

41 Li, Z., Guo, D., Wang, D. et al. (2021). Exploration of Metal/Ti3C2 MXene-derived composites as anode for high-performance zinc-ion supercapacitor. *Journal of Power Sources* 506: 230197.

3

Green Energy: Present and Future Prospectives

Irtiqa Amin[1], Quraazah Akeemu Amin[2], and Harpreet Kaur[3]

[1] *Department of Computer Application, Doctoral Scholar School of Computer Application Lovely Professional University, Phagwara, Punjab, India*
[2] *Division of Food Science and Technology, SKUAST Kashmir, Phagwara, Jammu and Kashmir, India*
[3] *DCA CGC Landran, Chandigarh, Punjab, India*

3.1 Introduction

Green energy, alternatively termed renewable or sustainable energy, denotes energy sources with a reduced environmental impact compared to conventional relic fuels such as ember, lubricant, and fracked gas. These energy sources are termed "green" because they emit less greenhouse gases and contribute less to climate change. As we know the population is growing, and so are the demands for energy sources and different parts of the world, therefore, different amounts of energy are also required. Compared to poorer nations, wealthier countries have a greater energy demand. People are increasingly concerned about renewable energy sources because they are clean, abundant, affordable, and readily available on Earth. It is essential that we utilize natural energy sources like solar, wind, tidal, biomass, and geothermal energy in renewable energy technologies. For the last two decades, people have shown a keen interest in clean energy sources due to their natural benefits for the environment. This focus aims to foster sustainable growth and development for various reasons. Climate change is currently the most pressing issue worldwide, emphasizing the importance of renewable energy sources. Utilizing these sources offers numerous advantages, including social and economic progress, enhanced energy security, mitigation of weather revolution, and better ecological and public well-being [1, 2].

Carbon-based Nanomaterials for Green Applications, First Edition. Edited by Upendra Kumar, Piyush Kumar Sonkar, and Suman Lata Tripathi.

Published 2025 by John Wiley & Sons, Inc.

There is a direct correlation between increased utilization of energy and GDP growth, making the energy sector one of the crucial factors for fiscal progress. The accessibility of sustainable power options varies between nations. In many developing countries, there is a growing interest in using renewable energy technologies to create job opportunities. Currently, people heavily depend on energy derived from fossil fuels, which tends to be costly and harmful to the environment. Since remnant energies are often in short supply, it is crucial to focus on exploring substitute energy sources. Interesting concepts for environmentally friendly houses involve everything from the utilization of waste materials to the reuse and recycling of gray water for watering and eliminating the utilization of PV cells to keep homes toasty and charge battery-powered automobiles [3, 4]. Other ideas include the use of reused components to build new structures out of old ones. Sustainable materials are those that are naturally occurring, renewable, or capable of continual reusing, recycling, and recovery. They should also be widely used, beneficial to the environment, and need less energy to extract. In many developing countries, there is a growing interest in using renewable energy technologies to create job opportunities. Currently, people heavily depend on energy derived from fossil fuels, which tends to be costly and harmful to the environment. Since fossil fuels are often in short supply, it is crucial to focus on exploring alternative energy sources.

Not every energy source used by the renewable energy industry is environmentally friendly because some produce pollution, like fossil fuels do. For example, while electricity generated from organic material in sustainable forests may be renewable, the burning process can still release carbon dioxide (CO_2), making it not always environmentally friendly. Below are some well-known forms of sustainable energy.

- **Solar Energy**: Solar energy is the process of capturing sunlight, turning it into electricity using PV cells, or producing heat using solar thermal systems.
- **Wind Energy**: Generating electricity by capturing the kinetic energy of the wind with wind turbines.
- **Hydroelectric Energy**: Utilizing the energy of moving water in dams or rivers to produce electricity is known as hydroelectric power.
- **Geothermal Energy**: Using heat from deep under the Ground's crust to produce power is known as geothermal energy.
- **Biomass**: Utilizing organic materials like garbage, leftover food, and wood to generate heat or power through combustion or other processes is known as biomass energy.
- **Ocean Energy**: Using the tides, waves, and temperature variations of the ocean to produce electricity.

The classification and types of green energy sources are mentioned in Figure 3.1.

Solar power has received a lot of attention as a renewable, healthy, cost-free, and promising method to address ecological damage and conversion of energy. Of all the renewable energies, wind power is predicted to increase its absolute generation

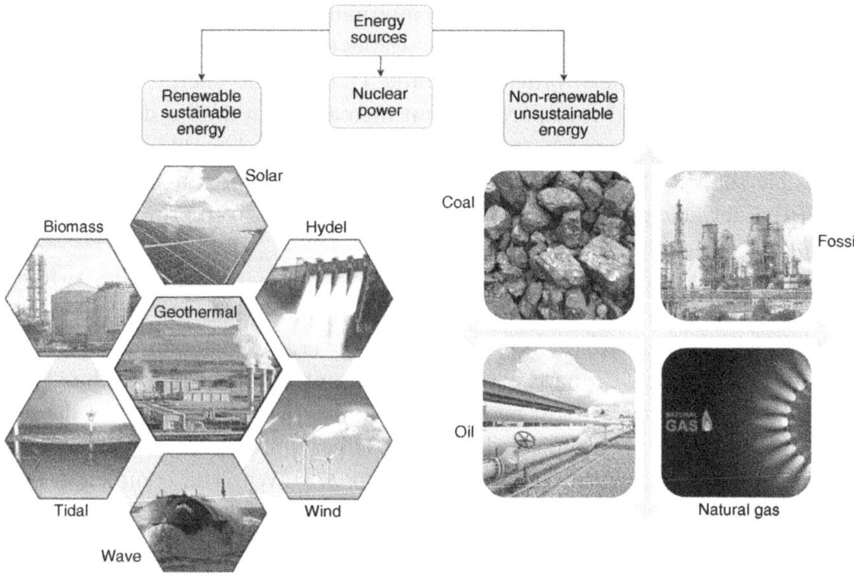

Figure 3.1 Classification of energy sources. *Source:* (a) Chawran/Adobe Stock. (b) Alex Mit/Adobe Stock Photos. (c) P123/Wikimedia Commons. (d) Wajan/Adobe Stock. (e) Pedphoto36pm/Adobe Stock. (f) Stockadrik/Adobe Stock. (g) Gretar Ívarsson/Wikimedia Commons. (h) Photosoup/Adobe Stock Photos. (i) Aikon/Adobe Stock. (j) chitsanupong/ Adobe Stock. (k) Sebos/Adobe Stock Photos.

the greatest. Greenlight catalytic nanomaterials are a prospective contender that can be utilized to produce the element hydrogen, degrade biological contaminants, reduce CO_2, change organic substances, create sensors, and other things by utilizing resources. While consumption of other uses of energy from renewable sources has decreased, renewable power usage has increased [5, 6].

3.1.1 Systematic Review Survey Reports

As consumption of alternative powers decreased, the use of green power increased by 3% in 2020. The production of electrical power from natural sources increased by nearly 7%, despite challenges like decreasing power demand and disruptions in source cables and construction in many parts of the world. This growth in green power was supported by long-term agreements, better access to the grid, and the ongoing construction of new facilities. Consequently, the proportion of clean energy in global electricity production rose from 27% in 2019 to 29% in 2020. However, there was a decrease in bioenergy due to reduced oil demand, which offset the 3% increase in biomass usage in industry because blended biofuels were also less commonly used.

The fastest-growing energy source both internationally and in the United States is renewable energy. In 2019, modern renewables (i.e. biomass, geothermal, solar, hydro, wind, and biofuels) provided around 11.2% of the energy used globally for heating, power, and transportation, an increase from 8.7% a decade earlier. By the end of 2020, just over 29% of the electricity produced worldwide came from renewable sources. More than 256 GW of capacity, or approximately 10% more installed renewable power capacity, was added in 2020, driven primarily by wind and solar PV (Figure 3.2).

The highest year-over-year rise in solar energy production since the 1970s is expected to occur in 2023 when it will increase by more than 11% to 8300 TWh. In 2021, most of the increase in renewable energy is anticipated to be from solar PV and wind sources. China is expected to account for about half of the global increase in renewable electricity generation, followed by the United States, the EU, and India. Wind energy is projected to see the largest growth, with a 17% increase in generation compared to 2020, equivalent to 275 TWh. The completion of a significant amount of capacity by developers in China and the United States at

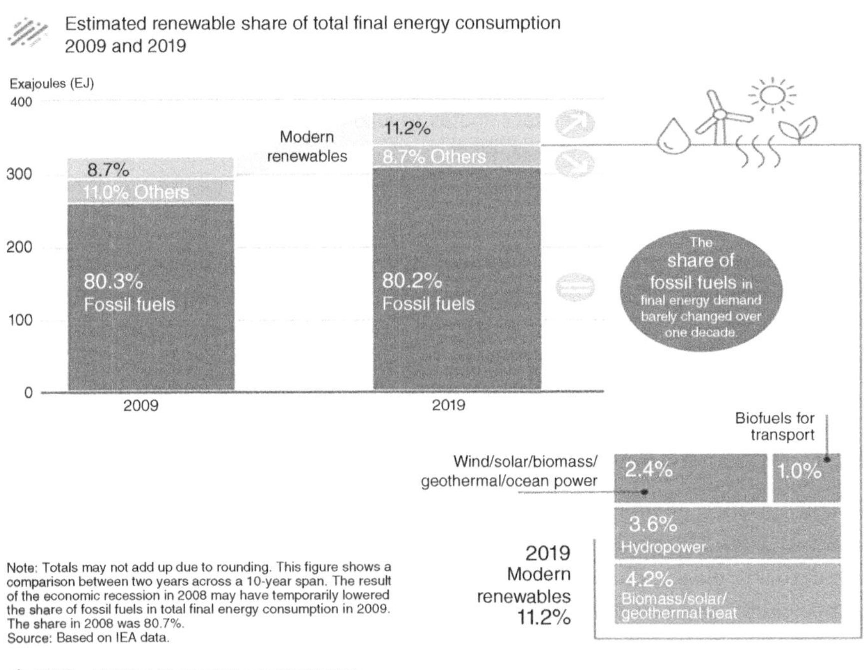

Figure 3.2 Approximated proportion of renewable energy in total final energy usage from 2009 to 2019. *Source:* REN21. 2021. Renewables 2021 Global Status Report (Paris: REN21 Secretariat).

the end of 2020, driven by policy deadlines, led to a substantial increase in generation at the beginning of 2021.

The ongoing worldwide energy crisis has sped up the urgency for transitioning to clean energy and underscored the crucial role that renewable energy plays, according to a poll by Renewable Energy Market Update, 2022 titled "Highlights Outlook for 2022 and 2023." Their updated forecast shows that before the recession regulations produced greater demand for green power. Especially, by 2023, wind and solar PV could potentially decrease the EU's reliance on natural gas from Russia. In the meanwhile, we cannot yet determine how the recently announced goals, after the Russian invasion of Ukraine, might affect our predictions for 2022 and 2023. This is because policies have not been put into action quickly enough. Countries like China, Japan, Thailand, and India will probably show a growth of 45% in green energy till 2023.

In 2021, India's solar market bounced back swiftly after experiencing a significant decline in new solar installations in 2020 due to COVID postponements. Meanwhile, Brazil and Vietnam are seeing increased electricity generation, driven by robust policy backing for circulated solar claims. Overall, universal solar PV electricity production is expected to surge by 145 TWh, surpassing 18%, and reaching 1000 TWh in 2021.

Renewable sources accounted for over 22% of the EU's total final energy supply in 2021. Although it fluctuates significantly across the EU, the proportion of sources of clean energy in the energy mix is roughly 60% in Sweden; over 40% in Denmark, Estonia, Finland, and Latvia; and between 10% and 15% in Belgium, Hungary, Ireland, Luxembourg, Malta, and the Netherlands. In the EU in 2020, wind and hydropower collectively produced 36% and 33% of all electricity from renewable sources, respectively, as per data from Eurostat. The remaining third came from various other renewable sources (8%), hard biofuels (8%), and solar energy (14%).

According to the survey study "World Economic Forum" [7] of the years 2030 and 2050 in the collection, The Green New Deal Road outlines a plan for 143 countries to tackle problems such as global warming, airborne pollution, and energy uncertainty. By 2050, these countries, responsible for 99.7% of global CO_2 emissions, aim to rely solely on clean, renewable energy with at least 80% of their energy from these sources by 2030.

The study breaks down countries into 24 regions that can work together on grid stability and energy storage technologies to ensure a balanced supply and call for energy from 2050 to 2052. Following this, the world could run on renewable energy sources.

By switching to renewable energy sources like wind, water, and solar, on a global scale, we can save 4–7 million premature deaths per year due to air pollution, slow down and then reverse the impacts of global warming, and stabilize the world's energy market (Figure 3.3).

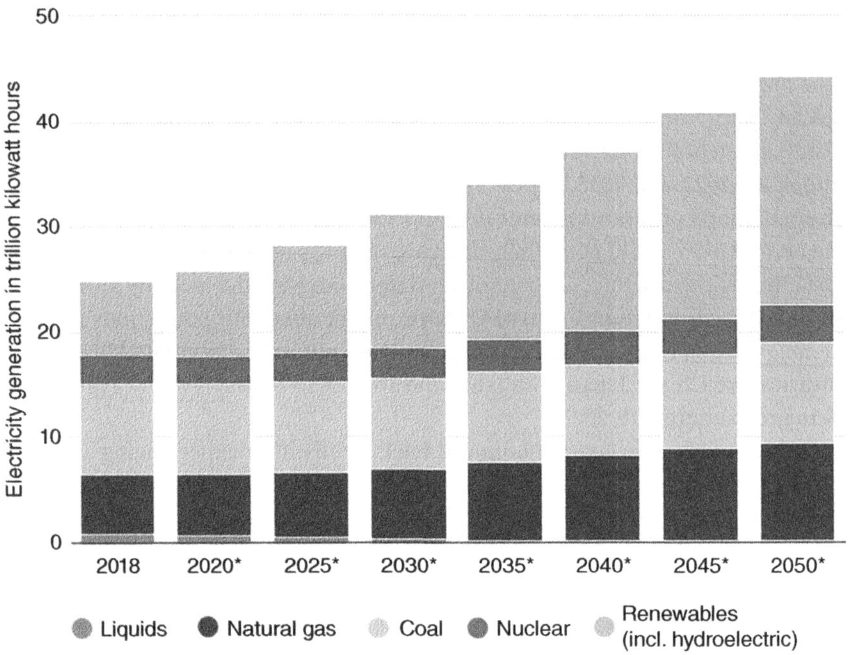

Figure 3.3 Projected electricity generation via various green energy sources: a survey report till 2050. *Source:* World Economic Forum [7]/Arindam Sengupta/OMC Power Private Limited.

3.2 Sustainable Energy Resources

Green and clean energy that does not rely on fossil fuels, such as renewable energy, is usually significantly less damaging to the environment. Through the selection of a more environmentally friendly energy source, there are several chances to influence environmental improvement. Through the assistance of power substitutes, green energy solutions are available to clients, helping them minimize their carbon impact. To reduce the effect of the day-to-day now, we just add green to your natural gas or electricity plan. As technology drives lower costs and begins to fulfill the potential of a clean energy future, the renewable energy sector is expanding.

Since the beginning of the business era, fossil fuels have dominated the power composition in most nations worldwide. Both the welfare of people and the overall atmospheric conditions are significantly impacted by this. Burning fossil fuels for energy production contributes to 75% of all worldwide greenhouse gas emissions.

Moreover, it is a major cause of local air pollution, leading to at least five million premature deaths each year. To combat this, we urgently need to shift toward cleaner, low-carbon energy sources such as nuclear and renewables. Over the coming years, transitioning to renewable energy will play a crucial role in reducing both CO_2 emissions and local air pollution [8, 9]. Because resources are finite, relying on fossil fuels isn't sustainable. Furthermore, burning fossil fuels depletes oxygen and releases CO_2, contributing to global warming and climate change. While using wood instead of coal presents a slightly better option, it comes with challenges. On the bright side, if sourced from responsibly managed forests, wood is renewable. Excess timber processing is repurposed into wood capsules and briquettes, effectively recycling. Additionally, compressed biomass fuels provide more energy compared to raw logs.

3.2.1 Wind Energy

With the power of wind turbines playing an increasingly significant role in supplying power to the National Grid, wind farms are becoming increasingly common across the United Kingdom. To generate electrical power from these systems, the turbine's power generators subsequently supply power to the local electricity grid. While there are options available for generating off-grid or household electricity, not every property is suitable for installing a residential wind turbine. Typically, wind energy is produced by turbines harnessing air and generating electricity using mechanical energy. As we know the wind is a form of solar energy. Factors such as the Earth's rotation, irregularities on its surface, and the warmth generated by solar radiation all play a role in the formation of winds. It is possible to use renewable electric power generated by the sun and wind to transmit electricity throughout the electrical grid if they are there. The most powerful wind turbine in the world, with a current annual output of 6000 MW, is typically located in China. It has been notified that a total of 534.5 TWh of wind energy has been produced globally as of 2017 (Pappas). In India, wind power construction increased by 31% to 2.07 GW in the fiscal year (FY) 2019–2020 from 1.58 GW in the FY 2018–2019. When contrasted with relic and wind energies usually has a smaller effect on the environment [10, 11].

Wind turbine systems are of two types: (1) horizontal-axis wind turbine systems and (2) vertical-axis wind turbine systems. Horizontal-axis wind turbine systems have typically three blades and work "upwind," turning at the highest point of the structure in such a way that their blades face the wind (Figure 3.4). Several kinds of vertical-axis wind turbines are available. Because of their 360° omnidirectional design, these turbines do not require special positioning so that they face the breeze to work (Figure 3.5).

Figure 3.4 Horizontal-axis wind turbine systems. *Source:* Dewis Schroeder/NREL 25897.

Figure 3.5 Vertical-axis wind turbine systems. *Source:* United States Department of Energy/ Mike VanBavel–42795.

3.2.1.1 Applications of Wind Turbine Systems

Numerous uses for wind energy may be found in daily life, which helps provide sustainable energy and lessen our reliance on fossil fuels. Some typical ways that wind energy is used in daily life are mentioned below:

- **Electricity Production**: Both large- and small-scale wind turbines are utilized to produce electricity. In windy areas, wind farms made up of numerous turbines are built to feed electricity into the grid, supplying households, companies, and industries.
- Residential wind turbines are installed by some homeowners and small enterprises to produce their electricity. This can assist in lowering carbon emissions and energy costs (Figure 3.6).
- **Water Pumping**: For millennia, people have used windmills to move water for drinking and cultivation. In rural areas and on farms, wind-powered water pumps are still in use.
- **Off-Grid Power**: Where it is difficult to reach conventional power sources, wind turbines can be employed in off-grid or isolated places. For cottages, outlying residences, and communication towers, they offer a dependable supply of power.
- Wind turbines can recharge batteries, which makes them helpful for projects like powering electric vehicles (EVs), distant weather stations, and emergency communication systems.
- **Desalination**: Wind energy can be used to run desalination plants, which turn salt water into fresh drinking water. This is particularly useful in desert places with scarce freshwater supplies.
- **Wastewater Treatment**: Municipalities can lower their energy costs and environmental effects by using wind energy to power wastewater treatment facilities.
- **Street Lighting**: To generate electricity, lower energy costs, and raise public awareness of renewable energy sources, some towns have installed wind turbines into their streetlights.
- **Hybrid Energy Systems**: In hybrid energy systems, wind power can be combined with other resourceful energies similar to sun-based power or hydropower. This makes for a more reliable and continuous supply of electricity [12].
- **Battery Backup Systems**: To store surplus energy produced during windy periods for usage during calm periods or power outages, wind turbines can be connected to battery backup systems in houses or businesses.
- **Displays for Education and Recreation**: To promote renewable energy and sustainability, wind turbines are occasionally placed in classrooms and public areas.
- Wind energy can be used for a variety of agricultural applications, such as grain drying, greenhouse heating, and powering electric fences.

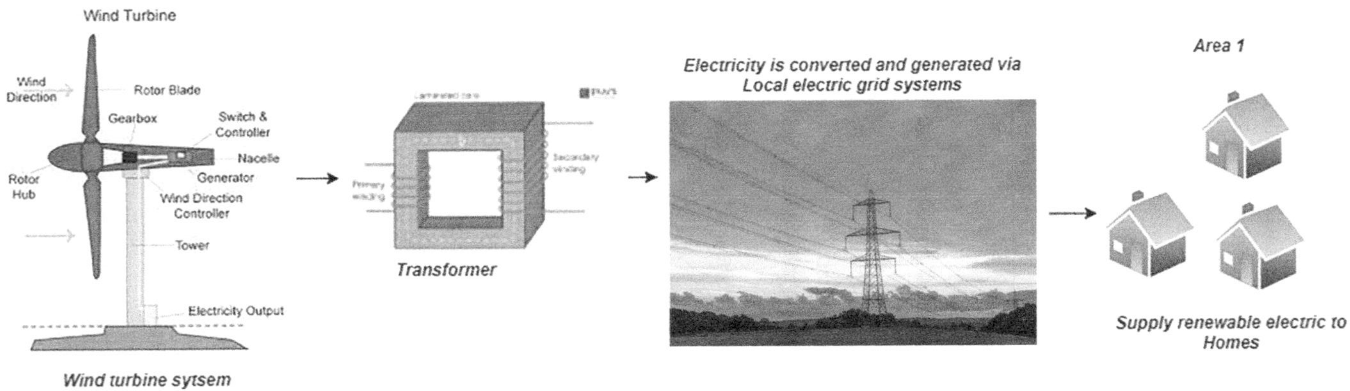

Figure 3.6 Wind turbine system home applications. *Source:* yelantsevv/Adobe Stock.

- **Community Projects**: To create local power and income and to foster an awareness of shared accountability for renewable energy, some towns have invested in community-owned wind turbines.
- Wind energy can be utilized to generate electricity for the charging of EVs, assisting in the shift to cleaner forms of transportation.
- These examples show how wind energy may be used for a variety of purposes in daily life, from supplying necessities like electricity and clean water to promoting sustainability and lowering greenhouse gas emissions.

3.2.1.2 Advantages of Wind Energy

A useful source of renewable energy, wind turbine systems have several benefits. Wind turbines employ the kinetic energy of the wind to produce electricity without emitting greenhouse gases or air pollution. Wind turbines use the energy from the wind to generate electricity without releasing greenhouse gases or causing air pollution which is considered fresh and reusable energy. They offer a pure, in addition to environmentally responsible, source of energy. A seemingly endless resource, wind is abundant. Wind energy is accessible in both developed and developing nations due to its widespread potential for exploitation. Wind energy helps improve energy security. It promotes energy independence and reduces exposure to fuel price fluctuations and supply disruptions by lessening reliance on fossil fuels. Once wind turbines are set up, they generally have lower operating and maintenance costs compared to fossil fuel power plants. They do not require fuel expenses, but regular upkeep such as lubrication and inspections are necessary.

The production, installation, maintenance, and operation of wind turbines are all jobs produced by the wind energy sector. This creates employment possibilities and boosts local economies. There are many different sizes of wind turbine systems, from small residential turbines to massive utility-scale installations. This scalability enables flexible deployment by energy requirements. The production of electricity with wind energy results in much lower CO_2 emissions, making it a crucial instrument in the fight against global warming. Wind turbines utilize a very small amount of water, making them less water-intensive than many traditional power plants, which need a lot of water for cooling. In comparison to some other types of energy infrastructure, wind projects can be created quickly. This may facilitate quicker satiation of rising energy demand. It is frequently possible to build wind turbines on agricultural land while still allowing farmers to use the land for grazing or growing crops. Landowners may receive additional money from this mixed land use. With improvements in weather forecasting allowing for greater energy planning and grid integration, wind energy is a predictable and reliable source of electricity in many places.

As wind turbine technology, components, and designs continue to progress, costs are coming down and efficiency is rising, making wind energy more competitive.

Wind energy has a lesser environmental impact than the mining and burning of fossil fuels, including less water pollution, habitat destruction, and land degradation. By supplying energy during times of high demand and preventing the need for fossil fuel power plants to run at partial capacity, wind turbines can help maintain the stability of the grid. Wind energy projects have the potential to benefit rural communities economically, including through increased tax income, lease payments to landowners, and local economic activity. Wind energy has many benefits, but when designing wind turbine installations, it's important to take into account things like wind variability, land use issues, and potential visual and acoustic consequences. Choosing the right location and involving the community is essential for managing potential issues and maximizing the advantages of wind energy installations.

3.2.1.3 Disadvantages of Wind Energy

Although wind turbine systems provide many benefits, several drawbacks and difficulties must be considered when planning their implementation. The intermittent energy source wind is one of these. Only when the wind is blowing within a specified range of speeds do wind turbines produce power. When compared to some other sources of electricity, wind energy may be less dependable due to its intermittency. The output of a wind turbine varies and is influenced by the speed and direction of the wind. The electrical grid may experience oscillations as a result of this unpredictability, necessitating backup power sources or energy storage techniques to keep the grid stable. Installing wind turbines can be expensive upfront, especially for large-scale projects. Turbines, infrastructure, and grid zconnections are all included in these expenses. Wind turbines may be considered visually obtrusive, particularly in scenic or residential regions. They may not be aesthetically pleasing to everyone. Additionally, noise from wind turbines might pose a problem for people residing in proximity. Large airstream farms require a lot of space, potentially displacing natural or agricultural ecosystems, and influencing how land and wildlife are used. Wind farms may also endanger birds and bats by causing crashes and habitat disruption.

Depending on where you are, you may or may not have access to wind resources that are strong enough to support the building of wind turbines. This may reduce wind energy's geographic viability. Wind turbines need routine maintenance and sporadic repairs, which can be expensive and call for specialized tools and experts. Large wind turbine components like blades and towers can be difficult to transport and install due to logistical issues, which necessitate substantial planning and infrastructure. Wind turbines may alter the surrounding scenery, which could have an impact on local real estate prices and tourists. Some individuals find the shadow flicker effect that occurs when the sun is hidden behind a moving wind turbine to be obtrusive or unpleasant. The production of wind energy does not always coincide with the electricity demand. Without sufficient energy storage, extra energy produced during windy conditions can be lost, and there might be shortages during calm conditions. It can be difficult to integrate wind energy into the grid since it necessitates balancing supply and demand, maintaining grid stability, and overcoming transmission issues. In cold climates, ice buildup can impair wind turbines, necessitating the installation of de-icing devices. They must also endure harsh weather, which may require downtime for maintenance or repairs.

Although there is still a lack of scientific agreement on these problems, some people who live close to wind turbines have reported experiencing stress and sleep disruptions. Compared to certain other energy sources, wind energy has a relatively low energy density, which means a sizable number of turbines and a sizable amount of land are frequently needed to generate a sizable amount of electricity.

Despite these drawbacks, wind energy is still a vital and expanding renewable energy source, and continual technology developments and smart project planning can help ease many of these difficulties. Addressing possible issues and ensuring the effective deployment of wind turbine systems requires careful site selection, community involvement, and environmental impact evaluations.

3.2.1.4 Future Prospectives and Challenges

As technology develops and the world's energy environment shifts in favor of renewable and sustainable sources, the future of wind turbine systems promises several exciting possibilities. Future wind turbines are probably going to be more effective in converting wind energy into electricity. Higher energy capture rates will be a consequence of advancements in aerodynamics, supplies, and controllers. In the future, wind turbines are probably going to get taller. Taller turbines provide more access to stronger, more reliable winds at higher altitudes, which increases the potential for energy production. Taller turbines are particularly advantageous for offshore wind projects. Fixed-bottom structures are not practical in deeper oceans, where floating wind turbines can potentially capture wind energy. Although it is still in its infancy, this technology shows promise for the growth of offshore wind. By making materials lighter and stronger, turbine components will be easier and less expensive to transport and install. Advanced composites and carbon fiber might be crucial in this. Integrating sophisticated energy storage technologies with wind turbines will increase their dependability and suitability for producing electricity even when the wind is not blowing. Examples of these technologies include better batteries or grid-scale solutions.

By better integrating wind turbines with smart grid systems, electricity generation and distribution can be managed. This will improve the efficiency and stability of the grid. Combining wind energy with solar and energy storage enhances reliability and consistency in power generation. Future wind turbines will be built to endure severe weather, which will lower maintenance costs and lengthen the life of the machines. The claim of artificial intelligence (AI) and sophisticated data analytics will allow for preventative maintenance, performance optimization of turbines, and downtime reduction. As small-scale, distributed wind energy systems for domestic and communal use proliferate, the need for centralized power grids may decline. Due to the enormous untapped wind resources in the world's oceans, offshore wind farms are anticipated to experience tremendous growth in the future. Improved installation methods and floating wind turbines will make this expansion possible. Future wind turbine designs will continue to prioritize limiting their environmental impact, including avoiding bird and bat collisions and addressing noise issues. Wind energy technology will continue to grow globally, bringing clean energy to more areas of the world as it develops and becomes more cost-effective.

Wind energy will likely become even more cost-competitive with fossil fuels thanks to innovations in design, manufacturing, and installation processes. By electrolyzing wind energy, hydrogen can be produced. In the evolution to a neutral carbon budget, this "green hydrogen" can be a useful energy carrier and storage medium. The wider modification to clean and maintainable power sources is intimately related to the future of wind turbine systems. Wind power is anticipated to play a big part in satisfying the world's energy needs while lowering greenhouse gas emissions and meeting the energy demands of the world via reducing the amount of glasshouse fumes as governments and companies increase their investments in renewable energy technology. With these future prospectives come challenges in terms of intermittency and variability, grid integration, energy storage, land use and space requirements, environmental impact, aesthetic concerns, noise pollution, shadow flicker, maintenance and repairs, extreme weather challenges, transport and installation, cost and financing, public acceptance, resource variability, and limited energy density. Despite these difficulties, the use of wind energy as a maintainable and pure source of electricity is on the rise, and continuous research and technology developments are aimed at resolving many of these problems. Successful wind energy project development depends on careful site selection, environmental impact analyses, and community involvement.

3.2.2 Solar Energy

Perhaps the most plentiful and inexpensive renewable resource on our planet is sunlight. Every hour, the amount of sunlight reaching the Earth's surface surpasses our entire annual energy requirements. However, the availability of solar energy fluctuates based on the time of day, the season, and where we live. Despite these variations, solar power is increasingly popular as a renewable energy option worldwide. Solar energy, emitted by the sun in the form of heat and radiation, is the source of this abundant resource. We incorporate the use of solar thermal energy, solar power energy, solar thermal energy, and artificial photosynthesis in this technology.

3.2.2.1 Applications of Solar Energy

Solar energy uses PV panels are used to produce electricity for industrial, commercial, and residential structures. Large arrays of solar panels make up utility-scale solar power facilities, which provide electricity for the grid (Figure 3.7). To produce electricity for their use, homeowners put solar panels on their rooftops. Extra energy can be used later or put back into the grid. The application of the solar energy is listed below:

- **Sun-powered Water Heating**: Water for domestic, commercial, and industrial uses, such as hot water for showers, swimming pools, and industrial processes, is heated using solar thermal systems.
- **Solar Drying and Cooking**: In locations with limited access to conventional fuels, solar cookers and dryers provide a sustainable alternative by heating and drying agricultural products or food using sunshine.
- **Solar-powered Automobiles**: EVs can use solar panels to increase their range or power auxiliary systems like entertainment and air conditioning.
- **Water Pumps Powered by the Sun**: Agricultural irrigation uses solar pumps, which offer a dependable energy source for pumping water from wells or other sources.
- **Solar Portable Chargers**: Portable electrical gadgets, such as cellphones, tablets, and camping gear, can be charged on the road using small solar chargers.
- **Streetlights Powered by Solar Energy**: Self-contained solar streetlights use the sun's energy to run LED lights in parking lots, streets, and public areas.
- **Sun-based Desalination**: Desalination plants can be powered by solar energy, transforming seawater into fresh drinking water, which is particularly useful in arid locations.

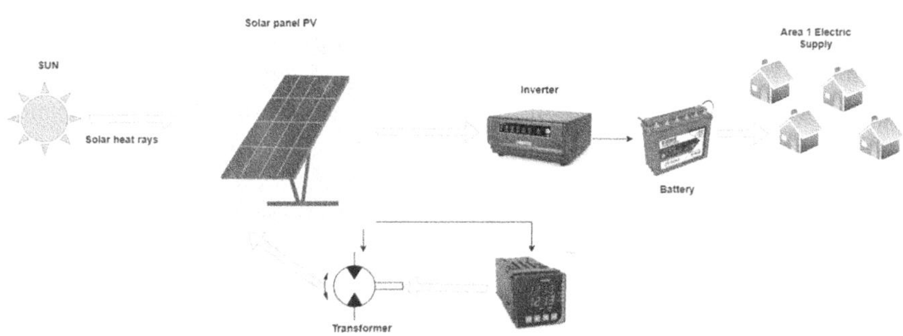

Figure 3.7 Solar energy systems.

- **Applications for Space**: To produce power from the sun's energy in space, solar panels are utilized in satellites and spacecraft.
- **Solar-powered Cooling and Ventilation**: Solar energy is used by solar fans and air conditioning systems to lower indoor temperatures and enhance ventilation in residential and commercial structures.
- **Refrigeration Powered by the Sun**: In off-grid and rural locations, perishable products including vaccinations, pharmaceuticals, and food are preserved using solar refrigeration devices.
- **Remote Monitoring Powered by Solar Energy**: Remote monitoring, communication, and environmental research devices are all powered by solar energy.
- **Purification of Water Using Solar Energy**: Water purification systems can be powered by solar energy, providing off-grid and rural communities with clean drinking water.
- **Solar Hybrid Systems**: To ensure a steady and uninterrupted supply of electricity, hybrid systems often incorporate energy storage along with solar energy and additional reusable resources such as airstream or hydropower to guarantee a consistent and uninterrupted electricity supply.
- **Stabilization of the Grid**: When outfitted with cutting-edge control systems, solar power plants can offer grid stabilization services including frequency regulation and voltage support.

As technology progresses and the globe looks for cleaner, more sustainable energy sources, solar energy's uses continue to grow. It is essential for combating climate change and lowering greenhouse gas emissions, and it also gives distant and marginalized areas access to electricity.

3.2.2.2 Advantages of Solar Energy

As a clean and renewable energy source, solar energy has several benefits which ensure not to emit pollutants or harmful discharges, making it friendly to the environment and lowering greenhouse gas emissions that cause climate change. The sun offers more energy in one hour than the entire planet consumes in a year, making it a plentiful and essentially infinite supply of power. Solar energy may be found all around the world, making it practical in a variety of settings. Solar panels can produce electricity for residences and commercial buildings, resulting in lower energy costs. These costs reduce expenses by frequently selling surplus energy back to the grid or storing it for future use. Additionally, solar panels entail minimal operating and maintenance costs once mounted. Cleaning and sporadic inspections are included in routine maintenance; there are no ongoing charges or fuel costs. Energy independence is increased and the vulnerability to changes in fuel prices and supplies is decreased because solar energy's capacity to help people, organizations, and communities depends less on fossil fuels. Jobs are created in manufacturing, installation, maintenance, and research and development by the solar sector. This may boost regional economies and create job possibilities.

There are many different types of solar energy systems, from small house installations to huge electric grids. Due to its scalability, energy needs can be met with flexibility. Distributed solar installations, such as rooftop solar panels, can lessen the load on the electrical grid during moments of high demand, assisting in preventing blackouts and grid congestion. By offering grid stabilization services like frequency regulation and voltage support, solar power plants can improve the steadiness and reliability of the electrical grids. Remote and off-grid places can use solar energy to generate electricity, giving those communities access to electricity that they would not have had before. When compared to some other types of energy infrastructure, solar projects can be created quickly. This may facilitate quicker satiation of rising energy demand. Solar energy is becoming more affordable and widely available due to continuous enhancements in solar panel efficiency, resources, and technology. Integrating energy storage into solar power systems enables the storing of extra energy produced during sunny times for use during cloudy days or at night. Energy storage devices include batteries. Solar panels on farmland can help farmers earn more money while still allowing them to use the land for grazing or cultivation. By situating solar energy production nearer to the point of consumption, transmission and distribution losses can be cut, refining the effectiveness of the entire energy organization. Solar energy has a major impact on reducing CO_2 emissions, which makes it a crucial instrument in combating climate change and lessening the environmental impact of producing power. To encourage the adoption of sun vigor, several governments provide rebates, tax credits, and incentives, making it more financially appealing for both individuals and enterprises. Overall, solar energy offers a sustainable and dependable source of electricity while benefiting people, communities, and the planet on an economic, environmental, and social level.

3.2.2.3 Disadvantages of Solar Energy

While solar energy offers many benefits, it also has several drawbacks and difficulties that should be taken into account when assessing its viability for certain uses and locales. Solar energy is intermittent and variable since it is influenced by the environment, the time of day, and the location. The generation of electricity may fluctuate as a result, making it less dependable to fulfill steady power demands. To combat solar power's sporadic nature, energy storage technologies like batteries are frequently needed. A solar energy system's overall cost will increase due to the potential expense of these storage solutions.

Although they have been declining in recent years, the initial expenses of buying and installing solar panels and related equipment can be high. The return on investment may not materialize for several years. Solar panels need a fair amount of room to spread out and create a lot of electricity. In regions with a high population density or in small spaces, this may be difficult. Cloudy or bad weather reduces the efficiency of solar panels. Solar energy production could be erratic in areas where the sky is often clouded. Solar panels can only produce electricity during the day, which limits their potential to supply electricity at night without the aid of energy storage systems. The manufacture and disposal of solar panels can affect the environment, releasing greenhouse gases during the manufacturing process and necessitating the management of electronic waste once the panels have reached the end of their useful lives. Land use conflicts could result from large-scale solar farms competing with natural or agricultural ecosystems for land usage, thereby causing habitat damage. Some people find solar panels unappealing, especially on residential rooftops, and aesthetic issues can operate as a barrier to adoption in some places. Solar resources are location-dependent, with some places receiving more sunlight than others. Solar energy may not be as beneficial in areas with scarce solar resources.

Solar energy is fully dependent on sunshine, which can be obstructed by shading from structures, trees, or other obstacles. Despite these drawbacks, improvements in solar technology and energy storage options, along with favorable government subsidies and falling costs, continue to make solar energy a desirable and more practical choice for the generation of clean and renewable electricity. Planning solar energy projects must take these disadvantages and the benefits into careful account.

3.2.2.4 Future Prospectives and Challenges

By capturing sunlight from both sides, bifacial solar panels may produce more electricity. More deployments of these panels are anticipated, particularly in industrial and utility-scale settings. By combining solar energy with cutting-edge energy storage technologies, such as better batteries, solar electricity will be available constantly, lowering dependency on the grid and enhancing the resilience of solar systems. The development of microgrids, which increase energy independence and grid resilience, is made possible by solar energy in conjunction with energy storage and smart grid technology. Emerging markets with inadequate access to power are predicted to adopt solar energy at a faster rate. For these areas, dependable power sources can be provided through off-grid and small-scale solar arrays. It reduces your dependency on centralized power plants and increases your energy resilience by decentralizing your energy generation through rooftop solar panels and small-scale solar installations. Water electrolysis using solar energy can make green hydrogen. Industries like transportation and heavy manufacturing can use green hydrogen.

Solar energy will be essential in helping to achieve global carbon reduction goals, which will aid in the fight against climate change. To reduce waste and advance a circular economy, it will be more crucial to recycle and reuse solar panels after they have served their purpose. Solar power is anticipated to become much more widely available and cost feasible as expertise develops, and the environmental advantages of solar energy become more obvious. It will be crucial in the shift to a future with lower carbon emissions and more sustainable energy sources. Other prospects will be in terms of increased efficiency, advanced materials, floating solar farms, building integrated solar systems, solar-powered transportation, community solar projects, and AI integration.

3.2.3 Biomass

This process entails transforming solid plant-based fuel into energy. While the fundamental concept of biomass involves burning organic materials to generate power, advancements are steadily making this process cleaner and more energy-efficient. Biomass provides electricity at reduced costs to both the environment and the economy by converting residential, marketable, and agricultural waste into dense, liquid, and gas fuels [13] (Figure 3.8). The living matter of bioenergy is derived from microorganisms or other byproducts. Chemical energy that comes from the sun is stored in biomass. Bioenergy is created by crops during photosynthesis. In addition to being directly consumed for warmth, biomass can also be processed into useful energy sources like CH4 gas, ethanol, or biodiesel.

When compared to fossil fuels, biomass energy has a lower impact on the environment since it produces fewer air pollutants such as CO, CO_2, NO_x, and particle substances. According to the current analysis, India has access to over 750 million metric tons of biomass annually. Pyrolysis and gasification are two processes used in the chemical transformation of biomass. Both thermal decomposition procedures involve heating biomass sources at elevated levels in sealed, pressure tanks known as gasifiers. They are influenced by the transformation process temperatures and the quantity of oxygen in the solution (Figure 3.9).

3.2.3.1 Applications of Biomass

The term "biomass," which describes organic materials originating from plants and animals, has several uses in a variety of industries.

- **Production of Bioenergy**: Processes like combustion, gasification, or pyrolysis can turn biomass into bioenergy. This bioenergy can be converted into biofuels, heat, or electricity.
- **Biofuels**: Bioethanol, biodiesel, and biogas are just a few of the biofuels that may be made from biomass as a feedstock. In heating and transportation applications, these biofuels can be used in place of or in addition to conventional fossil fuels.

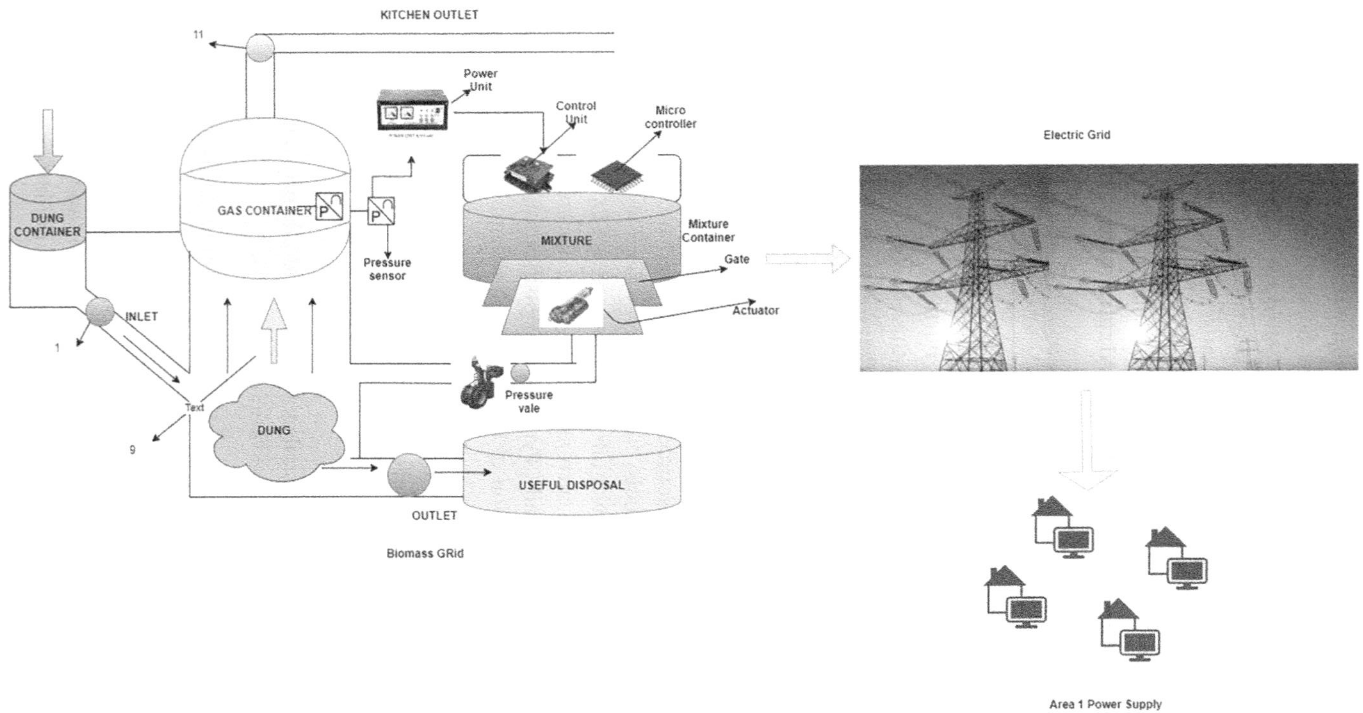

Figure 3.8 Biomass energy systems. *Source:* Patent 202111056299.

Figure 3.9 Biomass plant. *Source:* lusia83/Adobe Stock.

- **CHP (Combined Heat and Power) and Heat**: Direct combustion of biomass can generate heat for domestic, industrial, and commercial uses. Additionally, it is utilized in CHP systems to concurrently produce heat and electricity.
- **Creating Electricity**: Burning organic materials to produce electricity at biomass power plants makes for a sustainable and renewable energy source.
- **Production of Biogas**: Breaking down organic waste without oxygen like animal manure, agricultural residues, and dirt can result in the creation of biogas, which mostly contains methane. Biogas can be used to generate power, heat, or as a fuel for motor vehicles.
- **Bioplastics and Biochemicals**: To decrease the dependency on goods generated from petroleum, biomass can be used as a feedstock for the manufacturing of biochemicals, such as bioplastics, biobased chemicals, and biobased materials.
- **Biomass Pellets and Briquettes**: Briquettes and pellets made from compressed biomass are effective and practical fuels for industrial processes as well as for industrial heating and cooking.
- **Feeding Livestock and Bedding Them**: A useful resource for agriculture is provided by some agricultural waste products and crop by-products, which can be used as livestock bedding or as animal feed.
- **Fertilizer and Soil Amend**: Compost and manure are examples of organic products that can be utilized to enhance soil quality, add nutrients, and boost agricultural productivity.

- **Erosion Control and Land Restoration**: Land restoration and erosion control projects that involve the planting of cover crops and soil stabilization can benefit from the use of biomass.
- **Biomedicine and Pharmaceuticals**: Pharmaceuticals and biomedicine utilize some bioactive molecules generated from biomass, such as medicinal plants, for the creation of new drugs and natural cures.
- **Bioremediation**: Bioremediation is the technique of using specific plants and microorganisms to absorb or degrade pollutants from soil and water.
- **Food and Drinks**: Fruits, vegetables, cereals, and forage crops are all part of the biomass that is used as a primary source of food for people and other animals.
- **Fiber and Wood Products**: A typical biomass source used in the manufacture of timber, paper, furniture, and other wood-based goods is wood from trees.
- **Clothing and Textiles**: The production of textiles and garments uses natural fibers derived from plants, such as flax and cotton.
- **Crafts and Art**: Arts and crafts use a variety of biomass materials, including wood, leaves, and plant fibers.
- **Bioenergy Research**: Research on bioenergy, biotechnology, and bioprocessing relies heavily on biomass, which enables improvements in renewable energy and sustainable materials.
- **Climate Mitigation and Carbon Sequestration**: Afforestation and reforestation are two biomass activities that help with carbon sequestration and climate change mitigation.
- **Waste Minimization**: Utilizing biomass encourages sustainable waste management methods by reducing organic waste.
- **Green Products and Renewable Chemicals**: Chemicals and goods made from biomass are used to create renewable and ecologically friendly substitutes for traditional goods.
- Numerous industries, including energy, agriculture, and manufacturing, as well as others, employ biomass, which promotes sustainability, resource efficiency, and a decrease in greenhouse gas emissions.

3.2.3.2 Benefits and Disadvantages of Biomass

Biomass is a valuable and renewable resource since it gives advantages in a variety of industries. Biomass is a renewable resource since it can be renewed by natural processes like photosynthesis. It is also a sustainable resource. It is thus an environmentally friendly source of energy and materials. Biomass combustion can be carbon-neutral or even carbon-negative when it is managed responsibly. This indicates that the CO_2 absorbed by plants as they grow balances the CO_2 emitted during the combustion of biomass. Using biomass can help reduce overall greenhouse gas emissions and combat climate change. Using biomass lessens the environmental impact of landfills and trash disposal by managing organic wastes including food waste, forestry trimmings, and agricultural residues. There are many different types of biomass resources to choose from, including wood, crop leftovers, animal manure, algae, and more. The flexibility in procuring biomass resources is made possible by this diversity. The production of power, heat, and biofuels from biomass can help reduce dependency on fossil fuels and promote energy security and independence. Biomass industries, such as those that produce bioenergy and wood products, help boost local economies in rural and agricultural areas by adding jobs to the local labor force. The use of organic biomass materials, such as compost and crop leftovers, can enhance soil structure, fertility, and water retention, which is advantageous for farming and the health of the soil. By substituting biomass for fossil fuels, dependency on oil and natural gas can be less risky for the environment and the economy. Biodegradable materials and bioplastics can take the place of plastics that decompose slowly in the environment.

Unsustainable biomass harvesting from farms and forests can degrade ecosystems and deplete resources, which has an impact on biodiversity and soil health. While biomass combustion can be carbon-neutral or carbon-negative, inefficient conversion processes, poor land management, and transportation emissions can all lead to higher carbon emissions. When biomass is burned, air pollutants such as particulate matter, nitrogen oxides (NO_x), and volatile organic compounds (VOCs) can be produced. These pollutants are linked to poor air quality and health problems. Traditional biomass stoves used for cooking and heating in some developing regions can expose people to indoor air pollution, which can be harmful to their health, especially that of women and children. Infrastructure costs can be high, especially for large-scale operations, when building the infrastructure for biomass production, transportation, and storage. Some biomass crops, such as energy crops and specific bioenergy plantations, can demand substantial water resources for irrigation, thereby causing water scarcity difficulties. Food crops and biomass crops for energy production may compete for arable land and other resources, which may have an impact on food prices and security. Biomass conversion procedures, such as fermentation to produce biofuels or combustion for the production of heat and power, may be less effective than other energy-generation techniques, leading to energy losses.

3.2.3.3 Future Prospectives and Challenges

As a crucial element of a sustainable and renewable energy system (RES), biomass has a bright future. Despite obstacles, research and innovation are advancing biomass technology and techniques.

By advancing the efficiency of biomass-to-energy processes, such as gasification, pyrolysis, and torrefaction, it will be able to generate energy at a lower cost and with less environmental impact. These technologies are still being researched and developed. The creation of biorefineries, which can use different types of biomasses as feedstock to process a variety of bio-based products like biofuels, biochemicals, and biomaterials, will be essential to the efficient use of biomass. Biofuels: Advanced biofuels will become more commercially viable and competitive with fossil fuels, such as cellulosic ethanol and algae-based biofuels. These biofuels can aid in lowering transportation-related greenhouse gas emissions. Bioenergy with carbon capture and storage (BECCS) is a powerful instrument for mitigating climate change because it combines the generation of biomass energy with carbon capture and storage (CCS) technology. Sustainable biomass feedstock production will be aided by improved agriculture and forestry techniques, which will also reduce adverse effects on the environment and conflicts between different land uses.

By choosing high-yielding biomass species and doing research on genetically modified crops, it will be possible to boost biomass output while requiring less space for farming. To supply high-yield and low-input biomass feedstocks, the cultivation of specialized energy crops like switchgrass and miscanthus will increase. Biomass waste products, such as agricultural waste, forestry waste, and organic waste, will be used more and more to produce energy and recover resources.

3.2.4 Geothermal Energy

Geothermal energy, which gets its heat from the Earth's interior heat reservoirs is produced by utilizing the natural heat that is stored beneath the surface of the Earth. The creation of electricity, heating, and cooling are only a few uses for this energy. The interior of the Earth is extremely hot, with temperatures rising with depth. Both the original heat from the planet's formation and the radioactive isotopes' decay are the main sources of this heat today. Different parts of the Earth's crust experience different levels of heat flow. These areas are referred to as geothermal hotspots or reservoirs. Because the Earth's crust is thinner along tectonic plate boundaries and heat may escape more easily, geothermal energy is more readily available there. To access the hot reservoirs beneath the Earth's crust and extract geothermal energy, wells are drilled there. These wells can extract heat because they are deep enough where temperatures are high enough. To absorb heat from the Earth's subsurface, water or a solution and antifreeze are cycled down into the well. Production wells are used to transport the heated fluid to the surface. Heat is transmitted to a secondary fluid near the surface that has a lower boiling point than the primary fluid, such as water, or a working fluid like isobutane or ammonia. The secondary fluid evaporates and produces steam as it absorbs heat. To generate power, a turbine connected to an electricity generator can be driven by this high-pressure steam. To provide heating and cooling for structures, greenhouses, and other industrial activities, hot water or steam is supplied through a system of pipes in direct-use applications. To maintain the

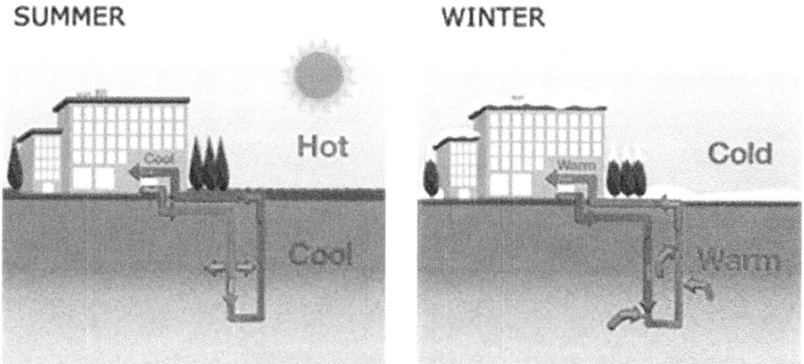

Figure 3.10 Geothermal energy application in homes.

pressure and long-term viability of the geothermal reservoir, the cooled fluid is frequently pumped back into the Earth (Figure 3.10).

Because it emits few greenhouse gases and leaves a relatively tiny environmental footprint in comparison to many other energy sources, geothermal energy is considered a pure source of energy. It is a crucial component of the switch to a more sustainable energy mix since it offers a continuous and dependable supply of power.

3.2.4.1 Applications and Future Prospectives

Geothermal energy has a bright future as a reliable and maintainable source of renewable energy. The development of these energies is being fueled by technological advancements, increasing investment, and growing public knowledge of its advantages. EGS technology, which involves constructing synthetic geothermal reservoirs by hydraulic fracturing, is anticipated to increase the use of geothermal energy. Geothermal resources in areas that were previously deemed unsuitable for conventional geothermal projects can now be accessed thanks to technology (Figure 3.11).

In residential, commercial, and industrial settings, GHP systems for heating and cooling buildings will become more prevalent. They are very effective and can aid in lowering the energy needed for air conditioning. Geothermal power plants that simultaneously generate electricity and useful heat (CHP) will proliferate, notably in industrial operations and district heating systems, improving energy efficiency. In areas with high cooling demand, geothermal cooling systems, which use the constant subterranean temperature for air conditioning, will become more popular.

Figure 3.11 Geothermal plants. *Source:* U.S. Energy Information Administration / https://www.eia.gov/energyexplained/geothermal/geothermal-power-plants.php / last accessed October 24, 2024.

This will assist in lowering peak power consumption and greenhouse gas emissions. Geothermal energy will continue to be used directly for heating in spa resorts, greenhouse agriculture, aquaculture, and district heating systems. As technology develops, new direct-use applications might be created. By combining geothermal energy with other sources like sun and airstream as well as storage, hybrid energy structures that generate power continuously and steadily will be created. To cut energy costs and environmental effects, geothermal energy can be used in industrial operations including drying, pasteurization, and sterilization. In addition to producing electricity, geothermal brine, which contains rich minerals and metals, can be utilized to extract and recover minerals.

Geothermal energy is a viable choice for reaching clean energy objectives and lowering greenhouse gas emissions because of its dependability and minimal environmental impact. Geothermal energy is set to take center stage in the global energy scene as technology develops and geothermal resources are more skillfully utilized.

3.2.5 Hydropower

Hydroenergy, usually stated as hydropower, is also a pure, maintainable source that uses the gravitational and kinetic energy of falling or flowing water to produce electricity. One of the oldest and most popular types of renewable energy exists today. Here is how hydroenergy functions in sustainable, renewable systems. Rivers, streams, and dams are common natural water sources near which hydroelectric power

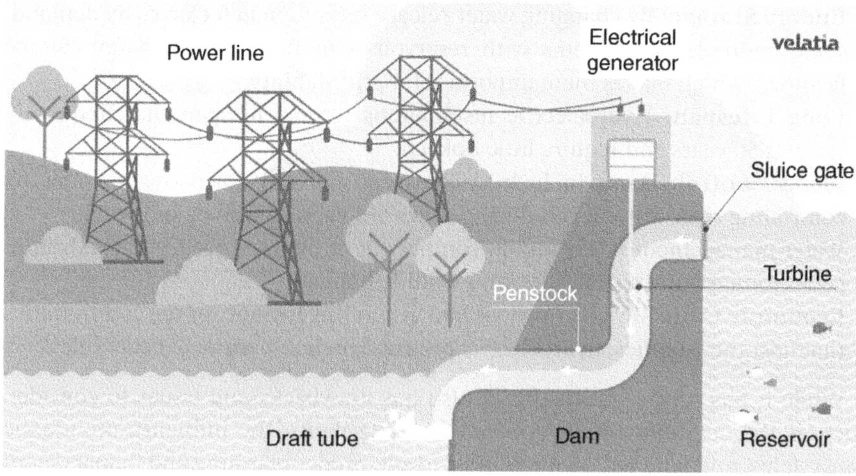

Figure 3.12 Hydel power. *Source:* Valetia/Velatia, S.L.

facilities are sited. The volume and flow rate of the water affects the hydropower system's ability to generate energy. A dam is built to produce a reservoir or impoundment in various hydroelectric schemes [14]. Run-of-river systems may not have a dam, but the river's flow is still changed by turbines. Water that has been diverted or discharged from a dam goes through turbines (Figure 3.12). Typically, turbines are installed at a powerhouse at the dam's base. The turbines are turned by the water's flow. Electricity is produced by generators that take the mechanical energy from the rotating turbines and turn it into current. After that, power lines carry this electricity to buildings, commercial establishments, and industries. Advanced control systems govern the water flow through the turbines, enabling operators to modify electricity generation in response to demand and environmental factors.

The key attributes and advantages of hydroenergy in renewable, environmentally friendly systems are:

- **Renewable**: Hydroenergy relies on the natural water cycle, which continuously refills water sources through precipitation and snowmelt. As a result, it is a renewable source of energy.
- **Clean and Green**: Hydroelectric power plants generate electricity without releasing air pollutants or greenhouse gases, making them environmentally friendly and assisting in the reduction of carbon emissions.
- Large hydropower plants can generate electricity consistently, reliably, and steadily, making them ideal base-load power sources since they can operate continuously.

- **Energy Storage**: By changing water release rates to match electricity demand, some hydroelectric systems with reservoirs can function as energy storage facilities, which makes them important for grid stability.
- **Long Lifespan**: Hydroelectric installations have a lifespan that frequently exceeds 50 years and require little upkeep.
- **Flood Control:** Dams in hydrosystems can prevent downstream floods by controlling river water levels during times of heavy rainfall or snowmelt.
- Water management is a common component of hydroprojects and is advantageous for agriculture, pleasure, and wildlife habitat.
- **Economic Gains**: By creating jobs and generating income for the communities that host the facilities, hydroelectric projects can boost regional economies.

While it has several profits but it also has drawbacks and issues to consider, such as the environmental impact of building dams, the influence on aquatic ecosystems, and land use conflicts. As a result, proper planning and management are essential for maximizing the advantages of hydroenergy while lessening its adverse impacts on the environment and society.

3.2.6 Tidal and Wave Energy

The kinetic and gravitational energies of tides and waves in the water are used to produce tidal energy. The movement of water in oceans and seas is used by these energy sources to create electricity. An overview of tidal and wave energies is given below.

3.2.6.1 Tidal Power

Tidal range energy harnesses the difference in water levels between high and low tides. Tidal plants or barrages are frequently used for this. During tidal cycles, tidal power plants regulate the flow of water into and out of an estuary or bay using sluice gates and turbines. Water runs through turbines, creating energy as the tide rises and falls. The kinetic energy of flowing water masses brought on by tidal currents is captured by tidal stream energy, sometimes referred to as marine current energy. Electricity is produced using underwater or ocean floor-mounted turbines. Water movement in enclosed coastal areas known as tidal lagoons can be used to produce power. Although they are built as enclosed structures, they operate on concepts that are comparable to those of tidal range energy systems (Figure 3.13).

3.2.6.2 Wave Power

Point absorbers are the floating, column-connected objects that link to the seabed at a point. The floating structure and the column move about each other as waves

Figure 3.13 Tidal power. *Source:* Alex Mit/Adobe Stock Photos.

pass by, creating mechanical energy that is later transformed into electricity. Oscillating water columns (OWCs) change the air pressure inside a chamber by using the motion of waves [15]. These pressure variations propel the airflow through a turbine, which produces power. Attenuators are lengthy, multi-segmented structures that bend and move in response to wave motion. With the help of the flexing motion, hydraulic pistons or pumps that are linked to generators generate electricity. These devices allow waves to enter a reservoir and create a water head. Turbines then release the water from the reservoir to produce energy (Figure 3.14).

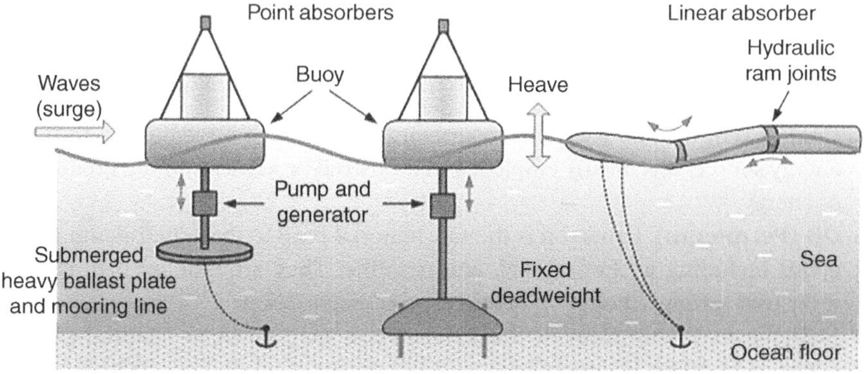

Figure 3.14 Wave power. *Source:* Chen (2016)/ with permission of Elsevier.

3.2.6.3 Benefits of Tidal and Wave Energy Systems

Tides and waves are reliable sources of energy since they are predictable due to gravitational forces and are renewable. Compared to certain other energy sources, tidal and wave energy systems emit very few greenhouse gases and cause very little visual and audible disturbance. These energy sources offer reliable energy production, which may help keep the system stable.

3.2.6.4 Challenges of Tidal and Wave Energy Systems

- **Location-Specific**: The geographic range of suitable sites for tidal and wave energy is constrained to areas with strong tidal currents or regular wave patterns.
- **High Capital Costs**: Tidal and wave energy systems can be expensive to build and install, requiring sizable upfront investments.
- Tidal barrages can change estuarine habitats and fish migration patterns, which can hurt the environment. Environmental mitigation and assessment must be done carefully.
- **Maintenance Obstacles**: Using and maintaining underwater equipment in abrasive marine environments can be difficult and costly.

Despite these obstacles, tidal skills are still being explored and developed to make them more affordable and widely available so they may help diversify renewable sources and reduce dependence on relic energies.

3.3 Non-Sustainable Energy Resources

Non-renewable energy sources are types of power produced from limited resources that can't be replaced naturally within a reasonable amount of time. These energy sources are limited in supply and will eventually run out when they are used up.

3.3.1 Fossil Fuels

- **Coal**: A sedimentary rock rich in carbon, coal has long served as a significant energy source. It is mostly employed in industrial processes and the production of energy.
- **Oil (Petroleum)**: Crude oil is the raw material used to make petroleum-based goods including jet fuel, diesel, and gasoline. They are employed in several industrial settings, including heating and transportation.
- Methane makes up most of the hydrocarbon gas known as "natural gas." It serves as a fuel for automobiles as well as a means of producing electricity, heat, and cooking.

3.3.2 Atomic Energy

Nuclear fission, which splits the nuclei of uranium or plutonium atoms, produces nuclear energy. Nuclear power facilities use the heat produced to produce electricity.

The following characteristics define non-renewable energy sources:

- **Limited Supply**: These energy sources have finite supplies, and as a result, there is a chance that they will eventually run out. These resources cannot be replaced in a human timeline once they are depleted.
- **Environmental Impact**: The extraction, refinement, and combustion of non-renewable energy sources frequently lead to environmental damage and the production of greenhouse gases, which fuel climate change.
- **Geopolitical Concerns**: As nations struggle for access to these priceless energy sources, non-renewable energy resources may be the source of geopolitical tensions and conflicts.
- **Economic Vulnerability**: Prices for non-renewable energy sources are susceptible to volatility due to shifts in supply and demand, geopolitical issues, and market dynamics. This has an impact on both the economy's stability and energy security.
- **Challenges of the Transition**: Changing from non-renewable to renewable energy sources can be a difficult and expensive process, but it is necessary to reduce negative environmental effects and combat climate change.

There is an increasing focus on switching to renewable and sustainable energy sources. This transition entails making investments in energy technologies that are more environmentally benign and have the potential to provide long-term energy sustainability (Figure 3.15).

3.4 Existing Green Energy Models

We have already discussed the green energy models including solar power, geothermal, wind, biomass, and hydroelectric. These green energy models are not only efficient and sustainable but have also reshaped national energy grids

Figure 3.15 Non-renewable resources. *Source:* (a) Photosoup/Adobe Stock Photos. (b) Eleonimages/Adobe stock. (c) Aikon/Adobe Stock.

worldwide over the past few decades. The comparison of the existing green energy models can be based on the search results which encompass a variety of green energy systems and models, including modeling sustainable energy systems, implementing RESs for homes, controlling RESs in buildings through modeling, analyzing RESs at different elevations, and reviewing advancements in hybrid RESs. These sources offer insights into diverse approaches, technologies, and applications of green energy systems (Table 3.1).

The survey, covering the years 2010 to 2024, compares available green energy systems or models, highlighting the growth and advancement of renewable energy sources, improvements in sustainable energy system modeling, and overall trends in renewable energy adoption (Table 3.2).

3.5 Conclusions

The development of the nation's economy depends heavily on traditional energies including coal, gas, and oil. The book chapter's primary developmental characteristics center on the use of clean energy for environmentally friendly growth. The use of green power is good for the natural world and human health where sustainable growth and the use of renewable power supplies are related. Energy technology focuses on moving energy from renewable technologies like solar energy, wind energy, tidal energy, geothermal energy, hydrogen energy, and biomass energy to store natural resources (fossil and nuclear). For the classification of renewable energy sources, all the aspects including greenhouse gas emissions, resource availability, land needs, water consumption, social implications, and cost of power generated are considered. As a result, the technical breakthrough reduces these civilizations' usage of green power. It is worth noting that research has demonstrated how green laws and technological advancement can affect consumer appetite for power in both developed and developing nations. Some important findings need to be noted.

i) **Growth of Renewable Energy**: Through the year 2020, the utilization of renewable energy surged by 3%, primarily driven by the installation of solar PV and wind systems, which comprised nearly 90% of all newly added renewable energy sources. Projections suggest that renewable electricity generation will grow by over 8% in 2021, with solar PV and wind driving two-thirds of this growth. Looking ahead, global renewable capacity is anticipated to soar by nearly 2400 GW (representing almost 75% growth) between 2022 and 2027, with solar PV and wind installations making up almost 95% of the total.

ii) **Growth of Solar PV and Wind**: Solar PV electricity generation is poised to rise by 145 terawatt-hours (TWh), marking an approximate 18% increase to

Table 3.1 Parameter-based comparison of existing green energy models.

Green energy models	Source/year	Comparisons
Sustainable energy systems modeling	doi: 10.3934/energy.2023014 2023	The initial outcome showcases a research endeavor proposing a strategy to achieve full carbon neutrality in the Swedish energy sector by 2050, employing computer-based simulations via EnergyPLAN software. The research considers the sustainable utilization of renewable resources across all economic sectors and cross-border energy exchange with neighboring countries under two scenarios, with the objective of establishing a completely carbon-free landscape. Findings indicate that achieving a 50% reduction in carbon emissions by 2035 and complete decarbonization by 2050 is feasible through this method.
Home renewable energy systems	https://www.energy.gov/energysaver/planning-home-renewable-energy-systems Survey-2023	The finding offers insights into the preparation stages for residential renewable energy setups, encompassing assessments of power consumption, comprehension of regional regulations, and selection of appropriate renewable energy technologies. Discussed technology options include compact solar electric setups, modest wind electric setups, micro-hydropower systems, and compact hybrid electric systems (combining solar and wind). The planning phase also entails deliberation over energy efficiency strategies, the choice between grid-connected or off-grid functioning, and integration of the system into the overall house design
Model-based control of renewable energy systems in buildings	DOI: 10.1080/10789669.2006.10391196 2006	The finding explores the simulation and regulation of renewable energy systems within buildings, employing a commercial simulation platform to replicate the current structure, facility, and control mechanism. The research emphasizes the alignment of the models with actual data collected from the building to guarantee precise depiction of the system.

(*Continued*)

Table 3.1 (Continued)

Green energy models	Source/year	Comparisons
Modeling and analysis of renewable energy systems at various altitudes	https://doi.org/10.1186/s40807-024-00103-5 2024	This discusses a research project that models and assesses a combined renewable energy system (utilizing both solar and wind sources) in India. The system aims to fulfil energy requirements across different elevations. The study suggests the integration of solar and wind technologies and evaluates its feasibility from both a technological and economic perspective, considering the intermittent availability of renewable energy resources. The findings indicate that the Net Present Cost (NPC) amounts to \$3.61 million, and the Levelized Cost of Electricity (LCOE) stands at \$0.255 per kilowatt-hour for a refined integrated renewable energy system deployed in a typical remote community. This community has a peak power consumption of 238.7 kW and an average daily load demand of 2230 kWh.
Overview of advancements in hybrid renewable energy systems	https://doi.org/10.1093/ijlct/ctad123 2024	The finding offers a summary of the current and anticipated cutting-edge developments in hybrid renewable energy systems (HRES). It explores the application of HRES in diverse settings, including rural and remote regions, contemporary urban infrastructure, island communities, industrial settings, academic institutions, and the tourism sector. The research also underscores the benefits of HRES, such as enhanced dependability, efficiency, and sustainability in comparison to conventional energy systems.

Table 3.2 Survey (2010–2050)-based comparison of existing green energy models.

Parameter	Survey year	Findings
Renewable energy growth	2010–2021	According to the IEA's Global Energy Review 2021, renewable energy consumption rose by 3% in 2020, despite a decline in demand for other fuels. Renewable electricity production is projected to grow by over 8% in 2021, marking the quickest annual growth rate since the 1970s. The proportion of renewables in global electricity generation surged to 29% in 2020, up from 27% in 2019, and is forecasted to reach a new peak of 30% in 2021.
Renewable energy sectors	2010–2020	The Statista report indicates a consistent rise in the proportion of renewables in overall energy usage worldwide across various sectors including electricity, heating, and transportation, spanning the period from 2010 to 2020. Wind and solar energy have experienced the most rapid growth among renewable sources, with leading companies like Vestas Wind Systems playing a prominent role in the industry.
Sustainable energy system modeling	2022–2050	Raza et al.'s research proposes a strategy to achieve full carbon neutrality in the Swedish energy sector by 2050, utilizing computer-based simulations conducted on the EnergyPLAN software. The findings suggest that a 50% reduction in carbon emissions could be attained by 2035, with complete decarbonization achievable by 2050, facilitated by the sustainable utilization of renewable resources and cross-border energy exchange with neighboring countries.
Comprehensive study of renewable energy sources	2010–2050	Ang et al.'s review offers a thorough examination of diverse renewable energy sources, encompassing their categorizations, obstacles, and recommendations for enhancement. The research delves into the patterns and fluctuations in renewable energy generation spanning the period from 2010 to 2020, shedding light on the progress and constraints of various renewable technologies
Total renewable energy usage	2010–2050	The data on ResearchGate illustrates how much renewable energy has been used to generate electricity from 2010 to 2020, revealing a consistent upward trend in usage throughout the years.

nearly reach 1000 TWh by 2021. Wind energy is anticipated to experience the most significant surge in renewable generation, with an expected growth of 275 TWh, equating to almost 17% in 2021. The addition of onshore wind installations is projected to rise from 74 gigawatts (GW) in 2021 to 109 GW by 2027.

iii) **Global Renewable Capacity**: It is forecasted to expand by nearly 2400 GW between 2022 and 2027, equivalent to China's entire installed power capacity. China is expected to lead in installing almost half of the new global renewable power capacity during 2022–2027.

iv) **Impact of Energy Crisis**: The energy crisis has expedited the adoption of renewable energy, as elevated prices of fossil fuels and electricity render renewable power technologies more financially appealing.

v) **Country-Specific Developments**: On a country-specific level, Chile has pledged to achieve carbon neutrality by 2050 and is actively pursuing projects to harness wind and solar energy. Meanwhile, Canada has outlined a clear roadmap for transitioning to renewable energy, including a phased increase in carbon emission taxes.

These findings underscore the swift expansion of renewable energy, particularly in solar PV and wind sectors, underscoring their growing significance in the global energy landscape.

References

1 Panwar, N.L., Kaushik, S.C., and Kothari, S. (2011). Role of renewable energy sources in environmental protection: a review. *Renewable and Sustainable Energy Reviews* 15 (3): 1513–1524. ISSN 1364-0321, https://doi.org/10.1016/j.rser.2010.11.037.

2 Williams, G.E. (2000). Geological constraints on the Precambrian history of Earth's rotation and the Moon's orbit. *Reviews of Geophysics* 38 (1): 37–59.

3 Huang, J., Lu, X., and McEloy, M.B. (2014). Meteorologically defined limits to a reduction in the variability of outputs from a coupled wind farm system in the Central US. *Renewable Energy* 62: 331–340. ISSN 0960-1481, https://doi.org/10.1016/j.renene.2013.07.022.

4 Paul, D.K. and Razdan, R. (1993). Space-fed optical beam-forming lens: a comparison with RF counterpart. *Proc. SPIE 1953, Photonics for Space Environments*. https://doi.org/10.1117/12.156584.

5 Lewis, M., Neill, S.P., Robins, P.E., and Hashemi, M.R. (2015). Resource assessment for future generations of tidal-stream energy arrays. *Energy* 83: 403–416. ISSN 0360-5442, https://doi.org/10.1016/j.energy.2015.02.038.

6 Bertani R. (2007). World geothermal generation in 2007. *Proceedings European Geothermal Congress*, 30.

7 World Economic Forum (2020). Renewable energy could power the world by 2050: Here is what that future might look like. Energy Translation.

8 Dincer, I. and Rosen, M.A. (1999). Energy, environment and sustainable development. *Applied Energy* 64 (1–4): 427–440. ISSN 0306-2619, https://doi.org/10.1016/S0306-2619(99)00111-7.

9 Meereboer, K.W., Misra, M., and Mohanty, A.K. (2020). Review of recent advances in the biodegradability of polyhydroxyalkanoate (PHA) bioplastics and their composites. *Green Chemistry* 22: 5519.

10 Ragaert, K., Delva, L., and Van Geem, K. (2017). Mechanical and chemical recycling of solid plastic waste. *Waste Management* 69: 24–58. ISSN 0956-053X, https://doi.org/10.1016/j.wasman.2017.07.044.

11 Wüstenhagen, R., Wolsink, M., and Bürer, M.J. (2007). Social acceptance of renewable energy innovation: an introduction to the concept. *Energy Policy* 35 (5): 2683–2691. ISSN 0301-4215, https://doi.org/10.1016/j.enpol.2006.12.001.

12 Amin, I. and Mithilesh, K.D. (2021). *"Artificial Intelligence Techniques in Smart Grids", Introduction to AI Techniques for Renewable Energy Sources*, vol. 1. CRC- Taylor Francis ISBN: 9781003104445.

13 Charanjit, I.A., Mithilesh, K., and Pooja, R. (2021). Early detection and prevention measure for the biogas plant. Patent of India, 202111056299.

14 Ormazabal_en Renewables (2023). Why is hydropower and how it works.

15 Wave Energy (2023). Alternative energy tutorial.

4

Carbon-Based 2D Materials: Synthesis, Characterization, and Their Green Energy Applications

Minakshi Sharma[1], *Varsha Yadav*[2], *Prachi Diwakar*[3], *Chandra Mohan Singh Negi*[3], *and Parvez Ahmed Alvi*[3]

[1] *Department of Physics, Om Sterling Global University, Hisar, Haryana, India*
[2] *School of Applied Sciences, Shri Venkateshwara University, Gajraula, Uttar Pradesh, India*
[3] *Department of Physical Sciences, Banasthali Vidyapith, Banasthali, Rajasthan, India*

4.1 Introduction

The expanding population and increased usage of fossil fuels raise people's awareness of energy needs and environmental consequences. The current high demand for fossil fuels has resulted in environmental devastation [1]. Therefore, developing solutions to promote sustainable green energy storage technologies is essential to meet critical energy demands. Novel nanomaterials are considered the most effective way to address these challenges. Several types of nanomaterials with unique features have been developed and employed for energy conversion and storage. Nuclear energy is another non-renewable energy that shares 8% of the world's total energy utilization [2]. Yet, in modern periods, the realization of finite fossil resource stocks has prompted a search for renewable energy sources. The surge in interest in generating important nanomaterials over the last several decades has sparked broad interest in the scientific community in developing a variety of innovative porous materials such as porous organic polymers, zeolites, and activated carbon [3]. Due to their outstanding chemical and physical characteristics, new 2D materials including graphene, transition metal dichalcogenides (TMDs), and topological insulators have received a lot of interest recently. Strong atomic interactions exist between atoms inside a layer of these layered materials,

Carbon-based Nanomaterials for Green Applications, First Edition. Edited by Upendra Kumar, Piyush Kumar Sonkar, and Suman Lata Tripathi.

Published 2025 by John Wiley & Sons, Inc.

whereas weak van der Waals forces exist among neighboring atomic layers [4]. Their unique electrical and optoelectronic characteristics are the result of this. Their strong non-linear optical (NLO) susceptibility, high carrier density, and mobility have made them useful as a variety of optical components, such as photodetectors, optical limiters, and passive saturable absorbers for Q-switching and mode-locking. Single-layered zeolites, layered double hydroxides, transition metal oxides, and perovskite materials are examples of other compounds that include O, S, and N atoms. These materials are in addition to single-atomic-thick graphene and nitride boron.

A nanoscale thickness of 2D materials, where the length and width are significantly greater than the thickness (resulting in a high length-to-thickness or width-to-thickness ratio). In 2004, 2D materials have been rapidly developing relative to the creation of single-layered graphene [5]. The 2D materials' ultrahigh surface-to-volume ratio is advantageous for mass and heat transport. They are often used in the objectives of heterogeneous catalysis, adsorption and electronic industries, analysis, and storage of energy because of their distinctive physical structure and adjustable chemical properties [6]. Carbon-based materials have drawn a lot of attention among different 2D nanostructures because of their widespread availability, superior chemical stability, electric and thermal conductivity, and relatively normal functionalization or hybridization. In a similar context, 2D carbon-based materials are made of carbon and other inorganic components with 2D structures. The first well-recognized, 2D substance is graphene, which is made up of a single sheet of carbon atoms arranged in a hexagonal lattice. Research on 2D material (graphene) has advanced swiftly compared to other carbon allotropes, as demonstrated by comparisons with fullerene as 0D and carbon nanotube as 1D material [7]. Carbon nanosheets are examples of similar types of materials like graphene and their functional derivatives like amine-doped graphene, phosphorus-doped graphene, reduced graphene oxide (rGO), GCN, etc. [8]. Rapid synthesis strategy advancement during the past 10 years has made it possible to create carbon nanosheets with adjustable thickness, specific surface functional groups, and varying pore size and surface area. There are different types of synthesis methods for developing 2D materials for different green energy device applications. Cutting-edge 2D materials have a significant influence on science and cutting-edge technology. They possess special mechanical, electrical, optical, and physical characteristics. Due to their potential applications in various fields, 2D materials have become a highly popular topic. Their uses include gas and chemical sensors, supercapacitors, batteries, fuel cells, wearable sensing technology, optoelectronics, photodetectors, catalysis, electrochemical sensing, biosensing, water and air purification, and advanced electronic devices [9]."

4.2 Synthesis of Graphene and Its Derivatives

The 2D substance known as TMDs can take on several crystal forms. The most prevalent trigonal symmetric structure with semiconductive properties is the 2H phase. These are among the most frequently used options for 2D transistors [10]. Bottom-up or top-down procedures are frequently used to manufacture 2D materials, and both are capable of producing nanomaterials with different lattice distributions, defects, vacancies, grain boundaries, and oxygen are a few of them. Stone-Wales flaws, functional groups, and dangling bonds with a certain level of functionalization as a result. Excellent reviews on the wide variety of techniques for producing two-dimensional materials have been investigated in recent years.

The intrinsic properties of 2D materials, considerably as the possibility for leading synthesis and processability, can be chosen in relation to the preparation technique.

The large concentration of functional groups that include oxygen that is dispersed throughout the surface of the GO flake results in an electrically insulating substance [12]. When oxidising chemicals are applied to graphite, epoxy, hydroxy, and carboxyl groups are produced, which increases the interlayer gap between the sheets. In-plane voids and fissures at the margins are caused by oxidative cleavage [13]. The oxygen residues are then removed by the following reduction of GO, which restores the sp^2-hybridized carbon system and its conductivity. The manufacturing of electrodes for sensors should be simple enough to allow for mass production, which is not the same as the process used to create individual graphene flakes from graphite using the renowned Scotch tape method. Even though they have many flaws compared to a perfect graphene flake, several additional production technologies have been created, and it was discovered by advanced material characterization processes that such two-dimensional (2D) carbon compounds are somehow comparable to graphene. The inherent material qualities are altered by all of the structural and chemical changes made during processing of such materials. This raises the query of how various exfoliation techniques impact the characteristics of "electrochemical sensors based on 2D carbon nanostructure-based materials. The step-by-step chemical modification of graphene and oxidized form are shown in Figure 4.1.

4.2.1 Graphene-Based 2D Materials

Since the isolation of the first 2D substance, graphene, examples of 2D materials are included. An isomorph of graphene, hexagonal boron nitride resembles graphene in terms of crystal structure [14]. However, in this special 2D substance, nitrogen and boron atoms take the place of carbon atoms. Beyond graphene, phosphorene

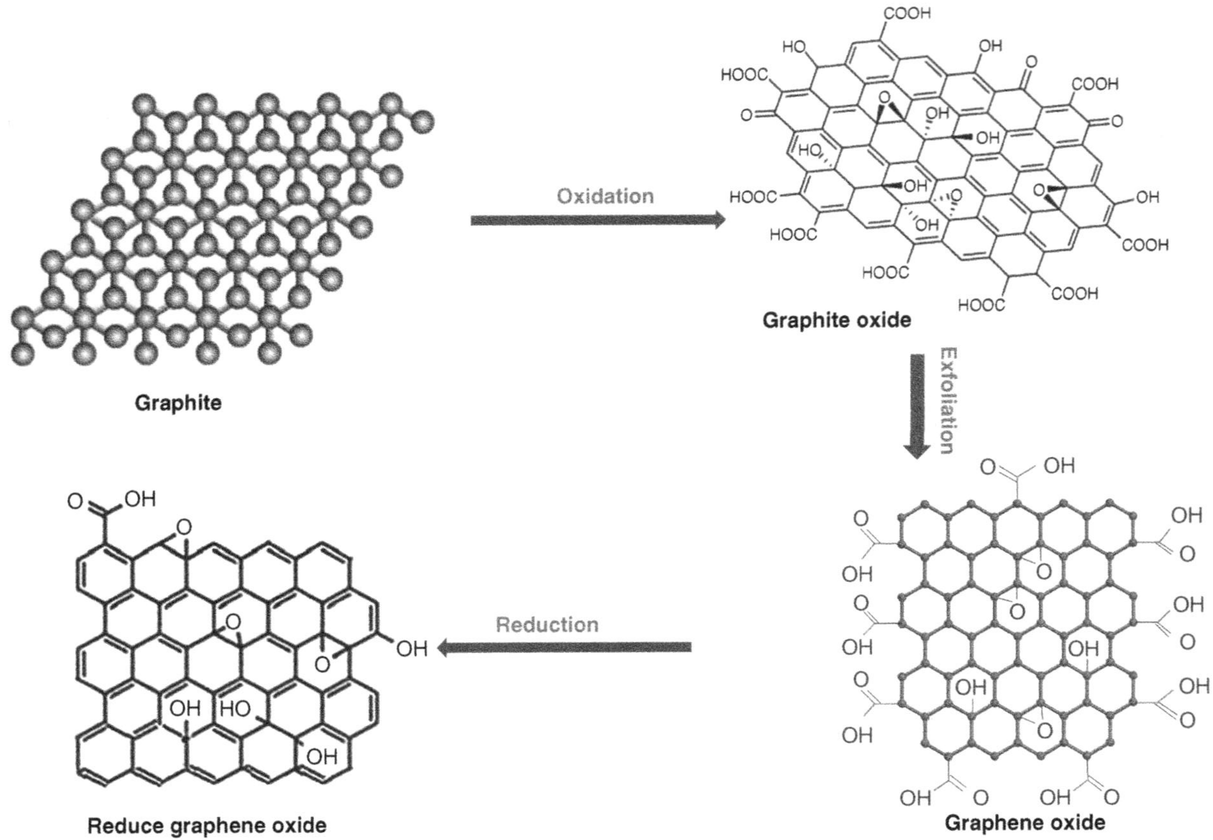

Figure 4.1 2D materials synthesis with different chemical routes.

is another example of a direct bandgap semiconductor with a structure resembling a puckered honeycomb. It is very beneficial for optoelectronic devices and transistors due to the strong charge mobility [15]. The hexagonal structure of xenes, which are monolayers of silicon, germanium, and tin, is identical to that of graphene. These 2D materials, despite being in their infancy, are thought to offer a lot of promise for usage in a variety of applications. The atoms of boron, nitrogen, and carbon make up the 2D nanomaterials known as borocarbonitrides [16]. These substances exhibit structural Stone-Wales flaws and a large surface area. Because of their combined ability to absorb CO_2 and CH_4 at a rapid rate and their special properties, borocarbonitrides are very effective for securing these gases.

The most significant manufacturing methods and functionalization tactics are assessed critically in the sections that follow. When coupled with graphene, other 2D crystals, including the TMDs (WS_2, MoS_2, and WSe_2), might allow new device topologies due to their metallic, insulating, and semiconducting properties [17]. Graphene is one of the materials that can be printed on rigid and flexible substrates and mass-produced in solution, exhibiting advanced hybrid topologies and electrical components including photodetectors and thin-film transistors (TFTs). Several techniques can be used to synthesize graphene, depending on the quality and quantity needed. One of the most widely used techniques for growing GO is chemical exfoliation using a modified Hummers' method [18]. This process is based on selecting appropriate oxidizing agents from graphite oxide. This process is inexpensive and produces a lot of graphene goods [19]. The highly oriented pyrolytic graphite (HOPG) or graphite rod that is submerged in electrolyte solutions is exfoliated using electricity in an electrochemical exfoliation procedure that forms graphene from the material. High-quality graphene products with controlled layers over large areas are produced using the chemical vapor deposition (CVD) process [20]. Typically, methane (CH4) and acetylene (C_2H_2) are used as carbon sources for graphene growth on copper (Cu) or nickel (Ni) foam at temperatures as high as 1000 °C [21]. The pattern of following the various roots for synthesis of the 2D materials is shown in Figure 4.2.

4.2.2 Graphene

Graphene synthesis refers to the process that is used to produce or synthesize graphene from graphite. The intended size, quantity, and purity determine the procedure to be used [22]. The method of synthesis influences the final product's structure and qualities. In several branches of science and technology, such as energy storage devices, biology, memory, electronics, and sensors, graphene layers come in variants from diverse processes such as a single layer, double layer, or multiple layers. Researchers employ various methods, particularly those involving graphene and its derivatives, for both applications and synthesis when large-scale

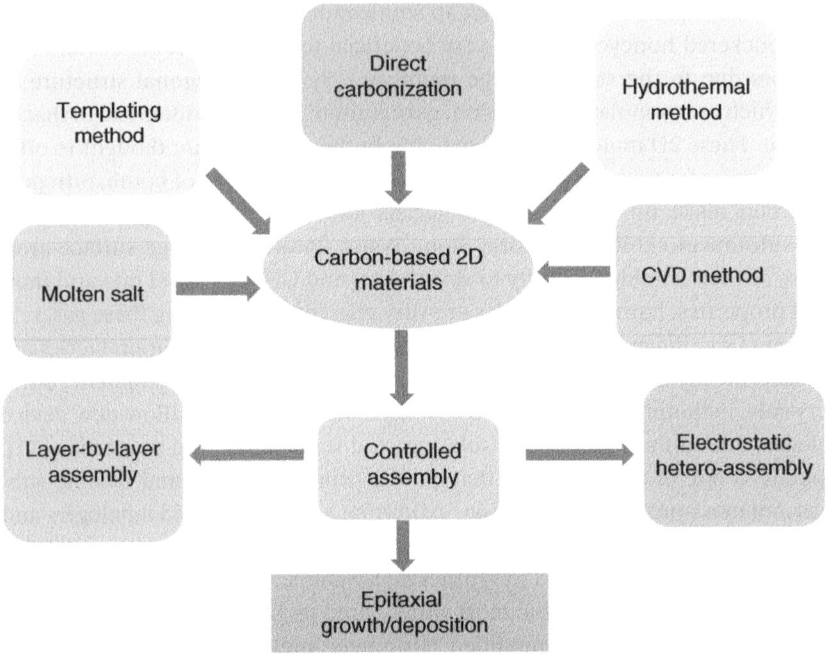

Figure 4.2 Strategic layout for the synthesis process of 2D materials. *Source:* Shelke and Late [11]/with permission of Elsevier.

production is required. We will next go through various synthesis methodologies and applications, their current state, their development to date, and their possibilities moving forward [23]. Ball milling and hydrothermal techniques are demonstrated to be effective in the production of graphene-based compounds. Also, there are different types of doped graphene like amine-doped graphene, phosphorus-doped graphene etc. These are commercially available and can also be synthesized using various chemical methods and have different applications in the field of energy storage [24].

4.2.3 Graphene Oxide

Graphite oxide may be converted into GO via a few different processes [25]. The most popular techniques are either stirring, sonication, or a mix of the two. Uses and applications of decreased GO sonication and GO application can result in a very successful and time-efficient exfoliation of graphene. It may, however, also seriously harm the graphene flakes, causing a large range of graphene platelet sizes and a reduction in surface size from microns to nanometers [26]. The quantity of layers is what distinguishes graphite oxide from GO. There are limited layers, and a monolayer of flakes present in the GO dispersion of graphite oxide, despite the fact that it is a multilayer system. The modified Hummers' technique was used to make GO from flake graphite [27]. First, 1 g of graphite powder and 0.5 g of sodium nitrate are added to 70 ml of concentrated H_2SO_4 in a round-bottomed, two-neck flask placed in an ice bath. Next, 3 g of $KMnO_4$ is gradually added. The mixture is stirred for two hours and then diluted with deionized water. Afterward, 30% of H_2O_2 was added until the mixture's color changed to brilliant yellow, suggesting graphite oxidation [28]. The product was dried, filtered, and scanned using the FESEM as shown in Figure 4.3.

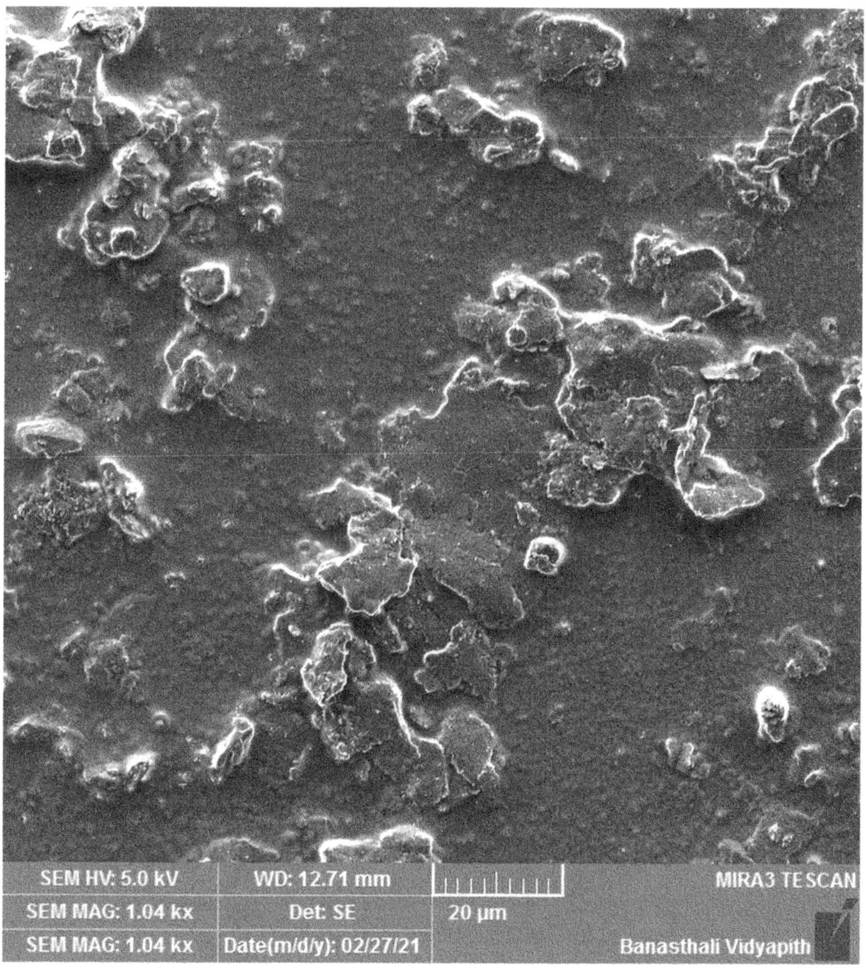

Figure 4.3 FESEM image of graphene oxide.

4.2.4 Reduced Graphene Oxide

Thermally and chemically treated GO is modified to rGO. Typically, oxidized GO and graphite sheets are used in the synthesis of rGO [29]. The reducing agents that are most frequently used to convert GO to rGO include hydrate, hydrazine, and l-ascorbic acid. Plants and plant extracts have been employed effectively in various research studies to generate rGO because they are easy to create and environmentally safe. Because of their interactions with enzymes, proteins, peptides, and DNA, the functional groups of rGO oxides facilitate their chemical production. Due to their affordability, biomolecules are used to prepare rGO, and they also have other advantages [30]. The reducing and capping agents required for the synthesis of nanoparticles are provided by plant extracts. They are easily chemically synthesized due to the functional groups of rGO oxides' interactions with DNA, proteins, peptides, and enzymes [31]. The preparation methods are displayed in the flow chart as shown in Figure 4.4

We have certain different methods to dissolve the rGO in order to obtain the homogeneous solution, which form the uniform thin film observed in the FESEM image as shown in Figure 4.5.

4.2.5 Graphitic Carbon Nitride

The most stable carbon nitride (CN) allotrope in ambient circumstances and comes with a plethora of surface features that make it an attractive material for many applications, including catalysis is g-C_3N_4 [32]. A conjugated polymer with just a CN bond assembly and no electron localization in the state is the optimal

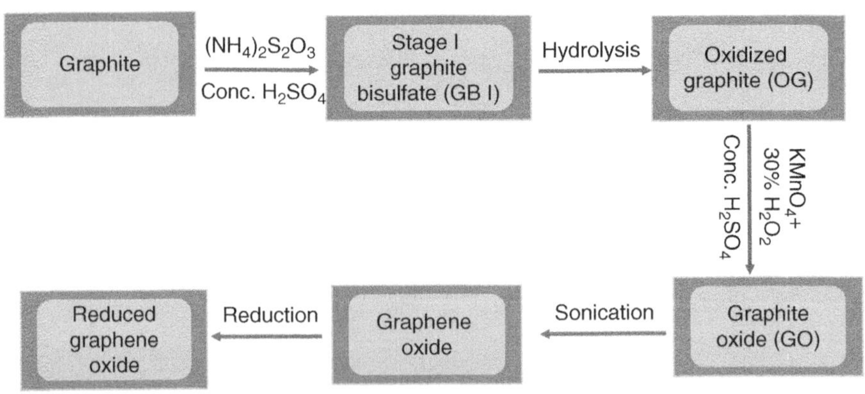

Figure 4.4 Diagram illustrating the steps involved in producing reduced graphene oxide.

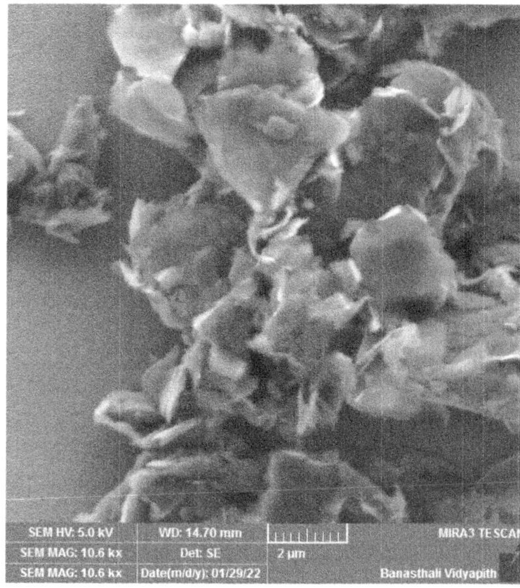

Figure 4.5 FESEM image of reduced graphene oxide.

form of g-C_3N_4. These are created by polycondensing cyanamide and have a tiny amount of hydrogen at their terminal edges in the form of primary and/or secondary amine groups. The presence of hydrogen indicates the existence of many surface imperfections and partial condensation of the actual g-C_3N_4 [33]. The Lewis-base character towards metal-free coordination is believed to be caused by these imperfections, which encourage electron relocalization on the surface.

The allotropes of C_3N_4, including —C_3N_4, have greatly surpassed due to g-C_3N_4 as it has a graphite-like structure and admirable stability [34]. It is made up of the three earth-abundant elements carbon, nitrogen, and hydrogen that are most frequently consumed in energy and storage applications. The most promising visible-light active photocatalyst is g-C_3N_4, which is polymeric and devoid of metals. g-C_3N_4 having 2.7 eV bandgap and low cost and environmental impact, may effectively replace the pricy and dangerous metal-holding photocatalysts with a narrow bandgap. Better mechanical characteristics, quick visible-light hydrogen evolution, and less recombination of photogenerated charges are all benefits of processing C_3N_4. It may be divided mostly into C_3N_4-based structures and those that are hybrids with other nanomaterials. The C_3N_4 hybrid structure operates quite well as a result of the combination of advantageous features of both materials. With this, C_3N_4-based thin films and ordered structures are included in the hybrid C_3N_4-graphene structures.

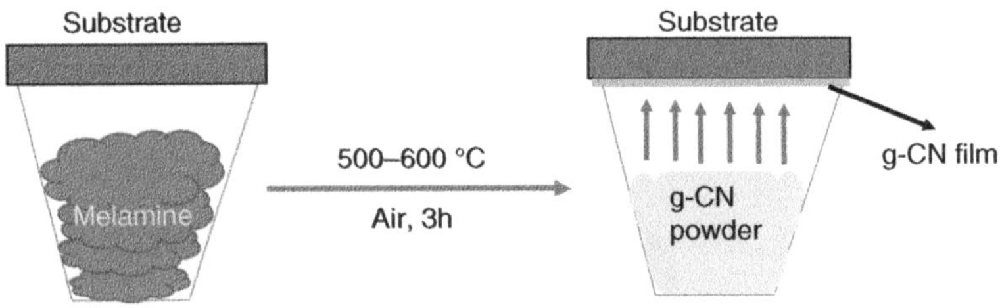

Figure 4.6 Preparation of graphitic carbon nitride film via thermal vapor condensation.

4.2.5.1 g-CN-Thin Film

There are primarily two processing methods: bottom-up and top-down for the preparation of C_3N_4-based thin films. In a crucible set in a muffle furnace melamine was heated to 600 °C at heating rates of 3 °C min^{-1} for two to three hours as a part of a bottom-up thermal vapor condensation method as shown in Figure 4.6.

4.2.5.2 Graphitic Carbon Nitride (g-CN)-Powder Form

GCN is a versatile material with numerous preparation methods tailored to different applications. Advances in synthesis techniques continue to enhance its properties, making it a promising candidate for various technological applications. By following Table 4.1, we have shown a comprehensive overview of the preparation methods of g-CN, highlighting both the technical aspects and their practical implications.

4.2.5.3 Thin Film of g-CN

Reactive compounds with prepended C—N core structures, such as variants of triazine and heptazine, are the most common precursors for the chemical synthesis of g-C_3N_4. The majority of these molecules are unstable, difficult to use, explosive, or hard to get. Creation of single-phase sp^3 makes it difficult to create hybridized CNs because of their insufficient thermodynamic stability. Thus, the creation of g-C_3N_4 with flaws is an intriguing subject.

In order to prepare bulk g-C_3N_4 firstly, 10 g of urea and melamine powder were ground using the mortar piston. The powder was then dissolved in a 1 : 1 combination of ethanol and water, which was heated to 110 °C while being agitated continuously. The crucible was heated for three hours at 550 °C. After another ethanol/water dissolved, the resulting bulk material was sonicated for eight hours. The finished product was oven-dried after centrifugation [40].

To make yellow g-C_3N_4, 5 g of melamine powder was put into a porcelain crucible, covered with a lid, and heated continuously at a rate of 12 °C per minute in a muffle furnace. The crucible was then held at 545 °C for two hours. The resultant yellow material was gathered and pulverized into powder when it had gradually

Table 4.1 Various methods to synthesize g-C_3N_4.

Method used for synthesis	Product	Material used	Synthesis process	Reference
Thermal condensation method	Bulk g-C_3N_4	DCDA, urea, melamine, thiourea, and cyanamide	The previous method involved heating a covered crucible—ideally constructed of alumina—in a furnace for up to four hours at a temperature of 500–600 °C. The recovered yellow residue was ground into a powder after allowing it to reach room temperature.	[32]
Method of fractional thermal polymerization	g-C_3N_4 powder	Melamine, guanidine carbonate, and DCDA	At a rate of 5 °C min^{-1}, heating the precursors in a covered crucible ranged from 20 to 360 °C. The temperature was maintained at 360 °C for 30 minutes, then elevated once again to a specified amount in the next 30 minutes, where it was maintained for 3 hours. There was cooling after that.	[35]
Method of Pyrolysis-assisted chemical vapor deposition	Free-standing g-C_3N_4 films	DCDA	The predecessor was kept within a quartz tube that had its intake connected to "Ar flow and its exit connected to vacuum in order to preserve the tube's pressure low. The" vapors entered the furnace (at 600 °C) due to this configuration. The vapors condensed thermally and covered the tube walls. The accumulation was then scrubbed off, and a uniform freestanding film was produced once the tube had reached room temperature.	[36]

(Continued)

Table 4.1 (Continued)

Method used for synthesis	Product	Material used	Synthesis process	Reference
Soft-template method with thermal polymerization	Exfoliated g-C_3N_4	Urea	A 1 : 1 ethanol to water combination containing 10 g of urea was heated to 110 °C while being continuously agitated. When the solvent heated, white precipitate was created. Then placed inside a crucible. Three hours of 550 °C heating were applied to the crucible. Following another ethanol/water dissolution, the resulting bulk material was sonicated for eight hours. The finished powder was oven-dried following centrifugation.	[37]
Post-calcination template-free solvothermal method"	Hierarchical nanoporous g-C_3N_4 microspheres	melamine and Cyanuric chloride	ACN served as the solvent and "the precursors were mixed in a specified ratio. This liquid was placed in a 100 ml in autoclave. For 12–24 hours, the system was heated to a certain temperature (between 120 and 200 °C). After being chilled, the objects were centrifuged, cleaned with 100% ethanol and ionised water, and then dried. The post-calcination procedure involves heating the sample to 550 °C for 4 hours in air or 2 hours in argon.	[38]

Solid reaction without solvent	powder of g-CN	melamine	Inside a 50 ml stainless steel autoclave 10.0 mmol melamine were weighed, it was sealed, heated to 400 °C for six hours, and the finished product was allowed to cool to ambient temperature before being cleaned and dried.	[39]
Chemical vapor deposition method	Ordered cubic mesoporous $g\text{-}C_3N_4$	Melamine	An alumina crucible of 50 ml containing 6 g of precursor was used, and 2 g of KIT-6 template was evenly distributed across the precursor layer. The closed crucible was heated to 320 °C for four hours in the furnace. When the precursor evaporates, it will deposit itself into the pores of the template. Then, for three hours, the temperature was increased to and maintained at 550 °C.	[40]

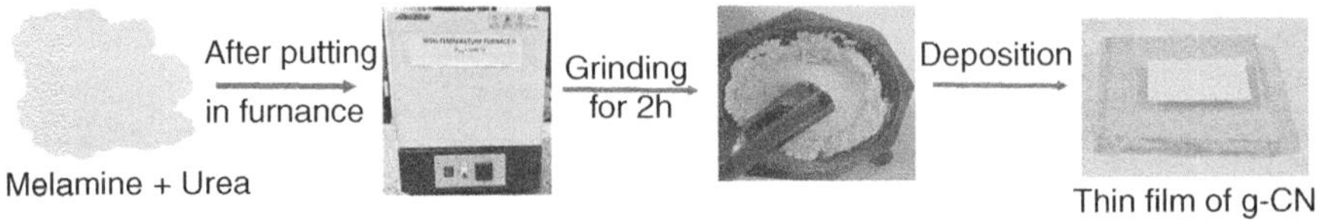

Figure 4.7 Thin film preparation of graphitic carbon nitride.

cooled to room temperature. After that, using the fine powder, the thin film is prepared on the glass substrate using doctor blade method as shown in Figure 4.7.

4.3 Properties of g-CN

Exfoliated graphene and g-C_3N_4 layered structures have been produced using a variety of methods. Graphite that is stacked is delaminated into a few or one sheet while maintaining its conjugated structure by the intercalation of chemical substances. The research's bulk graphitic CN (BCN) powder was created by heating melamine to 520 °C and condensing it in air [41]. Surface morphology g-C_3N_4 layered structures in two different forms as g-C_3N_4 bulk g-C_3N_4 porous as are exhibit in Figure 4.8a,b.

4.3.1 Morphologvical Properties

A picture of the freshly acquired g-C_3N_4 powder taken using a scanning electron microscope (FESEM). The FESEM image of g-CN was used to observe the pattern of the flakes and morphology of the thin film was examined at different

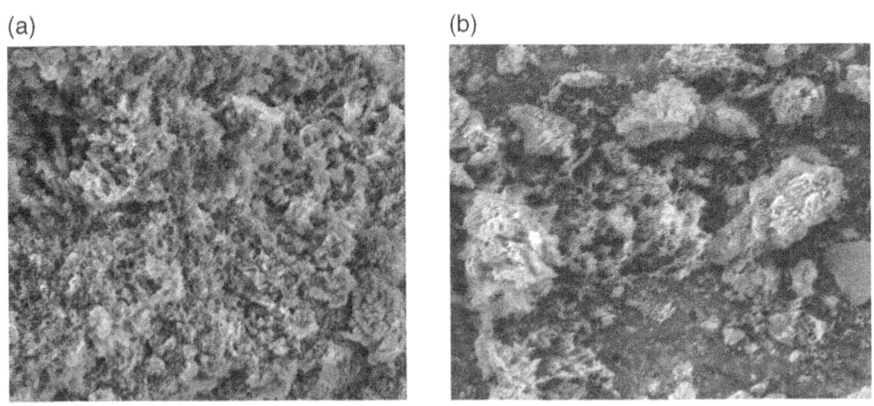

Figure 4.8 ESEM image of graphitic carbon nitride thin film (a) g-C_3N_4 bulk and (b) g-C_3N_4 porous.

magnifications. Thick, micron-sized particles that were irregularly dispersed over the whole sample's surface made up the bulk structure. The majority of the particles had rough surfaces and numerous stacked layers [42].

4.3.2 Band Gap

The optical properties of the bulk and several acid-treated g-C3N4 powders were examined using UV-visible absorbance spectroscopy. The absorbance spectra exhibit distinct absorption edges that indicate a change in bandgap energy from the bulk structure [43]. The absorption edge visibly moved for the acid-treated materials from a longer to a shorter wavelength (487–454 nm). The exfoliation and size decrease of the bulk g-C_3N_4 structure could have contributed to the band edge's elongation. Following acid treatment, the g-C_3N_4 particles' absorption edge was notably bluer than the bulk structure's. Compared to the bulk structure's 2.54 eV bandgap energy, the GCN showed the greatest bandgap energy at 2.72 eV [32].

4.3.3 Other Properties

Various techniques are often utilized to illustrate the surface chemistry, bonding states and composition of both g-C_3N_4 and its composites, in addition to the characterization methods stated above. The elemental makeup of a substance was also revealed by the findings of various spectroscopic investigations as each component's chemical environment. Still, when N and C's potential bonding configurations come together with the inevitable contact with the gases in the atmosphere (such as oxygen and hydrogen), an analysis of the structure is obtained.

Two contact approaches may be employed to evaluate the conductivity of materials based on C_3N_4. The comparison of conductivity of all C_3N_4-based materials synthesized at different temperatures and with varied forms has been done using the two-contact approach. Using Ohm's Law, electronic conductivity is determined. Additionally, electron paramagnetic resonance (EPR) or electron spin resonance (ESR) spectroscopy is a very sophisticated non-destructive technique for detecting free radicals, metallic ions, defects, photoelectrons, and conjugated structures. The electronic band structure of g-C_3N_4 has been examined, and the presence of vacancies in the structure has been confirmed, with ease, using this characterization technique.

4.4 Applications of g-CN

The porous 1D structure enhances charge transfer, which might boost the efficiency of the photovoltaic process. The photoexcited charge carriers' recombination processes, which can occur radiatively or non-radiatively, are what define the

PSC's efficiency. Eventually, surplus electrons and holes are recombined, and their energy is transmitted as a photon or a phonon, respectively. Due to its exceptional semiconducting qualities, g-C_3N_4 has drawn interest for its use in environmental and energy-related applications. Nevertheless, these materials' rapid electron-hole recombination reduces their effectiveness and limits their applications. In CN (C_3N_4)-based nanocomposites, this recombination is postponed, increasing efficiency. g-C_3N_4 materials are very appealing for photocatalytic applications because they are cheap, stable, sustainable, and metal-free. However, it is important to note that when g-C_3N_4 is combined with other types of nanoparticles, their applications are expanded beyond photocatalysis and include energy storage, bioimaging, sensing CN-based devices, and photocatalytic coenzyme regeneration [41].

While g-CN has been widely employed in solar energy applications, in order to draw readers' interest to less-explored fields including ecological and sensing applications, few are succinctly covered in this section.

4.4.1 g-CN Role in Organic Solar Cells

Owing to C_3N_4 large particle size and insoluble nature in most solvents, standard deposition techniques like spin coating and screen printing yield inadequate C_3N_4 coverage and poor adherence on widely utilized substrates, for example, transparent conductive oxides (FTO, ITO, etc.). Due to numerous optical and electrical properties of g-CN, it is able to be used as an a transporting layer as well as in the active layer of the organic solar cell (OSCs) [44]. The overall performance of the device is increased due to the role of g-CN in the photoactive layer as shown in Figure 4.9.

4.4.2 g-CN Role in Perovskite Solar Cells

With a validated PCE of 25.2%, the efficiency of Perovskite solar cells (PSCs) has skyrocketed in modern years. The efficiency of PSCs depends on the extraction of charge carriers as they can be destroyed by different radiative and non-radiative recombination. Eventually, the energy of surplus electrons and holes that have

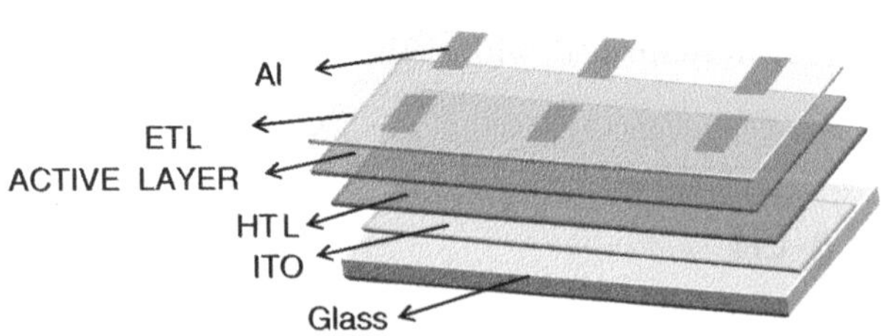

Figure 4.9 Schematic diagram of OSCs using g-CN in active layer.

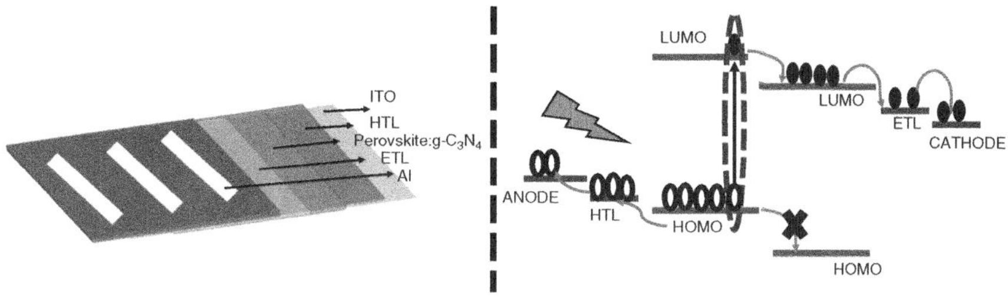

Figure 4.10 Structure and energy-level diagram of PSCs using g-CN in active layer.

recombined is transmitted as a phonon or photon, respectively [45]. Organic-inorganic halide perovskite materials have also demonstrated excellent passivation by amines (e.g., $-NH_2$) or nitrogen functionalities. Addition of urea in perovskite layer consequentially increases the PCE to 18.6% from 14.4% by the reduction of recombination at the interfaces and grain boundaries. Perovskite light absorber films' conductivity increased when $g\text{-}C_3N_4$ was added to improve the crystalline quality, which is advantageous for concerned charge extraction productivity [46]. The use of g-CN with perovskite enhanced the performance of the PSCs as shown in Figure 4.10.

4.4.3 g-CN Role in Dye-Sensitized Solar Cells

Using a straightforward ball-milling procedure, the composite of g-CN with conductive carbon black was used as a counter electrode to replace the costly Pt electrode to create dye-sensitized solar cells (DSSCs), with a 5.09% conversion efficiency. The use of an absorbent graphene/$g\text{-}C_3N_4$ composite as a counter electrode for DSSCs was subsequently described and studied, utilizing a simple hydrothermal process. By employing the porous $g\text{-}C_3N_4$/graphene composite as the counter electrode, the DSSC achieved a conversion efficiency of 7.13%. Since the photoanode is essential to DSSC, one of its hotspots is photoanode modification. The π-conjugated material $g\text{-}C_3N_4$ nanosheet accelerates the separation of charge carriers. After modifying the surface and controlling the morphology of $g\text{-}C_3N_4$, a TiO_2/$g\text{-}C_3N_4$ composite photoanode is created. The enhanced PCE of DSSCs is obtained due to the increased absorption of dye with the specific surface area of the TiO_2/$g\text{-}C_3N_4$ composite as photoanode [47].

4.4.4 g-CN Role as a Photocatalyst

Due to the weak van der Waals forces of attraction between the layers and the strong covalent bonds between the nitrogen and carbon atoms, exfoliation of the layers is possible in bulk g-CN. Various techniques, such as chemical heat exfoliation and liquid-phase ultrasonic exfoliation using strong acids, have been employed for material processing. The process of "peeling off" bulk g-C_3N_4 has been achieved using chemical blowing (using blowing agents). Because of its appropriate band gap, g-C_3N_4 can capture solar energy, which enables its use as a photocatalyst [48]. Although bulk g-C_3N_4 can only absorb blue light up to 450 nm in wavelength because of its huge bandgap of approximately 2.7 eV, tiny surface area, and reckless charge recombination rate, a variety of alterations have been performed to optimize and improve the absorption of light. These consist of dye-sensitization, doping, and copolymerization. CN graphitic traditionally, photocatalysts have been used in two primary contexts: for the degradation of contaminants as well as for separating water into H_2 and O_2.

4.4.5 g-CN-Sensing Applications

Despite being prime prospects for bioimaging and sensing, g-C_3N_4 nanostructures can be made more effective in these domains by modifying their architectures. Biomolecules like glucose and heparin as well as ions such as mercury, chromium, iron, copper, and other targets that the altered g-C_3N_4 nanocompounds could identify. It was discovered that the Pt/ZnO/g-C_3N_4 composite structure is a potentially useful sensor for the detection of gases like NO_2 and ethanol. Using a microwave to aid in manufacturing, exfoliated CNNSs were employed as a growing medium for ZnO nanorods through the Zn and N components bonding Subsequently, Pt nanoparticles were deposited onto the structure, resulting in the formation of Pt/ZnO/g-C_3N_4 nanostructures. [49].

4.4.6 g-CN Environmental Applications

To create "phosphorus-doped porous ultrathin g-C_3N_4 nanosheets (PCN-S)," thermal exfoliation and P element doping were combined. The advantages of PCN-S were shown by FTIR, photocurrent response (I–V), XRD, TEM, UV–vis DRS, and other characterization studies. These included strong visible light absorption, enhanced chemical resistance, a porous surface, and a large surface area. P-doping and the in-plane pores of the NSs were shown to have a beneficial effect on the photocatalytic effectiveness of PCN-S on Cr (VI) reduction, as they expanded the reaction region to light and increased the surface area with more active sites. The methods of the test may be applied to promote the production of gC3N4-based compounds with enhanced photocatalytic activity in order to eliminate more pollutants [50].

Due to their potential applications in electric vehicles and their relatively high specific energy and improved stability, several Li-based batteries, such as Li-ion, Li-O_2, and Li-S batteries, have received a lot of interest. Nevertheless, the slow kinetics of the processes on the electrodes limit the practical uses of these batteries. An effective catalyst should be created and used to address this issue. Because of its high nitrogen concentration, g-C_3N_4 could propose more active sites for electrode reactions. Nevertheless, the poor electronic conductivity places a constraint on g-C_3N_4. To boost the amount of electrons collected on the g-C_3N_4 surface, certain electrical conductive materials should be utilized [51]. Worldwide water body pollution has gotten worse due to extensive water pollution, mostly from the apparel, food service, and automotive sectors. Thus, novel approaches are needed for the identification and filtration of non-biodegradable contaminants as well as the treatment of water.

4.5 Conclusion

Carbon-based 2D materials have significantly impacted various technological fields due to their exceptional properties. Graphene, in particular, often reaches nearly 100% of its theoretical potential in key areas such as electrical conductivity, thermal conductivity, and mechanical strength. Its outstanding performance makes it a leading material in electronics, where graphene transistors outpace traditional silicon devices, and in green energy applications, where graphene-based batteries and supercapacitors achieve around 90% of their theoretical maximum capacities. Additionally, graphene sensors offer almost perfect sensitivity and selectivity, marking a near-total realization of their potential.

g-C_3N_4 also shows notable achievements, especially in photocatalysis, where it reaches up to 20% efficiency in applications like water splitting and pollutant degradation. While g-CN's electrical and mechanical properties are not as high as graphene's, it still performs well in composite materials used for green energy production and sensor applications, achieving efficiencies of around 20–40% and 30–50%, respectively. Even though g-C_3N_4 is known to have several advantages, more work has to be done to fully realize its potential. Creating complex g-C_3N_4-based nanocomposites and investigating how semiconductors affect g-C_3N_4 electrical characteristics are two interesting research directions. However, a general overview based on current trends and applications can be provided. Approximately, 40% of g-CN applications are in hydrogen production through photocatalytic water splitting. This is a major focus of current research due to the global interest in sustainable green energy sources. Similarly, the role of g-CN in solar energy conversion is 25% in photovoltaic devices and solar cells due to its effective light absorption and green energy conversion properties. Its use in this area is substantial but not as dominant as in photocatalysis. Next is g-CN's role in sensing application which is 15% for detecting gases and biomolecules. While this field is growing, it constitutes a smaller but significant portion of g-CN applications.

In batteries and supercapacitors, g-CN enhances performance and stability, contributing to advancements in energy storage technologies. However, its use in this area is only 10%, which is less compared to photocatalysis and solar energy conversion. Similarly, as in catalysis, g-CN applications are 10% in various chemical reactions, including CO_2 reduction and organic synthesis. Its role here is crucial but currently less widespread compared to its photocatalytic applications. These percentages provide a rough estimate of g-CN's participation in different applications, highlighting its predominant role in photocatalysis and significant contributions to green energy production, sensing, energy storage, and catalysis.

References

1 Paudyal, S., Cañizares, C.A., and Bhattacharya, K. (2011). Optimal operation of distribution feeders in smart grids. *IEEE Transactions on Industrial Electronics* 58 (10): 4495–4503. https://doi.org/10.1109/TIE.2011.2112314.

2 Lin, W. and Bitar, E. (2016). Decentralized stochastic control of distributed energy resources. *IEEE Transactions on Power Systems* 33 (1): 888–900. https://doi.org/10.1109/TPWRS.2017.2700472.

3 Power, A.C., Gorey, B., Chandra, S., and Chapman, J. (2018). Carbon nanomaterials and their application to electrochemical sensors: a review. *Nanotechnology Reviews* 7 (1): 19–41. https://doi.org/10.1515/ntrev-2017-0160.

4 Güryel, S., Alonso, M., Hajgató, B. et al. (2017). A computational study on the role of noncovalent interactions in the stability of polymer/graphene nanocomposites. *Journal of Molecular Modeling* 23 (2): https://doi.org/10.1007/s00894-017-3214-2.

5 Gopinath, S.C.B., Anbu, P., Theivasanthi, T. et al. (2018). Characterization of reduced graphene oxide obtained from vacuum-assisted low-temperature exfoliated graphite. *Microsystem Technologies* 24 (12): 5007–5016. https://doi.org/10.1007/s00542-018-3921-3.

6 Kumar, U., Sikarwar, S., Sonker, R.K., and Yadav, B.C. (2016). Carbon nanotube: synthesis and application in solar cell. *Journal of Inorganic and Organometallic Polymers and Materials* 26 (6): 1231–1242. https://doi.org/10.1007/s10904-016-0401-z.

7 Zhuge, F., Zheng, Z., Luo, P. et al. (2017). Nanostructured materials and architectures for advanced infrared photodetection. *Advanced Materials Technologies* 2 (8). https://doi.org/10.1002/admt.201700005.

8 Capezza, A., Andersson, R.L., Ström, V. et al. (2019). Preparation and comparison of reduced graphene oxide and carbon nanotubes as fillers in conductive natural rubber for flexible electronics. *ACS Omega* 4 (2): 3458–3468. https://doi.org/10.1021/acsomega.8b03630.

9 Liu, S., Zhang, X.D., Gu, X., and Ming, D. (2019). Photodetectors based on two dimensional materials for biomedical application. *Biosensors and Bioelectronics* 143: https://doi.org/10.1016/j.bios.2019.111617.

10 Díez-Pascual, A.M., Sánchez, J.A.L., Capilla, R.P., and Díaz, P.G. (2018). Recent developments in graphene/polymer nanocomposites for application in polymer solar cells. *Polymers* 10 (2): https://doi.org/10.3390/polym10020217.

11 Shelke, N.T. and Late, D.J. (2021). *Synthesis and Characterization of 2D Materials*. INC.

12 Subramanyam, B.V.R.S., Mahakul, P.C., Sa, K. et al. (2020). Investigation of improvement in stability and power conversion efficiency of organic solar cells fabricated by incorporating carbon nanostructures in device architecture. *JPhys Materials* 3 (4): https://doi.org/10.1088/2515-7639/abbf44.

13 Rajendran, R., Shrestha, L.K., Minami, K. et al. (2014). Dimensionally integrated nanoarchitectonics for a novel composite from 0D, 1D, and 2D nanomaterials: RGO/CNT/CeO2 ternary nanocomposites with electrochemical performance. *Journal of Materials Chemistry A* 2 (43): 18480–18487. https://doi.org/10.1039/c4ta03996c.

14 Xu, J., Brenner, T.J.K., Chabanne, L. et al. (2014). Liquid-based growth of polymeric carbon nitride layers and their use in a mesostructured polymer solar cell with v oc exceeding 1 v. *Journal of the American Chemical Society* 136 (39): 13486–13489. https://doi.org/10.1021/ja508329c.

15 Wang, S., Liao, J., Wu, M. et al. (2017). High rate and long cycle life of a CNT/rGO/Si nanoparticle composite anode for lithium-ion batteries. *Particle and Particle Systems Characterization* 34 (10). https://doi.org/10.1002/ppsc.201700141.

16 Singh, M.K., Pandey, R.K., and Prakash, R. (2017). High-performance photo detector based on hydrothermally grown SnO_2 nanowire/reduced graphene oxide (rGO) hybrid material. *Organic Electronics* 50: 359–366. https://doi.org/10.1016/j.orgel.2017.08.016.

17 Madaswamy, S.L., Wabaidur, S.M., Khan, M.R. et al. (2021). Polyaniline-graphitic carbon nitride based nano-electrocatalyst for fuel cell application: a green approach with synergistic enhanced behaviour. *Macromolecular Research* 29 (6): 411–417. https://doi.org/10.1007/s13233-021-9044-1.

18 Krishnamoorthy, K., Veerapandian, M., Yun, K., and Kim, S.J. (2013). The chemical and structural analysis of graphene oxide with different degrees of oxidation. *Carbon* 53: 38–49. https://doi.org/10.1016/j.carbon.2012.10.013.

19 Films, U.C. and Rise, T. (2021). Ultra-thin carbon films: the rise of," no. Figure 1, pp. 1–6.

20 Safa, S., Sarraf-Mamoory, R., and Azimirad, R. (2015). Ultra-violet photodetection enhancement based on ZnO–graphene composites fabricated by sonochemical method. *Journal of Sol-Gel Science and Technology* 74 (2): 499–506. https://doi.org/10.1007/s10971-015-3625-4.

21 Abid, P., Sehrawat, S., Islam, S. et al. (2018). Reduced graphene oxide (rGO) based wideband optical sensor and the role of temperature, defect states and quantum efficiency. *Scientific Reports* 8 (1): 1–13. https://doi.org/10.1038/s41598-018-21686-2.

22 Bîru, E.I. and Iovu, H. (2018). Graphene nanocomposites studied by raman spectroscopy. *Raman Spectroscopy* https://doi.org/10.5772/intechopen.73487.

23 Aydın, H., Kalkan, S.B., Varlikli, C., and Çelebi, C. (2018). P3HT–graphene bilayer electrode for Schottky junction photodetectors. *Nanotechnology* 29 (14): 145502.

24 Ferrari, A.C. and Basko, D.M. (2013). Raman spectroscopy as a versatile tool for studying the properties of graphene. *Nature Nanotechnology* 8 (4): 235–246. https://doi.org/10.1038/nnano.2013.46.

25 Subrati, M.S., Lonkar, S.P., and Abdala, A.A. (2014). Amine-functionalized graphene for natural gas sweetening. *Advanced Materials Research* 1064: 21–25. https://doi.org/10.4028/www.scientific.net/amr.1064.21.

26 Sung, S.J., Park, J., Cho, Y.S. et al. (2019). Enhanced gas barrier property of stacking-controlled reduced graphene oxide films for encapsulation of polymer solar cells. *Carbon* 150: 275–283. https://doi.org/10.1016/j.carbon.2019.04.120.

27 Chauhan, A.K., Gusain, A., Jha, P. et al. (2014). Graphene composite for improvement in the conversion efficiency of flexible poly 3-hexyl-thiophene:[6,6]-phenyl C_{71} butyric acid methyl ester polymer solar cells. *Applied Physics Letters* 104 (13). https://doi.org/10.1063/1.4869635.

28 Sharma, M., Negi, C.M.S., and Alvi, P.A. (2023). Carbon and its derivative-based ternary composites as a photoactive layer for optoelectronic application. *Indian Journal of Physics* https://doi.org/10.1007/s12648-022-02566-y.

29 Sharma, M., Alvi, P.A., Gupta, S.K., and Negi, C.M.S. (2021). The optoelectronic behavior of reduce graphene oxide-carbon nanotube nanocomposites. *Synthetic Metals* 281: 116892. https://doi.org/10.1016/j.synthmet.2021.116892.

30 Roy, R., Jha, A., Banerjee, D. et al. (2013). Edge effect enhanced electron field emission in top assembled reduced graphene oxide assisted by amorphous CNT-coated carbon cloth substrate. *AIP Advances* 3 (1): https://doi.org/10.1063/1.4789409.

31 Sharma, M., Sharma, N., Alvi, P.A. et al. (2022). P3HT-rGO composites for high-performance optoelectronic devices. *Optical Materials* 127: https://doi.org/10.1016/j.optmat.2022.112326.

32 Zheng, Y., Zhang, Z., and Li, C. (2017). A comparison of graphitic carbon nitrides synthesized from different precursors through pyrolysis. *Journal of Photochemistry and Photobiology A: Chemistry* 332: 32–44. https://doi.org/10.1016/j.jphotochem.2016.08.005.

33 Li, Z., Wu, S., Zhang, J. et al. (2020). Improving photovoltaic performance using perovskite/surface-modified graphitic carbon nitride heterojunction. *Solar RRL* 4 (3): https://doi.org/10.1002/solr.201900413.

34 Ayán-Varela, M., Villar-Rodil, S., Paredes, J.I. et al. (2015). Investigating the dispersion behavior in solvents, biocompatibility, and use as support for highly efficient metal catalysts of exfoliated graphitic carbon nitride. *ACS Applied*

Materials and Interfaces 7 (43): 24032–24045. https://doi.org/10.1021/acsami.5b06974.

35 Kumru, B., Antonietti, M., and Schmidt, B.V.K.J. (2017). Enhanced dispersibility of graphitic carbon nitride particles in aqueous and organic media via a one-pot grafting approach. *Langmuir* 33 (38): 9897–9906. https://doi.org/10.1021/acs.langmuir.7b02441.

36 Bayan, R. and Karak, N. (2019). Photoluminescent oxygeneous-graphitic carbon nitride nanodot-incorporated bioderived hyperbranched polyurethane nanocomposite anticounterfeiting attribute. *ACS Omega* 4 (5): 9219–9227. https://doi.org/10.1021/acsomega.9b00891.

37 Comeau, Z.J., Boileau, N.T., Lee, T. et al. (2019). On-the-spot detection and speciation of cannabinoids using organic thin-film transistors. *ACS Sensors* 4 (10): 2706–2715. https://doi.org/10.1021/acssensors.9b01150.

38 Fidan, T., Torabfam, M., Saleem, Q. et al. (2021). Functionalized graphitic carbon nitrides for environmental and sensing applications. *Advanced Energy and Sustainability Research* 2 (3). https://doi.org/10.1002/aesr.202000073.

39 Zhu, J., Xiao, P., Li, H., and Carabineiro, S.A.C. (2014). Graphitic carbon nitride: synthesis, properties, and applications in catalysis. *ACS Applied Materials and Interfaces* 6 (19): 16449–16465. https://doi.org/10.1021/am502925j.

40 Chetia, T.R., Ansari, M.S., and Qureshi, M. (2016). Graphitic carbon nitride as a photovoltaic booster in quantum dot sensitized solar cells: a synergistic approach for enhanced charge separation and injection. *Journal of Materials Chemistry A* 4 (15): 5528–5541. https://doi.org/10.1039/c6ta00761a.

41 Naseri, A., Samadi, M., Pourjavadi, A. et al. (2017). Graphitic carbon nitride (g-C3N4)-based photocatalysts for solar hydrogen generation: Recent advances and future development directions. *Journal of Materials Chemistry A* 5 (45): 23406–23433. https://doi.org/10.1039/c7ta05131j.

42 Alaghmandfard, A. and Ghandi, K. (2022). A comprehensive review of graphitic carbon nitride (g-C3N4)–metal oxide-based nanocomposites: potential for photocatalysis and sensing. *Nanomaterials* 12 (2): https://doi.org/10.3390/nano12020294.

43 Kumar, A., Singh, M.K., Sharma, M., and Kumar, U. (2022). Structural, dielectric and electrical properties of homovalent doped SrSn1-xTixO3 ($0 \leq x \leq 0.08$) system. *Indian Journal of Pure and Applied Physics* 60 (11): 909–913. https://doi.org/10.56042/ijpap.v60i11.65819.

44 Cao, Q., Kumru, B., Antonietti, M., and Schmidt, B.V.K.J. (2020). Graphitic carbon nitride and polymers: a mutual combination for advanced properties. *Materials Horizons* 7 (3): 762–786. https://doi.org/10.1039/c9mh01497g.

45 Sharma, M., Negi, C.M.S., Alvi, P.A., and Gupta, S.K. (2022). Bulk heterojunction PTB7-Th: ICBA-based optoelectronic devices. *2022 5th International Conference on Multimedia, Signal Processing and Communication Technologies,*

IMPACT 2022, Aligarh Muslim University, pp. 1–4, https://doi.org/10.1109/IMPACT55510.2022.10029022 (accessed 26–27 November 2022).

46 Shi, Z. and Jayatissa, A.H. (2017). The impact of graphene on the fabrication of thin film solar cells: current status and future prospects. *Materials* 11 (1): https://doi.org/10.3390/ma11010036.

47 Yang, X., Zhao, L., Wang, S. et al. (2021). Recent progress of g-C3N4 applied in solar cells. *Journal of Materiomics* 7 (4): 728–741. https://doi.org/10.1016/j.jmat.2021.01.004.

48 Andreas, R., Lesbani, A., and Akhmad Yusuf, F. (2019). The character istics (compositions, morphological, and structure) of nanocomposites polyaniline (PANI)/ZnO. *IOP Conference Series: Materials Science and Engineering* 509 (1): https://doi.org/10.1088/1757-899X/509/1/012126.

49 Kavinkumar, T. and Manivannan, S. (2016). Synthesis, characterization and gas sensing properties of graphene oxide- multi walled carbon nanotube composite. *Journal of Materials Science and Technology* 32: 626–632. https://doi.org/10.1016/j.jmst.2016.03.017.

50 Zhao, W., Yan, Z., and Qian, L. (2020). Graphitic carbon nitride: preparation, properties and applications in energy storage. *Engineered Science* 10: 24–34. https://doi.org/10.30919/es8d1008.

51 Liang, Z., Yan, L., Si, J. et al. (2021). Rational design and characterization of symmetry-breaking organic semiconductors in polymer solar cells: a theory insight of the asymmetric advantage. *Materials* 14 (21). https://doi.org/10.3390/ma14216723.

5

Exploring the Potential of Graphene in Sustainable Energy Solutions

M. Karthik[1], S. Allirani[2], G. Ilakkiya[2], and R. Adharsh[2]

[1] *Department of Electrical and Electronics Engineering, SRM Madurai College for Engineering and Technology, Pottapalayam, Tamil Nadu, India*
[2] *Department of Electrical and Electronics Engineering, Sri Ramakrishna Engineering College, Coimbatore, Tamil Nadu, India*

5.1 Introduction

Graphene is a hexagonal honeycomb chassis of carbon titles securely clicked together as shown in Figure 5.1. It's a carbon allotrope in the shape of an airplane of sp2 with a molecular bond length of 0.142 nm. Graphite is the result of stacking layers of graphene atop one another, with an interplanar distance of 0.335 nm. The cohesive forces between these layers, known as van der Waals forces, bind the distinct layers of graphene in graphite together. These forces can be overcome through processes like graphene exfoliation, which separates the layers from graphite.

Each carbon snippet in diamond is linked to four other imitations. Diamond is one of the hardest known accoutrements due to this veritably strong arrangement. Each snippet in graphite is connected to three others in hexagonal (six-sided) layers. Although the connections between the hexagonal wastes are strong, each subcaste is only noiselessly attracted to the coming, allowing the layers to glide past one another.

In 2004, two scientists hailing from the University of Manchester in the United Kingdom, Andre Geim and Konstantin Novoselov, utilized the related nature and list strength of graphite to produce graphene. Graphite flakes were uprooted from graphite using sticky bonds. The tenacious tape recording was also squeezed and

Carbon-based Nanomaterials for Green Applications, First Edition. Edited by Upendra Kumar, Piyush Kumar Sonkar, and Suman Lata Tripathi.

Published 2025 by John Wiley & Sons, Inc.

Figure 5.1 Graphene. *Source:* AlexanderAlUS/Wikipedia/CC BY-SA 3.0.

torn piecemeal. This system was continued until they attained a single snippet thick film of graphene.

The first Graphene samples attained were many square millimeters in size and were suitable for testing. Graphene, which is only one snippet thick, was allowed to be the first two-dimensional substance ever set up by humans.

In 2010, Andre Geim and Konstantin Novoselov were honored with the Nobel Prize in Physics for their groundbreaking work on discovering graphene. Graphene is one of the thinnest accoutrements known to humanity, with a consistency of 0.34 nm and only one snippet thick. It's also one of the most durable accoutrements. It's one of the topmost operators of electricity at room temperatures ranging from (4.840.44) 103 to (5.300.48) 103 Wm1K1. Graphene has the least electron smatter when compared to other accoutrements. The electron mobility in graphene has been determined to be less than 200,000 $cm^2V^1s^1$. One of graphene's major features is its harmonious immersion of light across the visible and near-infrared diapasons. It's worth noting that graphene is extremely well-acclimated for spin transfer. The chemical techniques for graphene from several carbon sources are schematically described in Figure 5.2.

The potential of wearable technology incorporating graphene is immense, poised to transform diverse sectors such as healthcare, fitness, fashion, and electronics. Graphene, consisting of a single layer of carbon atoms arranged in a two-dimensional lattice, boasts extraordinary properties, rendering it exceptionally well-suited for wearable devices.

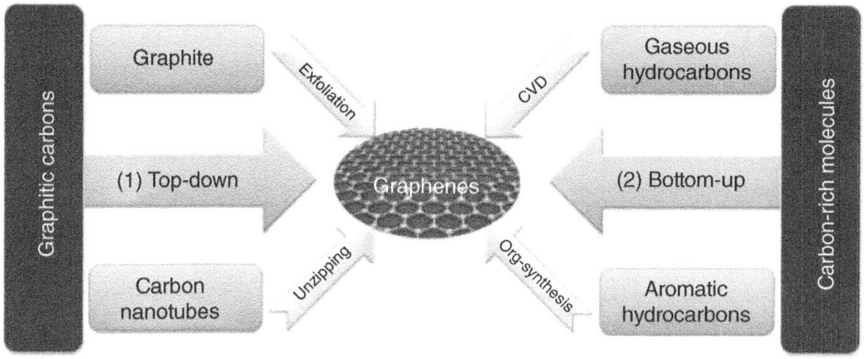

Figure 5.2 The chemical techniques for graphene from several carbon sources.

The salient features of using graphene in wearable technology are:

1) **Flexibility and Durability:** Graphene is an incredibly flexible material that can be integrated into wearable devices without compromising their mechanical properties. It allows for comfortable and long-lasting wear, making it suitable for applications such as flexible sensors and smart clothing.
2) **Lightweight:** Graphene is an ultralight material, adding minimal weight to wearable devices, which is crucial for comfort and portability.
3) **Conductivity:** Graphene exhibits excellent electrical conductivity, making it an ideal material for sensors, electrodes, and interconnects in wearable electronics.
4) **Thermal Properties:** Graphene has excellent thermal conductivity, which helps dissipate heat generated by wearable devices, preventing overheating and improving overall device performance.
5) **Transparency:** Graphene is transparent, making it suitable for integrating into transparent displays and smart windows in wearables.
6) **Biocompatibility:** Graphene is biocompatible, which means it is safe to use in contact with the human body. This property is especially important for medical and health-related wearable devices.

5.2 Usage of Graphene in Various Sectors

1) **Health Monitoring Devices:** Graphene-based sensors can be used to monitor various health parameters, such as heart rate, blood pressure, and body temperature, in real-time. These sensors can be integrated into wearable devices like smartwatches, fitness bands, and health patches.

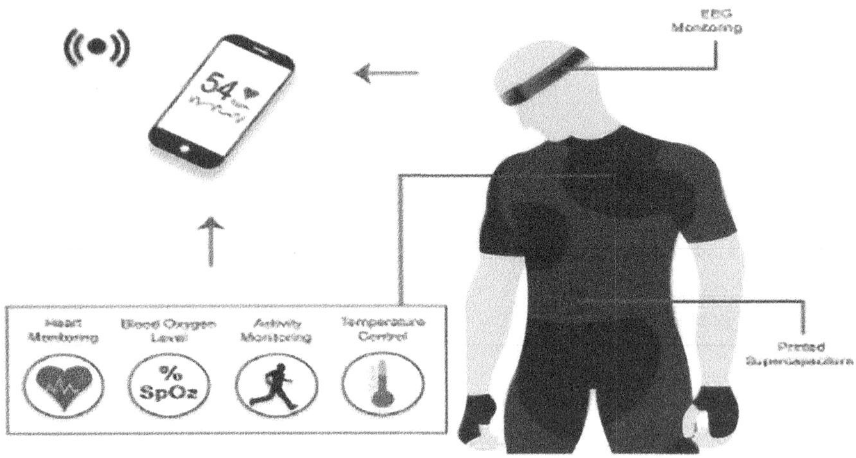

Figure 5.3 Graphene in textile industry.

2) **Smart Textiles:** Graphene can be woven into fabrics, creating smart clothing with built-in sensors for monitoring body movements, posture, and even vital signs as shown in Figure 5.3. These smart textiles can find applications in sports, healthcare, and military sectors.
3) **Energy Harvesting:** Graphene's electrical conductivity can be utilized to harvest energy from body movements or ambient sources, providing power to wearable devices without the need for batteries.
4) **Flexible Displays:** Graphene's transparency and flexibility make it an ideal candidate for transparent and flexible displays, allowing for more versatile and aesthetically appealing wearable devices.
5) **Brain-Computer Interfaces:** Graphene-based neural electrodes can be used to develop brain-computer interfaces, enabling seamless communication between the brain and external devices for medical or augmented reality applications.
6) **Crop Protection Using Graphene:** Crop protection using graphene is an emerging field that explores improving agricultural practices, enhance crop yield, and protecting crops from various threats. Graphene possesses unique properties that can be leveraged to address challenges in agriculture. The various applications of graphene in the protection of crops are:

   Pesticide Delivery: Graphene-based nanomaterials can act as carriers for targeted delivery of pesticides and other agrochemicals. By encapsulating agrochemicals in graphene nanoparticles or nanosheets, the release of these substances can be controlled, leading to more efficient and precise pesticide application.

Soil Remediation: Graphene-based materials have shown promise in environmental remediation. These materials can help in the removal of heavy metals and pollutants, reducing their harmful effects on crops and the environment.

Nanofertilizers: Graphene-based nanomaterials can be used as carriers for fertilizers, enabling controlled release and improving nutrient uptake by plants as shown in Figure 5.4. This can lead to more efficient use of fertilizers and reduce environmental pollution caused by excess nutrient runoff.

Disease Detection: Graphene-based biosensors can be used for early detection of plant diseases and pathogens. These sensors can detect specific biomolecules or pathogens associated with plant diseases, allowing farmers to take timely action to prevent disease spread.

Nanopesticides: Graphene-based nanomaterials can be engineered to have antimicrobial properties. These "nanopesticides" could potentially offer an alternative to traditional chemical pesticides, reducing the reliance on harmful chemicals in crop protection.

Smart Farming: Integration of graphene-based sensors into agricultural systems can provide real-time data on various environmental parameters such as temperature, humidity, soil moisture, and nutrient levels. This information can enable precision agriculture practices, optimizing resource usage and crop growth.

It's essential to consider the potential environmental and health impacts of using graphene-based materials in agriculture. Research is ongoing to

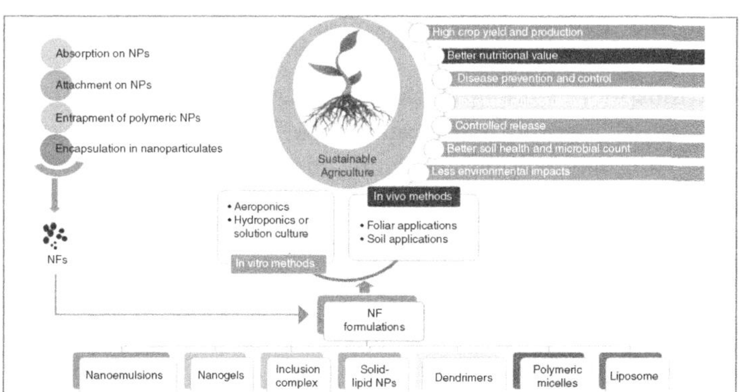

Figure 5.4 Role of nanofertilizers in the field of agriculture.

understand the long-term effects of these nanomaterials on the environment, including soil and water systems, as well as on human health.

As with any emerging technology, there are challenges to overcome in implementing graphene-based crop protection solutions. These challenges include large-scale production of graphene-based materials, cost-effectiveness, and ensuring the safety and regulatory compliance of these materials in agricultural applications.

Overall, the development of graphene-based technologies for crop protection holds promise in improving agricultural efficiency, sustainability, and resilience against various challenges faced by modern farming practices. Continued research and innovation in this area may lead to practical and safe applications in the agricultural industry.

7) **Reducing Food Waste Using Graphene:** Reducing food waste using graphene is an innovative approach that can address some of the key challenges associated with food spoilage and preservation. Graphene, with its unique properties, can be applied in various ways to extend the shelf life of food, improve packaging, and enhance food safety. The diverse applications of graphene to reduce food wastage are;

Graphene-based Sensors: Graphene sensors can be integrated into food packaging materials to monitor the freshness and quality of perishable goods. These sensors can detect changes in temperature, humidity, and gas concentrations (e.g. ethylene, a ripening agent) inside the packaging, providing real-time data on food conditions. This information can help consumers and retailers make informed decisions about food consumption and reduce waste by consuming food before it spoils.

Anti-Microbial Coatings: Graphene and graphene-based materials possess antimicrobial properties, which can be utilized to develop coatings for food packaging. These coatings can inhibit the growth of bacteria, molds, and fungi, extending the shelf life of packaged food products and reducing the chances of contamination.

Modified Atmosphere Packaging (MAP): Graphene films can be incorporated into modified atmosphere packaging (MAP) to control the gas composition surrounding the food. By regulating the levels of oxygen, carbon dioxide, and other gases, graphene-based MAP can slow down the degradation process and preserve the freshness of perishable foods for a longer time.

Temperature Regulation: Graphene's high thermal conductivity can be utilized to develop smart packaging that regulates the temperature of the food. By maintaining optimal storage conditions, the rate of food spoilage can be minimized.

Food Safety and Traceability: Graphene-based sensors can also be used to detect and monitor food contaminants, such as pesticides, heavy metals, and pathogens. By enhancing food safety and traceability, consumers can have more confidence in the food they purchase, reducing food waste caused by safety concerns.

Smart Labels: Graphene-based smart labels can be attached to food packages, providing real-time information on food quality, expiration dates, and storage instructions. This data can help consumers use food more efficiently and reduce the likelihood of discarding edible items prematurely.

It's important to note that while graphene offers promising solutions, the implementation of graphene-based technologies in food packaging and preservation requires thorough research to ensure safety, regulatory compliance, and cost-effectiveness. Additionally, sustainable and scalable production methods for graphene are necessary to make these solutions practical and accessible for the food industry.

Graphene-based technologies have the potential to play a vital role in reducing food waste by improving food packaging, enhancing food safety, and providing valuable data to consumers and retailers to make informed decisions about food consumption and storage. As research and development in this area continue, we may see more effective graphene-based solutions deployed in the food industry to combat food waste.

However, ongoing research and advancements are continuously improving the prospects of graphene-based wearable devices, and we can expect to see more innovative applications in the near future.

5.3 Implicit Operations of Graphene in the Renewable Energy Sector

5.3.1 Battery Technology

Over the last 20 times, the development of electronic factors has been extremely presto. The primary limiting factors in power storage options such as batteries and capacitors have been their size, power capacity, and efficiency. While most types of batteries face significant constraints, capacitors are comparatively less restricted. While accoutrements similar to lithium can now store enormous amounts of energy, the implicit volume decreases with each charge or recharge owing to electrode declination. In the case of lithium-ion batteries, for illustration, there's a tradeoff between energy viscosity and power viscosity. The salient features of graphene-based batteries are shown in Figure 5.5.

The power density of ray-seamed graphene (LSG) supercapacitors initially rivaled that of high-power lithium-ion batteries currently utilized. Additionally, LSG supercapacitors exhibit remarkable flexibility, lightweight construction, rapid charging capabilities, slim profile, and relatively cost-effective production.

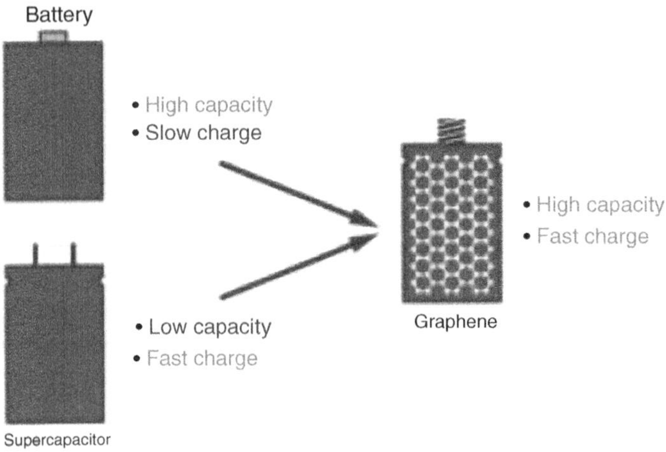

Figure 5.5 Graphene batteries.

Graphene is also being leveraged to enhance battery longevity, not just in terms of capacity and charging speed. For instance, with graphene oxide as an anode in lithium-ion batteries, battery lifespan between charges significantly increases (implicit capacity sees a tenfold boost), with minimal loss in storage capacity over time. This advancement makes technologies like electric vehicles a much more viable transportation solution for the future. Consequently, batteries (or capacitors) can now be engineered to last significantly longer and hold larger capacities than previously thought possible. Moreover, electronic devices can be charged within seconds rather than minutes or hours, and their battery life greatly extended.

Graphene's exceptional characteristics have unlocked numerous applications in battery technology. It serves as an ideal electrode material in both lithium-ion batteries and other advanced battery systems. Its high surface area and electrical conductivity make it a prime candidate for enhancing charge storage and transfer kinetics within battery electrodes. Incorporating graphene into the anode or cathode of lithium-ion batteries can significantly enhance battery performance. This translates to increased battery capacity and enhanced rate capabilities. Moreover, graphene can be incorporated into cathode materials for lithium-ion batteries. For example, it can be combined with transition metal oxides or sulfides to form composite cathodes. This integration enhances the cathode's electrical conductivity, thereby improving the overall performance of the lithium-ion battery system.

Graphene can be incorporated into solid-state batteries to enhance the electrical conductivity of solid-state electrolytes or as an additive to improve the performance of composite cathodes and anodes. It's important to note that while graphene holds great promise in battery technology, challenges remain in terms of large-scale production, cost-effectiveness, and scaling up to commercial applications. Nonetheless, ongoing research and development are likely to lead to further improvements and more widespread integration of graphene in various battery types.

5.3.2 Touchscreen

Graphene has several potential uses in touchscreens, offering improvements in various aspects of touchscreen technology. Graphene can serve as a replacement for traditional indium tin oxide (ITO) in the transparent conductive electrode layer of touchscreens. ITO has been widely used for its transparency and conductivity, but it has limitations, such as brittleness and high production costs. Graphene, on the other hand, provides similar or even better electrical conductivity and transparency while being more flexible and potentially cheaper to produce. Graphene's exceptional flexibility and mechanical strength make it an ideal material for flexible and foldable touchscreens. As touchscreens become more prevalent in wearable devices and foldable smartphones, graphene can play a significant role in enabling these advancements. Due to its high electron mobility, graphene-based touchscreens can exhibit improved sensitivity and responsiveness compared to conventional touchscreens. This means that the screen can detect touch inputs more accurately and with faster response times.

Graphene's high electrical conductivity allows for efficient charge transport, reducing power consumption in touchscreens. This can lead to longer battery life in touchscreen-enabled devices. Incorporating graphene into the touchscreen's structure can enhance its durability and resistance to scratches and abrasion. Graphene coatings can be applied to touchscreens to make them more resistant to fingerprints and smudges, improving the overall user experience. Graphene's unique electronic properties, particularly its ability to form quantum dots, can be leveraged to create quantum capacitive touchscreens. These touchscreens can offer higher resolution and better sensitivity to touch inputs. Graphene's scalability makes it suitable for large-area touchscreens, such as interactive displays, interactive whiteboards, and smart tables. By utilizing the mechanical properties of graphene, pressure-sensitive touchscreens can be developed, allowing devices to distinguish between different levels of touch pressure. In cold environments, graphene-based transparent heaters can be integrated into touchscreens to maintain the functionality of the display and improve user comfort.

Graphene's integration into touchscreens is still an area of ongoing research and development. While its potential benefits are clear, there are challenges related to large-scale production, cost-effectiveness, and compatibility with existing manufacturing processes. Nonetheless, as graphene technology advances and becomes more accessible, it is likely to play a significant role in shaping the future of touchscreens and related interactive technologies.

5.3.3 Integrated Circuits

Graphene has the potential to impact integrated circuits (ICs) in various ways, though its integration into mainstream IC technology is still a subject of ongoing research and development. Graphene's high electrical conductivity and low resistance make it an attractive candidate for replacing traditional metal interconnects in ICs. As transistors become smaller and more densely packed on chips, the resistance and heat dissipation issues associated with metal interconnects become more pronounced. Graphene-based interconnects could potentially offer lower resistance, improved heat dissipation, and reduced signal delays, leading to faster and more energy-efficient ICs. Graphene's exceptional thermal conductivity can be harnessed to enhance heat dissipation in ICs. As ICs become more powerful and generate more heat, effective thermal management becomes crucial to prevent overheating and maintain optimal performance. Graphene can be used as a heat spreader or integrated into thermal interface materials to improve heat dissipation. As mentioned earlier, graphene-based photodetectors can be integrated into ICs for efficient on-chip optical communication. Graphene's broadband response and high-speed response time make it suitable for high-speed data transmission within ICs and between chips.

Graphene's high electron mobility and frequency response enable its use in RF and wireless communication components. It can be utilized in RF transistors, amplifiers, and other high-frequency components, enabling faster and more efficient wireless communication in ICs. Graphene's high surface area and sensitivity to changes in its electronic properties make it well-suited for various sensor applications. In ICs, graphene-based sensors could be employed for temperature sensing, pressure sensing, gas sensing, and other environmental monitoring applications. Graphene's mechanical flexibility and transparency can be leveraged to create flexible and wearable ICs. These flexible circuits could be integrated into clothing, medical devices, or other wearable electronics, enabling new applications and form factors. Graphene's long electron spin lifetime makes it a promising material for spintronics applications in ICs. Spintronics leverages the electron's spin rather than its charge to store and process information, potentially leading to new types of memory and logic devices.

Despite the promising potential of graphene in ICs, several challenges need to be addressed before its widespread adoption. Large-scale production, compatibility with existing semiconductor fabrication processes, and cost considerations are some of the key hurdles that researchers and engineers are working to overcome. Nonetheless, as graphene technology continues to advance, its integration into ICs has the potential to revolutionize the semiconductor industry and enable next-generation electronic devices.

5.3.4 Flexible Memory

Graphene's exceptional properties have led to promising developments in flexible memory technologies. Flexible memory refers to memory devices that can withstand bending, stretching, and deformation without losing their functionality. Graphene-based flexible RAM can offer fast read and write access times, making it suitable for temporary data storage in various electronic devices. The high carrier mobility of graphene allows for quick charge transfer and data retrieval. Graphene's ability to function as a transparent and flexible electrode makes it an excellent candidate for use in flexible flash memory devices. Flexible flash memory retains data even when the power supply is turned off, enabling it to serve as non-volatile memory in portable and wearable electronics. Resistive RAM utilizes the resistance change of a material when an electric field is applied to it. Graphene-based ReRAM can be used in flexible memory applications due to its scalability, fast switching speed and compatibility with flexible substrates. FeRAM is a type of non-volatile memory that uses ferroelectric materials to store data. Graphene can be combined with ferroelectric materials to create flexible FeRAM, which retains data even without a continuous power supply and can withstand mechanical deformations.

PCM utilizes the reversible phase transition of material between amorphous and crystalline states to store data. Graphene's high thermal conductivity and electrical conductivity make it an attractive choice for use as a heater or electrode in flexible PCM devices. MRAM uses magnetic properties to store data. Graphene's high electrical conductivity and compatibility with spintronics make it a potential material for use in flexible MRAM applications. Graphene's versatility allows it to be integrated with other materials to create hybrid flexible memory devices, combining the advantages of different memory technologies. For example, graphene could be combined with organic materials or other 2D materials to create unique and efficient flexible memory solutions. The use of graphene in flexible memory is still an area of active research, and there are challenges to overcome, such as scalable manufacturing processes, long-term stability, and integration with other flexible components. However, ongoing developments and advances in graphene-based electronics show great promise for the realization of flexible memory solutions in the near future.

5.3.5 Solar Power Generation

Graphene holds great promise for various applications in solar power generation thanks to its unique combination of properties. As mentioned earlier, graphene can serve as a transparent and conductive electrode in solar cells. It can replace traditional materials like ITO, which is expensive and less flexible. Graphene's high electrical conductivity and transparency allow it to efficiently collect generated electrical charges while allowing light to pass through, making it an ideal candidate for next-generation transparent electrodes in solar panels. Graphene can be used in photocatalytic processes to harness solar energy for water splitting and other chemical reactions. By combining graphene with semiconductor materials, it enhances light absorption and charge separation, leading to more efficient photocatalytic reactions. This approach holds potential for producing clean hydrogen fuel and other renewable fuels using sunlight. Graphene's thin and flexible nature enables its integration into solar cells to enhance light trapping. By incorporating graphene layers into the design, it can increase the path length of light within the cell, leading to improved light absorption and higher energy conversion efficiency.

Graphene's high carrier mobility and conductivity make it an excellent material for facilitating charge transport and collection within solar cells. It can be used to form pathways for electron and hole transport, reducing recombination losses and improving overall device performance. Due to its flexibility and mechanical strength, graphene can be used to create flexible and lightweight solar panels. These panels can be integrated into various surfaces, such as clothing, windows, and curved structures, enabling new possibilities for solar power generation in unconventional settings. Graphene can be combined with other materials to form tandem solar cells. In these structures, multiple layers of different light-absorbing materials are stacked on top of each other to capture a broader spectrum of sunlight and increase energy conversion efficiency. Graphene can serve as a transparent contact layer between the different cell layers, ensuring efficient charge extraction. As mentioned earlier, graphene can enhance the performance of DSSCs by providing an efficient counter electrode or a scaffold for light-absorbing dyes. DSSCs are a low-cost alternative to traditional silicon solar cells and can be flexible and easy to manufacture. Graphene's hydrophobic nature and self-cleaning properties can help maintain the efficiency of solar panels over time. A graphene coating on the surface of solar panels can prevent dust and dirt from sticking to them, reducing the need for manual cleaning and ensuring better energy harvesting. Graphene's potential in solar power generation is still being explored, and research and development efforts are ongoing to optimize its integration and performance in various solar technologies. As these efforts progress, graphene-based solutions have the potential to improve the efficiency, cost-effectiveness, and versatility of solar energy systems, contributing to the global transition toward sustainable and renewable energy sources.

5.3.6 Photovoltaic Cells

Every time, eight thousand times more solar energy is created than is used encyclopedically; while part of the electricity reaching the Earth is captured and utilized, a significant portion is not. Solar power is growing more charming as energy demands rise, but its effectiveness remains low. Likewise, silicon-grounded solar cells are both precious to produce and install.

Perovskite solar cells (PSCs) are veritably implicit coming-generation solar power sources with extremely high effectiveness. Flagship experimenters made significant progress in enhancing the continuance and performance of PSCs while lowering the cost of PSC manufacture. The addition of a reduced graphene oxide spacer subcaste to a PSC redounded in the low-cost manufacturing of PSCs with 20 effectiveness, which was retained up to 95 after 1000 hours of operation. For the coming term, an airman manufacturing line and a 1 kW graphene-perovskite solar ranch are planned. Graphene might play a significant part in anti-reflection coatings for solar cells. According to Indian experimenters, similar cells reduce reflectance at the UV section of the diapason from 35 to 15.

5.3.7 Solar Cells

Graphene's outstanding electron transport capacity and remarkably high carrier mobility make it a promising candidate for highly efficient, flexible, and cost-effective solar systems. Recent advancements have led to the development of several graphene-based solar cells, where graphene serves various roles within the cells. Graphene's suitability for transparent and conductive electrodes in solar cells is a significant driving factor behind the interest in this material. Being a perfect two-dimensional material, graphene can be fabricated into film electrodes with outstanding transparency, high conductivity, and minimal roughness. Moreover, graphene holds promise in enhancing photovoltaic systems through its ability to be integrated into conjugated polymers. This integration enhances exciton dissociation and charge transport processes, further improving solar cell efficiency. Furthermore, graphene's properties can be customized to serve as a photoactive material by modulating its band gap through chemical functionalization or adjusting the size of graphene domains. The potential of graphene in solar energy is exemplified in graphene-enabled solar farms as shown in Figure 5.6, where its versatile properties contribute to enhanced energy harvesting and conversion.

5.3.8 Lithium-Ion Batteries

The reliability and efficiency of rechargeable lithium-ion batteries, extensively utilized in mobile devices like smartphones, laptops, and digital cameras, heavily rely on the physical and chemical characteristics of the electrode materials.

Figure 5.6 Graphene-enabled solar farm. *Source:* Courtesy of the Graphene Flagship.

Consequently, numerous research endeavors have focused on designing novel nanostructures and identifying innovative electrode materials to enhance battery capacity and accelerate charging rates. These efforts often involve leveraging graphene in various forms such as nanosheets, paper, carbon nanotubes, or fullerene nanotubes. Table 5.1 showcases the validation of graphene properties in conjunction with lithium-ion batteries, highlighting its role in improving battery performance and longevity.

5.3.9 Supercapacitors

Graphene has emerged as a highly promising material for supercapacitor applications due to its exceptional chemical stability, high conductivity, and extensive surface area. Unlike conventional supercapacitors utilizing activated carbons and carbon nanotubes, the effective surface area of a graphene-based capacitor electrode remains consistent regardless of pore distribution. However, the effective surface area of graphene-based materials does vary significantly based on the number of layers. Single- or few-layer graphene with minimal agglomeration is expected to exhibit a higher effective surface area, thus offering superior supercapacitor performance.

Graphene-based supercapacitors, often referred to as ultracapacitors, are characterized by their high power density and rapid charging and discharging capabilities. The large surface area and excellent electrical conductivity of graphene make it an ideal candidate for supercapacitor electrodes. This facilitates efficient charge storage at the interface between graphene sheets, resulting in high

Table 5.1 Property analysis of graphene battery with lithium-ion battery.

Property analysis of graphene battery with lithium-ion battery			
Property	**Al-graphene battery**	**Lithium-ion battery**	**Benefit**
Dimensions (cm)	7 × 200 × 100 for 8 modules	10.16 × 284.48 × 177.8 for 16 modules	54% smaller
Power	100 kWh	85 kWh	15% more power
Estimated Manufacturing cost	A$20/kWh	~A$200/kWh	Reduced manufacturing cost
Range	3500 km	426 km	8× range
Energy density	2000 Wh kg^{-1}	265 Wh kg^{-1}	7.5× stored energy
Weight	387 kg	544 Kg	30% lighter
Estimated cost	A$11,450	A$22,200	50% less
Recharge time	15 minutes	4.5 hours	not much longer than refueling with petrol
Durability	15 years (est.)	8 years (est.)	Lasts almost 100% longer
Accumulated battery loss	1% initially, then 1% per 3.000 km	5% initially, then 1% per 3.000 km	Retains peak efficiency longer

capacitance and fast charge-discharge cycles. Additionally, graphene serves as an effective catalyst support in metal-air batteries such as zinc-air or lithium-air batteries. By enhancing the kinetics of oxygen reduction reaction (ORR) and oxygen evolution reaction (OER) at the cathode, graphene enhances the overall efficiency and performance of these batteries.

Beyond its role in metal-air batteries, graphene is being investigated as an electrode material for sodium-ion batteries, presenting an alternative to lithium-ion batteries. Sodium-ion batteries hold promise for large-scale energy storage applications, and graphene's unique properties contribute to improving their capacity, cycling stability, and rate performance.

5.3.10 Graphene Transistors

Graphene transistors are electronic devices that utilize graphene as a channel material instead of traditional semiconductor materials like silicon. Graphene, characterized by a honeycomb lattice of carbon atoms arranged in a two-dimensional layer, possesses outstanding electrical properties. This makes it an appealing choice for constructing high-speed, low-power transistors. Among the advantages of graphene transistors are:

1) **High Electron Mobility**: Graphene exhibits high electron mobility, meaning that electrons can move rapidly through the material. This property allows for high-speed operation of transistors.
2) **High Carrier Mobility**: Graphene has high carrier mobility for both electrons and holes, enabling efficient charge transport.
3) **Low Resistance**: Graphene has extremely low resistance, resulting in minimal power dissipation and high energy efficiency.
4) **Atomically Thin**: Being only one atom thick, graphene transistors can be made very compact, leading to the possibility of building ultra-small and dense ICs.
5) **High Thermal Conductivity**: Graphene also has excellent thermal conductivity, which helps dissipate heat efficiently, a crucial factor for high-performance transistors.

The various drawbacks of graphene transistors are:

1) **Zero Band Gap**: Graphene has a zero bandgap, which means that it cannot be easily turned off in a transistor configuration. This limitation makes it challenging to achieve sharp on-off switching characteristics required for digital logic.
2) **Scalability**: While individual graphene transistors have shown great potential, scaling up production and creating reliable large-scale ICs using graphene remains a significant challenge.

The possible steps taken to overcome the drawbacks are:

1) **Engineering Band Gap**: Researchers are investigating ways to introduce a band gap in graphene or combine it with other materials to create a composite structure that exhibits both high mobility and a discernible bandgap.
2) **Graphene Nanoribbons**: Narrow strips of graphene known as graphene nanoribbons are being studied for their potential to introduce a bandgap and improve transistor performance which is shown in Figure 5.7.
3) **Hybrid Structures**: Combining graphene with other two-dimensional materials, such as transition metal dichalcogenides, can help overcome the bandgap limitation and enable more versatile transistor designs.

While graphene transistors are not yet widely used in commercial electronics, ongoing research and advancements in materials science and nanotechnology are steadily progressing toward developing practical graphene-based transistor technology. As a result, graphene transistors hold great promise for the future of high-speed, low-power, and compact electronic devices.

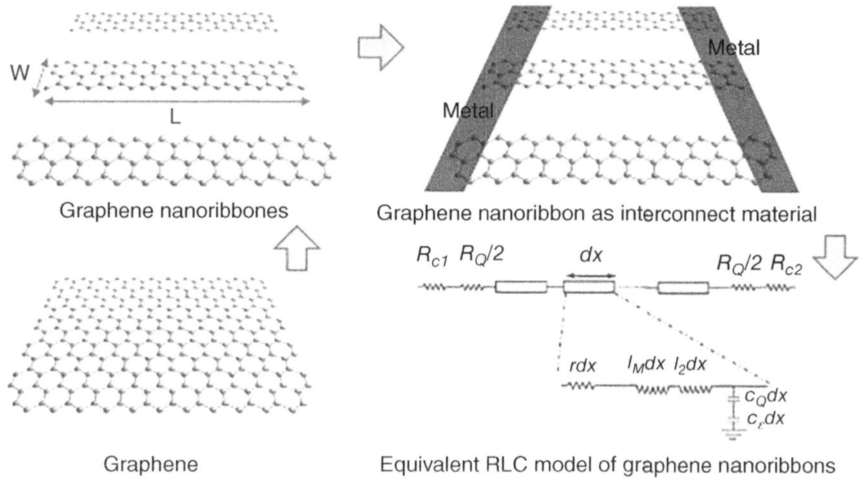

Figure 5.7 Graphene nanoribbons.

5.3.11 Graphene Semiconductors

Graphene, in its pure form, is not a semiconductor but rather an excellent conductor of electricity due to its unique electronic band structure. As mentioned earlier, it lacks a bandgap, which is a critical property for traditional semiconductors used in transistors and electronic devices.

However, researchers have been exploring various methods to modify the electronic properties of graphene to turn it into a semiconductor or to combine it with other materials to create graphene-based semiconductor devices.

Approaches to achieve graphene semiconductors:

1) **Quantum Dots**: Quantum dots are nano-scale structures that can be embedded into graphene to create localized electronic states and introduce a bandgap. By selectively patterning graphene with quantum dots, researchers can effectively tune its electrical properties, making it function as a semiconductor.
2) **Graphene Nanoribbons**: Graphene nanoribbons are narrow strips of graphene that can exhibit a tunable bandgap depending on their width and edge geometry. By engineering the width and edge structure of graphene nanoribbons, researchers can control their electronic properties and turn them into semiconductors.
3) **Chemical Functionalization**: Functionalizing graphene with certain chemical groups can modify its electronic structure, leading to the introduction of a bandgap. This process is known as "doping" and is being explored to create graphene-based semiconductors.

4) **Heterostructures**: Combining graphene with other two-dimensional materials, like transition metal dichalcogenides (TMDs), boron nitride, or other semiconducting materials, can lead to hybrid structures with tunable electronic properties. These heterostructures can exhibit semiconductor-like behavior.

The effective integration of graphene-based semiconductors is pivotal for advancing graphene-based electronics, including flexible displays, high-speed transistors, and other sophisticated electronic devices. Despite strides in developing graphene-based semiconductors, challenges persist, such as attaining high-quality, large-scale production and ensuring consistent device performance. Nevertheless, graphene research remains dynamic and vibrant, with ongoing progress holding the promise of future breakthroughs in practical graphene-based semiconductor devices.

5.3.12 Graphene Sensors

Graphene sensors are devices that utilize the unique properties of graphene to detect and measure various physical, chemical, or biological parameters. Graphene's outstanding electrical, mechanical, and chemical characteristics make it an excellent material for building sensitive and versatile sensors. The different types of graphene sensors are:

1) **Gas Sensors**: Graphene-based gas sensors can detect various gases, including nitrogen dioxide, ammonia, carbon dioxide, and volatile organic compounds (VOCs). Changes in the electrical properties of graphene occur when gas molecules adsorb onto its surface, allowing for the detection of specific gases.
2) **Biosensors**: Graphene biosensors can detect biomolecules, such as DNA, proteins, and antibodies. When the target biomolecule binds to the graphene surface, it leads to a change in the electrical conductivity, allowing for sensitive and label-free detection.
3) **Strain Sensors**: Graphene's electrical resistance can be significantly altered when subjected to mechanical strain or deformation, making it suitable for strain sensing applications.
4) **Temperature Sensors**: Graphene's electrical conductivity is highly temperature-sensitive, allowing it to be used as a sensitive material in temperature sensors.
5) **Humidity Sensors**: Graphene's electrical conductivity changes with humidity levels, making it suitable for humidity sensing applications.

The various applications of graphene-based sensors are:

1) **High Surface Area**: Graphene has an incredibly high surface area, providing ample sites for interaction with target molecules or analytes, making it suitable for gas sensing and biosensing applications.

2) **High Sensitivity**: Graphene is highly responsive to changes in its environment, enabling it to detect even minute variations in temperature, pressure, or analyte concentration.
3) **Fast Response Time**: Due to its excellent electrical conductivity and low mass, graphene sensors display swift response and recovery times, well-suited for diverse applications. real-time monitoring.
4) **Biocompatibility**: Graphene is biocompatible, making it suitable for biosensing applications, such as detecting biomolecules, pathogens, or other biological entities.
5) **Mechanical Flexibility**: Graphene can be integrated into flexible substrates, allowing for the development of wearable and conformal sensors.

The development of graphene sensors has attracted significant interest in various industries, including healthcare, environmental monitoring, aerospace, and consumer electronics. However, challenges remain in large-scale production, stability, and reproducibility of graphene sensors. Researchers and engineers are actively engaged in addressing these challenges, striving to optimize the design and production of graphene sensors for practical utilization. As a result, we can expect to see more widespread adoption of graphene-based sensors in the future, driving advancements in sensor technology and applications.

5.4 Catalysis

Experimentalists in catalysis have increasingly turned their attention to graphene owing to its unique two-dimensional structure, expansive surface area, and remarkable ballistic and electron transport properties. Graphene serves as an excellent support material for nanocatalysts, augmenting their stability and effectiveness. Its application spans across catalytic processes for energy generation, environmental cleanup, and chemical synthesis.

5.5 Renewable Energies

Experimenters at the University of Manchester revealed that graphene membranes may be illuminated by the sun and conduct protons at a faster pace. The print proton effect may be utilized to directly capture solar energy and replicate the photosynthesis process. This gas might also be utilized in electric autoenergy cells. As the cost of processing falls, such a strategy would help cut the cost of renewable hydrogen energy and make hydrogen energy stations more wide along roadways.

5.6 Nanotechnology

Graphene's exceptional properties have opened up numerous possibilities for its use in nanotechnology. Graphene's high carrier mobility and electrical conductivity make it a promising material for use in nanoelectronic devices. It can be employed in transistors, interconnects, and logic devices, enabling the development of faster and more efficient electronic components.

Graphene can be incorporated into various nanocomposites to enhance their mechanical, electrical, and thermal properties. When added to polymers, ceramics, or metals, graphene can improve their strength, stiffness, and electrical conductivity, this renders them appropriate for a diverse array of applications. Graphene-based nanosensors can be designed to detect a variety of substances, including gases, biomolecules, and chemicals. Its high surface area and sensitivity to changes in electrical properties make it an excellent material for ultrasensitive and selective sensors. Graphene holds promise in nanomedicine applications, including drug delivery systems, imaging agents, and biosensors. Its biocompatibility and ability to carry therapeutic agents make it a potential candidate for targeted drug delivery to specific cells or tissues.

Graphene's atomic-scale pores and high permeability make it suitable for nanofiltration applications. It can efficiently filter out nanoparticles, ions, and organic molecules, making it valuable in water purification and desalination processes. Graphene's exceptional mechanical properties, combined with its electrical conductivity, allow it to be used in NEMS devices, such as resonators, switches, and sensors. These devices operate at the nanoscale and have applications in communication systems and precision instruments.

Graphene's high thermal conductivity makes it an excellent material for nanoscale heat transport applications, such as thermal management in nanoelectronics and thermoelectric devices. The combination of graphene's high surface area and electrical conductivity positions it as an optimal material for nano-electrodes in electrochemical applications, including fuel cells, supercapacitors, and electrochemical sensors.

Graphene's low friction and wear properties make it useful for reducing friction and wear in nanoscale systems, potentially leading to improved performance and durability in mechanical applications.

5.7 Conclusion

Numerous fundamental challenges persist, particularly in the precise structural manipulation of graphene, essential for tasks such as bandgap correction and bulk functionalization. Graphene chemistry seems to be one of the best options for solving these problems. Graphene is also becoming a popular material for high-demand operations similar to graphene batteries and graphene supercapacitors. However, manufacturing processes and cost constraints continue to hinder mass production.

By integrating graphene with active phase substitutions, akin to carbon nanotubes, conductive polymers, and metal oxides, it is possible to significantly improve the performance through synthetic products, mixtures graphene-based represents a scientific and artificial interest and could become a competitive device for active-related energy.

Changing the elegant approach to obtain the required graphene sample remains the same—a serious difficulty. It seems that 3D printing media offers an attractive manufacturing option for three-dimensional graphene structures. Graphene 3D printing produces effects capable of allowing any size or shape of graphene to be created. More about 3D graphene can be implemented in our composition.

GRM can make a difference in solar cells, batteries, supercapacitors, hydrogen storage, and energy cells. They can be used to create new separate utilities or to improve the performance of the system. For example, activated graphene enables the use of supercapacitors to store energy while increasing the lifespan, energy capacity, and charging speed of lithium-ion batteries. GRM, similar to molybdenum disulfide, can be used to improve the lifespan of PSCs for power generation. Graphene is a wonder substance. In the long run, graphene will help humanity adopt a more sustainable lifestyle. It will support humanity in the ever-expanding renewable energy sector by developing products that are more efficient than existing bones.

Despite recent progress, the exploration of the connection between graphene-based nanomaterials and their potential for enhancing energy-related operations is still in its early stages. However, there is much anticipation for further advancements in this field in the near future.

Bibliography

1 Liu, Y., Xie, B., Zhang, Z. et al. (2012). Mechanical properties of graphene papers. *Journal of the Mechanics and Physics of Solids* 60 (4): 591–605.

2 Song, W.-L., Fan, L.-Z., Cao, M.-S. et al. (2014). Facile fabrication of ultrathin graphene papers for effective electromagnetic shielding. *Journal of Materials Chemistry C* 2 (25): 5057–5064.

3 Compton, O.C. and Nguyen, S.B.T. (2010). Graphene oxide, highly reduced graphene oxide, and graphene: versatile building blocks for carbon-based materials. *Small* 6 (6): 711–723.

4 Iwan, A. and Chuchmała, A. (2012). Perspectives of applied graphene: polymer solar cells. *Progress in Polymer Science* 37 (12): 1805–1828.

5 Agresti, A., Pescetelli, S., Taheri, B. et al. (2016). Graphene–perovskite solar cells exceed 18% efficiency: a stability study. *ChemSusChem* 9 (18): 2609–2619.
6 El-Kady, M.F., Shao, Y., and Kaner, R.B. (2016). Graphene for batteries, supercapacitors and beyond. *Nature Reviews Materials* 1 (7): 1–14.
7 Alhajji, E., Zhang, F., and Alshareef, H.N. (2021). Status and prospects of laser-induced graphene for battery applications. *Energy Technology* 9 (10): 2100454.
8 Luo, Z., Wang, X., Chen, D. et al. (2021). Ultrafast Li/fluorinated graphene primary batteries with high energy density and power density. *ACS Applied Materials & Interfaces* 13 (16): 18809–18820.
9 Bulbula, S.T., Lu, Y., Dong, Y., and Yang, X.-Y. (2018). Hierarchically porous graphene for batteries and supercapacitors. *New Journal of Chemistry* 42 (8): 5634–5655.
10 Xiong, Z., Liao, C., Han, W., and Wang, X. (2015). Mechanically tough large-area hierarchical porous graphene films for high-performance flexible supercapacitor applications. *Advanced Materials* 27 (30): 4469–4475.
11 Tsang, C.H.A., Huang, H., Xuan, J. et al. (2020). Graphene materials in green energy applications: recent development and future perspective. *Renewable and Sustainable Energy Reviews* 120: 109656.
12 Bak, S., Kim, D., and Lee, H. (2016). Graphene quantum dots and their possible energy applications: a review. *Current Applied Physics* 16 (9): 1192–1201.
13 Li, Q., Mahmood, N., Zhu, J. et al. (2014). Graphene and its composites with nanoparticles for electrochemical energy applications. *Nano Today* 9 (5): 668–683.
14 Karthick, R. and Chen, F. (2019). Free-standing graphene paper for energy application: progress and future scenarios. *Carbon* 150: 292–310.
15 Hughes, Z.E. and Walsh, T.R. (2015). Computational chemistry for graphene-based energy applications: progress and challenges. *Nanoscale* 7 (16): 6883–6908.

6

Fullerene for Green Hydrogen Energy Application

Manish Kumar and Sunil Kumar

Department of Chemistry, L.N.T. College, B.R.A. Bihar University, Muzaffarpur, Bihar, India

6.1 Introduction

Fullerene was first discovered in 1985 by H. Kroto, Robert Curl, and his coworkers. Fullerene has a closed hollow cage consisting of carbon atoms interconnected in pentagonal and hexagonal rings. C_{60} is the most common and the first discovered fullerene, which is made of a 6C molecule of carbon due to which it gains a shape like a soccer ball. There is a weak intermolecular force existing in a molecule of fullerene. So, it is necessary to functionalize to increase their reactivity and then it is used in storage of hydrogen. Ionization potential of fullerene is 7.61 eV while electron gain enthalpy of fullerene is 2.6–2.8 ev. The fullerene structure is unique in that the molecules are borderless, uncharged, and have no boundaries, no dangling bonds, and no unpaired electrons [1]. It is chemically reactive and can be added to a polymer structure to form a new copolymer with specific physical and chemical properties [2]. Hydrogen has been widely recognized to be an ideal alternative vitality instead of fossil fuel due to its non-pullulating tendency and high abundance in nature [3, 4]. Functionalized fullerenes are widely used in solar cells, drug delivery, and in materials science [5]. Fullerenes are tremendous conductors of heat and electricity. They have exceptional tensile strength. Due to high reactivity of fullerenes, they work as catalysts. Their derivatives are used in *n*-type materials [6]. They are also used in the manufacturing of anti-aging and cosmetic creams due to their antioxidant properties. Fullerenes do not get dissolved in water. However, they are soluble in many organic solvents like toluene and carbon disulfide. This dissolves with a

Carbon-based Nanomaterials for Green Applications, First Edition. Edited by Upendra Kumar, Piyush Kumar Sonkar, and Suman Lata Tripathi.

Published 2025 by John Wiley & Sons, Inc.

deep color solution [7]. The relevant storage media should be lightweight and inexpensive and must have the capacity to store a high density of H_2 by physisorption. In the last few decades, the use of carbon nanostructures as a potentially inexpensive, plentiful, and lightweight storage system has gained increasing attention [8, 9]. Theoretically, fullerene has the maximum capacity to contain 58 hydrogen atoms at 0 K, which is 7.5% by weight. The practical storage capacity of fullerene is lower than that of theoretical storage. In the chemical storage method, hydrogen is generally generated through a chemical reaction [10, 11]. The well-defined structures and multiple unsaturated bonds of fullerene enable easy modification, potentially yielding diverse compounds that could address the variability issues in morphology and composition that plague many carbon materials used in batteries [12]. Fullerene can be of 0D, 1D, 2D, 3D. Different dimensional fullerenes have different properties and uses. The zero-dimensional fullerene can adsorb substrate easily on its surface via non-covalent bonding, such as $\pi-\pi$ interaction. Fullerene can withdraw electrons easily from the local substrate position, where it is anchored. Lots of 2D materials can also be used as substrates to support the fullerene [12]. Boron carbon nitride (BCN) can be used as a 2D material, which exhibits multifunctional electrolytic capabilities toward HER, ORR, and OER. When CoNi LDH nanosheets are taken with fullerene quantum dots, they form hybrid catalysts, which show tremendous catalytic activity toward HER, OER, and urea oxidation reaction [13]. Fullerene can self-assemble into micro- or nanostructure that possesses the properties of both homogeneous and heterogeneous catalysts. Due to surface modification of fullerene, it rises to numerous derivatives, which inherit the feature of parental fullerene as well as also improve the physicochemical properties such as superconductivity, appropriate wettability, and on-demand surface functionality [14]. Due to high electron acceptance capacity of fullerene, it shows high redox reactivity during battery recycling. Due to its good electronic properties, it is widely used in solar cells. Fullerene has high electronic conductivity and great redox activity. It can't serve as an effective anode material in lithium-ion batteries but if it undergoes some modification or hybridization, it achieves a tremendous performance. Fullerene contains anode material [15]. There are some problems found in fullerene such as in visible light or in the presence of oxygen, it may break down. Fullerene can undergo several deactivation processes in the triplet excited state, such as ground state cooling by molecular oxygen and atomic electron transfer [16].

Fullerenes are emerging chemicals which are highly susceptible to hydrogen for future utilization. Structural functionalization improves hydrogen storage efficiency. Therefore, chapters review selective organic or inorganic materials, which can improve structural stabilization and unsaturation for hydrogen. Thanks to various organic groups on the surface of fullerene and chemical dopants, which attract researchers and scientists for further modification and assessment of energy supply for global demand, sustainability, and friendly ecostructure.

6.2 Green Hydrogen Energy

Hydrogen fuel refers to hydrogen used as a fuel source, often burned with pure oxygen. It can be a zero-carbon fuel if produced through carbon-free process. However, most hydrogen is currently produced from fossil fuels, leading to carbon dioxide emissions. [17] Depending on its production method and environmental impact, hydrogen is often categorized using color-coded terms such as white, blue, gray, black, and green hydrogen. Green hydrogen is produced through the electrolysis of water using renewable energy sources, and not through coal or lignite gasification, steam methane reforming, or natural gas methane, which are methods associated with gray hydrogen. Blue hydrogen is produced by reforming natural gas or coal gasification, coupled with carbon capture and storage or utilization technologies to minimize carbon emissions. Green hydrogen is produced by electrolyzing water with electricity sourced from renewable energy. The carbon intensity is ultimately determined by the carbon neutrality of the electricity used. In the context of reversible hydrogen storage, Li and Na atoms attach to the C—C bridge position of C_{20} through non-covalent closed shell interaction (Figure 6.1) [18]. This can be utilized in fuel cells or internal combustion engines. Hydrogen is now being used in commercial fuel-cell vehicles like passenger cars and has been employed in fuel-cell buses for several years.

Hydrogen is also employed as a fuel in spacecraft propulsion and is being considered for hydrogen-powered aircraft (Figure 6.2). Green hydrogen is seen as a promising alternative for facilitating this transition. It can be used for long-term

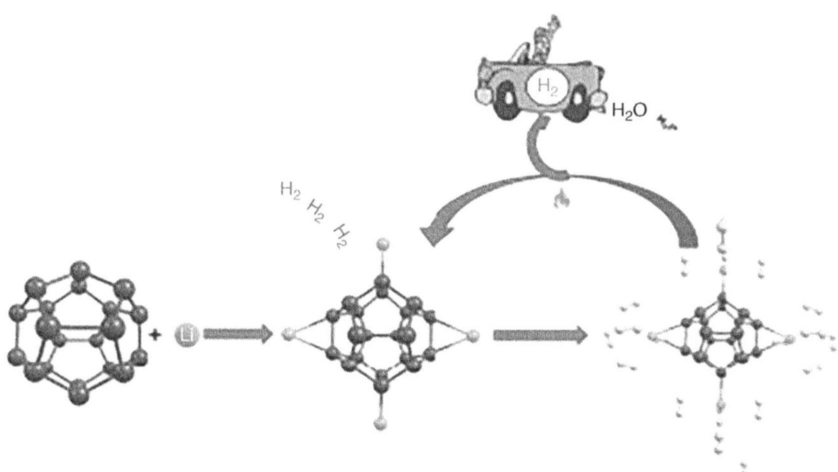

Figure 6.1 Reversible hydrogen storage alkali metal. *Source:* Sahoo et al. [18]/with permission of Elsevier.

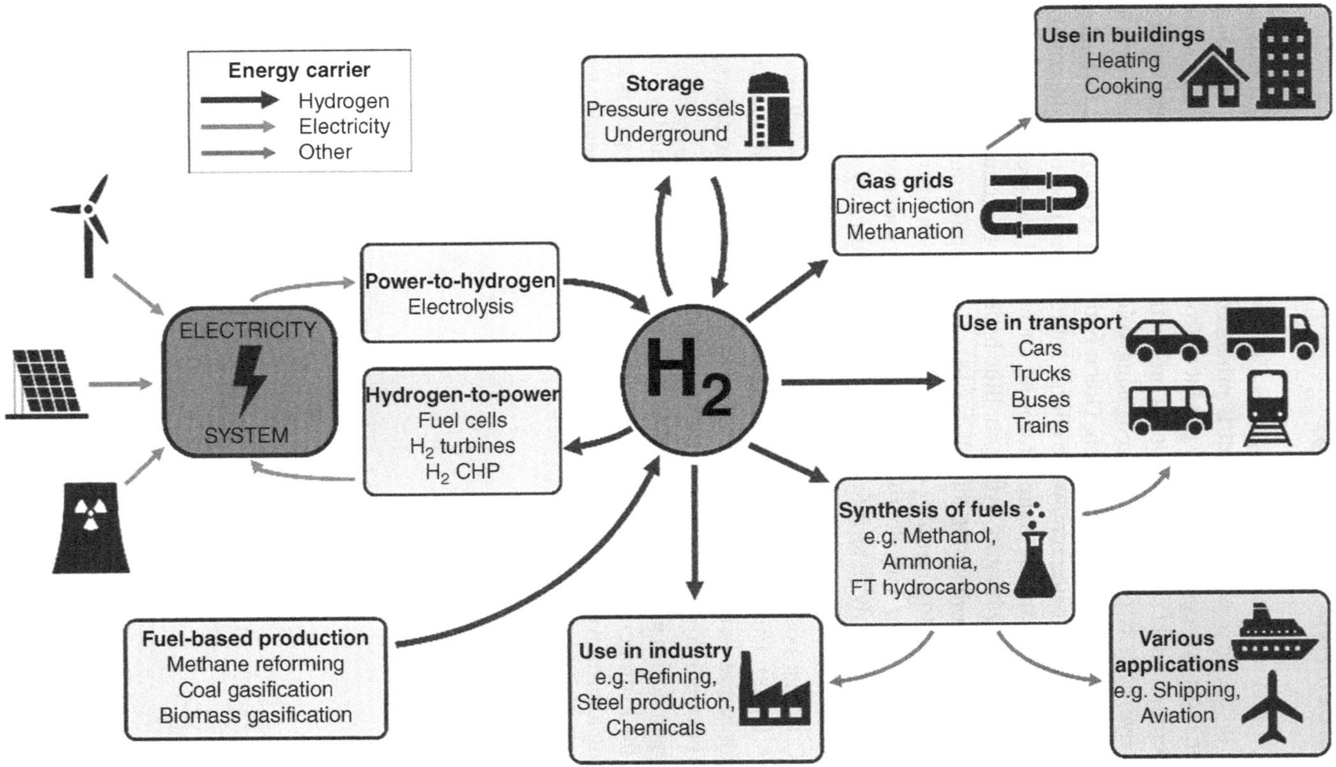

Figure 6.2 Uses of hydrogen as an energy source. *Source:* Quarton et al. [19] Royal Society of Chemistry/CC BY 4.0.

storage of renewable energy or as a replacement for fossil fuels. Green hydrogen is also utilized in the chemical industry for the production of ammonia and fertilizers. Additionally, it is employed in the petrochemical industry to manufacture petrochemical products [19].

6.3 Fullerene as a Hydrogen Storage Material

Physicochemical properties of fullerene decide capture capacity of hydrogen energy. There will be a maximum of 23, 24, and 25 hydrogen molecules that can be stored in C_{60} to form a stable hydrogen C_{60} nanocomposite material. There was a claim that there was not enough space in the C_{60} fullerene for adsorption of more than one molecule of hydrogen. Another confusion scientists have is that there is no possibility of chemisorption of hydrogen in the inner wall surface of fullerene. Hydrogenation of fullerene generally occurs at 12MPa and a temperature range of 200–600 °C. At the temperature range of 425–455 °C, the total hydrogen content is about 8.2% by weight. Scandium-doped, especially C_{24} fullerene structure possesses suitable adsorption and desorption energies for the hydrogen molecules. At desorption temperature, the diffusion energy of scandium atom significantly exceeds typical thermal energy levels [20]. These theories prove that scandium-decorated C_{24} fullerene is a potential hydrogen storage device. Theoretically, fullerene C_{60} has a maximum 58 hydrogen atom storage capacity, which is 7.5% by weight at 0 K. This produces a meta-stable shape having an internal pressure of 1.3 Mbar. Their practical storage capability is condescend but it is still promising. van't Hoff equation is adopted to recognize the surface assimilation behavior of hydrogen gas.

$$T_d = \left(\frac{E_{ad}}{k_B}\right)\left(\frac{\Delta S}{R} - \ln P\right)^{-1}$$

where T_d signifies desorption temperature, E_{ad} shows the average adsorption energy of the H_2 adsorbed host, k_B(Boltzmann constant), P (pressure) taken 1 atm, R indicates gas constant, and ΔS (change in entropy) of H_2 when it goes from gas to liquid medium. For the production of $C_{24}B_{24}$ out of 12 carbon atoms, six identical C—C bonds of C_{60} are removed and 24 carbon atoms in six rings are substituted with boron atoms (Figure 6.3) [21].

6.4 Size Effect of Fullerene and Hydrogen Storage Efficiency

The integrated unsaturated fullerene (cage-structured) may be readily functionalized to yield a variety of compounds with suitable volume or space. In general, smaller fullerenes have high surface area to volume ratio, providing more

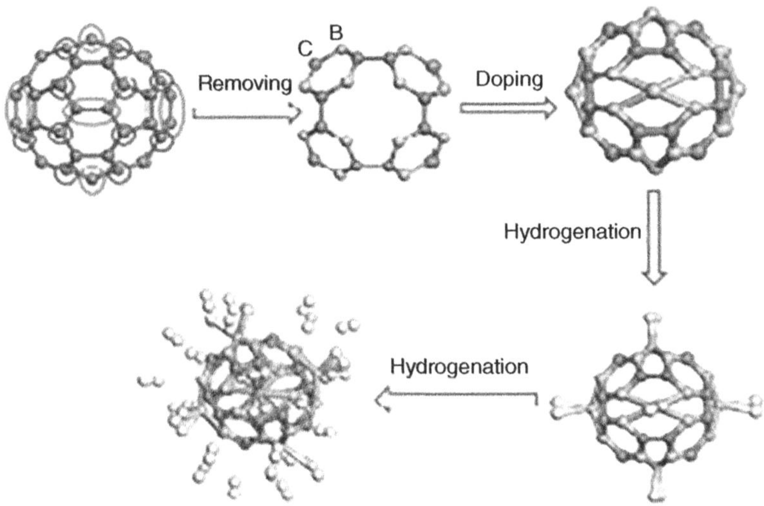

Figure 6.3 Hydrogenation on the surface of fullerene. *Source:* Tang et al. [21]/with permission of Elsevier.

accessible sites for hydrogen adsorption. This increased surface area can elevate the hydrogen storage capability of fullerenes. Boron nitride-doped C_{24} fullerene can store up to six hydrogen atoms. Ti-decorated B38 fullerene is recorded as having a strong ability to store hydrogen. A maximum of 58 molecules of hydrogen can be stored by fullerene C_{60}. Its spherical shape and high symmetry provide a large surface area for hydrogen adsorption [22].

6.5 Functionalized Fullerene, Chemical Structure, and Its Hydrogen Storage Performance

Due to functionalization of fullerenes, their hydrogen storage nature is enhanced. Pure fullerene has weak binding energy due to which it has low hydrogen storage capacity. Due to functionalization of fullerene by transition metal, it increases binding energy and also increases the adsorption capacity [21]. Ti-decorated fullerene has excellent adsorption energy and also has tremendous binding energy in the direction of hydrogen molecules. The binding energy of each hydrogen molecule is approximately 0.132 eV. If 14 Ti is undergoing doping on the C_{60} fullerene, then its storage capacity is approximately 7.5% by weight. C_{60} fullerene shows greater electron gain enthalpy and when mixed with alkali metal, it donates partially valence electrons, which results in strong bonding [23]. Boron-doped heterofullerene can store hydrogen up to 9% by weight. Lots of research found that metal-organic frameworks got much more attention due to their high hydrogen uptake nature at low temperatures. Isoreticular metal-organic framework can uptake 1.6% hydrogen by weight at 77 K. If temperature increases to 25 °C and taking pressure is 10 bar, hydrogen can be stored up to 2% by weight. The novel metal-organic framework which is coordinated with Mn^{2+} is capable of uptake of hydrogen by 6.9% by weight at 77 K temperature and 90 bar pressure [22]. Hydrogen storage performance has been elaborated through variations of surface groups, dopants, encapsulates, nanomaterials, and layered ecostructure in the surface of fullerenes.

6.5.1 Boron

The major preparation methods of boron-functionalized fullerene are laser ablation electron beam irradiation and chemical methods. Boron clusters can be obtained by laser ablation of boron nitrides with a frequency of 532 nm [24, 25]. A structure of boron-functionalized aromatic molecule ($B_{12}H_{32}$) was proposed a long time ago, which has icosahedral symmetry. It has been observed that the geometrical dualism between boron and carbon fullerenes is generally consistent. This mapping between boron and carbon fullerenes can be achieved using the Descartes-Euler formula:

$$P - C + F = 2$$

where, in a given polyhedral molecule, P indicates the number of vertices, C indicates the number of contacts, and F indicates the number of faces. Boron is known to be electron-deficient and when it is doped with fullerene, it improves the redox properties of fullerene. The most effective way to enhance redox potential $C_{59}B$ as compared to pure C_{60} is on cage dopping of C_{60} with boron [26]. The intensification of redox potential of $C_{59}B$ as compared to pure C_{60} occurs due to a singly occupied molecular orbital. Boron-functionalized fullerene has high binding energy, which can increase by replacing another element from the periodic table either inside or outside the cages. Boron fullerene is also used as a molecular device application as a sensor [27]. Boron fullerene is found in a well-balanced molecule, which possesses both acidic and basic sites. Boron-doped fullerene also shows stable optoelectronic properties. Boron fullerene has been remarked much more stable with a significant energy gap of 3.13 eV between HUMO AND LUMO. Boron fullerene has 431R Raman spectra and 73 Raman active modes [28].

6.5.2 Phosphorene or Black Phosphorus

Phosphorene demonstrates high electrical conductivity, excellent mobility, superior thermal conductivity, and significant tensile strength. Relevant phosphorene structure processes excellent electrical conductivity (2105 $cm^2\ V^{-1}\ s^{-1}$), good carrier mobility ($\approx$1000 cm^2 V^{-1} s^{-1}), and superior thermal conductivity (4840–5300 W m^{-1} K^{-1}). Phosphorene exhibits remarkable ability to bind organic and inorganic molecules by covalent or non-covalent interactions, primarily driven by electrostatic forces. Experimentalists and theoreticians have proposed phosphorene with fullerene (C_{60}) nanostructure with tremendous properties for technological applications, such as serving as a high-specific capacity battery electrode material. C_{60} can't be bonded with phosphorene via a covalent bond until breakage of C—C\C—C bond. X-ray, low-temperature scanning tunneling microscopy, and ultraviolet photoelectron show that fullerene molecules are generally undergoing physical adsorption on the surface of honeycomb structure of blue or black phosphorus at room temperature. It is synthesized by interfacial charge transfer by phosphorene to fullerene. Fullerene and phosphorene-doped fullerene are widely used in solar cells and electron acceptors due to their exciting chemical, electronic, and physical properties. The electronic transproperty of phosphorene-based fullerene may differ according to the interaction and also differ due to the contacting position of the electrode [29]. Phosphorene is a good 2D structure for C_{60} adsorption. The interaction between fullerenes and phosphorene varies with molecular size, forming either covalent or non-covalent complexes, and these complexes show significant stability even in the solution. There are two types of covalent complexes found by cycloaddition-like reactions. The first are short-range effects, which occur due to charge transfer and polarization, which determine the stability, and the second where short-range effects decay to avoid steric repulsion and for balancing long-range forces like electrostatic and dispersion, which favors stability [30]. Generally, high-size fullerene (C_{50}—C_{70}), when doped with mild phosphorene, forms only non-covalent complexes. Phosphorene increases the polar character of fullerene and also increases the ability to acquire induced dipole moments [31]. The metallic character of phosphorene increases due to small energy band gap fullerene.

6.5.3 Hexagonal Boron Nitride

The boron and nitride are abundant in earth's crust and can exist in fullerene structure with size range of 0.7–1.4 nm. Boron nitride fullerene is generally composed of four-, six-, or eight-membered rings. The boron nitride-doped fullerene nanostructure is chemically stable due to its non-ionic bond and due to its non-cytotoxic nature, it is ideal for drug delivery in human bodies. Boron nitride-doped fullerene has good adsorption and drug delivery capability. It is used as an anticancer drug hydroxyl urea (HU) as a drug vehicle [32]. The optimized structure suggested that the most favorable site of HU for adsorption on the fullerene is boron atoms in the four-membered rings. The O atom of carbonyl group adsorbs chemically HU drug on the local pyramidal site of boron nitride fullerene. The ionic character of bond is in boron nitride fullerene and it has a local pyramidal structure. Due to this nature, it has a high tendency to bind with other molecules physically and chemically. In ($B_{116}N_{124}$), boron nitride fullerene exhibits homonuclear bond. This feature enhances the fundamental properties of material [33]. It is well known that homonuclear bonds in certain systems can result in unique alterations in both the structure and electronic properties and when chemical composition is changed, this effect will happen. The $B_{116}N_{124}$ fullerene has a high electronic conductivity due to having a combination of hetero- and homonuclear bonds in the entire structure. The study of structural, electronic, and optical nature of a specific boron nitride fullerene can be carried out by using DFT and TD-DFT. The boron nitride fullerene material can be obtained in a nitrogen gas atmosphere by arc-melting method using boron-based powder and tungsten electrode [34]. Boron nitride fullerenes are expected to be a good nanomaterial, which provides good stability at high temperatures with electronic insulation in air. It is difficult to adsorb hydrogen gas inside or outside the surface of BN fullerene by physisorption, which is due to too weak bonding of physisorbed hydrogen at room temperature [35]. Hence chemisorption is necessary for hydrogen storage in BN fullerene at room temperature. Therefore, if the energy of chemisorption is lowered, then hydrogen storage capacity of BN fullerene would be improved.

6.5.4 Silicene

Silicene is a compound of hexagonal rings even though they are not planned due to symmetric buckling and their buckled honeycomb structure. Silicene shows a tremendous chemical reactivity among any other substance and also shows very strong adsorption of atoms and molecules. Silicene does not show good electronic and optoelectronic properties because it has zero band gaps. It shows poor performance in field-effective transistor application. Therefore, it is necessary to improve Silicene into good stabilized Silicene with a large band gap [36]. Silicene can be improved by using different methods using application of strain, incorporation of spin-orbit coupling, and also by using chemical adsorption. It can also be improved by decoration of hydrogen on the surface of Silicene. Silicene-doped fullerene enhances the performance of energy storage devices, including batteries and supercapacitor. Silicene's high surface area combined with fullerene conductivity can lead to improved charge storage and faster electron transport, resulting in energy storage system with higher efficiency and shorter charging times [37].

6.5.5 Carbon Nanotubes

Carbon nanotube and fullerene have excellent mechanical properties, which are used in many potential applications. A novel hybrid nanomaterial consisting of fullerene can be made by covalent attachment of fullerene on the surface of a carbon nanotube in the presence of Fe catalyst by a hot wire generator method and particles grow in situ via ferrocene vapor decomposition in the presence of CO and traces amount of H_2O and CO_2 [38]. This hybrid shows various electronic properties due to electron field emission efficiency. Functionalization of carbon nanotubes with fullerene has wide applications in electronic, medicinal, and light energy. Indeed, exploring the non-covalent operationalization of carbon nanotubes can be encouraging. It not only preserves the electronic structure but also minimizes various forms of damage to the tube including chemical, thermal, and sonication damage. This approach enhances the stability and opens avenues for various applications in nanotechnology and materials science. Functionalized fullerene with carbon nanotube can be used in drug delivery systems. Their unique structure allows for efficient encapsulation of drugs and targeted delivery to specific cells, reducing effects and enhancing treatment effectiveness. These hybrids can be utilized as sensors for detecting gases, biomolecules, or environmental pollutants. The functionalized structure provides a stable platform for immobilizing sensing elements, enhancing the sensitivity and selectivity of the sensors. It can be photovoltaic devices [39]. Hybrid structure of carbon nanotubes and fullerene derivatives is used in organic solar cells. The unique electronic properties of these materials enhance the efficiency of converting sunlight into electricity, making them valuable for renewable energy applications. Carbon nanotubes are renowned for their exceptional mechanical strength and stiffness. Integrating fullerenes with CNTs can reinforce the structural integrity of materials, resulting in composites with enhanced mechanical properties. This makes them suitable for applications requiring strong and lightweight material. A 3D carbon nanotube fullerene interconnected framework (NFIF) structure is designed at 300 K theoretically and its geometric stability is tested by molecular dynamic calculation [38]. The doping of Li on boron-doped carbon nanotube is shown in Figure 6.4.

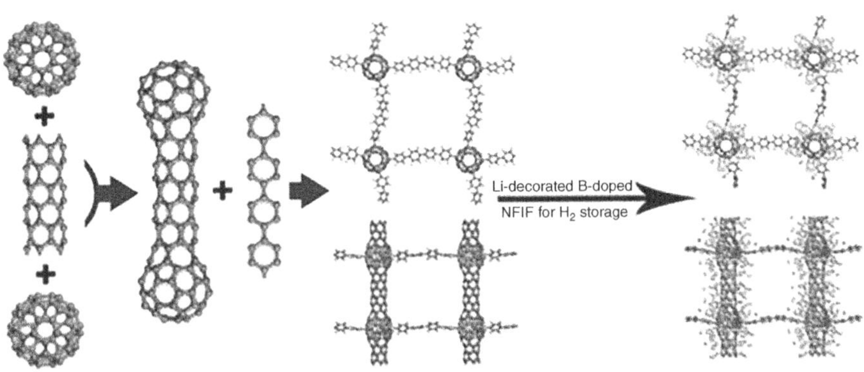

Figure 6.4 Structural decoration of Li on boron-doped NFIF and storage of hydrogen. *Source:* Shi et al. [38]/with permission of Elsevier.

6.5.6 Graphene

It is a single-layered carbon atom organized in a hexagonal lattice, making it the basic structural element of other carbon allotropes like graphite, carbon nanotubes, and fullerene. It has remarkable properties, including great electrical conductivity, thermal conductivity, mechanical strength, and flexibility. Graphene has various potential applications in electronics, materials science, energy storage, and many more due to its unique characteristics. The preparation of hybrid fullerene using graphene is done by chemical vapor deposition (CVD), solution-based method, mechanical exfoliation, and chemical synthesis [40]. Chemical methods can be employed to covalently attach fullerenes to graphene sheets. Functional groups on graphene can react with functionalized fullerene, forming strong covalent bonds between the graphene and fullerene molecules. Graphene-doped fullerene materials have been studied for their potential application in hydrogen storage. Graphene and fullerene materials both have high surface areas, providing ample sites for hydrogen molecules to adsorb. The structure of graphene-doped fullerene materials can be engineered to have specific pore sizes. This tenability is crucial for optimizing the interaction between hydrogen molecules and the material, enhancing storage efficiency. Graphene-doped fullerene materials can exhibit stability and reversibility during the hydrogen adsorption and desorption processes. This means that they can store hydrogen securely and release it when needed without significant degradation of the material. Graphene-doped fullerene materials can possess excellent mechanical strength and chemical stability, ensuring the durability of the storage system over multiple hydrogen adsorption and desorption cycles [41].

6.5.7 Ferrocene

Ferrocene is a chemical compound in two cyclopentadienyl anions and a central iron cation. It has sandwich-like arrangement. Ferrocene is a member of the class of compounds in which central metal is sandwiched between organic ring systems. Ferrocene was discovered by Nobel laureates Pauson, Fischer, and Kealy. Their unique structure was determined by X-ray crystallography. The preparation of ferrocene-doped fullerene involves integrating ferrocene, a metallocene compound, with fullerene molecules. This integration can be achieved through various methods ensuring proper interaction and bonding between the ferrocene and fullerene components. Ferrocene-doped fullerene materials (Figure 6.5) have been studied for their potential applications in hydrogen storage due to their unique properties [42]. These materials combine the advantages of both fullerene and ferrocene making them promising candidates for efficient hydrogen storage. The unique structure of ferrocene-doped fullerene material can offer kinetic advantages, allowing for faster adsorption and desorption

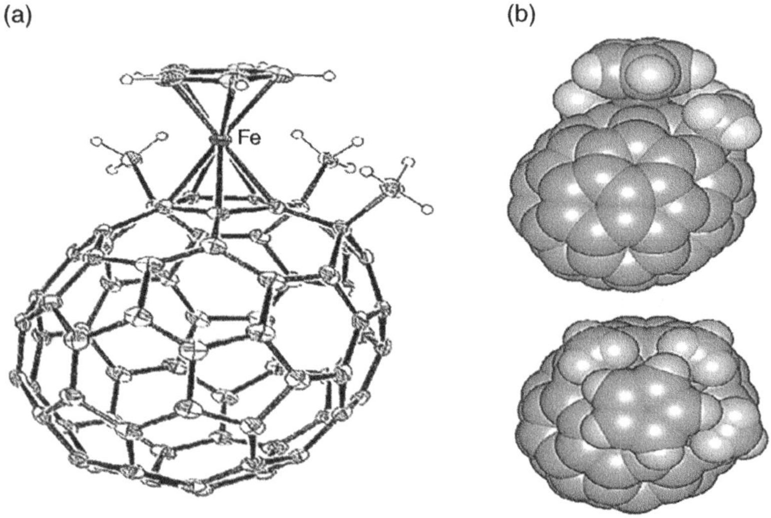

Figure 6.5 (a) Hybrid structure of fullerene and ferrocene, (b) bucky ferrocene. *Source:* Sawamura et al. [42]/American Chemical Society.

kinetics [43]. These rapid adsorption and release of hydrogen are essential for practical application. The presence of ferrocene in the material enhances the adsorption properties. The iron atoms in ferrocene can form strong interactions with hydrogen molecules, promoting adsorption and leading to higher storage capacity. Ferrocene is known for its magnetic properties due to the presence of iron. Introducing fullerene into fullerene could result in a material with interesting magnetic behavior depending on the interaction between ferrocene molecules and carbon structure. Doped fullerene can be explored for energy storage applications, including batteries and supercapacitors. The enhanced electrical conductivity and stability of marriage make it potentially valuable for energy storage [44].

6.5.8 MoS_2

It belongs to a class of materials known as transition metal dichalcogenides. MoS_2 has gained great attention in the scientific community due to its unique properties and various potential applications, especially in the field of nanotechnology and electronics [45]. Doped MoS_2 fullerene hybrid can be utilized in solar cells and photovoltaic devices. The combination of MoS_2's semiconductor properties with fullerene's excellent electron transport capabilities can enhance the efficiency of solar energy conversion. The mechanical flexibility of doped MoS_2 fullerene materials makes them suitable for flexible and wearable electronics [46]. They can be integrated into bendable electronic devices, sensors, and displays. MoS_2 fullerene hybrid composites can be used in thermal management applications, such as thermal interface material. Their excellent thermal conductivity properties can help dissipate heat deficiency in electronic devices. The combination of MoS_2 and fullerene provides a high surface area for hydrogen adsorption, potentially allowing for the storage of a significant amount of hydrogen gas. One of the key challenges in hydrogen storage is ensuring that the process is reversible, allowing for the efficient uptake and release of hydrogen gas. Doped MoS_2 fullerene material can potentially facilitate reversible hydrogen adsorption and desorption, making them suitable for on-demand hydrogen storage applications. MoS_2 catalytic activity can also enhance the kinetics of the hydrogen adsorption and release [47, 48]. This means that the process of storing and releasing hydrogen can occur at a faster rate, improving the efficiency of the hydrogen storage system. The stability of this hybrid fullerene composite is crucial for long-term hydrogen storage applications. The photovoltaic heterojunction between fullerene (C_{60}) and molybdenum disulfide is an interesting approach in the development of organic-inorganic hybrid solar cells. This heterojunction utilizes the unique properties of both materials to enhance the efficiency of solar energy conversion. Upon light absorption, excitation is created in MoS_2 [49]. This excitation can then migrate to the interface. The excitation can be dissociated into free electrons and holes due to the energy-level mismatch between MoS_2 and C_{60}. The electron moves toward the C_{60} layer, while the holes remain in the MoS_2 layer. The impact of doping semiconductor hollow closed fullerene-like nanoparticles of MoS_2 and WS_2 has been widely disregarded. However, introducing rhenium doping (Figure 6.6) results in significantly enhanced nanoparticle conductivity, minimized agglomeration, and a substantial reduction in friction and wear, approaching superlubricity [50].

6.5.9 Organometallic Framework

Organometallic framework (OMF) is a class of crystalline materials that combine organic ligands with metal-containing nodes to create a porous structure. Doping OMFs can lead to unique materials with enhanced properties and functionalities. Doped OMFs can be used for gas storage and separation due to the combining properties of OMFs (high surface area and tunable pores) and fullerene (high gas adsorption capacity). The fullerene dopants with OMFs can enhance the catalytic activity of the OMFs making them efficient catalysts for organic transformations. Doped OMFs can be tailored for specific sensing applications. The presence of fullerene molecules can improve the selectivity and sensitivity of the frameworks in detecting target molecules, making them useful for chemical and biological sensing. Doped OMFs can be utilized for drug delivery systems. The porous

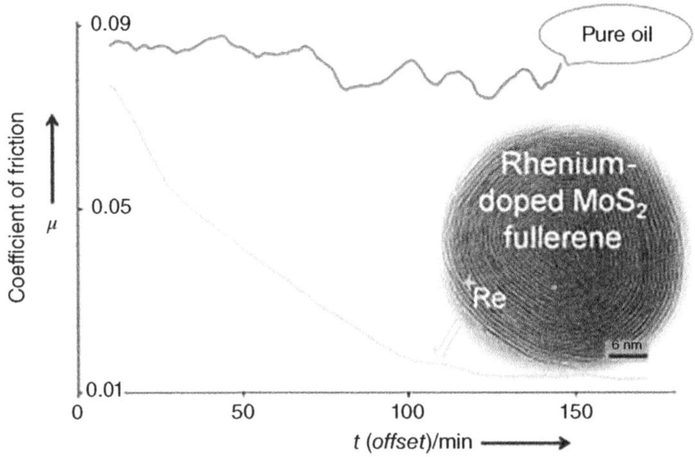

Figure 6.6 Effect of doping into semiconductor hollow closed fullerene. *Source:* Yadgarov et al. [50]/John Wiley & Sons.

structure of OMFs allows for the encapsulation of the drugs, and the presence of fullerene (Figure 6.7) molecules can aid in controlled drug release and targeting specific cells or tissues [51]. Doped OMFs can be employed in environmental cleanup process. They can adsorb and remove pollutants from air and water sources, contributing to environmental remediation efforts. OMFs are synthesized first and then fullerene molecules are introduced into the pores or onto the surface of the OMF [52–54]. This can occur through a chemical reaction that allows the fullerene molecules to attach to specific sites within the OMF structure.

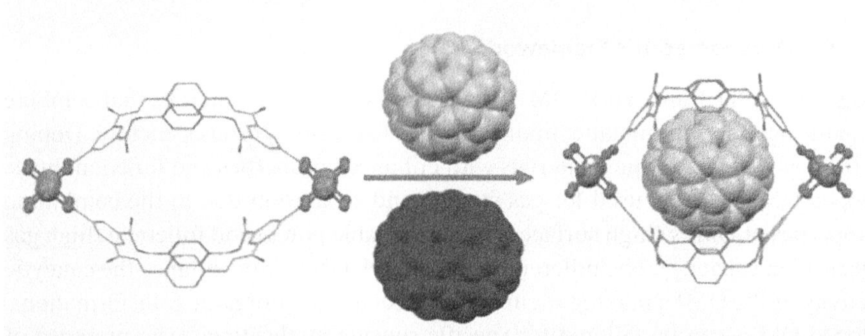

Figure 6.7 Encapsulation of fullerene by organometallic framework. *Source:* Adapted from Saura-Sanmartin et al. [51].

6.6 Charged Fullerene as Hydrogen Storage System

Charged fullerenes, especially those with extra electrons or missing electrons, have unique properties that can influence their potential application in hydrogen storage. Charged fullerenes can enhance their hydrogen adsorption due to presence of extra electron or electron deficiencies, which can create additional binding sites for hydrogen molecules. Charged fullerenes can exhibit higher stability, reducing the risk of hydrogen leakage and improving the overall safety of the storage system [55]. The enhanced binding sites are delocalized with nature surrounding the whole part of the charged fullerene and it is polarized by hydrogen molecules by a strong electric field, which enhances the storage capacity of fullerene by up to 8% by weight. The binding energy of molecular hydrogen to a bulky ball structure is approximately 0.02 ev. There are many ways to control charging state of fullerene, such as electrochemical doping, ion/electron impact, laser desorption, and chemical doping [56]. The charged state of fullerenes can be tuned, allowing scientists to optimize their properties of specific applications including hydrogen storage. Hydrogen stored in charged fullerene can be used as a source of fuel for hydrogen fuel cells. The controlled release of hydrogen from charged fullerenes can provide a stable and efficient fuel supply for fuel cell applications. First-charged fullerene is obtained by using electrochemical doping, which leads to the formation of cations or anions by irreversible charging by up to six electrons. It is generally generated in gas phase. When fullerene is highly charged, due to presence of strong electric field, hydrogen molecules will dissociate during structure optimization. Quantum mechanics demonstrate the origin of high electric field produced by fullerene [20].

6.7 Hydrogen Storage in Hydro- or Hydrogenated Fullerene

Hydrofullerenes are the class of fullerenes in which hydrogen is encapsulated inside the carbon cage structure. These hydrogen-filled fullerenes are also known as endohedral fullerenes. During encapsulation, hydrogen molecules are introduced to the fullerene structure under specific conditions, allowing them to trap inside the cage formed by carbon atoms [57, 58]. The encapsulation process can be complex and requires precise control of temperature, pressure, and other environmental factors to ensure the hydrogen molecules are securely held within the fullerene cage. Hydrofullerenes can potentially store a high density of hydrogen due to large internal volume of the carbon cage, allowing for a high number of hydrogen molecules to be encapsulated. The properties of hydrofullerenes including their size and charged state can be engineered to optimize the hydrogen storage capacity and release kinetics. The encapsulation of hydrogen inside a stable carbon structure enhances the safety of hydrogen storage, minimizing the risk associated with handling and transportation [59, 60]. Hydrofullerene reacts in terms of molar heat capacity as temperature varies from 5 to 340 K. Hydrogenated fullerenes are better alternatives in reversible hydrogen storage and renewable energy generation through next-generation photovoltaics [61–63]. Volumetric cavity enhancement results in capture of more H_2 molecules. In this way, proper functionalization and selection of structural units leads to producing a new era in green energy and environmental remediation. An important consideration is that increasing the surface area of a solar panel leads to a higher current output. Solar energy plays a crucial role in the separation of charge carriers, the recombination of electron-hole pairs, which leads to their annihilation, and the transmission of solar energy [64, 65].

6.8 Conclusions and Future Outlook

In summary, clean hydrogen is a potential energy source for future economy. Caged fullerene derivatives exhibit versatile properties due to the presence of functional groups. We have demonstrated the fullerene as green energy application by different functionalization or doping on fullerene surface. Hydrogen storage efficiency is a function of structure, organic groups, and dopants available on the surface of fullerene. Due to doping or functionalization of fullerene, absorption capacity of fullerene increases due to improvement in physicochemical properties of fullerene. We note some general trends in interaction of extra atoms or groups on the surface fullerene. This interaction is often used for tuning the electronic and absorption properties of fullerene. This also increases the catalytic activity of fullerene that is responsible for more absorption of hydrogen on the surface of fullerene. The present chapter inspires us to develop fullerene-based functional material to store green hydrogen energy for futuristic applications. In a broad sense, functional fullerenes are highly efficient candidates for advanced green energy storage and conversion systems, i.e. battery, fuel cells, and solar cells. Fullerene can play a role in carbon capture technologies that can be integrated with hydrogen production process, thus enabling a more sustainable and carbon-neutral production cycle.

Acknowledgments

The author (Manish Kumar) is thankful to the Department of Chemistry, L.N.T. College, B.R.A. Bihar University, and Muzaffarpur for providing favorable research ecosystem to collect and compile the data for the publication of the scientific chapter.

References

1 Boo, W.O.J. (1992). An introduction to fullerene structures. Geometry and symmetry. *Journal of Chemical Education* 69 (8): 605.

2 Lai, Y.Y., Cheng, Y.J., and Hsu, C.S. (2014). Applications of functional fullerene materials in polymer solar cells. *Energy & Environmental Science* 7 (6): 1866–1883.

3 Santos, A.L., Cebola, M.J., and Santos, D.M. (2021). Towards the hydrogen economy—a review of the parameters that influence the efficiency of alkaline water electrolysers. *Energies* 14 (11): 3193.

4 Li, M., Li, Y., Zhou, Z. et al. (2009). Ca-coated boron fullerenes and nanotubes as superior hydrogen storage materials. *Nano Letters* 9 (5): 1944–1948.

5 Paukov, M., Kramberger, C., Begichev, I. et al. (2023). Functionalized fullerenes and their applications in electrochemistry, solar cells, and nanoelectronics. *Materials* 16 (3): 1276.

6 Shinar, J. (ed.) (1999). *Optical and Electronic Properties of Fullerenes and Fullerene-based Materials*. CRC Press.

7 Toropov, A.A., Leszczynska, D., and Leszczynski, J. (2007). QSPR study on solubility of fullerene C60 in organic solvents using optimal descriptors calculated with SMILES. *Chemical Physics Letters* 441 (1–3): 119–122.

8 Durbin, D.J., Allan, N.L., and Malardier-Jugroot, C. (2016). Molecular hydrogen storage in fullerenes—a dispersion-corrected density functional theory study. *International Journal of Hydrogen Energy* 41 (30): 13116–13130.

9 Durbin, D.J., Allan, N.L., and Malardier-Jugroot, C. (2013). Physisorption of molecular hydrogen in curved carbon nanomaterials: a computational study. *WIT Transactions on Engineering Sciences* 77: 149–157.

10 Joseph, J., Sivasankarapillai, V.S., Nikazar, S. et al. (2020). Borophene and boron fullerene materials in hydrogen storage: opportunities and challenges. *ChemSusChem* 13 (15): 3754–3765.

11 Kuc, A., Zhechkov, L., Patchkovskii, S. et al. (2007). Hydrogen sieving and storage in fullerene intercalated graphite. *Nano Letters* 7 (1): 1–5.

12 Jiang, Z., Zhao, Y., Lu, X., and Xie, J. (2021). Fullerenes for rechargeable battery applications: recent developments and future perspectives. *Journal of Energy Chemistry* 55: 70–79.

13 Golberg, D., Bando, Y., Stéphan, O. et al. (1999). New fullerenes in the BCN system: synthesis and analysis by an electron beam. *Microscopy* 48 (6): 701–709.

14 Brant, J., Lecoanet, H., Hotze, M., and Wiesner, M. (2005). Comparison of electrokinetic properties of colloidal fullerenes (n-C60) formed using two procedures. *Environmental Science & Technology* 39 (17): 6343–6351.

15 Qiao, L., Sun, X., Yang, Z. et al. (2013). Network structures of fullerene-like carbon core/nano-crystalline silicon shell nanofibers as anode material for lithium-ion batteries. *Carbon* 54: 29–35.

16 Mi, D., Kim, J.H., Kim, H.U. et al. (2014). Fullerene derivatives as electron acceptors for organic photovoltaic cells. *Journal of Nano Science and Nanotechnology* 14 (2): 1064–1084.

17 Oliveira, A.M., Beswick, R.R., and Yan, Y. (2021). A green hydrogen economy for a renewable energy society. *Current Opinion in Chemical Engineering* 33: 100701.

18 Sahoo, R.K., Chakraborty, B., and Sahu, S. (2021). Reversible hydrogen storage on alkali metal (Li and Na) decorated C20 fullerene: a density functional study. *International Journal of Hydrogen Energy* 46 (80): 40251–40261.

19 Quarton, C.J., Tlili, O., Welder, L. et al. (2020). The curious case of the conflicting roles of hydrogen in global energy scenarios. *Sustainable Energy & Fuels* 4 (1): 80–95.

20 Mahamiya, V., Shukla, A., and Chakraborty, B. (2022). Exploring yttrium doped C24 fullerene as a high-capacity reversible hydrogen storage material: DFT investigations. *Journal of Alloys and Compounds* 897: 162797.

21 Tang, C., Chen, S., Zhu, W. et al. (2015). Transition metal Ti coated porous fullerene C24B24: potential material for hydrogen storage. *International Journal of Hydrogen Energy* 40 (46): 16271–16277.

22 Liu, P., Zhang, H., Cheng, X., and Tang, Y. (2016). Ti-decorated B38 fullerene: a high capacity hydrogen storage material. *International Journal of Hydrogen Energy* 41 (42): 19123–19128.

23 Das, S.K., Hossain, R., Hossain, K. et al. (2021). Alkali metal doped fullerene as hydrogen storage: a dft study. *Journal of Sustainable Energy* 12 (1).

24 Bi, L., Yin, J., Huang, X. et al. (2019). Lithium decoration of boron-doped hybrid fullerenes and nanotubes as a novel 3D architecture for enhanced hydrogen storage: a DFT study. *International Journal of Hydrogen Energy* 44 (5): 2934–2942.

25 Venkataramanan, N.S., Belosludov, R.V., Note, R. et al. (2010). Theoretical investigation on the alkali-metal doped BN fullerene as a material for hydrogen storage. *Chemical Physics* 377 (1–3): 54–59.

26 Zahedi, E. and Mozaffari, M. (2014). DFT study of hydrogen storage on Li-and Na-doped C 59 B heterofullerene. *Surface Review and Letters* 21 (04): 1450047.

27 Mahdy, A.E., Taha, H.O., Kamel, M.A., and El Shemy, F. (2016). Density functional theory study of hydrogen storage on Ni-doped C59X (X= B, N) heterofullerene. *Molecular Physics* 114 (9): 1539–1558.

28 Kaur, H., Kaur, J., and Kumar, R. (2022). A review on all boron fullerene (B40): a promising material for sensing and device applications. *Materials Today Proceedings* 48: 1095–1102.

29 Garara, M., Benzidi, H., Lakhal, M. et al. (2019). Phosphorene: a promising candidate for H2 storage at room temperature. *International Journal of Hydrogen Energy* 44 (45): 24829–24838.

30 Cortés-Arriagada, D. and Ortega, D.E. (2021). Fullerene–phosphorene–nanoflake nanostructures: modulation of their interaction mechanisms and electronic properties through the size of carbon fullerenes. *Carbon* 182: 354–365.

31 Baskar, A.V., Benzigar, M.R., Talapaneni, S.N. et al. (2022). Self-assembled fullerene nanostructures: synthesis and applications. *Advanced Functional Materials* 32 (6): 2106924.

32 Soliman, K.A. and Aal, S.A. (2021). Theoretical investigation of favipiravir antiviral drug based on fullerene and boron nitride nanocages. *Diamond and Related Materials* 117: 108458.

33 Juárez, A.R., Ortiz-Chi, F., Pino-Ríos, R. et al. (2020). The boron nitride (B116N124) fullerene: stability and electronic properties from DFT simulations. *Chemical Physics Letters* 741: 137097.

34 Oku, T. (2014). Hydrogen storage in boron nitride and carbon nanomaterials. *Energies* 8 (1): 319–337.

35 Chkhartishvili, L. (2010). Boron nitride nanostructures: molecules, sheets, tubes, fullerenes (an overview). *Nano Studies* 2: 139–174.

36 Hussain, T., Kaewmaraya, T., Chakraborty, S., and Ahuja, R. (2013). Functionalization of hydrogenated silicene with alkali and alkaline earth metals for efficient hydrogen storage. *Physical Chemistry Chemical Physics* 15 (43): 18900–18905.

37 Zhang, Y., Liu, P., and Zhu, X. (2021). Li decorated penta-silicene as a high capacity hydrogen storage material: a density functional theory study. *International Journal of Hydrogen Energy* 46 (5): 4188–4200.

38 Shi, M., Bi, L., Huang, X. et al. (2020). Design of three-dimensional nanotube-fullerene-interconnected framework for hydrogen storage. *Applied Surface Science* 534: 147606.

39 Srinivasan, S., Demirocak, D.E., Kaushik, A. et al. (2020). Reversible hydrogen storage using nanocomposites. *Applied Sciences* 10 (13): 4618.

40 Yadav, T.P. and Awasthi, K. (2022). Carbon nanomaterials: fullerene to graphene. *Transactions of Indian National Academy of Engineering* 7 (3): 715–737.

41 Azpeitia, J., Otero-Irurueta, G., Palacio, I. et al. (2017). High-quality PVD graphene growth by fullerene decomposition on Cu foils. *Carbon* 119: 535–543.

42 Sawamura, M., Kuninobu, Y., Toganoh, M. et al. (2002). Hybrid of ferrocene and fullerene. *Journal of the American Chemical Society* 124 (32): 9354–9355.

43 Yadav, S.P. (2016). Synthesis of carbon nanotubes using ferrocene by thermal CVD technique. Doctoral dissertation.

44 Rabti, A., Raouafi, N., and Merkoçi, A. (2016). Bio (sensing) devices based on ferrocene–functionalized graphene and carbon nanotubes. *Carbon* 108: 481–514.

45 Luo, C.Y., Huang, W.Q., Xu, L. et al. (2016). Electronic properties and photoactivity of monolayer MoS 2/fullerene van der Waals heterostructures. *RSC Advances* 6 (49): 43228–43236.

46 Chhetri, M., Gupta, U., Yadgarov, L. et al. (2016). Effects of p-and n-type doping in inorganic fullerene MoS2 on the hydrogen evolution reaction. *ChemElectroChem* 3 (11): 1937–1943.

47 Min, K. (2015). Quantum and atomistic simulation of mechanical and electronic properties of nano-materials.

48 Prakash, R., Kumar, S., and Maiti, P. (2022). Carbon nanotube based nanomaterials for solar energy storage devices. In: *Current and Future Developments in Nanomaterials and Carbon Nanotubes: Applications of Nanomterials in Energy Storage and Electronics* (ed. G. Manik and S.K. Sahoo), 1–18. Bentham Science Publishers.

49 Singh, E., Kim, K.S., Yeom, G.Y., and Nalwa, H.S. (2017). Atomically thin-layered molybdenum disulfide (MoS2) for bulk-heterojunction solar cells. *ACS Applied Materials & Interfaces* 9 (4): 3223–3245.

50 Yadgarov, L., Rosentsveig, R., Leitus, G. et al. (2012). Controlled doping of MS2 (M= W, Mo) nanotubes and fullerene-like nanoparticles. *Angewandte Chemie International Edition* 51 (5): 1148–1151.

51 Saura-Sanmartin, A., Martinez-Cuezva, A., Marin-Luna, M. et al. (2021). Effective encapsulation of C60 by metal–organic frameworks with polyamide macrocyclic linkers. *Angewandte Chemie* 133 (19): 10909–10914.

52 Kumar, S., Prakash, R., and Maiti, P. (2022). Advanced batteries and charge storage devices based on nanowires. In: *Current and Future Developments in Nanomaterials and Carbon Nanotubes: Applications of Nanomterials in Energy Storage and Electronics* (ed. G. Manik and S.K. Sahoo), 159–175. Bentham Science Publishers.

53 Singh, L., Dhavala, L., Bhimireddi, R. et al. (2023). Low-cost flame synthesized La2/3Cu3Ti4O12 electro-ceramic and extensive investigation on electrical, impedance, modulus, and optical properties. *Ceramics International* 49 (13): 21795–21803.

54 Gao, Y., Wu, X., and Zeng, X.C. (2014). Designs of fullerene-based frameworks for hydrogen storage. *Journal of Materials Chemistry A* 2 (16): 5910–5914.

55 Yoon, M., Yang, S., Wang, E., and Zhang, Z. (2007). Charged fullerenes as high-capacity hydrogen storage media.

56 Savoie, B.M., Rao, A., Bakulin, A.A. et al. (2014). Unequal partnership: asymmetric roles of polymeric donor and fullerene acceptor in generating free charge. *Journal of the American Chemical Society* 136 (7): 2876–2884.

57 Saraswat, A. and Kumar, S. (2022). A topical study of electrochemical response of functionalized conducting polyaniline: an overview. *European Polymer Journal* 182: 111714.

58 Gaboardi, M., Amadé, N.S., Aramini, M. et al. (2017). Extending the hydrogen storage limit in fullerene. *Carbon* 120: 77–82.

59 Lebedev, B.V., Bykova, T.A., and Lobach, A.S. (2000). Thermodynamic properties of hydrofullerene C60H36 from 5 to 340 K. *Journal of Thermal Analysis and Calorimetry* 62: 257–265.

60 Saraswat, A. and Kumar, S. (2023). Cutting-edge applications of polyaniline composites towards futuristic energy supply devices. *European Polymer Journal* 200: 112501.

61 Kumar, S. and Maiti, P. (2023). Review on functional electrolyte, redox polymers, and solar conversions in 3G emerging photovoltaic technologies: progress and outlook. *Energy & Fuels* 37 (19): 14473–14511.

62 Kumar, S., Rai, R.N., Singh, D. et al. (2023). Future perspectives on QDs embedded nano-fibrous materials as high capacity sustainable anode for Na-ion batteries technology. *MRS Energy & Sustainability* 10: 1–9.

63 Kumar, S., Chaurasia, P.K., and Patel, S.K. (2023). Improvement in the efficiency of organic semiconductors via molecular doping for OLEDs applications. In: *Organic Light Emitting Diode (OLED) Toward Smart Lighting and Displays Technologies* (ed. L. Singh, R. Dubey, and R.N. Rai), 203–220. CRC Press.

64 Tripathi, S.L. and Padmanaban, S. (ed.) (2020). *Green Energy: Solar Energy, Photovoltaics, and Smart Cities*. Wiley.

65 Wu, H., Fan, X., and Kuo, J.L. (2012). Metal free hydrogenation reaction on carbon doped boron nitride fullerene: a DFT study on the kinetic issue. *International Journal of Hydrogen Energy* 37 (19): 14336–14342.

7

Graphyne-Based Carbon Nanomaterials for Green Energy Applications

Kulsum Hashmi[1], Mohammad Imran Ahmad[2], Saman Raza[1], Nidhi Mishra[3], Seema Joshi[1], and Tahmeena Khan[2]

[1] *Department of Chemistry, Isabella Thoburn College, Lucknow, Uttar Pradesh, India*
[2] *Department of Chemistry, Integral University, Lucknow, Uttar Pradesh, India*
[3] *Department of Applied Sciences, Indian Institute of Information Technology, Allahabad, Uttar Pradesh, India*

7.1 Introduction

Carbon forms the basis of life. The atom exists in different hybridization states like sp, sp^2, and sp^3 forming diverse bonding. Carbon has several allotropes like graphene, fullerenes, carbon nanotubes (CNTs), nanorings, etc. [1, 2]. In 1987, Baughman et al. proposed the GYfamily members (GFMs) consisting of sp and sp^2 hybridized C atoms where some of the C—C bonds get substituted with C—C triple bonds, which reduces the cohesive energy. Among these, GY is the most stable and has also been associated with potential nanoelectronics and energy storage applications. GY has better electrical conductivity than graphene [3]. The presence of sp^2 and sp^3 hybridized carbon provides rich π-conjugation to graphene. It also possesses evenly distributed pores permitting water molecules to pass through it, a feature that has been useful in designing desalinators based on GY. GDY, a family member of graphene has been used in separation membranes for H_2 purification [4]. Metal-doped GYs are also found to be well-suited for H_2 storage and Li-ion batteries [5]. GYs and GDYs have been used in electronics, photovoltaics, and catalysis.

Carbon-based Nanomaterials for Green Applications, First Edition. Edited by Upendra Kumar, Piyush Kumar Sonkar, and Suman Lata Tripathi.

Published 2025 by John Wiley & Sons, Inc.

7.1.1 Structural Aspects of Graphyne

Graphene serves as the building block of several carbon-containing frameworks and holds the upper hand among all the allotropes of carbon [6]. Graphene has a high carrier mobility [7] and is used in ultrahigh-speed radiofrequency electronics [8]. However, the zero band gap holds limitations in field effect transistors (FETs) making it less useful for nanoelectronics. Therefore, the scientists worked on finding a new system having the intrinsic energy gap, which could be developed as a realistic candidate. Among 2D materials having structural similarity with graphene, GY [9] (Figures 7.1 and 7.2) and GDY [10] (Figure 7.3) were created from graphene and found to be even better in some aspects than graphene. As discussed earlier, the discovery of GY was made in the year 1987 [11] and the material got attention because of its direct band energy. It has D6h/mmm symmetry and is considered a hybrid system of graphene (sp^2C) and carbyne (spC) in which the hexagonal rings are joined by the acetylenic linkage. It possesses rich optoelectronic and elastic characteristics due to these linkages [12]. GY is classified as α, β, and γ graphyne based on the number of these acetylenic linkages to the different hexagonal rings [5]. α Graphyne has 100% acetylenic bonds present in each C—C bond of graphene, while in β and γ graphyne, they are at select C—C bonds. In β-GY, two-thirds of the C—C bonds are acetylenic,whereas in γ-GY, only one-third of such acetylenic bondsare present. The γ-GY acts as a direct band gap semiconductor having a tuneable band gap in varying circumstances [13]. α and β-GYs are

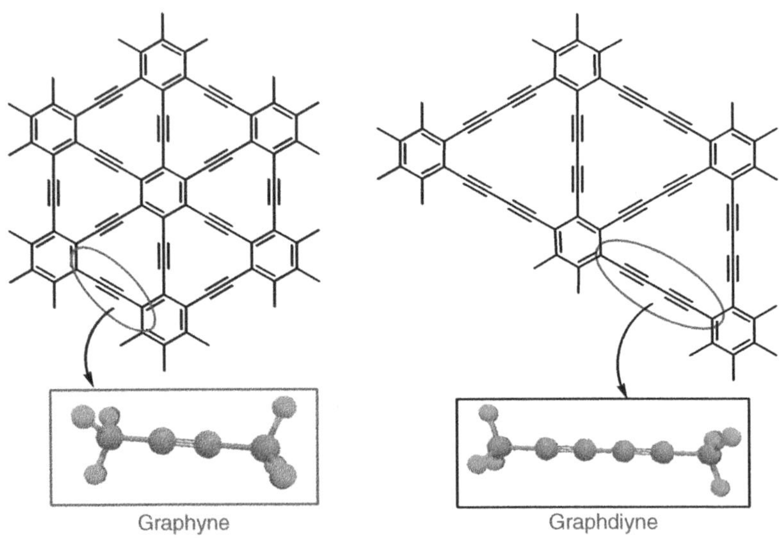

Figure 7.1 Structure of graphyne.

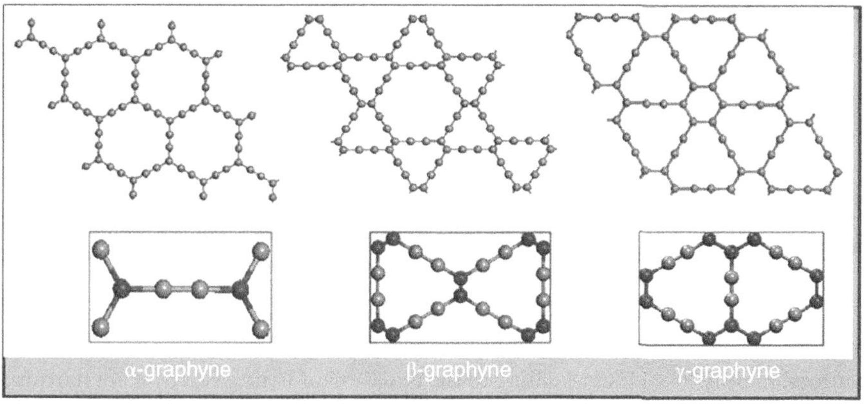

Figure 7.2 Different types of graphynes.

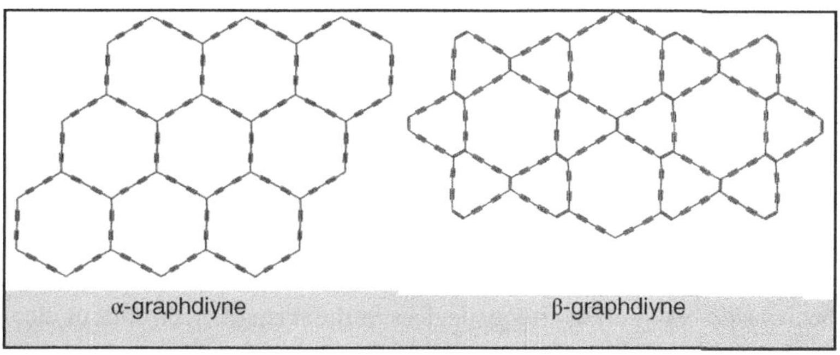

Figure 7.3 Different types of graphdiynes.

semi-metallic and have Dirac cones [14]. The directional electrical conductivity of GY makes it superior to graphene. The structure also has a higher degree of π-conjugation. GY also has mechanical stability and possesses high tensile strength and stiffness [15]. Diederich initially attempted the synthesis of GY [16]. Subsequently, the molecular fragments of GY and GDY molecular fragments were synthesized by M.M. Haley [17]. Despite the attempts, only trace amounts of GYs have been prepared in the laboratory. Large-area-multilayer GY films have been successfully synthesized [18] and also GY sheets [19]. GY has a higher degree of π-conjugation owing to *sp* as well as sp^2 hybridized C present. The carbon atoms involved in the hexagonal ring are sp^2 hybridized, whereas sp hybridized carbon atoms link the two adjacent rings. The C—C bond length in the hexagonal ring has been calculated to be 1.427 A. The px, py, and s orbitals are involved in σ bond formation between the neighboring atoms in GY, whereas the p_z orbitals are involved in π-bond formation, resulting in the π-conjugation. The bond distance between a sp^2 and sp hybridized C is 1.412 Å and is shorter than a normal σ bond (~1.470 Å).

7.2 Graphyne-Based Carbon Nanomaterials for Green Energy Applications

With the growing population and modernization worldwide, there is a consequent increase in the consumption of energy which mainly comes from non-renewable sources, mainly fossil fuels, leading to the emission of toxic gases that are harmful to not just our health but also the environment. It is, therefore, essential to search for sustainable and renewable energy sources. This objective has led to the development of various renewable energy production and storage systems, like fuel cells, SCs, LIBs, and solar cells [19–21]. These devices are portable, have high power and long life, and are also easily fabricated and have, therefore, gained much popularity and interest in the field of science and technology. The development of materials for use in the above-mentioned clean and green energy devices has been extensively researched. However, the future of such devices will depend on high energy efficiency, small cost, and ease of manufacture. This has led to considerable research to identify and discover materials that would increase energy efficiency at a low cost.

Over the past decades, the use of carbon-based materials like graphene and CNTshas been vastly explored in energy devices in the form of electrodes or electrocatalysts. Graphene has been found to be highly promising for use in energy storage and conversion devices, where it proves useful in the production of clean and sustainable energy. It can be used:

- in solar cells to increase the efficiency of energy conversion.
- in batteries to improve the charge-discharge cycle and efficiency.
- in supercapacitors to improve power density and efficiency.

Due to its structure, graphene has special electrical properties that are quite useful in energy-related applications, making it a material of choice in the preparation of such devices [22].

The success of graphene has prompted further interest in similar nanomaterials to improve the efficiency of the above-mentioned energy production and storage systems, together with reducing the cost of production, which would make their use more popular. Carbon-based nanomaterials are mostly used as electrodes in fuel cells, LIBs, and SCs as they have higher surface area and excellent electrical and mechanical properties.

GY is considered more useful than graphene in energy storage devices due to its larger surface area, greater electronic mobility, and intrinsic band gap. GDY, a novel carbon allotrope, has a two-dimensional carbon framework; its unique feature is the co-hybridization of sp and sp^2 hybridized carbon atoms. Such a hybridization gives GDY a 2D planar network structure, which imparts numerous advantages in precise growth and assembly. Several studies find it can be useful in the development of renewable energy storage and energy conversion devices. In recent years, GDY has been subjected to extensive research, which has helped in understanding its differences from and advantages over traditional carbon-based materials, like controlled growth and their physical and chemical properties. It has gradually developed into an important material, with promising applications in the energy sector [23].

The structural features in GDY that impart enhanced electrical properties are the catalytic active sites such as C_1 and C_2 atoms that bear a positive and negative charge, respectively; due to their positive charge, the C_1 atoms are able to readily bind with oxygen to accelerate reaction pathways like the oxygen reduction reaction (ORR) in an electrochemical cell [24]. Other than this, the conjugated π-bonds of GDY can enhance electron transmission, and the pores in the framework of GDY are useful in gaseous product transportation, such as in electrocatalytic reactions like ORR. Adding heteroatoms to GDY produces doped GDY, which increases its catalytic activities by changing the electronic configuration and also increases the number of active sites by the addition of heteroatom defects, which are useful for carrying the metal ions in batteries and SCs [25]. GY GDY and doped GY and GDY are used as anodes in lithium/sodium ion batteries due to their superior electrocatalytic and charge transport properties [26, 27]. The applications of GY and GDY in green energy storage and conversion devices (Figure 7.4) have been discussed in detail in the following sections.

7.2.1 Mechanisms Involved in Growth, Doping, Energy Storage, and Conversion Involving Graphyne

Fuel cells, batteries, and SCs are some of the most reliable options available for storage and conversion of electrical energy [28]. GY- and GDY-based materials have miscellaneous performances. In GYs, the sp and sp^2 hybridzed carbon framework structure is formed by linkages between benzene rings and acetylenic groups. The GYs usually involve cross-coupling reactions of reactants having benzene ring and acetylene groups in growth mechanisms. In the process, the terminal acetylene moieties of the reactants bind to the Cu complex catalysts and the reaction between these Cu complexes causes the cross-coupling of the terminal acetylene moieties to form GDY [29] on the surface of the template having different structures. A catalyst is not necessary for the synthesis of γ-GY. The

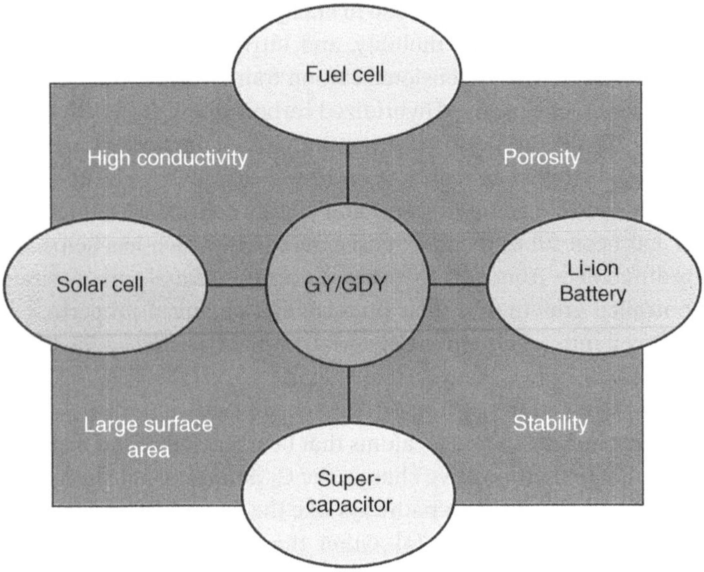

Figure 7.4 Properties and green energy applications of Graphyne and Graphdiyne.

theoretical findings suggest that GYs have higher electron density than graphene. The monolayer GDY possesses an intrinsic band gap of 0.5 eV and high electron mobility [30, 31] making it suitable for semiconducting applications in nanoscale devices like FETs [32, 33]. As mentioned earlier, in GDY the C_1 and C_2 carry a positive and negative charge respectively. Due to its positive charge, C_1 adsorbs O_2 and enhances the rate ofORRs, making them good catalysts in electrochemical catalysis [26]. The porous structure can facilitate the electrocatalysis process. Upon doping, the heteroatoms change the electronic assembly of GDY and the number of active sites increases by the induction of heteroatom defects [27]. Electrons and ions diffuse to the surface of GY electrodes readily leading to increased electrochemical performance [34]. The special electronic characteristics of GYs endow high specific capacity and cycling stability in the batteries. In lithium/sodium ion batteries, where GYs are used as anode materials, the metal ions are constantly added and removed. The enlarged in-plane pores enclosed within the butadiyne bonds and benzene ring make storage and diffusion of metal ions possible in GDY. Li/Na ions can diffuse in both directions, i.e. parallel as well as vertical to the plane of GDY and the doping with heteroatoms can improve its electrochemical capacity. The agglomeration of heteroatoms is also prevented by the formation of chemical bonds between GDY and heteroatoms. Doping with N and F that have high electronegativity can cause heteroatomic defects, which creates more active sites for Li^+ storage [28, 29]. Doped GYs are usually prepared by two methods. In the first, the synthesized GYs react with the heteroatom-containing reactants at high temperature through the acetylenic groups of GYs to form C-heteroatom-C linkages of GDY or the C atoms in GDY are substituted to form the doped GDY like S-GDY, N-GDY [5, 35], or Co-NGDY [36]. In the other methodology or the bottom-up approach, the heteroatom-substituted precursor is directly used in the preparation of doped GDY by cross-coupling [37, 38].

7.3 Fuel Cells

Fuel cells are a type of electrochemical devices that efficiently convert chemical energy to electrical energy without toxic waste production [39]. They have three main components, namely the anode, the cathode, and the electrolyte. Depending on the electrolyte used, fuel cells can be of different types like proton-exchange membrane fuel cells, solid acid fuel cells, phosphoric acid fuel cells, alkaline fuel cells, solid oxide fuel cells, etc., that have different working temperatures and energy efficiency [40]. Fuel cells convert the chemical energy from the redox reactions between a fuel and an oxidizing agent, into an electric current that can be used as an energy source. The fuel used is usually hydrogen and the oxidizing agent is oxygen. The two reactions taking place are the ORRat the cathode and the hydrogen oxidation reaction (HOR) at the anode, respectively, in the presence of a Pt catalyst [40]. Hydrogen gas is oxidized at the anode to form H^+ ions and release electrons that flow to the cathode where they reduce oxygen into water, which is released as the green waste product (Figure 7.5).

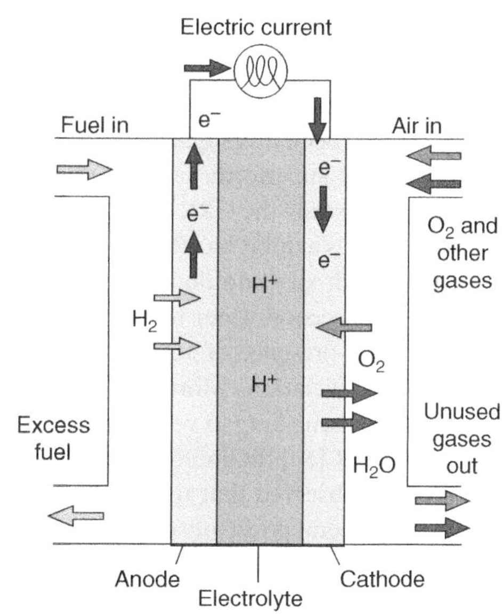

Figure 7.5 The general mechanism of a fuel cell. *Source:* [41]/Wikipedia/Dervisoglu/CC BY 1.0.

If a hydrocarbon is used as fuel, for example, in direct alcohol fuel cells, which are a type of proton-exchange membrane fuel cells, carbon dioxide is another byproduct formed. So, the emission of toxic wastes, as seen with the combustion of fuels, is minimized in a fuel cell. However, due to the use of costly components and other issues like lesser durability, heating up and low energy efficiency in some cases, their commercial use is still not widespread.

Presently, carbon-based materials are being explored to improve the efficiency of fuel cells in the form of electrocatalysts. The rate of ORR is slow due to multi-electron reactions and large overpotential and therefore requires a catalyst [42]. Presently, noble metals like platinum, iridium, and ruthenium are used as catalysts for these reactions [43, 44] but as these metals are expensive, they increase the cost of production of fuel cells and they need to be replaced by cheaper yet efficient materials. Carbon-based materials like graphene, graphyne, and CNTs can be used as electrocatalysts due to two reasons:

- their high electrical conductivity and chemical stability make them an excellent choice to be used in electrodes [45].
- doping them with heteroatoms, alters their chemical properties, imparting them with catalytic ability.

Consequently, heteroatom-doped carbon materials have been extensively studied for this purpose in recent years [46, 47]. Heteroatom-doped graphene has been found to show good electrocatalytic activity due to the 2D structure, greater surface area, and numerous active sites and several such materials have been prepared and used as electrocatalysts in fuel cells and metal-air batteries (MABs). For example, the ORR for Ruthenium electrocatalyst that is used in PEMFCs is 7.5 times faster when used as a Ru-N/G-750 composite of graphene, along with higher durability and lower cost [48]. Graphene has also been shown to reinforce the polymer electrolyte membranes (PEMs) of PEMFCs [49].

GY has also been investigated as an electrocatalyst in fuel cells and several studies suggest its superiority over graphene. For example, in a density functional theory (DFT) study the ORR for three GY models was studied and it was observed that the models exhibit an efficient reaction involving four electrons, in which the initial two-hydrogen addition gives a water molecule, followed byother two hydrogens giving a second water molecule with the recovery of GY planarity. This can be due to the presence of acetylenic bonds present in g GY that help in oxygen molecule dissociation. Moreover, the 18-membered ring structurespresent in α- and β-graphyne help in very efficient O_2 penetration, which enhances the efficiency of ORR [50]. In another DFT study on ORR using GY models in acid fuel cells, it was observed that the charge distribution across GY is not uniform but there exist several positively charged carbon atoms that are useful in the dissociation of the O—O bond. They act as catalysts and bind to oxygen molecules and when acid is used, the O—O bond breaks and water is formed. This increases the rate of ORR, thereby increasing the efficiency of the fuel cell [51]. Doping of GY and GDY with heteroatoms has also been found to increase their electrocatalytic efficiency in fuel cells and MABs. For example, GDY was doped with N by using ammonia and pyridine and the N-GDY prepared showed catalytic activity for ORR like that of Pt/C electrocatalyst used commercially and was stable in alkaline solution [35]. In another study, GDY doped with Fe and N (Fe-N-GDY) showed remarkable catalytic ability for ORR and was also stable in both acidic and alkaline electrolytes [31]. The promising results obtained in such studies highlight the potential of GY and GDY as electrocatalysts in making cheaper and more efficient fuel cells.

7.3.1 Oxygen Reduction Reaction (ORR) Catalyst for Hydrogen Fuel Cells or Metal-Air Batteries (MABs)

Increasing the battery system's energy density for electric vehicles (EVs) has drawn a lot of attention from researchers in the field of electrocatalysis [52–54]. The advancement of catalysts that effectively promote the ORR has led to significant progress in commercial applications within the field. Fuel-cell vehicles, like the Toyota Mirai, are now on the market as a result of the recent use of fuel-cell technologies in EVs [55]. This EV can start in cold conditions, offers an appealing travel distance of 500 km, and quicker refuelling times of just three minutes as compared to battery-based EVs [56]. However, the vehicle's price remains high as a result of the use of the Pt-based catalyst and low production volumes [57, 58]. Although some people might be able to afford this EV, it is unlikely to come into common use at its current price. Apart from addressing the cost aspects of vehicle development, the establishment of a nationwide or citywide hydrogen infrastructure is a significant challenge that poses a major obstacle to the extensive interest in hydrogen fuel cells in the field of transport. Unlike fuel cells, MABsare considered a good option between hydrogen fuel cell devices and LIBs. Li-air batteries (LABs), which are MABs, are supposed to have significantly better energy densities than LIBs. In theory, MABs like LAB are expected to exhibit significantly higher energy densities than LIBs. While estimations based on the volume of some non-active material resulted in relatively small volumetric energy density (smaller than $LiCoO_2$-based LIBs), the gravimetric energy density is estimated to be two times compared to LIBs [59]. Despite significant progress in commercial fuel cell application in EVs, most MABs are still in their initial prototype stages. The MABs, also known as semi-fuel cells, have a surface that resembles that of fuel cells [60, 61].

ORR plays an important role in various energy conversion devices, like fuel cells and MABs. But this reaction includes a multistep proton-linked electron transfer mechanism that is kinetically slow. To increase the rate and efficiency of the reaction, appropriate catalyst selection is required. Developing low-cost ORR catalysts is a big obstacle. GDY offers a large surface area with numerous catalytically active sites on its own, indicating significant potential for applications in ORR [62]. Kang *et al.* [63] employed DFT calculations to investigate the catalytic performance of nitrogen-doped graphdiyne (NGDY) in ORR. The synergy between sp-N and sp^2-N was discovered to be responsible for the excellent catalytic performance of NGDY, and the effect of NGDY on ORR was shown to be similar to that of Pt catalysts. N doping is the most popular and efficient method of modifying GDY in ORR [25]. A high catalytic activity can be achieved by combining it with Fe to produce more efficient catalytic sites. One such study used Fe and N co-doped graphdiyne (Fe–N-GDY) as an ORR catalyst in 0.1 M KOH [29]. In the carbon structure of GDY, N atoms interchanged carbon atoms, whereas Fe atoms were linked via Fe—N or Fe—C bonds. The unique acetylenic linkers in GDY provided stable bonding sites for Fe atoms and made the catalyst more stable.

To generate additional active sites, the researchers synthesized Co and N co-doped GDY (Co–NGDY) [36], possessing an active surface area of 1350 $cm^2\ g^{-1}$ determined through electrochemical surface area calculations. It was expected that the strong binding of the Co atoms to GDY would enhance the catalyst's stability. This Co-NGDY has been identified to be a potential catalyst for both hydrogen evolution reactions (HERs) and ORR. Lv et al. [64] employed a simple cross-coupling to regulate N bonding arrangement. They also synthesized a pyridinic nitrogen-doped GDY (PyN-GDY) where the PyN-containing active sites demonstrated excellent electrocatalytic activity for ORR in Zn–air batteries. PyN-GDY consists of acetylenic linkers and a pyridine ring, where each benzene ring of GDY had one carbon atom replaced by PyN. GYs, as ORR catalysts, possess an extensively conjugated π-electron system, and planar structure having similar hexagonal pores that promote gas diffusion. Furthermore, the number of active ORR sites may rise due to the vast specific area of GYs. The presence of heteroatoms in doped GDYs influences the electronic configuration of GDYs as well. These heteroatoms cause a redistribution of charge, leading to increased positive charges in the carbon atoms surrounding them, which enhances the adsorption of O_2. Consequently, doped GDYs exhibit more favorable catalytic activity for the ORR [62].

7.3.2 Lithium-Ion and Lithium-Metal Batteries

The rapid growth of EVs and portable electronics requires the development of new rechargeable batteries that are more powerful,are charged quickly, and also have higher energy density [65–68]. The development of high energy density primarily depends on technological advancements in electrode materials with markedly enhanced specific capacities. For example, the Li metal anode has a high specific capacity of $3860\,mAh\,g^{-1}$ and a low potential of −3.040V when compared to standard hydrogen electrodes, making it a prospective choice for the anode in rechargeable batteries havinghigher energy density [69–72]. LIBs are rechargeable batteries that use lithium ions as carriers to store electric charge and have shown rapid development [73, 74]. Lithium ions move from the anode to the cathode when LIBs are charged, and when discharged, they flow in reverse direction. LIBs have been successfully used in the past few decades to power portable electronic devices, EVs, and smart electric grids. Akira Yoshino, M. Stanley Whittingham, and John B. Goodenough have significantly contributed to developing the ground-breaking LIBs for energy storage for which they were awarded the Nobel Prize (2019) in Chemistry [75, 76]. Nowadays, graphite has gained extensive usage as an anode due to its high diffusivity of Li ions, providing high power, and its stability during lithium intercalation, resulting in minimal volume change. The highest possible arrangement of Li-intercalated graphite is one lithium atom for every six carbon atoms. This results in a composite material called C_6Li, whose specific capacity can reach up to $372\,mAh\,g^{-1}$ [77–79]. Li ions flow through the electrolytes across the separator during the charge and discharge processes, where they are reversibly exchanged between the two electrodes. Typically, the anode materials are usually graphitic carbon, while cathode materials are often lithiated transition metal oxides with a layered structure. When the battery is charged, Li ions are liberated from the cathode, leading to the oxidation of the active material. These ions then transfer the electrolyte and undergo intercalation within the layered structure of the active material at the anode, where reduction occurs. To balance the reaction, the compensating electrons are received by the host while they simultaneously flow via the external circuit. The opposite process takes place after discharge, as presented in Figure 7.6.

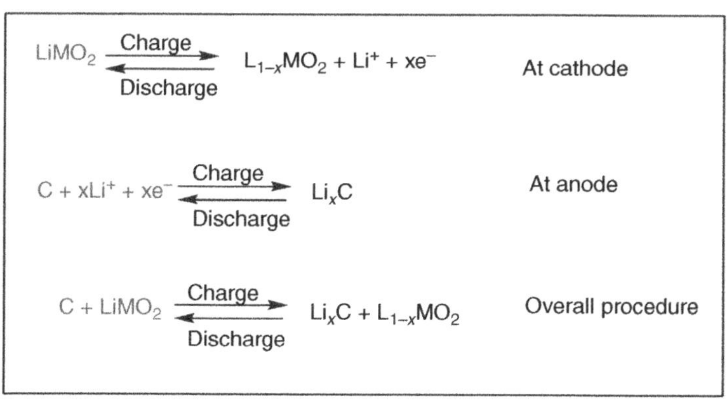

Figure 7.6 The charge and discharge processes of lithium-ion batteries involve reversible reaction pathways.

When the two electrodes are charged and discharged, Li^+ ions move back and forth between them in the reversible mechanism described above. Because of this, Li-ion cells are known as "rocking chairs" [80], "swing," or "shuttlecock" [81]. GDYs have a lot of potential applications in LIBs because of their huge specific surface area, wide spacing between layers, and outstanding capability to transport electrons [62]. The unique properties of GDY include its semiconducting behavior, porous 2D structure, asymmetrically dispersed π-electrons [34, 82, 83]. High conductivity and a huge specific surface area may provide an innovative possibility for lithium storage applications [84–86]. The specific capacities and lithiation potentials of GDY are much higher than graphite-based electrodes [86, 87]. DFT calculationsshow that the theoretical capacity, achieved when lithium is adsorbed on the surface due to the acetylene bonds in GDY, is double the value observed in graphene (744 mAh g^{-1}, LiC_3) [88]. The lithiation potentials of GDYare between 2.7 and 2.1 V. Furthermore, because of the huge pores in GDY, Li ions travel across the layers to produce an out-of-plane diffusion with a diffusion energy of 0.18–0.48 eV [89]. Using first-principle theory, the energy and kinetics of this process were investigated within a single-layer GDY [90, 91]. The findings revealed that lithium diffusion occurs both parallel to the plane of the GDY layer and within the pores. This characteristic renders GDY a suitable candidate for making the anode in LIBs [87]. A study reported a GDY-based electrode on the Cu surface without the use of a binder, that demonstrated excellent electrochemical characteristics, such as high specific capacities, efficient rate enhancement, and durability [92]. The researchers used GDY with different thicknesses to clarify the process behind the Li atom storage in multilayer GDY. Li atoms were mainly intercalated via the processes of surface absorption or desorption and interlayer insertion or extraction. Thus, further effort is required to improve the capabilities of GDY in storing Li. For example, it is important to optimize the arrangement of pores present between the GDY monolayer and sheet layer to enhance the rate of ion diffusion [93].

7.3.3 Supercapacitors

SCs, a novel class of energy conversion and storage devices, exhibit rapid charging and discharging capabilities, along with prolonged cycle life. They combine the quick energy exchange features of batteries with high-cycle durability. Energy is stored in SCs by generating a dual-layer boundary between the electrolyte and electrode. GYs are excellent anode materials for SCs owing to their large surface area and pore size [62]. There are two types of SCs: symmetric and asymmetric. A symmetric supercapacitor (SSC) has two identical electrodes, but an asymmetric supercapacitor (ASC) is generated when the two electrodes are made of different materials and electrolytes. Depending on the charge/discharge process of the

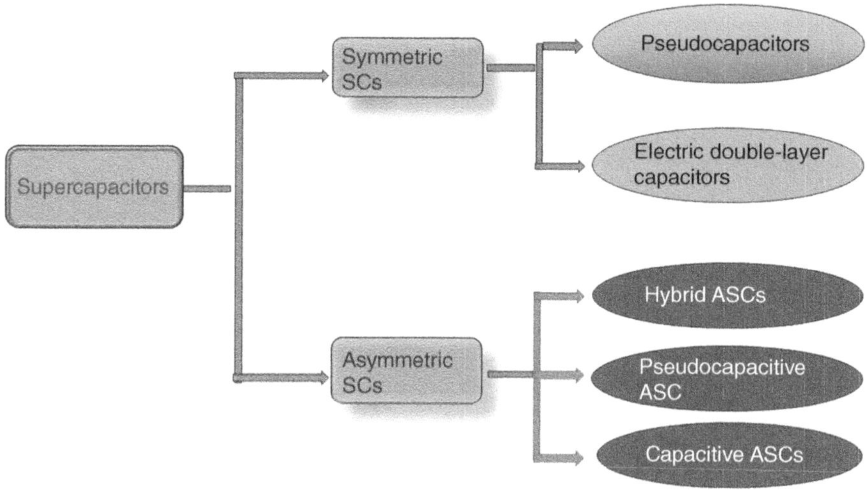

Figure 7.7 Classification of SCs.

electrodes, SSCs are classified as pseudocapacitors and electric double-layer capacitors (EDLCs). ASCs are an electrochemical energy storage device that releases energy through cycles of ion adsorption and desorption at the interface between electrolyte-electrode. With greater capacities, increased energy densities, higher power, and faster rate of charge-discharge, SCs are considered to be more effective devices than traditional capacitors. The literature categorized ASCs as either hybrid or capacitive ASCs [94]. The classification of SCs is illustrated in Figure 7.7.

SCs have become viable supplementary components in various applications, and in certain scenarios, they can function independently due to their exceptional electrochemical properties, such as extended life, greater power density, and reliable stability. Extensive studies and efforts are required to boost the energy densities of SCs, making them more suitable for broader commercial applications by aligning or approaching the performance level of comparable counterparts like rechargeable batteries. To increase its energy storage capacity, several techniques have been employed, like using conducting polymers, carbons as electrodes, and oxides of transition metal [95]. There is a growing demand for higher standards of energy and power supply in terms of durability, quality, and safety. In response to this demand, a new power grid called a microgrid has arisen, which smoothly incorporates distributed energy production. In this scenario, SCs have developed as an inventive energy storage device, offering short-term power supply and energy buffering capabilities. This ultimately contributes to improving the overall power quality of microgrids. Therefore, SCs are now one of the favored energy

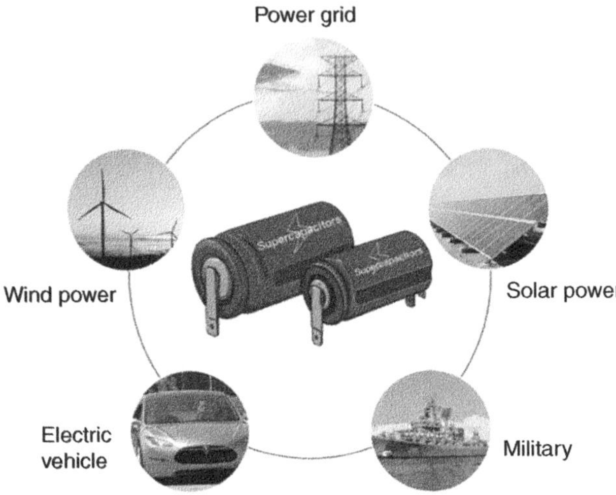

Figure 7.8 Application of SCs. *Source:* Ma et al. [100]/MDPI/CC BY 4.0.

storage devices in microgrid applications [96, 97]. SCs extend the life of batteries in electric or hybrid vehicles by acting as power sources. Furthermore, these are frequently used in distributed power production, photovoltaic, wind, and large-scale power storage systems [98–101]. Various applications of SCs are illustrated in Figure 7.8.

When GDY is used in SCs, its working process is different. For example, electrostatic double-layer capacitors (EDLCs) segregate charge at the interface between the electrolyte and conductive electrode surface in a Helmholtz double layer. Thus, the two key factors in enhancing its effectiveness are its potent adsorption capacity and huge specific surface area [102]. Mashhadzadeh et al. [103] used DFT to analyze the GDY adsorption capabilities of heavy metal ions. Their findings demonstrated that GDY with metal ions exhibited excellent chemisorption interactions. Researchers also investigated the adsorbing capacity of transition metals like Sc and Ti using DFT calculations [104]. The theoretical results provide a strong basis for GDY-based EDLCs, highlighting their strong adsorption ability for ions in EDLCs [89].

7.3.4 Wind Energy

Wind energy is a rich and renewable source and is used for green energy harvesting. The proper usage of wind energy is critical and often in traditional devices, there is a restriction of energy supply units. A triboelectric nanogenerator (TENG) is a passive device which uniquely strategizes energy production and detection of signals [105]. Passive devices are cost-effective, involve simple fabrication, and are highly energy-efficient. Carbon-based materials and their allotropes having unique physical properties have been used to fabricate nanostructures like carbon-involved TENG based on CNT and graphene [106] with substantial energy harvesting capacity.

A comparative study was done to know the effect of graphene sheet-embedded carbon (GSEC) and amorphous carbon films on the performance of TENG [107]. GDY-based materials due to their unique structure or specific morphology are expected to be of importance as an electrification material for TENG [108]. The structural design of the precursor during the growth process can alter the charge transfer ability of GDY [109] making it easy to design charged electrodes. Carbon material methyl-graphdiyne (M-GDY)-based nano thorn arrays as TENG core element have been designed and with its help, a new high-performance wind speed sensor having resistance toward water and cold is being developed, which can be applied even in typically extreme weather. M-GDY provides a convenient path for fast charge transfer due to its uniformly distributed methyl groups in a 2-D plane.The in-situ growing of M-GDY with nano thorn arrays on the surface of the Cu foil provides increased contact area and superhydrophobic properties endowing it with a wide applicability even in humid conditions [3]. It can also operate at different temperatures within the range of–20–30 °C.

7.4 Solar Energy

By using photovoltaic solar cells, the solar energy is transformed into electrical energy [110]. Some of the materials have a tendency to absorb light, which generates electron-hole pairs and facilitates the flow of these charge carriers, which is essential for efficient solar energy conversion [111]. In recent years, GY has attained a lot of importance after the first successful synthesis of GDY [112]. Due to the specific characteristics and tuneable electronic properties of GY or its analogues, it has been suggested that GY and GFMs are suitable for solar energy applications as they depend on their electronic band structure, conductivity, and other material properties [113].

Although graphene has been enormously explored for its possible application in solar cells due to its exceptional electrical conductivity and transparency, it isn't band-gap free, which is essential for effective solar energy conversion [114]. As the GY family is a modified version of graphene, it has the upper hand in their band-gap engineering and electrical characteristics.

GY and GFMs have been explored and used in solar cells [115]. Following are the important properties of graphyne and GFMs, which make them useful for solar cells [116].

- **High Electrical Conductivity**: These have good electrical conductivity, which is beneficial for charge transport in solar cells.
- **Tunable Band Gap**: The band gap of GY and GFMs can be tuned by modifying their structures, making it possible to optimize their absorption properties for solar energy conversion.
- **Light Absorption**: GY and GFMs have shown potential for efficient light absorption in the visible and near-infrared regions, which is crucial for solar energy conversion.
- **Flexibility and Lightweight**: GY and GFMs are lightweight and have good mechanical flexibility, which could be advantageous for the fabrication of flexible and lightweight solar cells.

In a study, it was presented that single-layer GY and graphene material is determined by various computational techniques are illustrated in Table 7.1. The one form of GY, a hexagonal geometry of 6,6,12-graphyne, lacks the symmetry found in polygons. The structure additionally exhibits anisotropy. The mobility of GR on it is much smaller than GY when the selected direction is fixed. Thus, it was concluded that GY has superior electrical properties than that of GR [117].

It has been investigated that GY and GY-nanoribbons exhibit band gaps within the semiconductor range of 0.59–1.25 eV. It was also shown that the mechanical

Table 7.1 The bandgap and intrinsic hole and electron mobilities (at 300 K) of a single-layer graphyne material determined using various computational techniques.

Carbon allotropes	Hole mobility μh ($10^4 cm^2 V^{-1} s^{-1}$)	Electron mobility μ^e($10^4 cm^2 V^{-1} s^{-1}$)	Bandgap (eV)	Reference
α-Graphyne	3.316^a, 2.960^b, 0.96	3.327^a, 2.716^b, 1.03	0	[117, 118]
β-Graphyne	1.076^a, 0.856^b	0.892^a, 0.798^b	0	[118, 119]
γ-Graphyne	0.35	1.62	0.454, 0.47, 0.471, 0.48, 0.52, 1.32, 2.23	[118–121, 10]
6,6,12-Graphyne	42.92^a, 12.29^b	54.10^a, 24.48^b	0	[117]
γ-Graphdiyne	1.97^a, 1.91^b	20.81^a, 17.22^b	0.44–1.47	[30, 34, 121–125]
Graphyne	32.17^a, 23.49	33.89^a, 32.02^b, 20.05	0	[126]

[a] Zigzag direction.
[b] Armchair direction.

strain affects band gap of infinite sheets of graphyne.The band gap increases by 1 eV under a tensile strain of 0.15, while a compressive strain of 0.1 decreases it by 0.3 eV.Thus, this provides a driving force to use GY-nanoribbons in solar cells [115, 127].

In another approach, surface modification was done and some GY-based nanostructures, especially GDY, were synthesized, which may have better properties for solar cell construction [128, 129].

In a theoretical study, it was reported that due to GDY's strong charge transport capacity and the creation of effective percolation pathways in the active layer, doping P3HT/PCBM solar cells can increase their short circuit current (Jsc) and power conversion efficiency (PCE) (Figure 7.9). This can lead to better and more efficient solar cell production [130, 131].

In an experimental study, it was reported that GDY linked with a benzene ring showed excellent optoelectronic properties. This is due to extensive conjugation between π-bonds of the ring and —C C-linkages. It was further investigated that it is self-doped and has two distinct Dirac cones close to the Fermi level and,

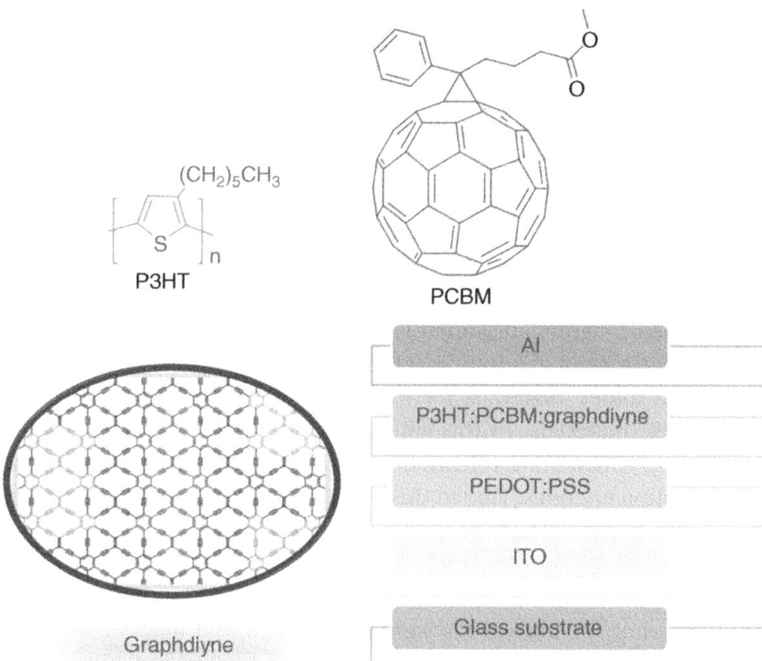

Figure 7.9 Chemical structures of P3HT, PCBM, and GDY and the structure of a GDY-containing photovoltaic device.

therefore, can be an excellent semiconductor material for electronic devices and solar energy cells [132]. In a theoretical study based on DFT simulation, a semiconductor-based device utilizing GY and GFMs films for electrical property measurement demonstrated a conductivity of 2.516×10^{-4} Sm^{-1}. Furthermore, it was also suggested that the devices mesh well with GY high conjugation with carbon-rich organic molecules can show optoelectronic and tunable structural properties and hence can be good candidates for the development of next-generation electronic and optoelectronic devices [133]. In another theoretical approach based on quantum mechanics, due to the unique atomic configuration in GY, it was described as the nonlinear stress-strain behavior; additionally, the acetylenic linkages provided tunable resilience. Thus, there is a possibility that a mechanically induced phase transition from Gra to GDY may be caused by in-plane. Hence, these systems may be promising materials for hydrogen storage units, battery anodes, energy-storage modules, and nanoelectronic devices [134].

In an approach through a cross-coupling reaction, HsGDY was synthesized. Due to an extended π-conjugated carbon skeleton with a porous structure and an aromatic hydrogen (Ar-H) group, HsGDY has a large number of active ion-binding sites, good electrochemical conductivity and a high specific surface area. Moreover, it was also investigated that this can be used in making the flexible free-standing electrode having a remarkably enhanced reversible and stable capacity of $1050\,mAh\,g^{-1}$ for LIBs, surpassing that of organic molecules and intrinsic carbon allotropes. Thus, it was proposed that the SIBs have the highest reversible capacity of any reported flexible electrode, reaching up to $650\,mAh\,g^{-1}$ after 100 cycles. In conclusion, HsGDY can be a promising option for LIBs and SIBs because of its superior rate capability, high reversible capacity, and excellent cycle performance [135].

In another study, it was reported that the band gap in the conjugated GY and GFMs can be tuned by changing the —C C-linking chain length. Thus, these materials can accommodate significant amounts of lithium with significantly higher lithiation potentials (in comparison to Li/Li^{+}) and specific capacities. It was further suggested that anode materials for lithium batteries could be effectively designed using these materials. In this study, these materials' lithium sites are found to be cationic, which allows them to adsorb molecular hydrogen with the appropriate adsorption energies. Due to this, electrons transfer from the lithium metal to the carbon surface. These properties make GFMs a candidate for solar energy cells [136].

In a recent approach, single-atom vacancy-induced changes in magnetic and electronic properties of GY and GFMs were investigated and it was concluded thatthe presence of extensive (—C C—) linkage enhances the capability of Li storage and also reduces the diffusion barrier of Li. Thus, GY can increase the performance of Li-ion batteries [137, 138].

7.5 Wastewater Treatment

Although the earth has an abundance of water, approximately 98% of the earth's accessible water resources are salty/saline. Due to rising water demands and the deteriorating quantity and quality of available water resources, as well as global population growth, water will become one of the most valuable resources in the future [139]. Water (H_2O) is an integrated part of life; without water, life is not possible. Due to various human activities, many pollutants are released into the soil, water, and atmosphere, endangering human health and lowering quality of life. To get fresh and drinkable water, desalination is one of the best methods. In desalination, salt ions are separated from saline water and thus convert into clean water, and it is suggested that this technique can be used to effectively address the shortage of water. In this instance, a membrane's function is essential to any desalination procedure. GY and GFMs are nanoporous structures for a variety of uses, including catalysis, gas separation, water desalination, and energy storage [111].

A few membranes based on GY were built to desalinate seawater. It was suggested that graphyne-3 membranes are capable of withstanding a wide range of salts and pollutants, including KCl, NaCl, $CuSO_4$, C_6H_6, and CCl_4. These membranes are therefore excellent choices for filtering contaminated water and desalinating seawater. It was suggested that graphyne-4 makes an excellent purification membrane when rejection rate and permeability are taken into account [112]. Furthermore, it has been proposed that negative charging can enhance water permeability and salt rejection even more. In addition to the GY-n structures, α- and β-GYs exhibit remarkable water desalination performance, exhibiting high water permeability and a 100% rejection rate [140].

In another study by a simple one-pot hydrothermal method, sulfur-doped graphene (SGDY) nanosheets were synthesized (Figure 7.10), which showed piezoelectric and peroxidase-like activities. Studies on the effects of SGDY on bacteria and dyes showed that it has great potential for use in wastewater treatment [141].

In another approach, it was proposed that due to —C=C— and —C— in GY and GFMs,they create holes, which are enough to permit the passage of water but not permeable for Na^+ and Cl^- ions. By taking inspiration, pristine-GY was made which has remarkably high water permeability. It can achieve 100% rejection of ions in seawater, such as Cl^-, Ca^{2+}, Na^+, Mg^{2+},and K^+ ions [115]. In wastewater treatment, radioactive ion removal is rarely taken into consideration; solar-driven steam generation is thought to be an environmentally friendly and sustainable strategy. Some GFMs were fabricated with hollow multishelled structures (HoMS) using the bottom-up synthesis approach, in which GDY-HoMS has achieved a stable evaporation rate of $2.7\,kg\,m^{-2}\,h^{-1}$. Thus, the heavy metallic elements were

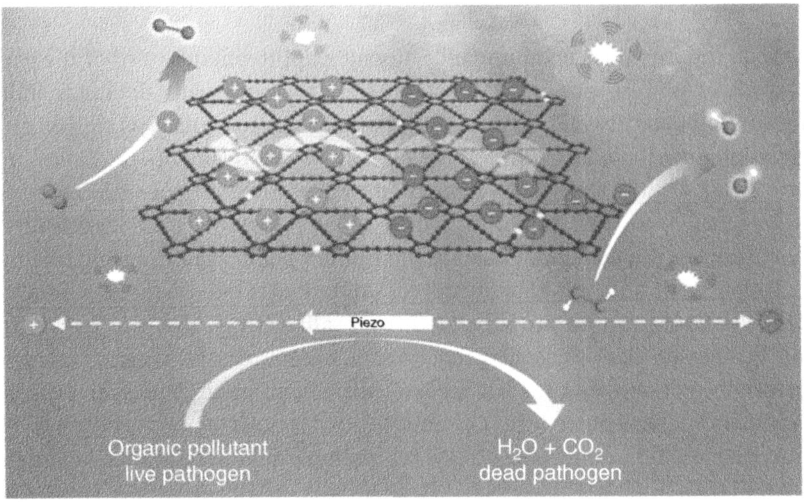

Figure 7.10 Sulfur-doped graphdiyne (SGDY) nanosheet showing dye and antibiotic degradation with antibacterial efficiency.

intercepted in the GDY film due to the interaction between the sp-C of GDY and the empty orbit of metallic atoms, resulting in an extremely effective ability to remove ions (Figure 7.11) [142].

Water bodies are constantly being contaminated by humans due to their ill activities. The main causes of water pollution are agriculture, inadequate water supplies or sewage treatment facilities, and industrialization. Industrial activities like nuclear power plants, distilleries, textile and food industries, tanneries, iron and steel industries, pulp and paper mills, and so forthare the primary sources of water pollution [143].

In this regard, one of the major causes of contamination of water is the production of fiber in the textile industry, which evolves a lot of contaminants including heavy metals, fibers, detergents, sulfides, nitro compounds, and dyes (such as anthraquinone, azo, triarylmethane, phthalocyanine, and indigo and sulphur dyes) [144].

Due to acetylenic linkages, GY and GFMs can interact with heavy metals and thus can be a good candidate for the removal of heavy metals as they can be adsorbed on the surface with acetylene linkages. In a study, it was reported that γ-GY-2 shows strong interaction with metal ions and has the capability of adsorption, which helps in the removal of heavy metal ions [117, 145].

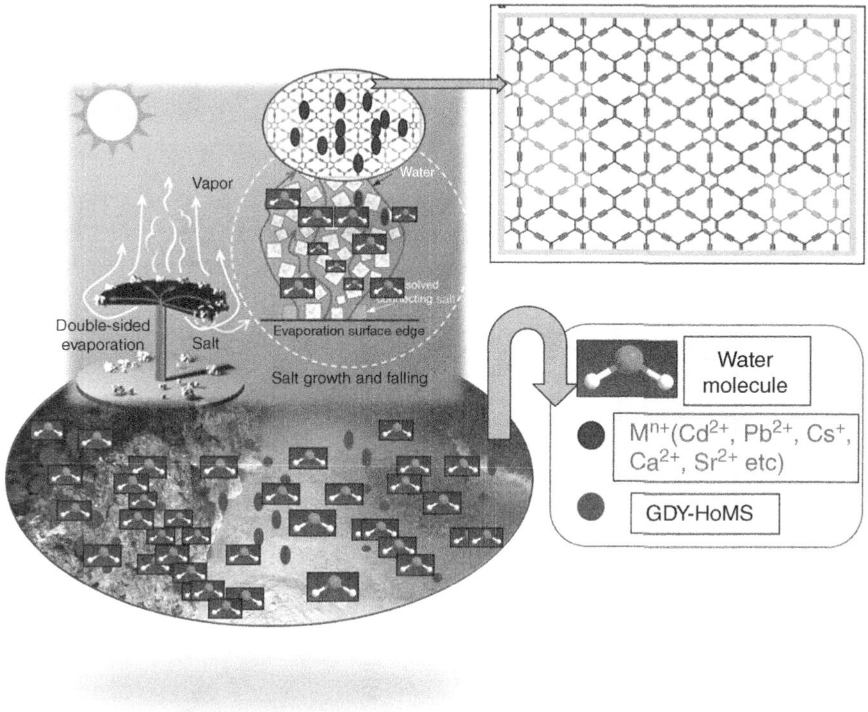

Figure 7.11 Schematic illustration of GDYs, the strong interaction between metallic ions and GDY contributes to GDY-HoMS's ultra-effective ion-removing ability. *Source:* Zhan et al. [142]/with permission of Elsevier.

In another approach, it was proposed that because of the presence of butadiyne and aromatic hydrogen (Ar–H) group, hydrogen-substituted graphdiyne (HsGDY) has good lead ion adsorption performance. The calculated Langmuir isotherm model has a maximum adsorption capacity of $2390\,mg\,g^{-1}$, which is almost five times greater as compared to GDY's reported results. According to DFT modelling, lead is primarily adsorbable through the adjacent acetylenic bond sites in the HsGDY plane, and the bonding strength with lead can be strengthened by the presence of hydrogen atoms [146, 147].

In a study, it was reported that the diacetylene linkage of graphyne dye (GD) can undergo hydrothermal reactions and partially transform into a 2D π-conjugated structure that is advantageous for electronic transmission. This property makes GD useful as an electron transport material in photodegradation processes. Further, it was presented that using graphyne oxide (GO) as a cross-linkage,

Table 7.2 Some reported studies on GFM membranes for water purification and desalination.

GFM membrane material type	Molecular medium	Rejected ions/salts	Permeability	Rejection (%)	Pressure applied (MPa)	Computational methods
$\alpha-/\beta-$GY, GDY,	Water	Na^+, Cl^-, Mg^{2+}, K^+, Ca^{2+}	~10 l cm^{-2} day^{-1} MPa^{-1}	100% rejection for GY-3 and a little lower for GY-4	0–250	MD
GY-3/-4/-5/-6	Water	NaCl, $CuSO_4$, C_6H_6, CCl_4	2.5–4.5 × 10^{-9} MPa^{-1} s^{-1}	75% above rejection for all the graphyne-*n* tested	50	MD
GY-3/-4/-5/-6	Water	Na^+, Cl^-	39.15 l cm^{-2} h^{-1}	100% rejection for GY-3 and lower for other GY-*n* membranes tested	NS	MD
Pristine/charged GY-3/-4/-5	Water	Na^+, Cl^-	13.5 l cm^{-2} day^{-1} MPa^{-1}	100% rejection for GY-3 and lower for other GY-*n* membranes tested	0–150	MD
Bare and hydrogenated α-GY and GY-2/-3/-4	Water	Na^+, Cl^-	85 l cm^{-2} day^{-1} MPa^{-1}	>90% rejection for all tested membranes except GY-4	0–1000	MD
Functionalized ¥-GY-1	Water	Na^+, Cl^-	8953 l m^{-2} h bar	100% rejection for all the functionalized γ-GY-1 membranes tested at <7.5 MPa	0–50	MD
Anisotropically nanoporous GY and GY-3/-4/-5	Water	Na^+, Cl^-	7.98–47.14 l cm^{-2} day^{-1} MPa^{-1}	100% rejection for GY-3 and ANGM-*n* ($n = 1, 2, 3$) membranes and others except GY-5, could reject >98.2% of ions up to 150 MPa	50–250	MD
Functionalized graphenylene	Water	Na^+, Cl^-	11,032 l m^{-2} h bar	99.4% rejection at 10 MPa for fluorinated functionalized pore	0–100	MD

GFMs, graphyne family members; GY, graphyne; MD, molecular dynamics; NS, Not supplied.
Source: Lasisi et al. [149]/Frontiers Media S.A./CC BY 4.0.

GDY can also be easily hybridized with Ag/AgBr. Thus, according to the results obtained, it was concluded that Ag/AgBr/GO/GD shows improved photocatalytic performance toward the degradation of methyl orange pollutant under visible light irradiation, in comparison to Ag/AgBr, Ag/AgBr/GO, and Ag/AgBr/GD [148]. In Table 7.2, some GFMs for water purification and desalination from computational and theoretical studies are presented [149].

7.6 Perspectives and Conclusion

The graphyne-based nanomaterials have been shown to possess several physical, chemical, and mechanical applications. GY and its associate GDY have unleashed useful properties. The different bonding patterns in GY and electronic properties impart rich properties even in the pristine state and the properties can be further modified by introducing vacancies, atoms adsorbed on the surface, mechanical strains and external electric fields that provide opportunities for further properties like band gap. There are still a few gaps in research like the bulk synthesis of GY and clarification on the band gap. GY and GDY also provide opportunities by doping through adsorption and substitution of other structures. GY's properties could make it more applicable to electronics than graphene. Although these properties have been mainly simulated, and need exploration through experimentation.

Acknowledgments

The authors are thankful to Professor Abdul Rahman Khan, Head, Department of Chemistry, Integral University for his support.

References

1 Kong, X.Y., Ding, Y., Yang, R., and Wang, Z.L. (2004). Single-crystal nanorings formed by epitaxial self-coiling of polar nanobelts. *Science* 303 (5662): 1348–1351. https://doi.org/10.1126/science.1092356.

2 Hirsch, A. (2010). The era of carbon allotropes. *Nature Materials* 9 (11): 868–871. https://doi.org/10.1038/nmat2885.

3 Li, X., Li, Y., Zhang, M. et al. (2021). Carbon nano thorn arrays based water/cold resisted nanogenerator for wind energy harvesting and speed sensing. *Nano Energy* 90: 106571. https://doi.org/10.1016/j.nanoen.2021.106571.

4 Cranford, S.W. and Buehler, M.J. (2012). Selective hydrogen purification through graphdiyne under ambient temperature and pressure. *Nanoscale* 4 (15): 4587–4593. https://doi.org/10.1039/C2NR30921A.

5 Zhang, S., Du, H., He, J. et al. (2016). Nitrogen-doped graphdiyne applied for lithium-ion storage. *ACS Applied Materials & Interfaces* 8 (13): 8467–8473. https://doi.org/10.1021/acsami.6b00255.

6 Geim, A.K. and Novoselov, K.S. (2007). The rise of graphene. *Nature Materials* 6 (3): 183–191. https://doi.org/10.1038/nmat1849.

7 Yamoah, M.A., Yang, W., Pop, E., and Goldhaber-Gordon, D. (2017). High-velocity saturation in graphene encapsulated by hexagonal boron nitride. *ACS Nano* 11 (10): 9914–9919. https://doi.org/10.1021/acsnano.7b03878.

8 Liao, L., Lin, Y.C., Bao, M. et al. (2010). High-speed graphene transistors with a self-aligned nanowire gate. *Nature* 467 (7313): 305–308. https://doi.org/10.1038/nature09405.

9 Malko, D., Neiss, C., Vines, F., and Görling, A. (2012). Competition for graphene: graphynes with direction-dependent dirac cones. *Physical Review Letters* 108 (8): 086804. https://doi.org/10.1103/PhysRevLett.108.086804.

10 Srinivasu, K. and Ghosh, S.K. (2012). Graphyne and graphdiyne: promising materials for nanoelectronics and energy storage applications. *Journal of Physical Chemistry C* 116 (9): 5951–5956. https://doi.org/10.1021/jp212181h.

11 Baughman, R.H., Eckhardt, H., and Kertesz, M. (1987). Structure-property predictions for new planar forms of carbon: Layered phases containing sp 2 and sp atoms. *The Journal of Chemical Physics* 87 (11): 6687–6699. https://doi.org/10.1063/1.453405.

12 Bhattacharya, B. and Sarkar, U. (2016). Graphyne -graphene (nitride) heterostructure as Nanocapacitor. *Chemical Physics* 478: 73–80. https://doi.org/10.1016/j.chemphys.2016.05.004.

13 Bhattacharya, B., Sarkar, U., and Seriani, N. (2016). Electronic properties of homo-and heterobilayer graphyne: The idea of a nanocapacitor. *Journal of Physical Chemistry C* 120 (47): 26579–26587. https://doi.org/10.1021/acs.jpcc.6b07092.

14 Lu, R., Rao, D., Meng, Z. et al. (2013). Boron-substituted graphyne as a versatile material with high storage capacities of Li and H 2: a multiscale theoretical study. *Physical Chemistry Chemical Physics* 15 (38): 16120–16126. https://doi.org/10.1039/C3CP52364K.

15 Pei, Y. (2012). Mechanical properties of graphdiyne sheet. *Physica B: Condensed Matter* 407 (22): 4436–4439. https://doi.org/10.1016/j.physb.2012.07.026.

16 Diederich, F. (1994). Carbon scaffolding: building acetylenic all-carbon and carbon-rich compounds. *Nature* 369 (6477): 199–207. https://doi.org/10.1038/369199a0.

17 Haley, M.M. and 2008 (2009). Synthesis and properties of annulenic subunits of graphyne and graphdiyne nanoarchitectures. *Pure and Applied Chemistry* 80 (3): 519–532. https://doi.org/10.1351/pac200880030519.

18 Li, G., Li, Y., Liu, H. et al. (2010). Architecture of graphdiyne nanoscale films. *Chemical Communications* 46 (19): 3256–3258. https://doi.org/10.1039/B922733D.

19 Li, Q., Li, Y., Chen, Y. et al. (2018). Synthesis of γ-graphyne by mechanochemistry and its electronic structure. *Carbon* 136: 248–254. https://doi.org/10.1016/j.carbon.2018.04.081.

20 Lu, J., Chen, Z., Pan, F. et al. (2018). High-performance anode materials for rechargeable lithium-ion batteries. *Electrochemical Energy Reviews* 1: 35–53. https://doi.org/10.1007/s41918-018-0001-4.

21 Khan, M.A., Zhao, H., Zou, W. et al. (2018). Recent progresses in electrocatalysts for water electrolysis. *Electrochemical Energy Reviews* 1: 483–530. https://doi.org/10.1007/s41918-018-0014-z.

22 Yang, G., Li, L., Lee, W.B., and Ng, M.C. (2018). Structure of graphene and its disorders: a review. *Science and Technology of Advanced Materials* 19 (1): 613–648. https://doi.org/10.1080/14686996.2018.1494493.

23 Zheng, X., Chen, S., Li, J. et al. (2023). Two-dimensional carbon graphdiyne: advances in fundamental and application research. *ACS Nano* 17 (15): 14309–14346. https://doi.org/10.1021/acsnano.3c03849.

24 Kang, B., Wu, S., Ma, J. et al. (2019). Synergy of sp-N and sp 2-N codoping endows graphdiyne with comparable oxygen reduction reaction performance to Pt. *Nanoscale* 11 (35): 16599–16605. https://doi.org/10.1039/C9NR05363H.

25 Lv, Q., Si, W., He, J. et al. (2018). Selectively nitrogen-doped carbon materials as superior metal-free catalysts for oxygen reduction. *Nature Communications* 9 (1): 3376. https://doi.org/10.1038/s41467-018-05878-y.

26 Yang, Z., Shen, X., Wang, N. et al. (2018). Graphdiyne containing atomically precise N atoms for efficient anchoring of lithium ion. *ACS Applied Materials & Interfaces* 11 (3): 2608–2617. https://doi.org/10.1021/acsami.8b01823.

27 He, J., Wang, N., Yang, Z. et al. (2018). Fluoride graphdiyne as a free-standing electrode displaying ultra-stable and extraordinary high Li storage performance. *Energy & Environmental Science* 11 (10): 2893–2903. https://doi.org/10.1039/C8EE01642A.

28 Gao, X., Dong, Y., Li, S. et al. (2020). MOFs and COFs for batteries and supercapacitors. *Electrochemical Energy Reviews* 3: 81–126. https://doi.org/10.1007/s41918-019-00055-1.

29 Si, W., Yang, Z., Wang, X. et al. (2019). Fe, N-codoped graphdiyne displaying efficient oxygen reduction reaction activity. *ChemSusChem* 12 (1): 173–178. https://doi.org/10.1002/cssc.201802170.

30 Narita, N., Nagai, S., Suzuki, S., and Nakao, K. (1998). Optimized geometries and electronic structures of graphyne and its family. *Physical Review B: Condensed Matter and Materials Physics* 58 (16): 11009. https://doi.org/10.1103/PhysRevB.58.11009.

31 Pan, Y., Wang, Y., Wang, L. et al. (2015). Graphdiyne–metal contacts and graphdiyne transistors. *Nanoscale* 7 (5): 2116–2127. https://doi.org/10.1039/C4NR06541G.

32 Zhang, M., Li, Y., Li, X. et al. (2020). Graphdiyne ink for ionic liquid gated printed transistor. *Advanced Electronic Materials* 6 (7): 2000157. https://doi.org/10.1002/aelm.202000157.

33 Li, Y., Zhang, M., Hu, X. et al. (2020). Light and heat triggering modulation of the electronic performance of a graphdiyne-based thin film transistor. *Journal of Physical Chemistry Letters* 11 (6): 1998–2005. https://doi.org/10.1021/acs.jpclett.0c00113.

34 Long, M., Tang, L., Wang, D. et al. (2011). Electronic structure and carrier mobility in graphdiyne sheet and nanoribbons: theoretical predictions. *ACS Nano* 5 (4): 2593–2600. https://doi.org/10.1021/nn102472s.

35 Lv, Q., Si, W., Yang, Z. et al. (2017). Nitrogen-doped porous graphdiyne: a highly efficient metal-free electrocatalyst for oxygen reduction reaction. *ACS Applied Materials & Interfaces* 9 (35): 29744–29752. https://doi.org/10.1021/acsami.7b08115.

36 Wang, X., Yang, Z., Si, W. et al. (2019). Cobalt-nitrogen-doped graphdiyne as an efficient bifunctional catalyst for oxygen reduction and hydrogen evolution reactions. *Carbon* 147: 9–18. https://doi.org/10.1016/j.carbon.2019.02.033.

37 Shang, H., Zuo, Z., Zheng, H. et al. (2018). N-doped graphdiyne for high-performance electrochemical electrodes. *Nano Energy* 44: 144–154. https://doi.org/10.1016/j.nanoen.2017.11.072.

38 Zhou, W., Shen, H., Wu, C. et al. (2018). Direct synthesis of crystalline graphdiyne analogue based on supramolecular interactions. *Journal of the American Chemical Society* 141 (1): 48–52. https://doi.org/10.1021/jacs.8b09945.

39 Mekhilef, S., Saidur, R., and Safari, A. (2012). Comparative study of different fuel cell technologies. *Renewable and Sustainable Energy Reviews* 16 (1): 981–989. https://doi.org/10.1016/j.rser.2011.09.020.

40 Iqbal, M.Z., Siddique, S., Khan, A. et al. (2020). Recent developments in graphene based novel structures for efficient and durable fuel cells. *Materials Research Bulletin* 122: 110674. https://doi.org/10.1016/j.materresbull.2019.110674.

41 Loyselle, P. and Prokopius, K. (2011). Teledyne Energy Systems, Inc., Proton Exchange Member (PEM) Fuel Cell Engineering Model Powerplant. Test Report: Initial Benchmark Tests in the Original Orientation (No. NASA/TM-2011-216225).

42 Dai, L., Xue, Y., Qu, L. et al. (2015). Metal-free catalysts for oxygen reduction reaction. *Chemical Reviews* 115 (11): 4823–4892. https://doi.org/10.1021/cr5003563.

43 Lim, B., Jiang, M., Camargo, P.H. et al. (2009). Pd-Pt bimetallic nanodendrites with high activity for oxygen reduction. *Science* 324 (5932): 1302–1305. https://doi.org/10.1126/science.1170377.

44 Lee, Y., Suntivich, J., May, K.J. et al. (2012). Synthesis and activities of rutile IrO2 and RuO2 nanoparticles for oxygen evolution in acid and alkaline solutions.

Journal of Physical Chemistry Letters 3 (3): 399–404. https://doi.org/10.1021/jz2016507.

45 Zhai, Y., Dou, Y., Zhao, D. et al. (2011). Carbon materials for chemical capacitive energy storage. *Advanced Materials* 23 (42): 4828–4850. https://doi.org/10.1002/adma.201100984.

46 Duan, J., Chen, S., Jaroniec, M., and Qiao, S.Z. (2015). Heteroatom-doped graphene-based materials for energy-relevant electrocatalytic processes. *ACS Catalysis* 5 (9): 5207–5234. https://doi.org/10.1021/acscatal.5b00991.

47 Wang, X., Sun, G., Routh, P. et al. (2014). Heteroatom-doped graphene materials: syntheses, properties and applications. *Chemical Society Reviews* 43 (20): 7067–7098. https://doi.org/10.1039/C4CS00141A.

48 Zhang, C., Sha, J., Fei, H. et al. (2017). Single-atomic ruthenium catalytic sites on nitrogen-doped graphene for oxygen reduction reaction in acidic medium. *ACS Nano* 11 (7): 6930–6941. https://doi.org/10.1021/acsnano.7b02148.

49 Su, H. and Hu, Y.H. (2021). Recent advances in graphene-based materials for fuel cell applications. *Energy Science & Engineering* 9 (7): 958–983. https://doi.org/10.1002/ese3.833.

50 Kang, B. and Lee, J.Y. (2014). Graphynes as promising cathode material of fuel cell: improvement of oxygen reduction efficiency. *The Journal of Physical Chemistry C* 118 (22): 12035–12040. https://doi.org/10.1021/jp502780y.

51 Wu, P., Du, P., Zhang, H., and Cai, C. (2012). Graphyne as a promising metal-free electrocatalyst for oxygen reduction reactions in acidic fuel cells: a DFT study. *Journal of Physical Chemistry C* 116 (38): 20472–20479. https://doi.org/10.1021/jp3074305.

52 Cheng, F. and Chen, J. (2012). Metal–air batteries: from oxygen reduction electrochemistry to cathode catalysts. *Chemical Society Reviews* 41 (6): 2172–2192. https://doi.org/10.1039/C1CS15228A.

53 Fu, J., Cano, Z.P., Park, M.G. et al. (2017). Electrically rechargeable zinc–air batteries: progress, challenges, and perspectives. *Advanced Materials* 29 (7): 1604685. https://doi.org/10.1002/adma.201604685.

54 Wang, H.F., Tang, C., and Zhang, Q. (2018). A review of precious-metal-free bifunctional oxygen electrocatalysts: rational design and applications in Zn– air batteries. *Advanced Functional Materials* 28 (46): 1803329. https://doi.org/10.1002/adfm.201803329.

55 Yoshida, T. and Kojima, K. (2015). Toyota MIRAI fuel cell vehicle and progress toward a future hydrogen society. *Electrochemical Society Interface* 24 (2): 45–49. https://doi.org/10.1149/2.F03152if.

56 Kojima, K. and Fukazawa, K. (2015). Current status and future outlook of fuel cell vehicle development in Toyota. *ECS Transactions* 69 (17): 213–219. https://doi.org/10.1149/06917.0213ecst.

57 Holewinski, A., Idrobo, J.C., and Linic, S. (2014). High-performance Ag–Co alloy catalysts for electrochemical oxygen reduction. *Nature Chemistry* 6 (9): 828–834. https://doi.org/10.1038/nchem.2032.
58 Whiston, M.M., Azevedo, I.L., Litster, S. et al. (2019). Expert assessments of the cost and expected future performance of proton exchange membrane fuel cells for vehicles. *Proceedings of the National Academy of Sciences of the United States of America* 11 (6): 4899–4904. https://doi.org/10.1073/pnas.1804221116.
59 Lu, Y.C., Gallant, B.M., Kwabi, D.G. et al. (2013). Lithium–oxygen batteries: bridging mechanistic understanding and battery performance. *Energy & Environmental Science* 6 (3): 750–768. https://doi.org/10.1039/C3EE23966G.
60 Bai, P., Viswanathan, V., and Bazant, M.Z. (2015). A dual-mode rechargeable lithium–bromine/oxygen fuel cell. *Journal of Materials Chemistry A* 3 (27): 14165–14172. https://doi.org/10.1039/C5TA03335G.
61 Pointon, K., Lakeman, B., Irvine, J. et al. (2006). The development of a carbon–air semi fuel cell. *Journal of Power Sources* 162 (2): 750–756. https://doi.org/10.1016/j.jpowsour.2005.07.023.
62 Gong, Y., Shen, L., Kang, Z. et al. (2020). Progress in energy-related graphyne-based materials: advanced synthesis, functional mechanisms and applications. *Journal of Materials Chemistry A* 8 (41): 21408–21433. https://doi.org/10.1039/D0TA08521A.
63 Kang, H., Chen, Y., Xu, L. et al. (2019). Top-down strategy synthesis of fluorinated graphdiyne for lithium ion battery. *RSC Advances* 9 (54): 31406–31412. https://doi.org/10.1039/C9RA05974A.
64 Lv, Q., Wang, N., Si, W. et al. (2020). Pyridinic nitrogen exclusively doped carbon materials as efficient oxygen reduction electrocatalysts for Zn-air batteries. *Applied Catalysis, B: Environmental* 261: 118234. https://doi.org/10.1016/j.apcatb.2019.118234.
65 Chiang, Y.M. (2010). Building a better battery. *Science* 330 (6010): 1485–1486. https://doi.org/1126/science.1198591.
66 Colclasure, A.M., Dunlop, A.R., Trask, S.E. et al. (2019). Requirements for enabling extreme fast charging of high energy density Li-ion cells while avoiding lithium plating. *Journal of the Electrochemical Society* 166 (8): A1412. https://doi.org/10.1149/2.0451908jes.
67 Kim, N., Chae, S., Ma, J. et al. (2017). Fast-charging high-energy lithium-ion batteries via implantation of amorphous silicon nanolayer in edge-plane activated graphite anodes. *Nature Communications* 8 (1): 1–10. https://doi.org/10.1038/s41467-017-00973-y.
68 Yang, X.G. and Wang, C.Y. (2018). Understanding the trilemma of fast charging, energy density and cycle life of lithium-ion batteries. *Journal of Power Sources* 402: 489–498. https://doi.org/10.1016/j.jpowsour.2018.09.069.

69 Xu, W., Wang, J., Ding, F. et al. (2014). Lithium metal anodes for rechargeable batteries. *Energy & Environmental Science* 7 (2): 513–537. https://doi.org/10.1039/C3EE40795K.

70 Tarascon, J.M. and Armand, M. (2001). Issues and challenges facing rechargeable lithium batteries. *Nature* 414 (6861): 359–367. https://doi.org/10.1038/35104644.

71 Li, G., Gao, Y., He, X. et al. (2017). Organosulfide-plasticized solid-electrolyte interphase layer enables stable lithium metal anodes for long-cycle lithium-sulfur batteries. *Nature Communications* 8 (1): 850. https://doi.org/10.1038/s41467-017-00974-x.

72 Bruce, P.G., Freunberger, S.A., Hardwick, L.J., and Tarascon, J.M. (2012). Li–O2 and Li–S batteries with high energy storage. *Nature Materials* 11 (1): 19–29. https://doi.org/10.1038/nmat3191.

73 Liu, J., Bao, Z., Cui, Y. et al. (2019). Pathways for practical high-energy long-cycling lithium metal batteries. *Nature Energy* 4 (3): 180–186. https://doi.org/10.1038/s41560-019-0338-x.

74 Liu, Y., Zhu, Y., and Cui, Y. (2019). Challenges and opportunities towards fast-charging battery materials. *Nature Energy* 4 (7): 540–550. https://doi.org/10.1038/s41560-019-0405-3.

75 Scrosati, B. and Garche, J. (2010). Lithium batteries: status, prospects and future. *Journal of Power Sources* 195 (9): 2419–2430. https://doi.org/10.1016/j.jpowsour.2009.11.048.

76 Tomaszewska, A., Chu, Z., Feng, X. et al. (2019). Lithium-ion battery fast charging: a review. *ETransportation* 1: 100011. https://doi.org/10.1016/j.etran.2019.100011.

77 Dahn, J.R. (1991). Phase diagram of Li xC 6. *Physical Review B* 44 (17): 9170–9177. https://doi.org/10.1103/PhysRevB.44.9170.

78 Satoh, A., Takami, N., and Ohsaki, T. (1995). Electrochemical intercalation of lithium into graphitized carbons. *Solid State Ionics* 80 (3-4): 291–298. https://doi.org/10.1016/0167-2738(95)00149-Z.

79 Stura, E. and Nicolini, C. (2006). New nanomaterials for light weight lithium batteries. *AnalyticaChimica Acta* 568 (1–2): 57–64. https://doi.org/10.1016/j.aca.2005.11.025.

80 Scrosati, B. (1992). Lithium rocking chair batteries: an old concept? *Journal of the Electrochemical Society* 139 (10): 2776. https://doi.org/10.1149/1.2068978.

81 Ohzuku, T., Iwakoshi, Y., and Sawai, K. (1993). Formation of lithium-graphite intercalation compounds in nonaqueous electrolytes and their application as a negative electrode for a lithium ion (shuttlecock) cell. *Journal of the Electrochemical Society* 140 (9): 2490. https://doi.org/10.1149/1.2220849.

82 Nayebi, P. and Zaminpayma, E. (2017). Metal decorated graphdiyne: a first principle study. *Physica B: Condensed Matter* 521: 112–121. https://doi.org/10.1016/j.physb.2017.06.065.

83 Bu, H., Zhao, M., Zhang, H. et al. (2012). Isoelectronic doping of graphdiyne with boron and nitrogen: stable configurations and band gap modification. *The Journal of Physical Chemistry. A* 116 (15): 3934–3939. https://doi.org/10.1021/jp300107d.

84 Shang, H., Zuo, Z., Yu, L. et al. (2018). Low-temperature growth of all-carbon graphdiyne on a silicon anode for high-performance lithium-ion batteries. *Advanced Materials* 30 (27): 1801459. https://doi.org/10.1002/adma.201801459.

85 Lin, Z., Liu, G., Zheng, Y. et al. (2018). Three-dimensional hierarchical mesoporous flower-like TiO_2@ graphdiyne with superior electrochemical performances for lithium-ion batteries. *Journal of Materials Chemistry A* 6 (45): 22655–22661. https://doi.org/10.1039/C8TA08225A.

86 Mohajeri, A. and Shahsavar, A. (2016). Li-decoration on the edge oxidized graphyne and graphdiyne: a first principles study. *Computational Materials Science* 115: 51–59. https://doi.org/10.1016/j.commatsci.2015.12.048.

87 Hwang, H.J., Koo, J., Park, M. et al. (2013). Multilayer graphynes for lithium ion battery anode. *Journal of Physical Chemistry C* 117 (14): 6919–6923. https://doi.org/10.1021/jp3105198.

88 Sun, C. and Searles, D.J. (2012). Lithium storage on graphdiyne predicted by DFT calculations. *Journal of Physical Chemistry C* 116 (50): 26222–26226. https://doi.org/10.1021/jp309638z.

89 Wang, N., He, J., Wang, K. et al. (2019). Graphdiyne-based materials: preparation and application for electrochemical energy storage. *Advanced Materials* 31 (42): 1803202. https://doi.org/10.1002/adma.201803202.

90 Li, X., Gao, C., Duan, H. et al. (2013). High-performance photoelectrochemical-type self-powered UV photodetector using epitaxial TiO2/SnO2 branched heterojunction nanostructure. *Small* 9 (11): 2005–2011. https://doi.org/10.1002/smll.201202408.

91 Xue, M., Qiu, H., and Guo, W. (2013). Exceptionally fast water desalination at complete salt rejection by pristine graphyne monolayers. *Nanotechnology* 24 (50): 505720. https://doi.org/10.1088/0957-4484/24/50/505720.

92 Huang, C., Zhang, S., Liu, H. et al. (2015). Graphdiyne for high capacity and long-life lithium storage. *Nano Energy* 11: 481–489. https://doi.org/10.1016/j.nanoen.2014.11.036.

93 Wang, Y., Yang, P., Zheng, L. et al. (2020). Carbon nanomaterials with sp2 or/and sp hybridization in energy conversion and storage applications: a review. *Energy Storage Materials* 26: 349–370. https://doi.org/10.1016/j.ensm.2019.11.006.

94 Shao, Y., El-Kady, M.F., Sun, J. et al. (2018). Design and mechanisms of asymmetric supercapacitors. *Chemical Reviews* 118 (18): 9233–9280. https://doi.org/10.1021/acs.chemrev.8b00252.

95 Olabi, A.G., Abbas, Q., Abdelkareem, M.A. et al. (2022). Carbon-based materials for supercapacitors: recent progress, challenges and barriers. *Batteries* 9 (1): 19. https://doi.org/10.3390/batteries9010019.

96 Yang, H. (2018). A review of supercapacitor-based energy storage systems for microgrid applications. *2018 IEEE Power & Energy Society General Meeting (PESGM),* Portland, OR, USA. pp. 1–5, https://doi.org/10.1109/PESGM.2018.8585956.

97 Zhang, L., Hu, X., Wang, Z. et al. (2021). Hybrid electrochemical energy storage systems: an overview for smart grid and electrified vehicle applications. *Renewable and Sustainable Energy Reviews* 139: 110581. https://doi.org/10.1016/j.rser.2020.110581.

98 Kai, W., Baosen, R., Liwei, L. et al. (2017). A review of modeling research on supercapacitor. *2017 Chinese Automation Congress (CAC)*, Jinan, China, pp. 5998–6001. https://doi.org/10.1109/CAC.2017.8243857.

99 Huang, X., Zhang, X., Wei, T. et al. (2017). Development and applications status of supercapacitors. *Advanced Technologyof Electrical Engineering and Energy* 36 (11): 63–70.

100 Ma, N., Yang, D., Riaz, S. et al. (2023). Aging mechanism and models of supercapacitors: a review. *Technologies* 11 (2): 38. https://doi.org/10.3390/technologies11020038.

101 Zhai, C., Luo, F., and Liu, Y. (2020). Cooperative power split optimization for a group of intelligent electric vehicles travelling on a highway with varying slopes. *IEEE Transactions on Intelligent Transportation Systems* 23 (6): 4993–5005. https://doi.org/10.1109/TITS.2020.3045264.

102 Du, H., Yang, H., Huang, C. et al. (2016). Graphdiyne applied for lithium-ion capacitors displaying high power and energy densities. *Nano Energy* 22: 615–622. https://doi.org/10.1016/j.nanoen.2016.02.052.

103 Mashhadzadeh, A.H., Vahedi, A.M., Ardjmand, M., and Ahangari, M.G. (2016). Investigation of heavy metal atoms adsorption onto graphene and graphdiyne surface: a density functional theory study. *Superlattices and Microstructures* 100: 1094–1102. https://doi.org/10.1016/j.spmi.2016.10.079.

104 Chen, X., Gao, P., Guo, L. et al. (2017). High-efficient physical adsorption and detection of formaldehyde using Sc-and Ti-decorated graphdiyne. *Physics Letters A* 381 (9): 879–885. https://doi.org/10.1016/j.physleta.2017.01.009.

105 Chen, X., Ren, Z., Han, M. et al. (2020). Hybrid energy cells based on triboelectric nanogenerator: from principle to system. *Nano Energy* 75: 104980. https://doi.org/10.1016/j.nanoen.2020.104980.

106 Matsunaga, M., Hirotani, J., Kishimoto, S., and Ohno, Y. (2020). High-output, transparent, stretchable triboelectric nanogenerator based on carbon nanotube thin film toward wearable energy harvesters. *Nano Energy* 67: 104297. https://doi.org/10.1016/j.nanoen.2019.104297.

107 Zhang, W., Diao, D., Sun, K. et al. (2018). Study on friction-electrification coupling in sliding-mode triboelectric nanogenerator. *Nano Energy* 48: 456–463. https://doi.org/10.1016/j.nanoen.2018.04.007.

108 Huang, C., Zhao, Y., and Li, Y. (2019). Graphdiyne: the fundamentals and application of an emerging carbon material. *Advanced Materials* 31 (42): 1904885. https://doi.org/10.1002/adma.201904885.

109 Xie, C., Hu, X., Guan, Z. et al. (2020). Tuning the properties of graphdiyne by introducing electron-withdrawing/donating group. *Angewandte Chemie, International Edition* 132 (32): 13644–13648. https://doi.org/10.1002/ange.202004454.

110 Mellit, A. and Kalogirou, S. (2022). Solar radiation and photovoltaic systems: modeling and simulation. In: *Handbook of Artificial Intelligence Techniques in Photovoltaic Systems*, 1–41. Elsevier. https://doi.org/10.1016/B978-0-12-820641-6.00001-6.

111 Narang, G., Bansal, D., Joarder, S. et al. (2023). A review on the synthesis, properties, and applications of graphynes. *FlatChem* 40: 100517. https://doi.org/10.1016/j.flatc.2023.100517.

112 Kang, J., Wei, Z., and Li, J. (2019). Graphyne and its family: recent theoretical advances. *ACS Applied Materials & Interfaces* 11 (3): 2692–2706. https://doi.org/10.1021/acsami.8b03338.

113 Matsuoka, R., Toyoda, R., Shiotsuki, R. et al. (2019). Expansion of the graphdiyne family: a triphenylene-cored analogue. *ACS Applied Materials & Interfaces* 11 (3): 2730–2733. https://doi.org/10.1021/acsami.8b00743.

114 Bagade, S.S., Patel, S., Malik, M.M., and Patel, P.K. (2023). Recent advancements in applications of graphene to attain next-level solar cells. *Journal of Carbon Research* 9 (3): 70. https://doi.org/10.3390/c9030070.

115 Peng, Q., Dearden, A.K., Crean, J. et al. (2014). New materials graphyne, graphdiyne, graphone, and graphane: review of properties, synthesis, and application in nanotechnology. *Nanotechnology, Science and Applications* 1. https://doi.org/10.2147/NSA.S40324.

116 Torres-Pinto, A., Silva, C.G., Faria, J.L., and Silva, A.M.T. (2021). Advances on graphyne-family members for superior photocatalytic behavior. *Advanced Science* 8 (10): 2003900. https://doi.org/10.1002/advs.202003900.

117 Chen, J., Xi, J., Wang, D., and Shuai, Z. (2013). Carrier mobility in graphyne should be even larger than that in graphene: a theoretical prediction. *Journal of Physical Chemistry Letters* 4 (9): 1443–1448. https://doi.org/10.1021/jz4005587.

118 Enyashin, A.N. and Ivanovskii, A.L. (2011). Graphene allotropes. *Physica Status Solidi B* 248 (8): 1879–1883. https://doi.org/10.1002/pssb.201046583.

119 Xi, J., Wang, D., Yi, Y., and Shuai, Z. (2014). Electron-phonon couplings and carrier mobility in graphynes sheet calculated using the Wannier-interpolation approach. *The Journal of Chemical Physics* 141 (3): 034704. https://doi.org/10.1063/1.4887538.

120 Kim, B.G. and Choi, H.J. (2012). Graphyne: hexagonal network of carbon with versatile Dirac cones. *Physical Review B* 86 (11): 115435. https://doi.org/10.1103/PhysRevB.86.115435.

121 He, J., Ma, S.Y., Zhou, P. et al. (2012). Magnetic properties of single transition-metal atom absorbed graphdiyne and graphyne sheet from DFT+U calculations. *Journal of Physical Chemistry C* 116 (50): 26313–26321. https://doi.org/10.1021/jp307408u.

122 Jiao, Y., Du, A., Hankel, M. et al. (2011). Graphdiyne: a versatile nanomaterial for electronics and hydrogen purification. *Chemical Communications* 47 (43): 11843. https://doi.org/10.1039/c1cc15129k.

123 Bu, H., Zhao, M., Wang, A., and Wang, X. (2013). First-principles prediction of the transition from graphdiyne to a superlattice of carbon nanotubes and graphene nanoribbons. *Carbon* 65: 341–348. https://doi.org/10.1016/j.carbon.2013.08.035.

124 Luo, G., Qian, X., Liu, H. et al. (2011). Quasiparticle energies and excitonic effects of the two-dimensional carbon allotrope graphdiyne: theory and experiment. *Physical Review B* 84 (7): 075439. https://doi.org/10.1103/PhysRevB.84.075439.

125 Zheng, Q., Luo, G., Liu, Q. et al. (2012). Structural and electronic properties of bilayer and trilayer graphdiyne. *Nanoscale* 4 (13): 3990. https://doi.org/10.1039/c2nr12026g.

126 Castro Neto, A.H., Guinea, F., Peres, N.M.R. et al. (2009). The electronic properties of graphene. *Reviews of Modern Physics* 81 (1): 109–162. https://doi.org/10.1103/RevModPhys.81.109.

127 Lei, S., Cao, Q., Geng, X. et al. (2018). The mechanical properties of defective graphyne. *Crystals* 8 (12): 465. https://doi.org/10.3390/cryst8120465.

128 Li, J., Jiu, T., Chen, S. et al. (2018). Graphdiyne as a host active material for perovskite solar cell application. *Nano Letters* 18 (11): 6941–6947. https://doi.org/10.1021/acs.nanolett.8b02863.

129 Das, S., Pandey, D., Thomas, J., and Roy, T. (2019). The role of graphene and other 2D materials in solar photovoltaics. *Advanced Materials* 31 (1): 1802722. https://doi.org/10.1002/adma.201802722.

130 Li, Y., Xu, L., Liu, H., and Li, Y. (2014). Graphdiyne and graphyne: from theoretical predictions to practical construction. *Chemical Society Reviews* 43 (8): 2572. https://doi.org/10.1039/c3cs60388a.

131 Du, H., Deng, Z., Lü, Z. et al. (2011). The effect of graphdiyne doping on the performance of polymer solar cells. *Synthetic Metals* 161 (19–20): 2055–2057. https://doi.org/10.1016/j.synthmet.2011.04.015.

132 Zhao, Y., Guo, P., Li, X., and Jin, Z. (2019). Ultrafast photonics application of graphdiyne in the optical communication region. *Carbon* 149: 336–341. https://doi.org/10.1016/j.carbon.2019.04.075.

133 Jiang, P.H., Liu, H.J., Cheng, L. et al. (2017). Thermoelectric properties of γ-graphyne from first-principles calculations. *Carbon* 113: 108–113. https://doi.org/10.1016/j.carbon.2016.11.038.

134 Andrew, R.C., Mapasha, R.E., Ukpong, A.M., and Chetty, N. (2019). Erratum: Mechanical properties of graphene and boronitrene. *Physical Review B* 100, 20: 209901. https://doi.org/10.1103/PhysRevB.100.209901.

135 He, J., Wang, N., Cui, Z. et al. (2017). Hydrogen substituted graphdiyne as carbon-rich flexible electrode for lithium and sodium ion batteries. *Nature Communications* 8 (1): 1172. https://doi.org/10.1038/s41467-017-01202-2.

136 Zhao, F., Wang, N., Wang, K. et al. (2021). Nitrogen substituted graphdiyne as electrode for high-performance lithium-ion batteries and capacitors. *2D Materials* 8 (4): 044013. https://doi.org/10.1088/2053-1583/ac2592.

137 Septya, M.M., Asih, R., Arifin, R., and Darminto, D. (1951). Electronic and magnetic properties of single vacancy graphene with hydrogen adsorptions analyzed using density functional theory method. *Journal of Physics: Conference Series* (1): 012012. 2021, https://doi.org/10.1088/1742-6596/1951/1/012012.

138 Kang, B., Ai, H., and Lee, J.Y. (2017). Single-atom vacancy induced changes in electronic and magnetic properties of graphyne. *Carbon* 116: 113–119. https://doi.org/10.1016/j.carbon.2017.01.068.

139 Cosgrove, W.J. and Loucks, D.P. (2015). Water management: current and future challenges and research directions. *Water Resources Research* 51 (6): 4823–4839. https://doi.org/10.1002/2014WR016869.

140 Mehrdad, M. and Moosavi, A. (2019). An efficient graphyne membrane for water desalination. *Polymer* 175: 310–319. https://doi.org/10.1016/j.polymer.2019.05.054.

141 Zhang, J., Bai, Q., Bi, X. et al. (2022). Piezoelectric enhanced peroxidase-like activity of metal-free sulfur doped graphdiyne nanosheets for efficient water pollutant degradation and bacterial disinfection. *Nano Today* 43: 101429. https://doi.org/10.1016/j.nantod.2022.101429.

142 Zhan, S., Chen, X., Xu, B. et al. (2022). Hollow multishelled structured graphdiyne realized radioactive water safe-discharging. *Nano Today* 47: 101626. https://doi.org/10.1016/j.nantod.2022.101626.

143 Lin, L., Yang, H., and Xu, X. (2022). Effects of water pollution on human health and disease heterogeneity: a review. *Frontiers in Environmental Science* 10: 880246. https://doi.org/10.3389/fenvs.2022.880246.

144 Shindhal, T., Rakholiya, P., Varjani, S. et al. (2021). A critical review on advances in the practices and perspectives for the treatment of dye industry wastewater. *Bioengineered* 12 (1): 70–87. https://doi.org/10.1080/21655979.2020.1863034.

145 Ghosh, A., Orasugh, J.T., Ray, S.S., and Chattopadhyay, D. (2023). Prospects of 2D graphdiynes and their applications in desalination and wastewater remediation. *RSC Advances* 13 (27): 18568–18604. https://doi.org/10.1039/D3RA01370G.

146 Liu, R., Zhou, J., Gao, X. et al. (2017). Graphdiyne filter for decontaminating lead-ion-polluted water. *Advanced Electronic Materials* 3 (11): 1700122. https://doi.org/10.1002/aelm.201700122.

147 Xie, S., Pan, C., Yao, Y. et al. (2023). Ultra-high-efficiency capture of lead ions over acetylenic bond-rich graphdiyne adsorbent in aqueous solution. *Proceedings. National Academy of Sciences. United States of America* 120 (16): e2221002120. https://doi.org/10.1073/pnas.2221002120.

148 Xiao, S.X., Huang, C.S., and Li, Y.L. (2017). Carbon materials. In: *Modern Inorganic Synthetic Chemistry*, 429–462. Elsevier. https://doi.org/10.1016/B978-0-444-63591-4.00016-1.

149 Lasisi, K.H., Abass, O.K., Zhang, K. et al. (2023). Recent advances on graphyne and its family members as membrane materials for water purification and desalination. *Frontiers in Chemistry* 11: 1125625. https://doi.org/10.3389/fchem.2023.1125625.

8

Mesoporous Carbon for Green Energy Applications

Vikas Jangra[1,2], Narvadeswar Kumar[2], Harpreet Kaur[2], Lal Bahadur Prasad[1], and Piyush Kumar Sonkar[2]

[1] *Department of Chemistry, Banaras Hindu University, Varanasi, Uttar Pradesh, India*
[2] *Department of Chemistry, MMV, Banaras Hindu University, Varanasi, Uttar Pradesh, India*

8.1 Introduction

In recent decades, energy has become one of the most necessary and popular areas of study. Rapid consumption and exhaustion of fossil fuels for energy lead to environmental pollution [1]. It is critical to use energy wisely and to find sustainable, green energy sources that can replace fossil fuels. Solar energy, geothermal energy, wind energy, and tidal energy are few sources of sustainable and renewable energy [2]. However, all these sources are not sufficient for our day-to-day power consumption. Thus, developing energy conversion and storage systems with outstanding efficiency and excellent stability is required.

Additionally, there is a considerable need for energy storage devices due to the rapid rise of portable consumer electronics and electric automobiles [3]. Several energy conversion and storage methods have been used during the last 30 years to generate powerful, ecologically friendly energy sources. Fuel cells, lithium batteries, supercapacitors, and solar energy cells are a few examples of these systems [4]. Energy storage devices are beginning to play a bigger role in our lives because the sun doesn't always shine and the wind doesn't always blow. The design, production, and characterization of an array of new and unique energy materials and catalysts for energy storage and conversion depend primarily on nanomaterials that are used as electrode modifiers. Due to their low cost, chemical stability, and large specific surface area, carbon-based nanomaterials like graphene, carbon

Carbon-based Nanomaterials for Green Applications, First Edition. Edited by Upendra Kumar, Piyush Kumar Sonkar, and Suman Lata Tripathi.

Published 2025 by John Wiley & Sons, Inc.

nanotubes, graphite carbon nitride, carbon quantum dots, and carbon aerogel have received a lot of interest [5]. However, the majority of this carbon-based nanomaterial has an irregularly shaped porous structure with an excessive number of micropores that cause significant electrolyte ion mass transfer resistance [3]. On the other hand, ordered mesoporous carbon (OMC) with controllable pore sizes ranging from 2 to 50 nm are particularly advantageous for energy conversion as well as for storage because of their easy synthesis, high surface area, tunable framework composition, diverse morphologies, efficient mass transport, nanoconfinement effect, chemical inertness, and ideal electroconductivity as represented by Figure 8.1 [4, 6]. Their very porous architectures make it possible to add more functional groups such as inorganic or organic components on the surface of the carbon walls or within the porous channels, greatly enhancing their characteristics and functionality [7]. Ryoo et al. published the first study on the production of MC materials in 1999. They used mesoporous silica as a hard template and sucrose as a carbon source to create a mesoporous material with a large specific surface area while regulating pore diameters between 2 and 50 nm [4]. The groups led by Dai and Zhao have made significant progress in producing MC materials, mostly due to the assembly of copolymer molecular arrays and carbon precursors. It is possible to manufacture mesoporous materials with a high degree of structural control by using a variety of templates with different structures. For instance, a variety of different-structured mesoporous silica materials, including MCM-48, SBA-1, SBA-15, SBA-16, KIT-6, and KIT-5, are used as templates to create different-structured MC materials [6, 7]. The invention of modified OMCs, which have improved the chemical and physical characteristics of OMCs via the addition of inorganic components either enchased on the pore walls or bound within the channels, has played a major role in the development of the carbon-based electrode or catalyst in its potential applications [8].

Simple pyrolysis and high-temperature physical or chemical treatment of polymers, fruit shells, wood, and coal have been used to produce porous carbon materials on a large scale, including carbon molecular sieves and activated carbon. However, because of the heteroatoms, lack of structural control and constrained flow routes, the carbon materials made using these processes have poor conductivity, structural integrity, and mass transfer. The introduction of mesopores for possible applications has overcome these limitations [9].

IUPAC classified porous materials into three categories, namely microporous with pore diameter <2 nm, mesoporous with pore diameter 2–50 nm, and macroporous with pore diameter >50 nm) (Figure 8.2) [10]. Among these, MC has attracted enormous attention in a variety of fields due to its huge specific surface area, excellent thermal durability, favorable chemical and mechanical characteristics, homogeneous controlled porous structure, and enormous pore volume [7].

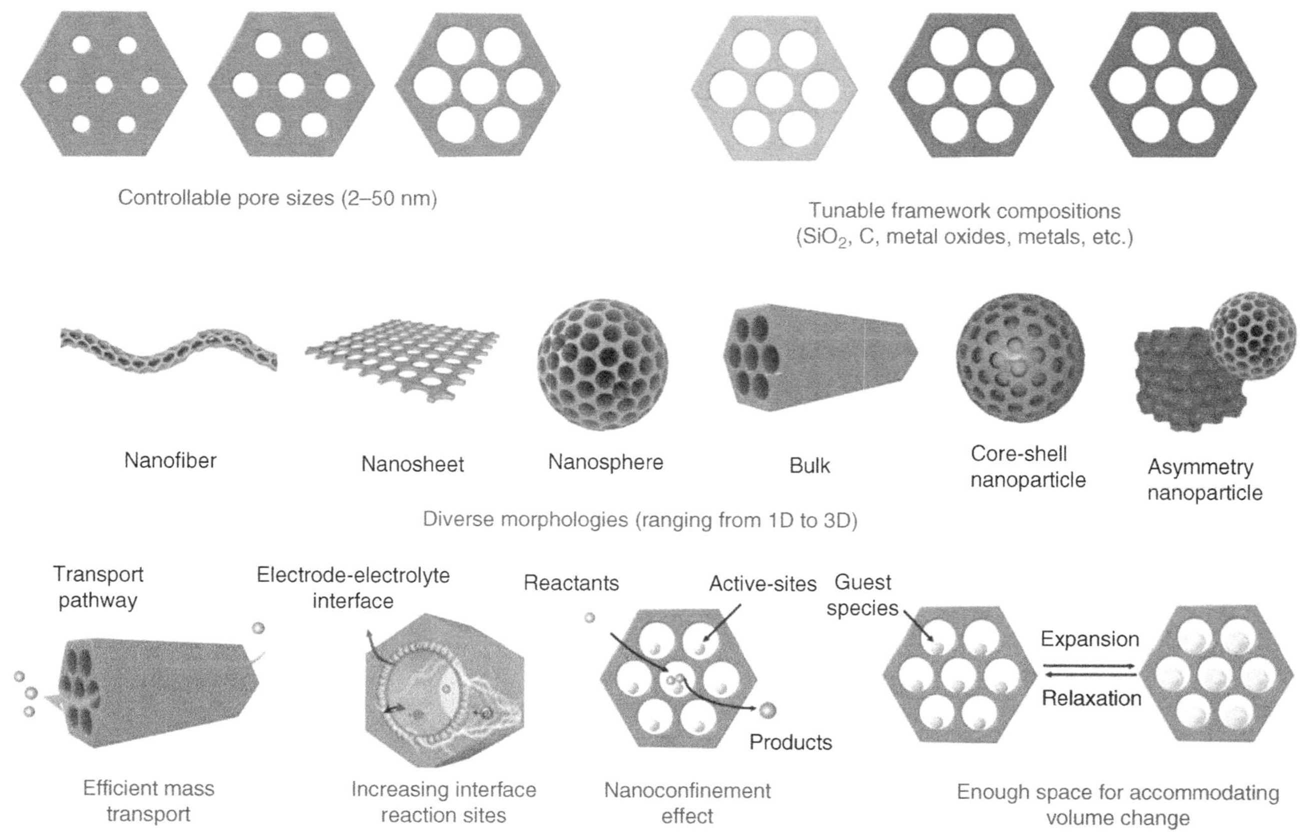

Figure 8.1 Mesoporous materials: their features and structural characteristics. *Source:* Zu et al. [6]/John Wiley & Sons.

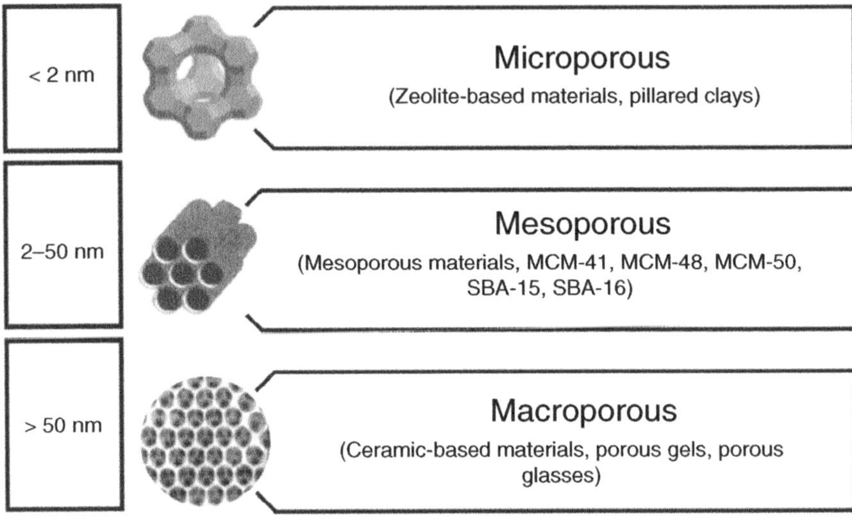

Figure 8.2 Classification of porous materials. *Source:* Azam et al. [10]/John Wiley & Sons.

This chapter describes the common procedures for producing MC structures. Different approaches to the modification of MC materials are then discussed. Considering recent studies in the last five years, green energy applications for these materials are also examined, including supercapacitors, Li batteries, fuel cells, and other devices. The challenges associated with producing MC materials for green energy applications are then discussed, along with recommendations for future research areas.

8.2 Recent Advances in Synthetic Techniques

Especially in the last two decades, there have been considerable advancements in the synthesis of MC and OMC. A summary of some of the synthetic processes that are utilized to produce MC and OMC is provided in this section. These are typically synthesized by casting method. A hard mold, typically composed of wax, plaster, metal, or another material, is required in a macroscopic casting process. The three basic steps of the nanocasting process to produce nanostructured materials are: (1) creating the template; (2) casting with target precursors, which includes converting the precursor typically a molecular one to a solid and (3) removing the template using highly acidic or alkaline solution as shown in Figure 8.3 [12]. There are various methods for the synthesis of MC, the hard-template and soft-template approaches are the most accepted fabrication methods, which will be addressed in the next section.

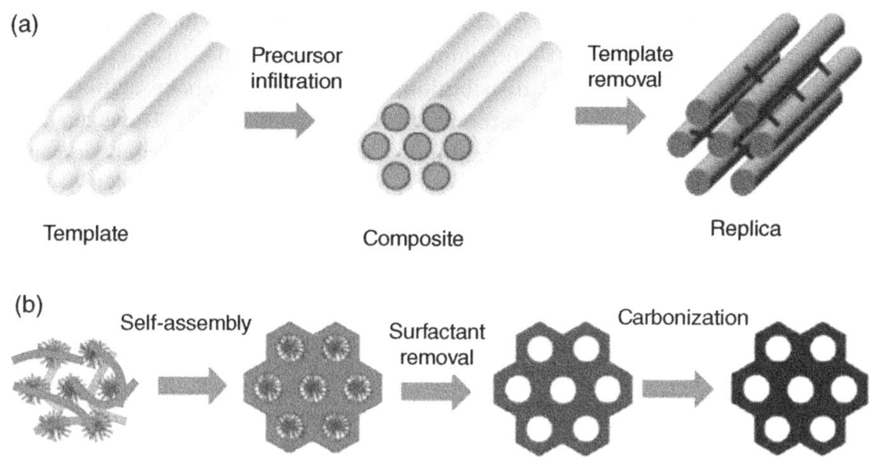

Figure 8.3 Methods for synthesizing OMC include: (a) Hard template and (b) Soft template. *Source:* Eftekhari and Fan [11]/Royal Society of Chemistry.

8.2.1 Hard Template Technique

One of the simplest methods of producing OMCs is the hard template approach, which consists of several fixed steps. These steps include:

i) Creating a controlled architecture, usually mesoporous silica or zeolites.
ii) Adding a suitable carbon precursor into the mesopores through chemical vapor deposition (CVD) or wet impregnation.
iii) Carbonizing the resulting composite at a high temperature through polymerization.
iv) Removing the silica mold by HF or alkaline dissolution as shown in Figure 8.3a [11, 13].

8.2.1.1 Carbon Precursors

Commonly used precursors can be divided into two groups: those with loose molecular structures, such as sugar, sucrose, furfuryl alcohol, and those with complex aromatic structures, such as pitch, pyrene, polyacrylonitrile, acenaphthene. By employing dense aromatic structures as precursors, the material can form a more accurate copy of the hard template, with minimal shrinkage, high strength, and very few tiny pores. On the other hand, if loose-structured precursors are used, framework shrinkage will result in an increase in mesopore size and the generation of more micropores in the framework, which would cause a two-modal distribution of pore size and a bigger specific surface area [11].

It has long been known that the hard-template approach provides superior control over the highly ordered structure and pore size. Using SBA-15 as a hard template, bimodal or monomodal distributions of pore sizes can be achieved by adjusting the stabilization and carbonization temperatures [4]. However, this method is more time-consuming, requires a longer synthesis period, presents more challenges in controlling pore size [7], and is more expensive. Also the elimination of the mesoporous silica hard templates by very poisonous hydrofluoric acid puts a barrier in the way of their commercialization [14]. As a result, the researchers discovered eco-friendly soft templating techniques to create OMCs without sacrificing textural qualities.

8.2.2 Soft Template Technique

Numerous mesoporous materials have been produced via the widespread use of the soft-templating technique. However, directly synthesizing OMCs using this approach is not easy. Various steps involved in this method are:

i) Creation of supramolecular arrangements.
ii) Thermopolymerization to create a highly cross-linked composite.
iii) Surfactant removal.
iv) Carbonization of the resulting material is shown in Figure 8.3b.

The soft-templating approach uses amphiphilic compounds, such as metal-organic frameworks, block copolymers, or surfactant molecules, as templates. In the beginning, the carbon precursor molecules in the templates assemble themselves through interactions such as hydrogen bonding. An organized mesostructured carbon is the result of these self-assembly processes. The elimination of the amphiphilic compounds can then be achieved by high-temperature calcination, which produces mesopore structures with highly controllable pore shape and size [14].

The soft templating techniques are appealing, but they mostly rely on the usage of carbon precursors based on phenolic compounds with aldehyde cross-linkers, which aren't necessarily environmentally friendly. As a result, the development of OMC nanomaterials has received a lot of attention by utilizing environmentally friendly and sustainable precursors. Recently, the traditional self-assembly triggered by solvent evaporation methodology—which uses natural biopolymers like lignin—has been applied to synthesize OMC in an environmentally friendly and sustainable manner [13].

Apart from these template methods, the template-free method is also extensively employed in the synthesis of random-distributed mesoporous materials, which are more common than ordered mesostructures. Mesopore voids in this path are mostly caused by structural flaws, nanoparticle aggregation, self-assembly of nanoblocks, and structure defects. But most of the time, it's challenging to pinpoint the exact pore structures of the resulting materials [6]. The three main components added to the solvent during the nanocasting process to create MCs are carbon sources, templates, and catalysts. Carbon precursors and templates have a major impact on the pore size and structure of MC materials [8]. Making the proper choice of thermosetting carbon sources and thermally degradable surfactants, solvents, and temperature is crucial to controlling the development of these mesostructures. So far, various synthetic conditions have been used. Some recent advances in synthetic techniques in the last five years and their potential uses are outlined in Table 8.1 [15–41].

8.3 Applications of Mesoporous Carbon

OMC provides enormous pore volumes, uniform pore size distribution, high specific surface areas, and highly organized porous structure. These interesting chemical and physical properties enable them to be successfully applied across an extensive array of uses. Many studies have investigated the use of nanoporous carbon materials including OMCs for a variety of possible uses, such as adsorption, separation, sensing, photocatalysis, drug delivery, fuel cell supports, ORR catalysts, post-Li and conventional Li-ion batteries, supercapacitors, H2 storage, and devices for field emission and capture of carbon dioxide (Figure 8.4) [14]. In this section, we will discuss the importance of OMC for advanced conversion and storage of energy only.

8.3.1 Applications in Lithium Batteries

In recent decades, electric cars and portable gadgets have made extensive use of rechargeable batteries, particularly lithium-ion batteries. The materials used to build electrodes in these devices are basically conductors. Using carbon as a conductor material is recommended worldwide. However, the difficulty of Li-ion penetration makes carbon inconvenient electrode material. As a result, the progressive solid-state dispersion presents the main challenge for lithium batteries [7]. Mass transport and diffusion processes, as well as charge transfer, can be significantly impacted by the properties of mesoporous frameworks. Li^+ transport in mesoporous materials is represented by the diffusion capacity of lithium ions, a critical kinetic factor for ion insertion and extraction. The term effective diffusion coefficient (D_{eff}) in electrolytes, which can be expressed as $D_{eff} = D\varepsilon/\tau$, represents the kinetic patterns in different porous electrodes. The sizes of the porous electrodes affect the tortuosity factor (τ), porosity (ε), and intrinsic diffusion coefficient (D) in electrolytes [6]. The efficient dispersion of ions and molecules is facilitated by mesoporous electrodes with high specific surface areas, linked porous channels, and mesopore walls at the nanoscale, as per the previously indicated formula. Some recent advancements in Li-ion batteries are represented in Table 8.2 [17, 42–50].

Table 8.1 Recent advancements in synthetic techniques and applications of OMC.

Material	Template	Precursor	Solvent	T (°C)	d pore (nm)	Morphology	Applications	Reference
MMC	KIT-6, NiO	Sucrose	DI water	900	4.6	Mesoporous structure with interconnected macropores	Li-ion batteries	[15]
S-OMC/Ge	*n*-butanol and P123	*n*-butanol, P123	DI water	600	500	Spherical ordered mesoporous carbon	Li-ion batteries	[16]
Si/SiO_2-OMC	P123	Sucrose	DI water	700	—	Highly ordered rod-like structure	Li- and Na-ion batteries	[17]
Ge/OMC-N-S	SBA-15	Sucrose, L-cysteine	DI water	900	—	Ordered mesoporous channels	Li-ion batteries	[18]
OMC	SBA-15	Sucrose	DI water	900	500–800	Well-ordered hexagonal arrangement	K-ion batteries	[19]
PMCs	Pluronic F127	Phenolic resol	Ethanol	850	6.6–14.2	Foam-like monolithic morphology with a 3D interconnecting macrocellular network	Li-ion batteries	[20]
N-DHMCSs	PS-b-PAA, TEOS	Dopamine	Ethanol	700–900	2–10	Double-shelled and hollow particle morphology	Li-ion batteries	[21]
MCOs	—	Cu-BTC	DI water	800	2–5	Porous octahedral skeleton	K-ion batteries	[22]
AC	—	Corn stalks	KOH	550	8.49	Slit-shaped pores	Li-ion batteries	[23]
P-MCN-1	SBA-15	g-C_3N_4	DI water	650	0.8–1.5	Platelet-like particles	Li-ion batteries	[24]
MCFC	—	Fullerene C70	Mesitylene	900	3.50	Mesoporous cubic crystal with narrow size distribution	Supercapacitors	[25]
HCS	TEOS	CTAB	Ethanol	800	40	Uniform spherical morphology	Supercapacitors	[26]
N-OMCS	F127	O-aminophenol and hexamine	Ammonia	800	300	Uniform and disperse nanospheres	Supercapacitors	[27]

MC	TPOS	Resorcinol formaldehyde	EtOH/H2O	800	3.9–5.7	Sheet-like morphology	Supercapacitors	[28]
MCNs	SBA-15	TEOS	HCl solution	550	201 ± 70	Nano-rod like morphology	Supercapacitors	[29]
NPMCs	F127	TEOS	EtOH/H2O	850	4.3	Stacked flake-like and 2D-hexagonal mesostructures	Supercapacitors	[30]
MCM	$Na_2O_3Si.9H_2O$	Sucrose	DI water	650	10–30	3D interconnected mesoporous network	Supercapacitors	[31]
N-MCS	Mica powder	Resorcinol-formaldehyde	EtOH/H2O	800	2.7	Uniform sheet-like structures	Supercapacitors	[32]
N/O-OMC	KIT-6	Polyvinylpyrrolidone		800	3–132	Ordered dual-mesoporous structure	Supercapacitors	[33]
PLG	F127	Lignin	DI water	900	5–8	2D-hexagonal pore structures	Supercapacitors	[34]
AOMCs	Pluronic F127	Mimosa tannin	DI water	900	4.6	2D hexagonally ordered mesoporous structure	Supercapacitors	[35]
N-OMCs	Pluronic F127	Terephthalaldehyde and m-aminophenol	DI water	750–900	2.5–3.0	Rough characteristic with highly ordered mesopores	ORR	[36]
Pt@s-OMC	P123	Sucrose	DI water	900	2.8	2D hexagonally ordered mesoporous structure	ORR	[37]
CMK-8	KIT-6	Sucrose	DI water	900	6.1–8.9	3D cubic surface morphology	ORR	[38]
$Co-C_3N_4$/OMC	SBA-15	Sucrose	DI water	550	200	Plate-like morphology and ordered structure	ORR	[39]
Fe,N-OMC	Colloidal silica	OPD, melamine, and $FeCl_3{\cdot}6H_2O$	DI water	900	25	Mesh-like structure with a number of uniform pores	ORR	[40]
SN-OMCs	SBA-15	Ferrocene	DI water	500	2.2–3.4	Rod-like morphology	ORR	[41]

Source: From [15–41].

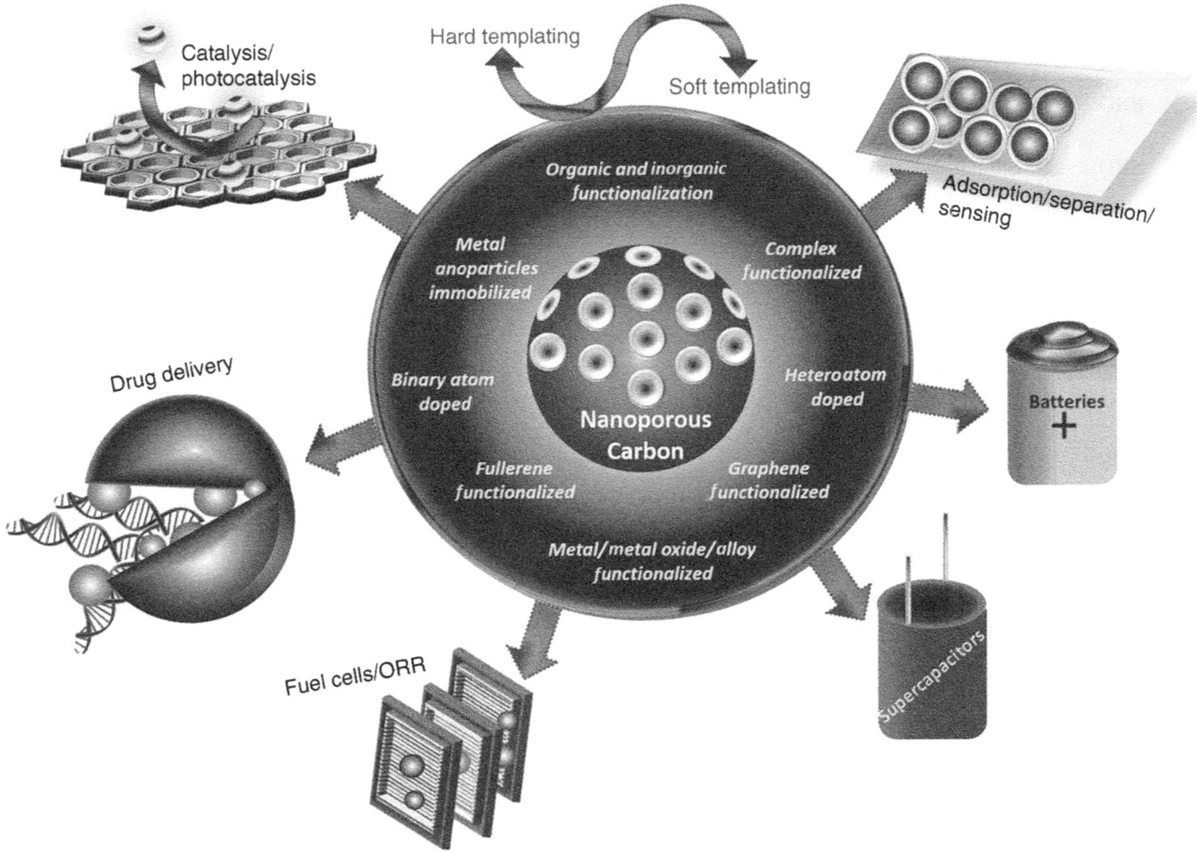

Figure 8.4 Mesoporous carbon synthesis, modification, and uses. *Source:* Benzigar et al. [14]/Royal Society of Chemistry.

Table 8.2 Recent advancements in Li-Ion batteries.

Material	OMMs	Electrode	[a]DC (mA h g^{-1})	[b]RC (mA h g^{-1})	[c]CE (%)	Cycles	Application	Reference
Si/SiO_2-OMC	OMC	Anode	595	958	98.60	100	Li-Ion batteries	[17]
SnO_2@OMC	OMC	Anode	440	769	98.70	60	Li-Ion batteries	[42]
FeO_x/OMC	OMC	Anode	1718	660	58.40	50	Li-Ion batteries	[43]
OMCNW/Fe_2O_3	OMC	Anode	774	819	98.80	1200	Li-Ion batteries	[44]
MoO_2/OMC	OMC	Anode	1641.8	1049.1	98.70	50	Li-Ion batteries	[45]
OMC/WO_3	OMC	Anode	1275	440	97.60	100	Li-Ion batteries	[46]
OMC/TiO_2	OMC	Anode	2676	540.9	98	60	Li-Ion batteries	[47]
Ge/GeO_2-OMC	OMC	Anode	517	1018	98	100	Li-Ion batteries	[48]
V_2O_3-OMC	OMC	Anode	535.70	536	99.40	180	Li-Ion batteries	[49]
Co_3O_4/OMC	OMC	Anode	1515	1025	98	100	Li-Ion batteries	[50]

[a] DC, discharge capacity.
[b] RC, reversible capacity.
[c] CE, Coulombic efficiency.
Source: From [17, 42–50].

Wang and co-workers synthesized Si/SiO_2-OMC using the magnesiothermic reduction method with NaOH etching and SiO_2-OMC as a precursor. It demonstrated exceptional reversible capacity (958 $mA\,h\,g^{-1}$ at 0.2 $A\,g^{-1}$) when used as an anode for lithium-ion batteries [17]. SnO_2@OMC nanocomposites were developed by Wang and colleagues using SnO_2 nanoparticles for use in lithium-ion battery anodes. An outstanding reversible capacity up to 769 $mA\,h\,g^{-1}$ is observed [42]. Li and colleagues successfully produced OMC modified by iron oxide nanoparticles using a two-step, controlled approach. The FeO_x/OMC hybrids show greater Coulombic efficiency (58.4%) and a reversible capacity of over 660 $mAh\,g^{-1}$ [43]. A soft-hard dual-template technique was presented by Hu and colleagues to create OMCNW/Fe_2O_3 composite nanowires. The OMC/Fe_2O_3 Nanowires (OMCNW/Fe_2O_3) composite maintains a high capacity of up to 819 $mAh\,g^{-1}$ even up to 1200th cycles, which is more than twice that of a conventional graphite anode [44]. Chen and his colleagues developed a new hybrid of MoO_2/OMC using a two-step solvothermal chemical reaction process. With an initial Coulombic efficiency of 63.6%, the MoO_2/OMC hybrid could have a first discharge capacity of 1641.8 $mAh\,g^{-1}$. Even after 50 cycles, it could still have a reversible capacity of 1049.1 $mAh\,g^{-1}$ at a current density of 100 $mA\,g^{-1}$ [45]. An MC matrix confinement growth technique was reported by Wang and colleagues to synthesize ultrasmall WO_3 nanocrystals for lithium storage. At a current density of 0.1 $A\,g^{-1}$, the resulting OMC/WO_3 composites have a high specific capacity of 440 $mAh\,g^{-1}$ after 100 cycles [46]. Ma and colleagues used a straightforward solvothermal technique to create OMC/TiO_2 (OMCT) nanocomposites, where TiO_2 nanocrystals were embedded within OMC. The produced composite can withstand up to 60 cycles at an intense current density of 100 $mA\,g^{-1}$ and has a large reversible storage capacity of 540.9 $mAh\,g^{-1}$, excellent rate capability, and a stable Coulombic efficiency of 98% [47].

MC was utilized as a nanoreactor by Zeng and colleagues to create the Ge/GeO_2-OMC nanocomposite. After 100 cycles at a current density of 0.1 $A\,g^{-1}$, a significant reversible capacity of 1018 $mA\,h\,g^{-1}$ was obtained in the resultant composite [48]. Zeng and colleagues prepared V_2O_3-OMC and used it as an anode material for Li-ion intercalation. After 180 cycles at a current density of 0.1 $A\,g^{-1}$, these materials demonstrate a substantial reversible capacity of 536 $mA\,h\,g^{-1}$ [49]. Park and his coworkers synthesized Co_3O_4/OMC and used it as anode material for Li-ion batteries. At a current density of 0.1 $A\,g^{-1}$, the resultant material has a high reversible capacity of 1025 $mA\,h\,g^{-1}$ after 100 cycles [50]. It indicates the remarkable energy storage capacity of the modified electrode.

8.3.2 Applications in Supercapacitors

Supercapacitors are devices that store energy and have a long cycling life, a quick charge/discharge operation and an improved power density of 410 $kW\,kg^{-1}$. The two electrodes of a supercapacitor are divided by a porous matrix or separator.

Supercapacitors are generally divided into two categories: pseudocapacitors and double-layer electrical capacitors. A double-layer electrical capacitor stores electrical energy using carbon electrodes by intercalating charges at the electrode-electrolyte interface to create a double layer of charges. A pseudocapacitor is a kind of supercapacitor that stores electrical energy via faradaic processes using conducting polymer or metal oxide electrodes [7]. The manufacturing procedure is also relatively straightforward for these materials [14]. This chapter provides a summary of the progress made over the last five years in using these nanostructures for supercapacitor applications (Table 8.3) [35, 51–59].

By using a universal method of graphitic carbon nitride (g-C_3N_4) assisted self-assembly of phenolic resin and F127 surfactant, Xie and colleagues synthesized nitrogen-doped OMC. An outstanding rate efficiency of 61.4%, a superior energy density of 29.8 $Wh\,l^{-1}$ at 990 $W\,l^{-1}$, an immense capacitance of 420 $F\,cm^{-3}$ (382 $F\,g^{-1}$ equivalent) at 0.5 $A\,g^{-1}$ and retaining 258 $F\,cm^{-3}$ at 20 $A\,g^{-1}$ were recorded, suggesting an appealing option for transportable power storage appliances [51]. High B-doped OMC was produced by Zhang and co-workers using the evaporation-induced self-assembly (EISA) method with four-hydroxyphenylboronic acid-modified phenolic resin as a boron and carbon precursor [52]. Chen and collaborators proposed a straightforward one-pot aqueous technique to create the precursor for the composite of sulfur-doped MCs and N-doped graphene (SG/NMC). At 1.0 $A\,g^{-1}$, SG/NMC demonstrates a high specific capacitance of around 304.2 $F\,g^{-1}$ [53]. Mohammadi and associates demonstrated the supercapacitive characteristics of a novel nanocomposite based on MnO_2 and flawed MC. The nanocomposite demonstrated the maximum specific capacitance of 292 $F\,g^{-1}$ at a current density of 0.5 $A\,g^{-1}$. Additionally, DMC/MnO_2 showed excellent rate efficiency (61% retention of capacity over 0.5–10 $A\,g^{-1}$) and remarkable cycling stability (21% loss of capacity after 2000 charge/discharge cycles) [54]. Using a simple, self-developed method, monodispersed $CoMoO_4$ nanoclusters were impregnated into OMCs by Li and co-workers. The developed hybrids enhance the electrochemical efficiency of supercapacitors with their larger specific surface area (>700 $m^2\,g^{-1}$), consistent mesoporous pathways (~4 nm) and enhanced wetting ability. With a 99.8% cycling efficiency, these uniformly distributed $CoMoO_4$ nanoclusters show strong reversibility and a notable specific capacitance of 367 $F\,g^{-1}$ in aqueous KNO_3 electrolyte [55]. CuO/MPC composite was used by Saravanakumar and colleagues to create an asymmetric supercapacitor cell, with activated carbon serving as the electrodes and gel polymer (PVA-KOH) serving as the electrolyte. This asymmetrical device exhibits improved electrochemical performance, with a power density of 438 $W\,kg^{-1}$ and an energy density of 26.6 $Wh\,kg^{-1}$. After 5000 cycles, this device preserves 69% of its original capacitance [56].

Zallouz et al. propose a one-pot bottom-up approach for designing cobalt oxide (Co_3O_4) nanoparticles enclosed in an MC. The phenolic resin, surfactant, and cobalt salt were assembled together in this method, which was then followed by a

Table 8.3 Recent advancements in supercapacitors.

Material	[a]S_{BET} ($m^2\,g^{-1}$)	[b]D (nm)	[c]Cg ($F\,g^{-1}$)	Principal findings and remarks	Application	Reference
NOMC973-C_3N_4	272	2.1	385 @ 0.5 $A\,g^{-1}$	N-doped OMC was produced by g-C_3N_4, phenolic resin and F127 as surfactant.	Supercapacitor	[51]
0.15-HPB-OMC	567.6	4.79	183 @ 1 $A\,g^{-1}$	With an increase in B-doped concentration, specific capacitance progressively rises.	Supercapacitor	[52]
SG/NMC	1039.65	3.61 ± 0.42	304 @ 1 $A\,g^{-1}$	SG/NMC yield a large specific surface area of 1039.65 $m^2\,g^{-1}$	Supercapacitor	[53]
DMC/MnO_2	185	13.1	264 @ 1 $A\,g^{-1}$	At 0.5 $A\,g^{-1}$, DMC/MnO_2 demonstrated maximum specific capacitance at 292 $F\,g^{-1}$.	Supercapacitor	[54]
$CoMoO_4$@C300	708	4	367 @ 1 $A\,g^{-1}$	The newly formed binary metal oxide nanoclusters exhibit strong retention and outstanding capacitance, making them ideal electrodes for ecofriendly supercapacitors.	Supercapacitor	[55]
CuO/MPC	25.33	2.4–11	616 @ 1 $A\,g^{-1}$	The extremely conducting carbon included in the CuO/MPC nanocomposite provides extra routes for transporting ions as well as numerous electro-active sites.	Supercapacitor	[56]
C/Co_3O_4-750	413	4.3	54 @ 0.5 $A\,g^{-1}$	Electronic conductivity, capacitance, and capacitive behavior all increased as temperature increased.	Supercapacitor	[57]
$FeNiCo_2S_4$@N, S-CMK-3	890	3–6	197.8 @ 1 $A\,g^{-1}$	Resultant materials demonstrate a partially delocalized charge in addition to an elevated density of state intensity near the Fermi level.	Supercapacitor	[58]
OPC-CDs-700	535.3	2.1	312 @ 0.1 $A\,g^{-1}$	Using high-temperature calcination, they combined proanthocyanidins with CDs to create a light-assisted rechargeable supercapacitor with enhanced capacitance.	Supercapacitor	[59]
AOMC	1222–2061	0.9–1.6	37 @ 0.2 $A\,g^{-1}$	CO_2 is used to physically activate OMCs, increasing their surface area from 500 to approximately 2000 $m^2\,g^{-1}$.	Supercapacitor	[35]

[a] S_{BET}, specific surface area.
[b] D, pore diameter.
[c] Cg, specific capacitance.
Source: From [35, 51–59].

reduced temperature oxidation phase (190–240 °C) and an extreme temperature combustion phase (600–800 °C). In a two-electrode setup employing 2 M KOH, the material annealed at 750 °C had a capacity of 54 $F g^{-1}$ at 0.1 $A g^{-1}$, an enhanced rate efficiency of 82% after 10,000 cycles and an extremely high-rate capability of 48.7% at 10 $A g^{-1}$ [57]. Lai et al. produced Fe-doped $NiCo_2S_4$ nanosheets coated over N, S-doped OMC (Fe-$NiCo_2S_4$@N, S-CMK-3). At a scan rate of 2 $mV s^{-1}$, these nanosheets show a higher specific capacitance of 197.8 $F g^{-1}$ in ionic liquid-based supercapacitors [58]. After high-temperature calcination, Wang and colleagues coupled proanthocyanidins with CDs (OPC-CDs-700) to produce a light-assisted reusable supercapacitor with significantly enhanced capacitance. The OPC-CDs-700 provides a specific capacitance of 312 $F g^{-1}$ at 0.1 $A g^{-1}$ under light luminescence, with a 54.4% rise in capacitance under dark conditions [59]. Castro-Gutierrez and colleagues used only tannin, water and Pluronic F127 as raw ingredients to create OMCs utilizing the simple, quick, and environmentally friendly surfactant-water-assisted mechanochemical mesostructuration (SWAMM) approach. CO_2 is used to mechanically stimulate OMCs, increasing their surface area from about 500 to over 2000 $m^2 g^{-1}$. The maximum cell capacitance values of certain stimulated OMCs in water and organic electrolyte solutions are 37 and 27 $F g^{-1}$, respectively, at 0.2 $A g^{-1}$ [35].

8.3.3 Applications in Fuel Cells

One of the most important processes in fuel cells is oxygen reduction reaction (ORR), which produces water as a byproduct while electrochemically reducing oxygen and oxidizing hydrogen to generate energy. The electrodes based on platinum (Pt) are the most appropriate electrocatalyst for ORR. Nevertheless, the main obstacles to the commercialization of Pt-based fuel cell electrocatalysts are their high cost and unavailability. In addition to being expensive, the Pt-based electrode is vulnerable to CO poisoning and fuel crossover in electrochemical settings. In order to find non-platinum ORR electrocatalysts for FCs that are both inexpensive and have a high ORR activity, an enormous amount of work has been done [14]. Carbon is a good catalyst support because of its broad surface area and high electrical conductivity. The ideal alternative is the mesoporous structure because when the catalyst is manufactured simultaneously, it allows the catalyst nanoparticles to be consistently dispersed throughout the mesoporous framework and facilitates the creation of homogeneous nanoparticles over the mesopores as a template. Making carbon/catalyst nanocomposites is trickier than it first appears since a lot of the catalyst is coated with carbon, making it electrically non-responsive [11]. Some recent progress made over last five years in using mesoporous nanostructures for fuel cell applications is presented in Table 8.4 [60–69].

Table 8.4 Recent advancements in fuel cells.

Material	$^{a}S_{BET}$ (m^2 g^{-1}]	^{b}D (nm)	Current density (mA cm^{-2})	Principal findings and remarks	Application	Reference
mC-TpBpy-Fe	362		5.92	Mesoporous carbon produced from COF in a Zn-air battery demonstrated that it could function competitively in ambient settings.	ORR	[60]
NMCNs	635	37	5.8	By modifying the superficial contacts between dopamine and F127 in water/ethanol systems as a whole the facilitator, TMB has been demonstrated to be essential in the production of carbon nanospheres.	ORR	[61]
Fe, N doped mesoporous carbon	800	3.8	4.6	In the electrochemical elimination of oxygen in polymeric electrolyte fuel cells, inexpensive mesoporous carbon loaded with Fe and N_2 shows promise as a catalytic substitute for expensive metal catalysts.	ORR	[62]
NSMC	666	8.8	2.8	NSMC displayed a smaller Tafel slope of 70 mV dec^{-1}, a small percentage of production of peroxide (81%), an average current density of 2.8 mA cm^{-2}, a two-electron pathway and a greater positive onset potential of 0.78 V and half-wave potential of 0.68 V.	ORR	[63]
P-MC-4	737	5.8	4.25	By using the simplest dehydrogenation pathway and the straight four-electron channel for oxygen reduction, each produced EC effectively stimulates the FAO process.	ORR	[64]
Pd-CeO_2/NMCS	464.7	3.5	1.5	In contrast to conventional carbon black supports, the NMCS's nitrogen doping and predominant mesopores offer a wealth of Pd deposition seeding sites.	Methanol oxidation	[65]
NiCu hybrid oxides @OMC	436	3.8	182.07	The large surface area, small pore size distribution and significant amount of mechanical ordering of OMC suggest that it holds possibility to be used as a foundation material for nanoparticles.	Methanol oxidation	[66]

1Ni1Ru/N-OMC	409	3.8	107.47	The composite 1Ni1Ru/NOMC shows enhanced electrocatalytic ability because of the combined effect of triple component, which enables methanol splitting along with complete oxidation at weaker potentials.	Methanol oxidation	[67]
PdePdO/OMC	743.6	3.7	1.13	With the use of Pd ePdO/OMC catalyst, an air-breathing microfluidic fuel cell achieves the highest possible power density of 63.0 $mW cm^{-2}$, which is nearly one-fold greater than similar previous experiments.	Formic acid oxidation	[68]
PdeCo/OMC	538.3	3.5	185	An improvement in electrocatalytic performance is achieved through the elimination of intermediate species generated during the FAOR through the use of an OMC structure.	Formic acid oxidation	[69]

[a] S_{BET}, specific surface area.
[b] D, pore diameter.
Source: From [60–69].

By using a novel p-toluenesulfonic acid-facilitated mechanochemical technique, Zhao and colleagues were able to synthesize a silicate-plated bipyridine-containing covalent organic framework. After carbonization and template removal, the resulting material was iron-nitrogen-doped MC (mC-TpBpy-Fe). The final product mC-TpBpy-Fe has a significant pore volume and surface area and exhibits a limiting current density of 5.92 mA cm^{-2} and an adequate half-wave potential of 0.845 V [60]. In order to create highly homogeneous N-doped MC nanospheres with both nanosized diameter and huge adjustable pore diameters, Peng and colleagues established a flexible nanoemulsion assembly technique. Due to these special qualities, dendritic MC nanospheres have a higher current density (5.8 mA cm^{-2}), better durability (~98% after 10,000 cycles), and superior methanol tolerance for ORR [61]. The production, characterization, and electrocatalytic properties of MC materials coated with Fe and N_2 atoms for the catalytic reduction of oxygen gas at fuel cell cathodes are reported and compared by Kim and coworkers. The resulting product exhibits large surface areas about 800 m^2 g^{-1} strong electric conductance approximately 19 S cm^{-1} making it desirable to be utilized as an excellent electrocatalyst [62]. Duraisamy and colleagues utilized two-dimensional Santa Barbara fifteen amorphous (SBA-15) silica as a foundation and L-cysteine as an N and S dual precursor for the production of NSMC. In comparison to RHE, NSMC-0.4 displayed a higher positive onset potential of 0.78 V, a half-wave potential of 0.68

V, a current density of 2.8 mA cm^{-1} and an 81% peroxide generation in a basic electrolyte [63]. Using phenolic resin as the carbon source, F127 as the soft template and dialkylphosphinic acid as the P source, Zhao and co-workers produced P-doped MCs. Trace amounts of iron species were introduced to act as a catalyst for the graphitization process. In a basic electrolyte, the resultant P-doped MCs (P-MC-4) follow a ~four electron route and have impressive ORR activity [64]. High dispersion Pd nanoparticles encircled by CeO_2 dots were deposited by Tan and colleagues on a nitrogen-doped, core-shell structured MC sphere (NMCS) support. The NMCS demonstrated CO tolerance, electrocatalytic activity and stability for the MOR in alkaline environments. The anode for Pd(20%)-CeO_2(20%)/NMCS showed an open-circuit voltage of 0.86 V and produced current densities of 20 mA cm^{-2} at 0.5 V [65]. Theres and colleagues used a non-conventional soft-template method to create OMC with a large surface area. By using the wet chemical reduction process, 15% weight of three distinct Ni:Cu compositions (3 : 1, 1 : 3, and 1 : 3) and their monometallic counterparts have been created. Among all the tested samples, NiCu hybrid oxides@OMC electrocatalyst had the largest specific surface area (436 m^2 g^{-1}), greatest current density (182.07 mA cm^{-2}), a minimum onset potential (0.34 V) and charge transfer resistance (R_{ct}) value of 0.355 Ω with 87.68% persistence in current density over 7200 seconds. This was achieved when Ni and Cu were combined in equal weight% [66].

For efficient methanol electro-oxidation, Gubert and colleagues produced a ternary hybrid catalyst made up of nitrogen-doped OMC (catalyst/co-catalyst/scaffold) doped with B, Ru, and Ni. The precise ratio of Ni to Ru was 3 : 1 or 1 : 1 or 1 : 3. When the two metal components (Ni and Ru) combined in a 1 : 1 weight percent ratio (1Ni1Ru/N-OMC), the maximum current density, lowest onset potential, and significant two-hour stability were observed [67]. In order to produce the unique OMC-supported palladium-palladium oxide catalyst (Pd-PdO/OMC), Zhou and colleagues developed a new synthesis method. The catalyst was then used in the anode of a direct formic acid microfluidic fuel cell. In addition to demonstrating excellent activity for the electro-oxidation of formic acid, the PdePdO/OMC catalyst also demonstrated excellent stability due to a stronger anchoring effect of PdePdO catalyst on OMC substrate [68]. The polyol technique of PdeCo/OMC synthesis and its electrochemical evaluation for the FAOR were published by Cazares-Avila and coworkers. The anodic mass current density (185 mA mg^{-1} Pd) of the FAOR is delivered by PdeCo/OMC at 0.5 mol l^{-1} HCOOH, whereas PdeCo/MWCNT only delivers 94.3 mA mg^{-1} Pd [69]. It is clear from the above discussion that the modified electrocatalysts have a variety of the applications in fuel cell to improve their viability.

8.4 Further Directions, Opportunities, and Challenges

Although functionalized OMCs exhibit remarkable performance in energy-storage applications, a workable approach for mass-producing affordable electrode materials and electrocatalysts that are extremely steady and trustworthy still has to be discovered. During synthesis, this material could get mixed up with the nanoporous material. Thus, the next step will be production of contamination-free functionalized ordered mesoporous material. The main challenge is to more economically synthesize very pure functionalized MC materials without any contamination utilizing inexpensive biodegradable carbon or nitrogen precursors instead of using extremely hazardous and expensive chemicals. During high-temperature pyrolysis synthesis, uncontrollable doping and irreversible molecular agglomeration present more challenges that must be resolved for improved use. Additionally, from an industrial perspective, the reduced production of MC using existing technology continues to be an issue, which should be considered during material synthesis.

In order to address these issues, the following information must be taken into account:

1) Selecting an OMC with an appropriate form—1D, 2D, or 3D—while combining OMC with other materials is a prudent practice. The OMC framework can

take on a variety of shapes, including spherical, cage-like and honeycomb-like forms.

2) The variety of sizes of pores is extended by extending the bounds of the controllable mesoporous size and preserving an exceptionally durable, ordered mesoscopic framework.
3) Make the most of the unique properties of the material, consider its structure and be aware of alterations in the physical properties of the pore structure.

8.5 Conclusions

This chapter provides a thorough understanding of the synthesis techniques for MC materials and the range of uses for them in energy storage devices, including fuel cells, lithium-ion batteries and supercapacitors. The latest developments in MC synthesis techniques, including soft and hard templates, were thoroughly examined. In conclusion, MC materials, with their distinct mesoporous structure, high surface area, high carbonaceous content and greater biocompatibility, are thought to be an upcoming class of inorganic substances for use in renewable energy applications. Even though several studies have shown that mesoporous structures are appropriate materials for energy storage, the field of study is still in its very early stages. More research is needed to determine practical and affordable ways to prepare MCs for widespread use in both commercial and scientific contexts. In order to further improve OMC for green energy, we anticipate that new ideas will arise from the transfer of information from earlier studies.

References

1 Yu, S., Ng, V.M.H., Wang, F. et al. (2018). Synthesis and application of iron-based nanomaterials as anodes of lithium-ion batteries and supercapacitors. *Journal of Materials Chemistry A* 6 (20). Royal Society of Chemistry: 9332–9367. https://doi.org/10.1039/c8ta01683f.

2 Liu, C., Li, F., Lai-Peng, M. et al. (2010). Advanced materials for energy storage. *Advanced Materials* 22 (8): https://doi.org/10.1002/adma.200903328.

3 Gao, Y., Wang, Q., Ji, G. et al. (2021). Doping strategy, properties and application of heteroatom-doped ordered mesoporous carbon. *RSC Advances* 11 (10). Royal Society of Chemistry: 5361–5383. https://doi.org/10.1039/d0ra08993a.

4 Wang, J., Xin, H.L., and Wang, D. (2014). Recent progress on mesoporous carbon materials for advanced energy conversion and storage. *Particle and Particle Systems Characterization* 31 (5): 515–539. https://doi.org/10.1002/ppsc.201300315.

5 Liu, C.J., Burghaus, U., Besenbacher, F. et al. (2010). Preparation and characterization of nanomaterials for sustainable energy production. *ACS Nano* 4 (10): 5517–5526. https://doi.org/10.1021/nn102420c.

6 Zu, L., Zhang, W., Qu, L. et al. (2020). Mesoporous materials for electrochemical energy storage and conversion. *Advanced Energy Materials* 10 (38): https://doi.org/10.1002/aenm.202002152.

7 Mehdipour-Ataei, S. and Aram, E. (2023). Mesoporous carbon-based materials: a review of synthesis, modification, and applications. *Catalysts* 13 (1). MDPI: https://doi.org/10.3390/catal13010002.

8 Xin, W. and Song, Y. (2015). Mesoporous carbons: recent advances in synthesis and typical applications. *RSC Advances* 5 (101). Royal Society of Chemistry: 83239–83285. https://doi.org/10.1039/c5ra16864c.

9 Rahman, M.M., Ara, M.G., Alim, M.A. et al. (2021). Mesoporous carbon: a versatile material for scientific applications. *International Journal of Molecular Sciences* 22 (9). MDPI: https://doi.org/10.3390/ijms22094498.

10 Azam, M.A., Ramli, N.S.N., Nor, N.A.N.M. et al. (2021). Recent advances in biomass-derived carbon, mesoporous materials, and transition metal nitrides as new electrode materials for supercapacitor: a short review. *International Journal of Energy Research* 45 (6). John Wiley and Sons Ltd: 8335–8346. https://doi.org/10.1002/er.6377.

11 Eftekhari, A. and Fan, Z. (2017). Ordered mesoporous carbon and its applications for electrochemical energy storage and conversion. *Materials Chemistry Frontiers* 1 (6). Royal Society of Chemistry: 1001–1027. https://doi.org/10.1039/c6qm00298f.

12 Lu, A.H. and Schüth, F. (2006). Nanocasting: a versatile strategy for creating nanostructured porous materials. *Advanced Materials* 18 (14): 1793–1805. https://doi.org/10.1002/adma.200600148.

13 Collins, G., Kasturi, P.R., Karthik, R. et al. (2023). Mesoporous carbon-based materials and their applications as non-precious metal electrocatalysts in the oxygen reduction reaction. *Electrochimica Acta* 439: https://doi.org/10.1016/j.electacta.2022.141678.

14 Benzigar, M.R., Talapaneni, S.N., Joseph, S. et al. (2018). Recent advances in functionalized micro and mesoporous carbon materials: synthesis and applications. *Chemical Society Reviews* 47 (8). Royal Society of Chemistry: 2680–2721. https://doi.org/10.1039/c7cs00787f.

15 Kim, D.Y., Jin, X., Lee, C.H. et al. (2018). Improved electrochemical performance of ordered mesoporous carbon by incorporating macropores for Li–O_2 battery cathode. *Carbon N Y* 133: 118–126. https://doi.org/10.1016/j.carbon.2018.03.016.

16 Zhao, M., Zhao, D.L., Han, X.Y. et al. (2018). Ge nanoparticles embedded in spherical ordered mesoporous carbon as anode material for high performance lithium ion batteries. *Electrochimica Acta* 287: 21–28. https://doi.org/10.1016/j.electacta.2018.08.050.

17 Zeng, L., Liu, R., Han, L. et al. (2018). Preparation of a Si/SiO_2–ordered-mesoporous-carbon nanocomposite as an anode for high-performance lithium-ion and sodium-ion batteries. *Chemistry - A European Journal* 24 (19): 4841–4848. https://doi.org/10.1002/chem.201704780.

18 Fang, Y., Liu, R., Zeng, L. et al. (2019). Preparation of Ge/N, S co-doped ordered mesoporous carbon composite and its long-term cycling performance of lithium-ion batteries. *Electrochimica Acta* 318: 737–745. https://doi.org/10.1016/j.electacta.2019.06.123.

19 Wang, W., Zhou, J., Wang, Z. et al. (2018). Short-range order in mesoporous carbon boosts potassium-ion battery performance. *Advanced Energy Materials* 8 (5): https://doi.org/10.1002/aenm.201701648.

20 Wang, J., Xia, Y., Liu, Y. et al. (2019). Mass production of large-pore phosphorus-doped mesoporous carbon for fast-rechargeable lithium-ion batteries. *Energy Storage Materials* 22: 147–153. https://doi.org/10.1016/j.ensm.2019.01.008.

21 Xu, C., Niu, D., Zheng, N. et al. (2018). Facile synthesis of nitrogen-doped double-shelled hollow mesoporous carbon nanospheres as high-performance anode materials for lithium ion batteries. *ACS Sustainable Chemistry & Engineering* 6 (5): 5999–6007. https://doi.org/10.1021/acssuschemeng.7b04617.

22 Xia, G., Wang, C., Jiang, P. et al. (2019). Nitrogen/oxygen co-doped mesoporous carbon octahedrons for high-performance potassium-ion batteries. *Journal of Materials Chemistry A* 7 (19): 12317–12324. https://doi.org/10.1039/c8ta12504j.

23 Li, Y., Li, C., Qi, H. et al. (2018). Mesoporous activated carbon from corn stalk core for lithium ion batteries. *Chemical Physics* 506: 10–16. https://doi.org/10.1016/j.chemphys.2018.03.027.

24 Kesavan, T., Partheeban, T., Vivekanantha, M. et al. (2020). Design of P-doped mesoporous carbon nitrides as high-performance anode materials for Li-ion battery. *ACS Applied Materials & Interfaces* 12 (21): 24007–24018. https://doi.org/10.1021/acsami.0c05123.

25 Bairi, P., Maji, S., Hill, J.P. et al. (2019). Mesoporous carbon cubes derived from fullerene crystals as a high rate performance electrode material for supercapacitors. *Journal of Materials Chemistry A* 7 (20): 12654–12660. https://doi.org/10.1039/c9ta00520j.

26 Zhang, F., Zong, S., Zhang, Y. et al. (2021). Preparation of hollow mesoporous carbon spheres by pyrolysis-deposition using surfactant as carbon precursor. *Journal of Power Sources* 484: https://doi.org/10.1016/j.jpowsour.2020.229274.

27 Wang, J.G., Liu, H., Sun, H. et al. (2018). One-pot synthesis of nitrogen-doped ordered mesoporous carbon spheres for high-rate and long-cycle life supercapacitors. *Carbon N Y* 127: 85–92. https://doi.org/10.1016/j.carbon.2017.10.084.

28 Liu, B., Liu, L., Yu, Y. et al. (2020). Synthesis of mesoporous carbon with tunable pore size for supercapacitors. *New Journal of Chemistry* 44 (3): 1036–1044. https://doi.org/10.1039/c9nj05085j.

29 Nazari, M., Rahmanifar, M.S., Noori, A. et al. (2021). The ordered mesoporous carbon nitride-graphene aerogel nanocomposite for high-performance supercapacitors. *Journal of Power Sources* 494: https://doi.org/10.1016/j.jpowsour.2021.229741.

30 Xin, X., Kang, H., Feng, J. et al. (2020). A novel sol-gel strategy for N, P dual-doped mesoporous carbon with high specific capacitance and energy density for advanced supercapacitors. *Chemical Engineering Journal* 393: https://doi.org/10.1016/j.cej.2020.124710.

31 Zhu, S., Taberna, P.L., Zhao, N. et al. (2018). Salt-template synthesis of mesoporous carbon monolith for ionogel-based supercapacitors. *Electrochemistry Communications* 96: 6–10. https://doi.org/10.1016/j.elecom.2018.09.003.

32 Mo, Y., Du, J., Lv, H. et al. (2021). N-doped mesoporous carbon nanosheets for supercapacitors with high performance. *Diamond and Related Materials* 111: https://doi.org/10.1016/j.diamond.2020.108206.

33 Liang, K., Ma, C., Liu, L. et al. (2019). Nitrogen and oxygen co-doped ordered dual-mesoporous carbon for supercapacitors. *Journal of Alloys and Compounds* 805: 859–867. https://doi.org/10.1016/j.jallcom.2019.07.129.

34 Herou, S., Ribadeneyra, M.C., Madhu, R. et al. (2019). Ordered mesoporous carbons from lignin: a new class of biobased electrodes for supercapacitors. *Green Chemistry* 21 (3): 550–559. https://doi.org/10.1039/c8gc03497d.

35 Castro-Gutiérrez, J., Díez, N., Sevilla, M. et al. (2019). High-rate capability of supercapacitors based on tannin-derived ordered mesoporous carbons. *ACS Sustainable Chemistry & Engineering* 7 (21): 17627–17635. https://doi.org/10.1021/acssuschemeng.9b03407.

36 Liu, X., Zhou, Y., Wang, C.L. et al. (2022). Solvent-free self-assembly synthesis of N-doped ordered mesoporous carbons as effective and bifunctional materials for CO_2 capture and oxygen reduction reaction. *Chemical Engineering Journal* 427: https://doi.org/10.1016/j.cej.2021.130878.

37 Liao, Y., Wang, Y., Zhang, R. et al. (2020). Anti-sintering Pt particles confined in short ordered mesoporous carbon with rapid mass transport for superior and robust oxygen reduction. *ChemCatChem* 12 (7): 1958–1962. https://doi.org/10.1002/cctc.201902282.

38 Duraisamy, V., Palanivel, S., Thangamuthu, R. et al. (2018). KIT-6 three dimensional template derived mesoporous carbon for oxygen reduction reaction: effect of template removal on catalytic activity. *ChemistrySelect* 3 (42): 11864–11874. https://doi.org/10.1002/slct.201802539.

39 Hassan, M., Liu, T., Bo, X. et al. (2019). Cobalt-doped carbon nitride supported on ordered mesoporous carbon as noble metal-free oxygen reduction electrocatalysts. *Journal of Physics and Chemistry of Solids* 131: 111–118. https://doi.org/10.1016/j.jpcs.2019.03.023.

40 Liu, X., Zou, S., Chen, S. et al. (2016). Ordered mesoporous carbons codoped with nitrogen and iron as effective catalysts for oxygen reduction reaction. *Nanoscale* 8 (46): 19249–19255. https://doi.org/10.1039/c6nr05884a.

41 Xu, J., Zhao, Y., and Shen, C. (2013). Sulfur and nitrogen-doped, ferrocene-derived mesoporous carbons with efficient electrochemical reduction of oxygen. *ACS Applied Materials & Interfaces* 5 (23): 12594–12601. https://doi.org/10.1021/am4039294.

42 Wang, X., Li, Z., Yin, L. et al. (2013). Nanocomposites of SnO_2@ordered mesoporous carbon (OMC) as anode materials for lithium-ion batteries with improved electrochemical performance. *CrystEngComm* 15 (37): 7589–7597. https://doi.org/10.1039/c3ce41256c.

43 Li, Z., Wang, X., Wang, C. et al. (2013). Hybrids of iron oxide/ordered mesoporous carbon as anode materials for high-capacity and high-rate capability lithium-ion batteries. *RSC Advances* 3 (38): 17097–17104. https://doi.org/10.1039/c3ra40682b.

44 Hu, J., Sun, C.F., Gillette, E. et al. (2016). Dual-template ordered mesoporous carbon/Fe_2O_3 nanowires as lithium-ion battery anodes. *Nanoscale* 8 (26): 12958–12969. https://doi.org/10.1039/c6nr02576e.

45 Chen, A., Yin, L., and Qi, Y. (2013). MoO_2 /ordered mesoporous carbon hybrids as anode materials with highly improved rate capability and reversible capacity for lithium-ion battery. *Physical Chemistry Chemical Physics* 15 (32): 13601–13610. https://doi.org/10.1039/c3cp51255j.

46 Wang, C., Zhao, Y., Zhou, L. et al. (2018). Mesoporous carbon matrix confinement synthesis of ultrasmall WO_3 nanocrystals for lithium ion batteries. *Journal of Materials Chemistry A* 6 (43): 21550–21557. https://doi.org/10.1039/c8ta07145d.

47 Ma, J., Xiang, D., Li, Z. et al. (2013). TiO_2 nanocrystal embedded ordered mesoporous carbons as anode materials for lithium-ion batteries with highly reversible capacity and rate performance. *CrystEngComm* 15 (34): 6800–6807. https://doi.org/10.1039/c3ce40642c.

48 Zeng, L., Huang, X., Chen, X. et al. (2016). Ge/GeO_2-ordered mesoporous carbon nanocomposite for rechargeable lithium-ion batteries with a long-term cycling performance. *ACS Applied Materials & Interfaces* 8 (1): 232–239. https://doi.org/10.1021/acsami.5b08470.

49 Zeng, L., Zheng, C., Xi, J. et al. (2013). Composites of V_2O_3-ordered mesoporous carbon as anode materials for lithium-ion batteries. *Carbon N Y* 62: 382–388. https://doi.org/10.1016/j.carbon.2013.06.021.

50 Park, J., Kim, G.P., Umh, H.N. et al. (2013). Co_3O_4 nanoparticles embedded in ordered mesoporous carbon with enhanced performance as an anode material for Li-ion batteries. *Journal of Nanoparticle Research* 15 (9): https://doi.org/10.1007/s11051-013-1943-4.

51 Xie, M., Meng, H., Chen, J. et al. (2021). High-volumetric supercapacitor performance of ordered mesoporous carbon electrodes enabled by the faradaic-active nitrogen doping and decrease of microporosity. *ACS Applied Energy Materials* 4 (2): 1840–1850. https://doi.org/10.1021/acsaem.0c02948.

52 Zhang, Y., Qi, F., and Liu, Y. (2020). Fabrication of high B-doped ordered mesoporous carbon with 4-hydroxyphenylborate phenolic resin for supercapacitor electrode materials. *RSC Advances* 10 (19): 11210–11218. https://doi.org/10.1039/d0ra00561d.

53 Chen, P., Yang, C., He, Z. et al. (2019). One-pot facile route to fabricate the precursor of sulfonated graphene/N-doped mesoporous carbons composites for supercapacitors. *Journal of Materials Science* 54 (5): 4180–4191. https://doi.org/10.1007/s10853-018-3122-6.

54 Mohammadi, N., Pourreza, K., Adeh, N.B. et al. (2021). Defective mesoporous carbon/MnO_2 nanocomposite as an advanced electrode material for supercapacitor application. *Journal of Alloys and Compounds* 883: https://doi.org/10.1016/j.jallcom.2021.160874.

55 Li, J., Zhao, C., Yang, Y. et al. (2019). Synthesis of monodispersed $CoMoO_4$ nanoclusters on the ordered mesoporous carbons for environment-friendly supercapacitors. *Journal of Alloys and Compounds* 810: https://doi.org/10.1016/j.jallcom.2019.151841.

56 Saravanakumar, B., Radhakrishnan, C., Ramasamy, M. et al. (2019). Copper oxide/mesoporous carbon nanocomposite synthesis, morphology and electrochemical properties for gel polymer-based asymmetric supercapacitors. *Journal of Electroanalytical Chemistry* 852: https://doi.org/10.1016/j.jelechem.2019.113504.

57 Zallouz, S., Réty, B., Vidal, L. et al. (2021). Co_3O_4 nanoparticles embedded in mesoporous carbon for supercapacitor applications. *ACS Applied Nano Materials* 4 (5): 5022–5037. https://doi.org/10.1021/acsanm.1c00522.

58 Lai, F., Feng, J., Ye, X. et al. (2019). Partially delocalized charge in Fe-doped $NiCo_2S_4$ nanosheet-mesoporous carbon-composites for high-voltage supercapacitors. *Journal of Materials Chemistry A* 7 (33): 19342–19347. https://doi.org/10.1039/c9ta06250e.

59 Wang, H., Cao, J., Zhou, Y. et al. (2020). Carbon dot-modified mesoporous carbon as a supercapacitor with enhanced light-assisted capacitance. *Nanoscale* 12 (34): 17925–17930. https://doi.org/10.1039/d0nr05532h.

60 Zhao, X., Pachfule, P., Li, S. et al. (2019). Silica-templated covalent organic framework-derived Fe-N-doped mesoporous carbon as oxygen reduction electrocatalyst. *Chemistry of Materials* 31 (9): 3274–3280. https://doi.org/10.1021/acs.chemmater.9b00204.

61 Peng, L., Hung, C.T., Wang, S. et al. (2019). Versatile nanoemulsion assembly approach to synthesize functional mesoporous carbon nanospheres with tunable pore sizes and architectures. *Journal of the American Chemical Society* 141 (17): 7073–7080. https://doi.org/10.1021/jacs.9b02091.

62 Kim, D., Zussblatt, N.P., Chung, H.T. et al. (2018). Highly graphitic mesoporous Fe,N-doped carbon materials for oxygen reduction electrochemical catalysts. *ACS Applied Materials & Interfaces* 10 (30): 25337–25349. https://doi.org/10.1021/acsami.8b06009.

63 Duraisamy, V., Venkateshwaran, S., Thangamuthu, R. et al. (2022). Hard template derived N, S dual heteroatom doped ordered mesoporous carbon as an efficient electrocatalyst for oxygen reduction reaction. *International Journal of Hydrogen Energy* 47 (95): 40327–40339. https://doi.org/10.1016/j.ijhydene.2022.03.250.

64 Zhao, H., Hu, Z., Zhu, Y. et al. (2019). P-doped mesoporous carbons for high-efficiency electrocatalytic oxygen reduction. *Chinese Journal of Catalysis* 40 (9): 1366–1374. https://doi.org/10.1016/j.ijhydene.2022.03.250.

65 Tan, Q., Shu, C., Abbott, J. et al. (2019). Highly dispersed Pd-CeO_2 nanoparticles supported on N-doped core-shell structured mesoporous carbon for methanol oxidation in alkaline media. *ACS Catalysis* 9 (7): 6362–6371. https://doi.org/10.1021/acscatal.9b00726.

66 Theres, G.S., Velayutham, G., Krishnan, P.S. et al. (2019). Synergistic impact of Ni–Cu hybrid oxides deposited on ordered mesoporous carbon scaffolds as non-noble catalyst for methanol oxidation. *Journal of Materials Science* 54 (2): 1502–1519. https://doi.org/10.1007/s10853-018-2884-1.

67 Gubert, S.T., Guruviah, V., and Kannan, S. (2020). Boron-doped Ni-Ru supported on nitrogen-doped ordered mesoporous carbon as effective electrocatalyst for methanol electrooxidation. *ChemistrySelect* 5 (28): 8455–8464. https://doi.org/10.1002/slct.202001842.

68 Zhou, Y., Zhu, X., Zhang, B. et al. (2020). High performance formic acid fuel cell benefits from Pd–PdO catalyst supported by ordered mesoporous carbon. *International Journal of Hydrogen Energy* 45 (53): 29235–29245. https://doi.org/10.1016/j.ijhydene.2020.07.169.

69 Cazares-Ávila, E., Ruiz-Ruiz, E.J., Hernández-Ramírez, A. et al. (2017). Effect of OMC and MWNTC support on mass activity of Pd–Co catalyst for formic acid electro-oxidation. *International Journal of Hydrogen Energy* 42 (51): 30349–30358. https://doi.org/10.1016/j.ijhydene.2017.08.156.

9

Green Synthesis of Carbon Dots and Its Application in Hydrogen Generation Through Water Splitting

Mandakini Gupta

Department of Chemistry, Sunbeam Women's College Varuna, Varanasi, Uttar Pradesh, India

9.1 Introduction

The worldwide increase in CO_2 emissions has experienced a significant surge since the Industrial Revolution leading to a major cause of global warming. As per the report of the United States Environmental Protection Agency, in 2021, the global average atmospheric carbon dioxide level reached the highest, recorded in the past 800,000 years (about 415.26 ppm). This rise in CO_2 levels has contributed to an average surface temperature increase of approximately 0.8–1.1 °C, posing a threat to the ecosystem [1, 2]. According to data, fossil fuels fulfill nearly all of the world's energy needs (approx. 86.4%) [3]. Thus, the main origins of artificial carbon dioxide emissions are closely linked to the burning of fossil fuels. Addressing this environmental issue requires urgent global action to shift our energy needs toward sustainable methods of energy generation, i.e. renewable energy.

Renewable energies such as wind, hydroelectric, solar, and geothermal are environmentally friendly. Solar power, which annually absorbs 3.85×10^{24} J in the Earth's atmosphere, can be harnessed globally for electricity and heat generation. Currently, only 2% of total solar energy can be utilized as green energy through different available methods [3].

In addition to the aforementioned renewable resources, hydrogen is identified as a promising, clean, and abundant fuel capable of replacing energy generated from fossil fuels. It possesses 2.75 times the energy content of hydrocarbon fuels and is 2.40 times more active as compared to methane [3]. Hydrogen can be

Carbon-based Nanomaterials for Green Applications, First Edition. Edited by Upendra Kumar, Piyush Kumar Sonkar, and Suman Lata Tripathi.

Published 2025 by John Wiley & Sons, Inc.

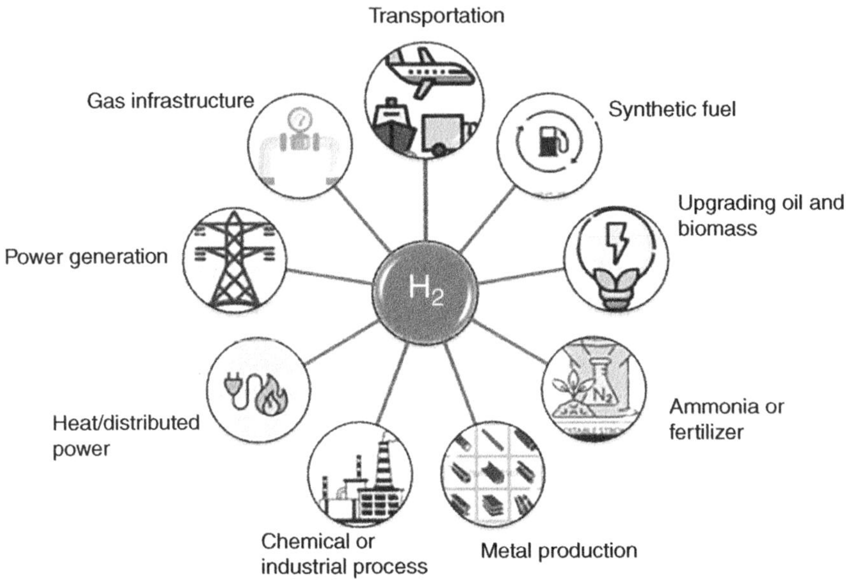

Figure 9.1 Hydrogen generation as a clean and renewable source of energy and its application as fuel in different sectors.

generated as a fuel from various sources and through diverse methods. The predominant approaches involve steam-methane reforming and electrolysis, which involves splitting water with electricity. However, scientists are actively investigating alternative methods for hydrogen production such as thermochemical procedures that transform biomass into gases or liquids and subsequently extract hydrogen. Biological methods utilizing microbes like microalgae and bacteria to generate hydrogen, and photolytic techniques utilizing solar energy to break up water into oxygen and hydrogen are among the diverse approaches explored for hydrogen production [4] (Figure 9.1).

Further, photocatalytic water splitting, a method for renewable energy production has become increasingly studied, offering a carbon dioxide emission-free alternative to fossil fuel combustion. Abundant water serves as a promising source through semiconductor-based photocatalysis as one of the ways for hydrogen production nowadays.

This approach involves utilizing the sun's energy to facilitate reactions between photoinduced protons and electrons, ultimately yielding molecules of hydrogen. It might offer a sustainable solution to the world's energy requirements.

In this chapter, our prime focus will be CDs as a novel class of carbon-based nanomaterials, their green synthesis, their properties, and their application toward hydrogen generation from photocatalytic water splitting.

9.2 Carbon Dots

CDs are carbon-based nanomaterials characterized by their zero-dimensional structure and a size of less than 10 nm. Their discovery traces back to 2004 when the single-walled carbon nanotubes (SWCNTs) were produced through arc-discharge methods during electrophoretic purification by Xu et al. [2]. Arc-synthesized SWCNTs were purified via electrophoresis method using matrixes of agarose gel and glass beads. Two major impurities were short tubular carbon and fluorescent carbon, which were found to be interesting nanomaterials. Analysis of these impurities was performed. As a novel class of nano9.2

materials based on carbon, CDs display distinctive optical and electrical properties and also exhibit tunable fluorescence properties, garnering increasing attention [5]. Figure 9.2a–g highlights significant milestones in the development of CDs. In 2006, surface passivation was first employed to significantly enhance emission by modifying surface chemistry by Sun et al. [6] and carbon quantum dots (CQDs) were separated as fluorescent carbon nanoparticles. By 2010, size-dependent photoluminescence (PL) had been discovered through the effective isolation of size-controlled CDs [7]. The synthesis of CDs using polymers occurred for the first time in 2013. Chiral CDs were first reported in 2016, expanding the

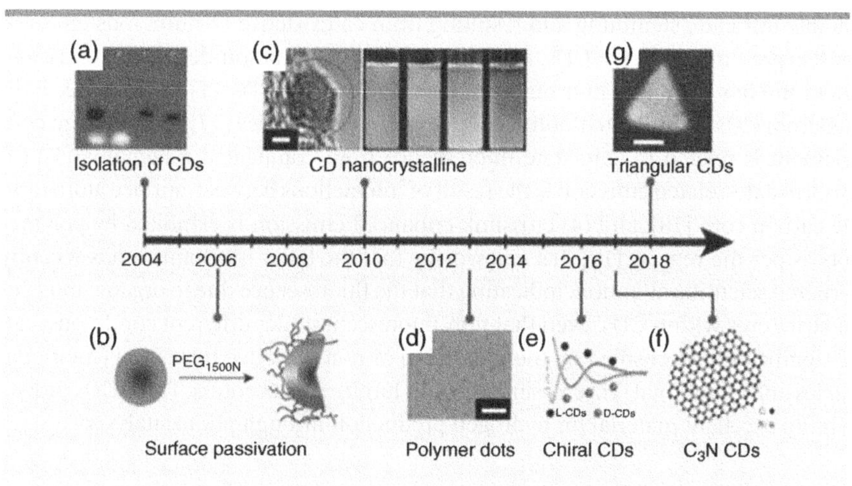

Figure 9.2 (a) Discovery of CDs during gel electrophoresis of CNTs, (b) passivation of CDs as more emissive material, (c) crystalline CDs with size-dependent fluorescent property, (d) fluorescent polymer dots, (e) circular dichroism spectrum of chiral CDs, (f) C_3N CDs' chemical structure, (g) CDs' controlled morphology with a triangular shape. *Source:* Bowen et al. (2019)/with permission of Elsevier.

applications of CDs [8]. In 2018, triangular-shaped CDs with an ultra-narrow emission bandwidth were synthesized [9]. Based on their specific carbon core structure and surface groups, CDs can be categorized into three main types: graphene or graphene oxide quantum dots (GQDs or GOQDs), carbonized polymer dots (CPDs) and CQDs [3].

Typical CDs are composed of an sp^2-/sp^3-hybridized carbon core inside and a layer consisting of plentiful functional groups such as carboxyl (—COOH), hydroxyl (—OH), and amine (—NH_2) outside. Their functionalized surfaces and small sizes also endow CDs with many eccentric properties such as easy-to-modify electronic structure, tunable PL, good dispersion, and excellent electron donor/acceptor properties [9]. Because of these novel properties, CDs find broad applications in photodynamic therapy [10], chemical sensing [11], drug delivery [12], and bioimaging [13] and also emerged as new candidates for photocatalysis due to their easily adjustable electronic structure, supplementing the traditionally studied inorganic semiconductors like CdS and TiO_2 [14, 15]. Furthermore, CDs exhibit high thermo- and photostability, showcasing efficient emission with quantum yields (QYs), which reach up to 90% as per the reports available in literature.

In comparison to classical semiconductor quantum dots, CDs differ significantly, not only in the energy structure but also in expected emission mechanisms. Quantum confinement of energy levels and related optical transitions, common in traditional quantum dots [16], is the one aspect reported for GQDs. The light emission in CDs is notably intricate, stemming and resulting from co-existence of numerous emissive centers present within the CDs, resulting in a broad, excitation-dependent emission. Literature has identified four major emission origins for CDs: (1) fluorescence radiated from CDs is similar to fluorescence of organic molecules [17]; (2) quantum confinement is related to core-state fluorescence from conjugated π-domains [5]; (3) surface-state-related emission is the result of interactions between surface atoms and the carbon core [18]; and (4) crosslink-enhanced emission is exhibited by polymer dots as per the reports [19]. It's noteworthy that the latter mechanism has recently garnered scientific attention, indicating that the fluorescence due to organic molecular structures within CDs, even that non-fluorescent under different conditions, can be significantly intensified by the repression of non-radiative relaxation (rotational modes and vibrational) via crosslinking with luminescent centers. Thus, CDs proved to be an excellent material for hydrogen production through photocatalysis.

9.3 Processes Used for Synthesis of CDs

Most CDs typically consist of an sp^2/sp^3 hybridized carbon core underneath and functional groups on the surface. For example, graphene quantum dots (GQDs) are made up of mono-/multi-layer nanometer-sized graphite with interlayer defects or edge/surface functional groups. The size of π-conjugated domains and the surface/edge structures are the major factors influencing the optical properties of GQDs, which have lateral dimensions greater than their height and are anisotropic [20, 21].

The trajectory of CDs' development underscores a growing demand for pure and size-dependent CDs with well-structured morphology for research and expanding applications. As discussed, CDs possess different functional groups such as carbonyl, amino, and hydroxyl groups, which enable them to become highly polar in nature and responsible for high solubility in water. Organic, inorganic, and even biological species can be used for surface functionalization of CDs [22]. In contrast to GQDs, CPDs and CQDs typically have spherical cores connected with surface groups. In CQDs, spherical cores consist of multiple-layer graphite structures, and their intrinsic state luminescence and the quantum confinement effect of size are mainly responsible for their unique PL emission properties.

The diverse structures of CDs can be achieved by various synthetic routes, such as "bottom-up" and "top-down" approaches [9] (Figure 9.3). Consequently, various synthesis routes can be employed to alter the surface functional groups of CDs, enabling them to show tunable light emission. Specifically, CDs with modified surface moieties contribute to exhibiting efficient up-converted PL behavior and expanding the light utilization range. This suggests that CDs can be designed to efficiently respond to visible light across whole spectrum of sunlight. Moreover, enhanced photocatalytic efficiency is benefited by photoexcited electron transfer and the light-harvesting ability of CDs-based composites.

9.3.1 Bottom-Up Synthesis Processes

The bottom-up method is advantageous for large-scale production of CDs and provides a wide choice of precursors for CD synthesis with tunable structures and properties. This approach utilizes small organic molecules or polymers as precursors for CDs. The synthesis may include polymerization, dehydration, and

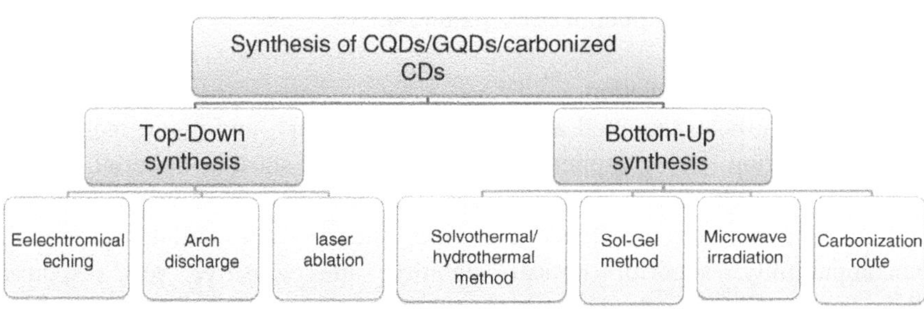

Figure 9.3 General synthesis methods of CQDs/GQDs/carbonized CDs.

carbonization processes, utilizing methods such as microwave pyrolysis, plasma treatment, thermal decomposition, and hydrothermal/acidic oxidation.

9.3.1.1 Solvothermal/Hydrothermal Method

Solvothermal and hydrothermal methods are wet chemical processes used for nanostructure generation. Nanoparticles are synthesized under high pressure in hot solvent/water in an autoclave. Solvothermal or hydrothermal carbonization (HTC) is an environmentally benign method for producing CQDs from various precursors. Ding et al. [23] used HTC to fabricate photoluminescent CQDs from coke powder with a QY of 34.27%, with sizes less than 5 nm. Amino-functionalized luminescent CQDs were produced by Yang et al. [24] using chitosan via HTC at low temperatures. Yang et al. [25] used citric acid with L-serine or monoethanolamine as initiators, deionized water, and N,N-dimethylformamide (DMF), were employed as solvents to manufacture two varieties of nitrogen-doped carbon dots (N-CDs) via the hydrothermal method. The deionized water CDs, with a yellow color, emitted blue fluorescence at 365 nm under UV rays while the DMF-CDs, displaying a deep green color, exhibited blue-green luminescence. TEM studies revealed uniform spherical shapes for DMF-CDs and DW-CDs with sizes of 6.7 and 4.7 nm, respectively. Jin et al. [26] utilized threonine to create CDs, which are pH-sensitive with a size of approximately 8.2 nm and a spherical shape using a solvothermal method. In another approach, Barati et al. [27] synthesized N-CDs using glucose and urea at 200 °C for up to 12 hours with sizes ranging from 4 to 6 nm via hydrothermal route. These N-CDs were employed as pH nanosensors that exhibited independence from concentration over a varied pH range.

9.3.1.2 Sol-Gel Method

This method involves three steps: (1) hydrolysis, (2) condensation (sol formation, i.e. nanoparticles distributed in a solvent through Brownian motion, using a metal precursor in an acid or basic medium), and (3) growth (gel formation). Liang et al. [28] utilized the sol-gel method to fabricate TiO_2@g-CQDs and used it for the degradation of nonylphenol ethoxylate [28].

9.3.1.3 Microwave Irradiation

Microwave synthesis emerged as an eco-friendly, rapid, and green method for CQD preparation. Using *p*-phenylenediamine as the starting material, N-CDs were produced via microwave synthesis, demonstrating great water solubility and emitting orange fluorescence at 592 nm. These N-CDs exhibited promising biocompatibility, superior optical stability, and sensitive pH response. Phenylenediamine has also been employed as a precursor by Sun et al., resulting in CDs with a 15% QY, excitation under 470–620 nm with an approximate diameter of 3.8 nm and high photostability. CDs prepared from this route exhibited pH response within the range of 5–10, with quenching achievable via ferric ions [29].

The use of 1-butyl-3-methylimidazoliumtetra-fluoroborate in the microwave method facilitated the large-scale production of carbon nanoparticles with surface modification agents, resulting in the synthesis of N-CQDs with elevated QY values of 30.2%. These N-doped CQDs demonstrated high sustainability and boundless potential for hydrogen evolution [30].

9.3.1.4 Carbonization Route

In the carbonization route, a carbon-containing precursor is heated to synthesize desired quantum dots using appropriate reagents. The chemical method of HTC plays a role in converting organic substrates into structured carbon at high pressure and low temperatures (180–250 °C). Examples include chitosan as a starting compound for CQD synthesis [24], molecules, [31], BBQ char [32], and atmospheric plasma-based synthesis [33]. Carbonization process can be achieved by following ways:

i) **Electrochemical Carbonization**: Electrochemical carbonization uses applied potential and electrodes to synthesize CQDs through electric charges. Deng et al. [34] used low molecular weight alcohol for CQD preparation via electrochemical carbonization, operating with working electrodes of Pt sheets along with two auxiliary and a calomel electrode. The resulting CQDs demonstrated excellent luminescence properties without intricate purification and modification, boasting a QY of 15.9%.

ii) **Microwave Irradiation**: Microwave irradiation emerged as an eco-friendly, rapid, and green approach for CQD preparation. The utilization of various synthetic routes for CQD synthesis offers distinct advantages and disadvantages, catering to diverse applications and employing different precursors. The selection of methods for synthesis depends on the desired properties and applications of the resulting CQDs.

9.3.2 Top-Down Synthesis Processes

This approach involves breaking down large carbon structures, like graphite, graphene oxide, or carbon nanotubes into nanorange having dimensions below 10 nm. Techniques such as arc discharge, laser ablation, and electrochemical methods are employed for this purpose.

9.3.2.1 Laser Ablation

This approach involves directing laser irradiation onto the bulky carbon structures in water to produce CQDs [35]. Initially, carbon nanoparticles are transformed into CQDs by dissolving them in water, followed by a 12-hour reflux with HNO_3.

Subsequently, a 28-hour reflux process is conducted in PEG-200 solvent. The procedure is further extended with a 31-hour reflux, incorporating the addition of N-acetyl-L-cysteine, resulting in a yellow-brown CQD solution. The extraction process utilizes ethyl acetate to eliminate any unreacted components. The use of organic solvents is carefully chosen to enhance the light emission from CQDs through surface characterization. The fluorescence characteristics are influenced by the presence of ligands on the outer surface of the CQDs.

9.3.2.2 Arc Discharge

In the process of arc discharge synthesis, the vaporization of source material occurs through an arc discharge between electrodes. Further, the subsequent steps involved are: (1) condensation, (2) nucleation, and (3) growth of nanoparticles. This method is employed for the generation of metal/metal oxide nanoparticles, fullerenes, carbon nanotubes, or graphene films. Xu and collaborators oxidized the arc discharge soot using 3.3 M HNO_3 for incorporating the carboxyl group. Subsequently, they extracted the chamber using sodium hydroxide, resulting in a robust black-colored slurry [36]. During gel electrophoresis, the highly fluorescent CDs with an approximate size of 18 nm are distinguishable from the fast-moving band [37]. Bottini et al. utilized the electronic flash method and separated photoluminescent nanomaterials from both nitric-acid-oxidized and clean carbon nanomaterials [38]. The luminescent nanomaterials derived from hydrophobic, narrowly dispersed carbon nanotubes exhibited exceptional characteristics. When oxidized nanotubes were dispersed in a solvent, there was an accumulation of fluorescent nanomaterials.

9.3.2.3 Chemical and Electrochemical Oxidation Methods

The electrochemical oxidation technique involves the electrolysis of carbon materials like carbon nanotubes and graphite to produce CQDs. This method provides advantage of adjustable size and luminescence properties of the CDs by varying the current intensity, with low preparation costs, and extraordinary aq. dispersibility, highly pure, and environmentally benign methods of synthesis of CQDs [39]. Zhou and colleagues introduced the electrochemical technique in 2007 as a means of synthesizing blue luminescent CQDs from multi-walled CNTs, employing an electrolyte (tetra-butyl ammonium perchlorate solution) [40]. The process involves the electrochemical oxidation of larger carbon precursors, facilitated by a reference electrode, to break them down into smaller components. Another study by Zhao et al. demonstrated the preparation of fluorescent carbon nanomaterial through electrochemical oxidation, utilizing a graphite rod as the working electrode [41]. Ming et al. [42] conducted a simple, one-step, and additive-free etching process to synthesize CQDs using only ultra-pure water, without the use of any other chemicals. A graphite rod served as the anode and was immersed in the water, while a graphite rod was used as a counter-electrode, and was positioned laterally with a separation (7.5 cm). Operating with DC supply at constant voltage ranging between 15 and 60 V, the reaction completed after 120 hours of continuous stirring, resulting in the corrosion of the graphite rod (anode) and the formation of a deep yellow solution. The CQDs were obtained through centrifugation and filtration, showcasing sizes ranging from 3 to 6 nm, an extremely small size distribution, and consequently high homogeneity. The CQDs produced from this process were reinforced with graphene oxide layers and exhibited excellent luminescence.

9.3.2.4 Ultrasonic Treatment

In ultrasonic treatment method, carbon precursors, along with alkali, acid, and other oxidants, are subjected to intense ultrasonic waves, causing the fragmentation of carbon structures into extremely tiny nanomaterials and continuous cavitations of molecules to occur during the process. The application of high-energy ultrasound waves eliminates the need for a complicated post-treatment process, enabling the straightforward synthesis of CQDs with a reduced size [43, 44]. In a comprehensive study, Li et al. produced water-soluble fluorescent CDs using activated carbon using ultrasonic treatment method, assisted by one-step H_2O_2 in the same year. TEM analysis revealed that the surface of the prepared CDs was abundant in —OH groups, bearing a detected size range of 5–10 nm [45]. Figure 9.4 shows a diagrammatic presentation of green preparation of CDs via oxidation and ultrasonication.

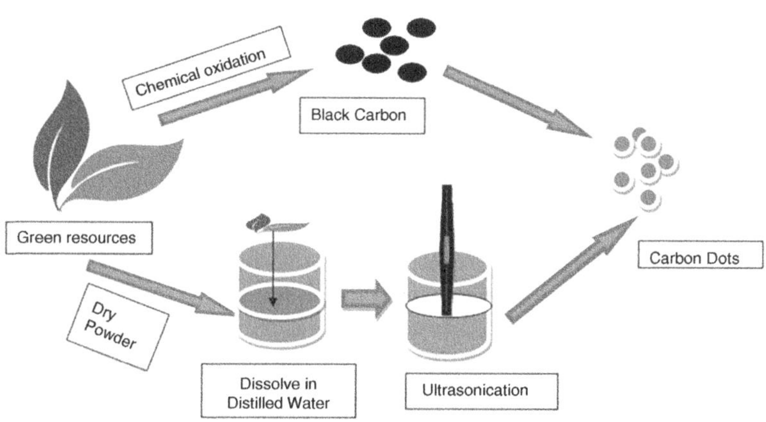

Figure 9.4 Chemical oxidation and ultrasonication methods for green synthesis of CDs.

9.4 Green Synthesis of Carbon Dots

In successful material applications, crucial factors include general, physical, and chemical characteristics, such as the type of chemical functionalities and abundance, morphology, and resulting material properties. The synthesis process plays a pivotal role in achieving these characteristics. The green synthesis process aims to eliminate or minimize the use of hazardous chemicals, extended time of reaction, high temperature and pressure, and intricate synthesis steps. This approach not only delivers economic advantages but also contributes to environmental benefits in comparison to traditional physical and chemical methods. Presently, green synthesis processes used for CD preparation primarily encompass abovementioned processes, such as microwave irradiation, HTC, ultrasound treatment, self-exothermic synthesis, and ozone/hydrogen peroxide oxidation. Recent efforts focus on sustainable synthesis aligned with the principles of green chemistry. This involves utilizing green carbon sources, such as biomass-based resources, and developing eco-friendly synthesis processes. Biomass, with its unique advantages of renewability, nontoxicity, low cost, and wide availability, is a promising carbon precursor for CD synthesis. Carbon sources based on Biomass can be further classified into animal waste and plant waste, or their extract-based resources.

9.4.1 Biomass-Based Green Synthesis of CDs

Within the context of choosing appropriate carbon precursors for CD preparation, biomass presents distinctive advantages, including renewability, nontoxicity, wide availability, and cost-effectiveness. The bio-economy model is supported by utilization of biomass, which revolves around deriving essential components for materials, energy and chemicals from renewable biological resources [46]. While transitioning from traditional precursors to biomass-based carbon sources marks a significant advancement, challenges such as prolonged reaction times or high temperatures may compromise sustainable idea of green synthesis of CDs. Nonetheless, the diverse nature of natural biomass supports the production of CDs with variable structures and properties.

Microwave-assisted HTC techniques and HTC are effective in eliminating the need for pre-drying biomass and converting wet-state biomass into CDs [47]. Pyrolysis carbonization is another widely used technique for CD synthesis which requires dry biomass feedstock [48]. Microwave irradiation has demonstrated efficacy in converting amino acids, carbohydrates, and ascorbic acid into CDs, proving suitable for biomass residues and industrial by-products like palm kernel shell [49], mint leaves [50], kelp [51], and empty fruit bunch biochar. Apart from these techniques, other green synthetic methods for CD production, including ultrasonic treatment and chemical oxidation, lack specific requirements for the type of CD precursor [52].

The biomass-based carbon sources are currently categorized into animal waste or plant waste, or their extract-based resources. This distinction recognizes the potential of utilizing diverse materials, such as residues from plants and animals or their extracts, in the synthesis of CDs.

9.4.1.1 Plant Waste-Based Green Synthesis of Carbon Dots

Efforts to transform biopolymers based on plants and waste products, including starch, cellulose, waste paper, lignin, and spent coffee grounds, into CDs similar to nanographene oxide caught attention of scientists recently. The resulting CDs demonstrated concentration-dependent self-association behavior and pH-dependent self-assembly behavior, offering opportunities to tailor the morphology of nanocomposites. Thus, the focus has shifted to synthesizing CDs from residues of plants, driven by environmental and economic considerations. Therefore, inexpensive waste products have become promising materials to generate high-value and low-order carbon nanomaterials, which can hold significant potential especially when the production is cost-effective. In contrast to other methods, the HTC technique, developed by Bergius in 1913, which involves raw biomass residues to be preheat-dried, enables the direct conversion of biomass into carbon-rich products, where water acts as the reagent as well as reactant medium. This process eliminates the need for pre-drying. Figure 9.5 showcases a schematic representation of green synthesis of CDs via banana peels with microwave assistance.

Under plant waste, HTC stands out for its ability to convert biomass directly into carbon-rich products without the need for preheat-drying treatment. Various plant residues, such as lotus, camellia, rapeseed flower pollens, rice husks, and Sago industry waste, have been successfully utilized for CD synthesis. Additionally, fruits and extracts from apple, banana, papaya, lemon, peach, mangosteen pulp,

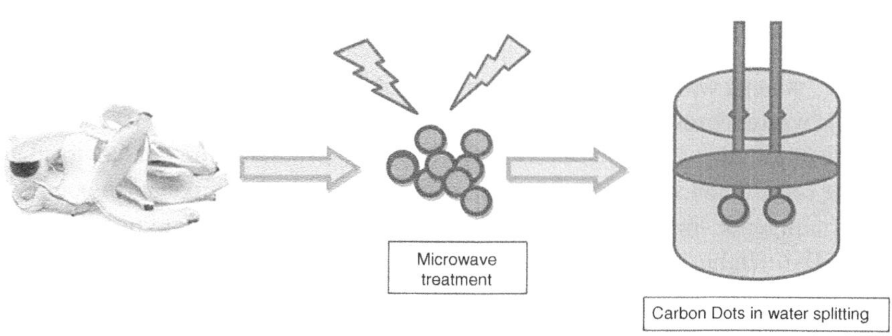

Figure 9.5 Utilization of microwave-assisted CDs prepared from banana peel for water-splitting application.

and cherry tomatoes have shown potential as suitable precursors for functional CDs. Carbon precursors such as camellia, rapeseed flower pollens, and lotus were employed in scaled-up HTC at 180 °C for 24 hours to synthesize N-CDs [53]. These CDs, ranging in size from 1 to 2 nm, exhibited excellent biocompatibility and PL intensity, suggesting promising applications in bio-imaging. Moreover, Sago industry wastes in Southeast Asia were utilized for producing fluorescent CDs and were used in detection of heavy metal ions [54]. Bamboo leaves [55] and Grass [56] were employed to generate precisely defined carbon nanoparticles through a series of processes including dehydration, hydrolysis, fragmentation, condensation, polymerization, and aromatization. Additionally, various fruits and extracts such as banana [57], apple [28] papaya [58, 59], cherry tomatoes [59], and mangosteen pulp [60] have been validated as suitable precursors for functional CDs.

In the above discussion, a range of biopolymers based on plant and waste products, including cellulose [61], agro-industrial residues [62], starch [63], lignin [64], waste paper [61], trapa bispinosa peel [65], and spent coffee grounds [66] were discussed to prepare nanographene oxide-type CDs (nGO-CDs) involving exfoliation-oxidation treatment and microwave-assisted HTC. In this continuation, reduced nanographene oxide-type (r-nGO-CDs) were produced via microwave-assisted HTC in superheated water, with or without caffeic acid as a reducing agent [67]. The recovery of sp^2 conjugated structures is suggested by the appearance of a π–π^* transition in the XPS spectrum of r-nGO-CDs. The sizes of nGO-CDs and r-nGO-CDs typically ranged a few nanometers, primarily influenced by the biomass source, temperature, and reaction time during the microwave-assisted HTC process. Furthermore, nGO-CDs demonstrated concentration-dependent self-association behavior and pH-dependent self-assembly behavior in aqueous solutions into polymer matrices, enabling the tuning of nanocomposite morphology [68]. Klongklaw et al. prepared CQDs through green synthesis using edible mushrooms [69]. Green synthesis has also been reported using precursors like sweet pepper [70], watermelon peel [71], and lemon peel.

9.4.1.2 Animal Waste-Based Green Synthesis of CDs

Animal waste represents a rich source for CD synthesis, and hair fibers, composed mainly of keratin, stand out as cost-effective and sustainable precursors for producing carbon nanomaterials. Keratin, comprising disulfide linkages in polypeptide chain, lends itself well to the synthesis of nitrogen and sulfur co-doped CDs (N,S-CDs). Higher temperatures favor the formation of smaller-sized N,S-CDs with high sulfur content, and longer emission wavelengths in PL. Additionally, N,S-CDs exhibit low toxicity, high solubility, and excellent biocompatibility. Successful production of N,S-CDs has been achieved using hair fibers from humans [72], pigs [73], and sheep [74], showcasing ionic tolerance, high pH stability, and varying fluorescence QYs of 29.0%, 20.1%, and 25.6%, respectively, depending on the hair precursor used.

A microwave-assisted hydrothermal method has been employed on goose feathers and resulting CDs, containing nitrogen, sulfur, and oxygen atoms, exhibit a QY of 17.1% [75]. Residues from seafood industries, including crab shells, fish scales, and prawn shells, also serve as viable sources for CD synthesis [76–78]. Nitrogen-doped fluorescent CDs derived from fish scales display a homogeneous size and a relatively high nitrogen content of 14.6%, surpassing values reported for most N-CDs [79].

Prawn, crab, or shrimp shells, rich in chitin, have been subjected to microwave-assisted hydrothermal/pyrolysis treatments for CD fabrication, resulting in magneto-fluorescent properties [80, 81]. Interestingly, chitosan (a derivative of chitin) led to CDs with a lower nitrogen content (approximately 6.1%) by utilizing the microwave-assisted hydrothermal treatment. Conversely, high-temperature pyrolysis of shrimp shells produced CDs with a higher nitrogen content of 11.8% [81].

9.5 Application of CDs in Water Splitting

The electrolysis of water holds promise for producing highly pure hydrogen in an environmentally friendly manner. Unfortunately, this process is currently hindered by its high cost and inefficiency, limiting its industrial applicability.

Photocatalytic water splitting [82] and photocatalytic improving organics [83] are two major ways to manufacture H_2. Photocatalytic water splitting deals with redox reaction by the use of electrons/holes. However, photocatalytic improving organics involves oxidation of organic substances yielding proton [82]. Next, electrons exterior of the photocatalyst contribute to the rehabilitation of protons into H_2 [84] (Figure 9.6).

Apart from this, the production of H_2 via water splitting process has also been highlighted, as we know that H_2 on combustion gives the only product H_2O, which facilitates cyclic use of the fuel. Sargin et al. [85] achieved the fabrication of CQDs as TiO_2 sensitizer through green synthesis using edible mushrooms, contributing to future fuel production (H_2 gas).

9.5.1 Hydrogen Generation via Water Splitting (Photoreduction)

Utilizing photocatalytic water splitting under sunlight as a green and sustainable energy source involves generating holes and electrons through light absorption, which are then employed in the production of H_2 and O_2. However, a significant challenge lies in preventing water formation through a reverse reaction, as

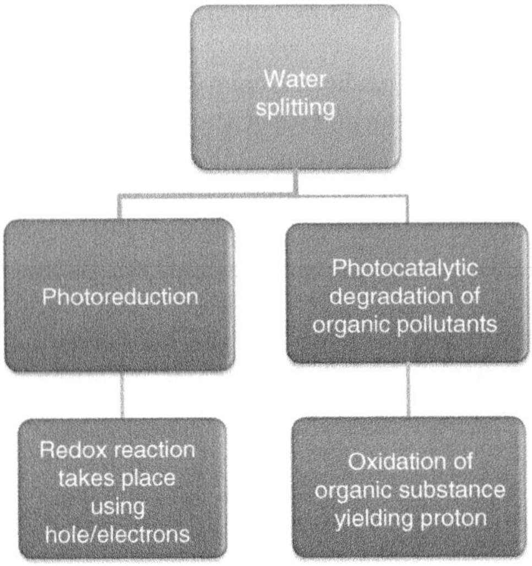

Figure 9.6 Photocatalytic water-splitting pathways.

discussed by Saravanan and colleagues [86]. The endothermic nature of this reaction, requiring about 237 kJ mol^{-1} of energy, is highlighted by Reddy et al. [1]. To optimize water splitting via photocatalytic route, the band gap of the photocatalyst must satisfy specific criteria, ensuring that the valence band is more positive than the redox potential of O_2/H_2O (1.23 V versus NHE) and the conduction band is more negative than the redox potential of H^+/H_2 (0 V versus NHE). This sets the minimum band gap requirement at an 1100 nm wavelength, equivalent to 1.23 eV for photocatalytic water splitting. The water-splitting reaction consists of two half-reactions: the O_2 evolution reaction (OER) and the H_2 evolution reaction (HER).

The typical 4 e$^-$ half-reactions are mentioned below:

$$2H_2O_{(l)} \rightarrow O_{2(g)} + 2H_{2(g)} \tag{9.1}$$

$$\textbf{Hydrogen generation site:}\ 4H^+ + 4e^- \rightarrow 2H_{2(g)} \tag{9.2}$$

$$\textbf{Oxygen generation site:}\ 2H_2O_{(l)} \rightarrow O_{2(g)} + 4H^+ + 4e^- \tag{9.3}$$

The typical 4 e$^-$ half-reactions are as follows:

The **two-step 2e$^-$/2e$^-$** process first generates H_2 and H_2O_2 from water, followed by the release of energy due to the disproportion of H_2O_2:

$$2H_2O_{(l)} \rightarrow H_2O_{2(l)} + H_{2(g)} \tag{9.4}$$

$$2H^+ + 2e^- \rightarrow H_{2(g)} \tag{9.5}$$

Hydrogen peroxide generation site:

$$2H_2O_{(l)} \rightarrow H_2O_{2(l)} + 2H^+ + 2e^- \quad (9.6)$$

$$H_2O_{2(l)} \rightarrow \frac{1}{2}2O_{2(g)} + H_2O_{(l)} \Delta G = -106.1 \text{ kJ mol}^{-1} \quad (9.7)$$

The process of photocatalytic water splitting involves the generation and transfer of electron and hole pairs inside the semiconductor, their motion toward the semiconductor surface, and the capturing of photogenerated electrons and holes by surface energetic location for subsequent utilization in the production of O_2 and H_2. Efficient electron-hole pair separation is crucial for semiconductor efficiency in sunlight-induced water splitting as the formation and recombination of excited electrons and holes occur within micro- to femtoseconds. To achieve high efficiency, charge recombination must be suppressed both on the surface and in the bulk of the catalyst (Figure 9.7).

While TiO_2 photocatalyst has been extensively investigated due to its valence band and conduction band positions, i.e. conduction band is more negative than the redox potential of H^+/H_2 (0V versus NHE) as discussed earlier, but its large band gap limits its effectiveness to the UV zone of sunlight (3.2eV) and it belongs to only 5% of total solar spectrum [87]. Thus, the main focus of researchers is to

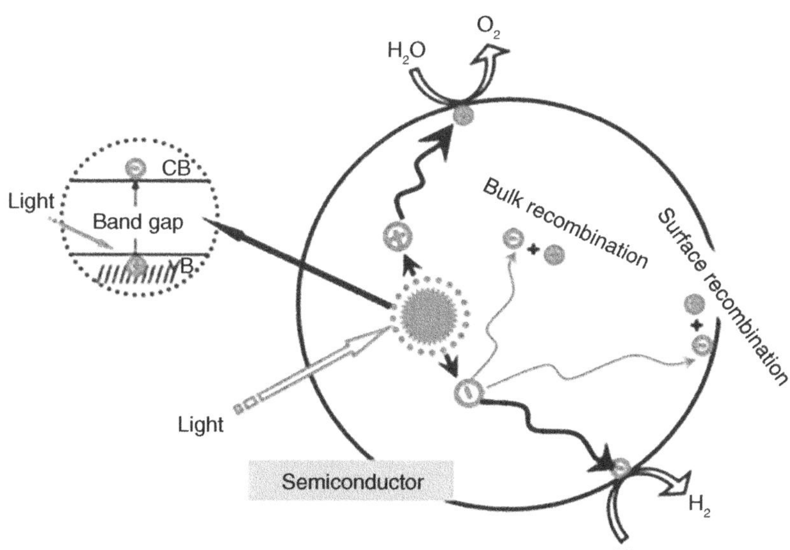

Figure 9.7 Schematic presentation of semiconductor-based photocatalytic water splitting.

design photocatalysts with convenient band gaps that operate in both NIR and visible light nowadays.

In a specific study by Zhang et al. [88], a hydrothermal process was employed to fabricate Fe_2O_3/CQD nanocomposites, demonstrating their effectiveness as semiconductor photocatalysts for the visible-light photodegradation of toxic fumes like methanol or benzene. The amalgamation of fluorescence and magnetic characteristics on one substrate is made possible by the distinct characteristics of CQDs, such as their magnetic architecture. The high electron storage capacity of CQDs enables electrons stimulated by photons from Fe_2O_3/CQD nanoparticles to move easily across the CQD-conducting system. As a result, when adsorbent reductants/oxidants interact with electron-hole pairs, more active oxygen radicals are produced that can eliminate hazardous gases. A comparison table has been presented here to discuss hydrogen energy production through photocatalytic water splitting using various photocatalysts (Table 9.1).

Table 9.1 Comparative study of hydrogen energy production through photocatalytic water splitting using various photocatalysts.

S. no.	Photocatalyst used for hydrogen generation	Solar to hydrogen efficiency (STH)/$\mu mol_{H2}\ h^{-1}$	Reference
1	Indium gallium nitride photocatalyst	9.2% STH	[89]
2	Al-doped SrTiO3 ($SrTiO_3$:Al) photocatalyst	10% STH	[90]
3	CQDs as a photosensitizer in combination with Ni catalyst	$398\ \mu mol_{H2}\ h^{-1}$	[91]
4	CQDs and Au@CQDs	260 μmol and $280\ \mu mol\ h^{-1}$	[92]
5	5% CDs/$BiVO_4$ QDs as photocatalysts	$0.92\ \mu mol\ h^{-1}$	[93]
6	5% CDs/p(DB) organic polymer photocatalyst	$140.88\ \mu mol\ g^{-1}\ h^{-1}$	[94]
7	Photoluminescent carbon nanoparticles (CNPs)	$4.89\ \mu mol\ h^{-1}$ over 2.1 mg CNPs.	[95]
8	Graphitic carbon nitride (g-C_3N_4)	$983.7\ \mu mol\ h^{-1}\ g^{-1}$	[96]
9	CdS quantum dots/ZnO	$22.12\ mmol\ h^{-1}\ g^{-1}$	[97]
10	Carbon ring conjugated TCN (C-TCN) organic photocatalysts	$204.6\ \mu mol\ g^{-1}\ h^{-1}$	[98]
11	g-$C_{54}N_6$-COFcovalent organic framework	$2518.9\ \mu mol\ g^{-1}\ h^{-1}$	[99]
12	Isomethylbytylamine (IMBA)/rGO nanosheets	$16.49\ \mu mol\ g^{-1}\ h^{-1}$	[100]

9.5.2 Photocatalytic Degradation of Organic Pollutants

Unquestionably, the advancement of technology has improved people's quality of life everywhere. However, it has concurrently given rise to potential pollutants, resulting in approximately 38,000 known chemicals in circulation by 2011, with an annual addition of an estimated 300 new compounds [101]. Each day, significant volumes of domestic and industrial pollutants find their way into waterways, posing threats to human health, the environment, and surrounding ecosystems [102]. Consequently, the availability of uncontaminated drinking water continues to diminish [103]. Key contributors to water pollution include industrial effluents, organic compounds, and heavy metals (e.g. cadmium and chromium). This section focuses on the latter, which encompasses small organic compounds (e.g. dyes or drugs) and longer-chained molecules (e.g. perfluorooctanoic acid or perfluorooctane sulfonate). This underscores the urgent need for efficient and cost-effective methods to remove these pollutants after use, disposal, or loss of functionality. To remove these contaminants, a number of alternatives—such as photocatalytic technologies—use sunlight and renewable energy sources [104].

A photocatalyst can change a pollutant that has been adsorbed onto its surface into a less hazardous breakdown product by absorbing the right kind of radiation. Photocatalysis provides several advantages over classical chemical oxidation, including environmental friendliness and the attainment of near-complete degradation during the reaction [105]. In a typical photocatalytic degradation process as indicated in Figure 9.8, suitable radiation is absorbed by semiconductor, bridging its band gap and creating an electron-hole pair by excitation an electron (e^-) from the valence band to the conduction band. These photogenerated e^- and h^+ are photoreactive, initiate a series of reactions and culminate degradation of pollutants,

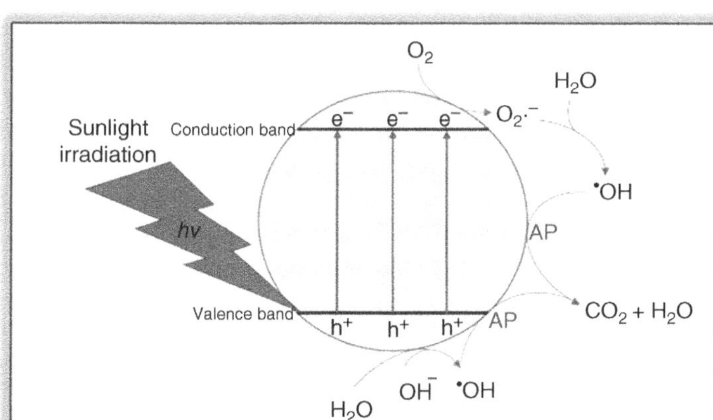

Figure 9.8 General photocatalytic degradation mechanism.

transforming them into CO_2 and H_2O. Therefore, the use of photocatalysts facilitates the conversion of sun's energy into a chemical reaction that degrades pollutants into a less harmful form [104]. However, this process is hampered by fast recombination of photogenerated e^- and h^+, which may release energy in the form of radiation or heat. PL, a property associated with CDs, is a competing process for photodegradation, potentially leading to a decline in efficiency.

9.6 Factors Affecting Characteristics of Nanomaterials of Carbon in Photocatalytic H_2 Production

Understanding the elements and challenges influencing the efficacy of photocatalysts based on carbon in hydrogen (H_2) production reactions is paramount. The performance of carbonaceous photocatalysts in H_2 production relies significantly on their dimensions, the number of layers, the presence of defects, the conductivity of the carbon material, chemical functionalities, and the interfacial contact area between semiconductors and carbon materials. To optimize photocatalytic H_2 generation using carbonaceous materials, systematic approaches are required to regulate the structural effects of property of carbon materials [106].

9.6.1 Doping

The introduction of heteroatoms into pristine carbon nanomaterials enhances their catalytic activity. Doping with heteroatoms introduces defect sites, altering the physicochemical properties of carbon nanomaterials and enhancing active site reaction. Heteroatom doping also affects the electronic structure, electrical conductivity, and semiconductor or metal-like properties of carbon materials. Despite recent progress in heteroatom doping technology, the detection and understanding of atomic-level composition and catalytic centers remain challenging. Non-metallic elements, particularly nitrogen, are the burning area of recent research on heteroatom doping in carbon nanomaterials for photocatalytic H_2 generation. Examples include N-GQDs and phosphorus-doped graphene, both exhibiting improved photocatalytic activities in H_2 production.

9.6.2 Defects

The presence of defects on the surface of carbonaceous catalysts, especially inherent defects, enhances their photocatalytic efficacy. Defects on multiple scales, such as point, plane, line, and volume defects, influence the surface and electrical properties of CNMs. The distribution of electron density and charge in carbon nanomaterials due to defects significantly impacts the photocatalytic reaction. The active sites created by defects serve as crucial centers for photocatalytic reactions. Although it is challenging to produce defect-free carbon nanomaterials, controlling defects is essential to understanding their impact on photocatalytic performance.

9.6.3 Dimensions

The size of carbon nanomaterials plays an important role in determining their photocatalytic efficiency. Zero-dimensional carbon nanomaterials, such as CQDs, exhibit dual photocatalytic and co-catalytic applications. However, descending the size of CQDs may increase charge recombination, limiting their photocatalytic efficiency. One-dimensional carbon nanomaterials, with their high specific surface area, offer numerous active centers for photocatalytic reactions and efficient charge transfer, positively affecting hydrogen evolution speed. However, practical applications of one-dimensional materials for photocatalytic H_2 production require essential modifications. Two-dimensional carbon nanomaterials possess advantages such as specific surface area, high photochemical stability, and optical and mechanical properties. They show potential in developing efficient photocatalytic H_2 systems but face challenges in improving mass transfer performance and H_2 generation efficiency. Two-dimensional carbon nanomaterials can efficiently absorb solar radiation and transfer charge in multiple directions, making them more efficient in H_2 evolution compared to zero- and one-dimensional counterparts. The thickness of two-dimensional carbon nanomaterials is a crucial parameter influencing photoinduced electron mobility and overall photocatalytic efficiency.

9.7 Conclusion

Hydrogen has been suggested as a future environmentally friendly and clean energy that has capability to replace fossil fuel. In fuel cell systems, hydrogen is used to produce power, with the only byproduct being pure water. Hydrogen gas must be collected from a variety of sources, including hydrocarbons, water, acids, and other molecules, as it is not naturally occurring on Earth as a gas. Currently, hydrocarbon fossil fuels are responsible for about 95% of hydrogen production using gas reforming and gasification coal processes, which have limited resources and are not clean processes. In contrast, just about 4% of hydrogen comes from rich water supplies.

This book chapter provides an overview of various methods of synthesis of CDs, their classification, properties such as fluorescence and factors affecting characteristics of carbon-based nanomaterials, such as doping and defects, in hydrogen production. As discussed above, cost-effective and metal-free CDs with exceptional electronic and optical properties are playing an important role in renewable energy production to address the energy crisis and environment-related problems such as photocatalytic degradation of pollutants.

Even though CDs have made great strides in the conversion of solar energy to hydrogen, there are still many important problems and obstacles that need to be resolved before this technology can be developed further and eventually commercialized. Enhancing the long-term stability of the CD-based photosystem is currently the top objective, particularly at high catalytic rates. Going forward, considerably more work should be put into developing hybrid photocatalysts based on CDs or solar-driven overall water splitting of individual CDs in order to do away with the requirement for costly sacrificial agents.

References

1 Reddy, C.V., Reddy, K.R., Harish, V.V.N. et al. (2020). Metal-organic frameworks (MOFs)-based efficient heterogeneous photocatalysts: synthesis, properties and its applications in photocatalytic hydrogen generation, CO2 reduction and photodegradation of organic dyes. *International Journal of Hydrogen Energy* 45 (13): 7656–7679. https://doi.org/10.1016/j.ijhydene.2019.02.144.

2 Xu, X. et al. (2004). Electrophoretic analysis and purification of fluorescent single-walled carbon nanotube fragments. *Journal of the American Chemical Society* 126 (40): 12736–12737. https://doi.org/10.1021/ja040082h.

3 Rasool, M.A., Sattar, R., Anum, A. et al. (2023). An insight into carbon nanomaterial-based photocatalytic water splitting for green hydrogen production. *Catalyst* 13 (1): 1–31.

4 Production of hydrogen—U.S. energy information administration (EIA). https://www.eia.gov/energyexplained/hydrogen/production-of-hydrogen.php (accessed 1 December, 2023).

5 Ozyurt, D., Kobaisi, M.A., Hocking, R.K., and Fox, B. (2023). Properties, synthesis, and applications of carbon dots: a review. *Carbon Trends* 12: 100276. https://doi.org/10.1016/j.cartre.2023.100276.

6 Sun, D. et al. (2019). Electrochemical dual-aptamer-based biosensor for nonenzymatic detection of cardiac troponin I by nanohybrid electrocatalysts labeling combined with DNA nanotetrahedron structure. *Biosensors & Bioelectronics* 134: 49–56. https://doi.org/10.1016/j.bios.2019.03.049.

7 Sahana, S., Gautam, A., Singh, R., and Chandel, S. (2023). A recent update on development, synthesis methods, properties and application of natural products derived carbon dots. *Natural Products and Bioprospecting* 13 (1): 1–21. https://doi.org/10.1007/s13659-023-00415-x.

8 Dong, J., Fang, L., Li, J. et al. (2023). Synthesis of aspartic acid-derived chiral carbon dots and their effect on bovine serum albumin amyloid fibrillation by multispectral and molecular interaction studies. *Journal of Molecular Structure* 1291: 136045. https://doi.org/10.1016/j.molstruc.2023.136045.

9 Wu, H., Lu, S., and Yang, B. (2022). Carbon-dot-enhanced electrocatalytic hydrogen evolution. *Accounts of Materials Research* 3 (3): 319–330. https://doi.org/10.1021/accountsmr.1c00194.

10 Sailaja Prasannakumaran Nair, S., Kottam, N., and Prasanna Kumar, S.G. (2020). green synthesized luminescent carbon nanodots for the sensing application of Fe3+ Ions. *Journal of Fluorescence* 30 (2): 357–363. https://doi.org/10.1007/s10895-020-02505-2.

11 Zheng, D.-W. et al. (2016). Carbon-dot-decorated carbon nitride nanoparticles for enhanced photodynamic therapy against hypoxic tumor *via* water splitting. *ACS Nano* 10 (9): 8715–8722. https://doi.org/10.1021/acsnano.6b04156.

12 Kuo, T.-R., Sung, S.-Y., Hsu, C.-W. et al. (2016). One-pot green hydrothermal synthesis of fluorescent nitrogen-doped carbon nanodots for in vivo bioimaging. *Analytical and Bioanalytical Chemistry* 408 (1): 77–82. https://doi.org/10.1007/s00216-015-9138-8.

13 Er, S. et al. (2021). Amino acids, peptides, and proteins: implications for nanotechnological applications in biosensing and drug/gene delivery. *Nanomaterials* 11 (11): 3002.

14 Brachi, P. (2020). Synthesis of carbon dots (CDs) through the fluidized bed thermal treatment of residual biomass assisted by γ-alumina. *Applied Catalysis B: Environmental* 263: 118361.

15 N. Pinna, K. Weiss, H. S. Kongehl, W. Vogel, J. Urban, M. P. Pileni, Triangular CdS nanocrystals: synthesis, characterization, and stability *Langmuir*, vol. 17, no. 26 pp.7982–7987, 2001. DOI: https://doi.org/10.1021/la010287t.

16 Bayat, A. and Saievar-Iranizad, E. (2018). Graphene quantum dots decorated rutile TiO2 nanoflowers for water splitting application. *Journal of Energy Chemistry* 27 (1): 306–310.

17 Shafi, A., Bano, S., Sabir, S. et al. (2020). Eco-friendly fluorescent carbon nanodots: characteristics and potential applications. In: *Carbon-Based Material for Environmental Protection and Remediation*. IntechOpen https://library.oapen.org/bitstream/handle/20.500.12657/43403/external_content.pdf?sequ#page=151.

18 Cui, L., Ren, X., Sun, M. et al. (2021). Carbon dots: synthesis, properties and applications. *Nanomaterials (Basel)* 11 (12): 3419. https://doi.org/10.3390/nano11123419.

19 Nayak, S., Guleria, K., Sen, A. et al. (2023). Chemically induced crosslinked enhanced emission of carbon polymer dots discerning healthy and cancer cells through pH-dependent tunable photoluminescence. *Journal of Materials Chemistry B* 11 (3): 594–605. https://doi.org/10.1039/D2TB01836E.

20 Li, L., Wu, G., Yang, G. et al. (2013). Focusing on luminescent graphene quantum dots: current status and future perspectives. *Nanoscale* 5 (10): 4015–4039. https://doi.org/10.1039/C3NR33849E.

21 Zhu, S. et al. (2011). Strongly green-photoluminescent graphene quantum dots for bioimaging applications. *Chemical Communications* 47 (24): 6858–6860. https://doi.org/10.1039/C1CC11122A.

22 Wu, Z.L., Liu, Z.X., and Yuan, Y.H. (2017). Carbon dots: materials, synthesis, properties and approaches to long-wavelength and multicolor emission. *Journal of Materials Chemistry B* 5 (21): 3794–3809. https://doi.org/10.1039/C7TB00363C.

23 Ding, L. et al. (2020). One-pot hydrothermal synthesis of carbon quantum dots with excellent photoluminescent properties and catalytic activity from coke powders. *Journal of Fluorescence* 30 (1): 151–156. https://doi.org/10.1007/s10895-019-02480-3.

24 Yang, Y. et al. (2011). One-step synthesis of amino-functionalized fluorescent carbon nanoparticles by hydrothermal carbonization of chitosan. *Chemical Communications* 48 (3): 380–382. https://doi.org/10.1039/C1CC15678K.

25 Yang, M., Li, B., Zhong, K., and Lu, Y. (2018). Photoluminescence properties of N-doped carbon dots prepared in different solvents and applications in pH sensing. *Journal of Materials Science* 53 (4): 2424–2433. https://doi.org/10.1007/s10853-017-1700-7.

26 Jin, X. et al. (2015). pH-sensitive carbon dots for the visualization of regulation of intracellular pH inside living pathogenic fungal cells. *Carbon* 81: 388–395. https://doi.org/10.1016/j.carbon.2014.09.071.

27 Barati, A., Shamsipur, M., and Abdollahi, H. (2016). Carbon dots with strong excitation-dependent fluorescence changes towards pH. Application as nanosensors for a broad range of pH. *Analytica Chimica Acta* 931: 25–33. https://doi.org/10.1016/j.aca.2016.05.011.

28 Liang, H., Tai, X., and Du, Z. (2020). Photocatalytic degradation of nonylphenol ethoxylate and its degradation mechanism. *Journal of Molecular Liquids* 302: 112567. https://doi.org/10.1016/j.molliq.2020.112567.

29 Ding, Y.-Y. et al. (2018). Facile preparation of bright orange fluorescent carbon dots and the constructed biosensing platform for the detection of pH in living cells. *Talanta* 189: 8–15. https://doi.org/10.1016/j.talanta.2018.06.060.

30 D. Xiao, D. Yuan, H. He, and M. Gao, "Microwave assisted one-step green synthesis of fluorescent carbon nanoparticles from ionic liquids and their application as novel fluorescence probe for quercetin determination," *Journal of Luminescence*, vol. 140, pp. 120–125, Aug. 2013, https://doi.org/10.1016/j.jlumin.2013.02.032.

31 Wang, J. et al. (2013). Versatility with carbon dots—from overcooked BBQ to brightly fluorescent agents and photocatalysts. *RSC Advances* 3 (36): 15604–15607. https://doi.org/10.1039/C3RA42302F.

32 Safavi, A., Sedaghati, F., Shahbaazi, H., and Farjami, E. (2012). Facile approach to the synthesis of carbon nanodots and their peroxidase mimetic function in azo dyes degradation. *RSC Advances* 2 (19): 7367–7370. https://doi.org/10.1039/C2RA20355C.

33 Wang, J., Wang, C.F., and Chen, S. (2023). Amphiphilic egg-derived carbon dots: rapid plasma fabrication, pyrolysis process, and multicolor printing patterns. *Angewandte Chemie International Edition* 51: 9297–9301. https://doi.org/10.1002/anie.201204381.

34 Deng, J. et al. (2014). Electrochemical synthesis of carbon nanodots directly from alcohols. *Chemistry – A European Journal* 20 (17): 4993–4999. https://doi.org/10.1002/chem.201304869.

35 Gonçalves, H., Jorge, P., Afonso Fernandes, J.R., and Silva, J. (2010). Hg(II) sensing based on functionalized carbon dots obtained by direct laser ablation. *Sensors and Actuators B: Chemical* 145: 702–707. https://doi.org/10.1016/j.snb.2010.01.031.

36 Zuo, P., Lu, X., Sun, Z. et al. (2016). A review on syntheses, properties, characterization and bioanalytical applications of fluorescent carbon dots. *Microchimica Acta* 183: 519–542.

37 Varenikov, A., Shapiro, E., and Gandelman, M. (2021). Decarboxylative halogenation of organic compounds. *Chemical Reviews* 121 (1): 412–484. https://doi.org/10.1021/acs.chemrev.0c00813.

38 Bottini, M. et al. (2005). Covalent decoration of multi-walled carbon nanotubes with silica nanoparticles. *Chemical Communications* 6: 758–760. https://doi.org/10.1039/b412876a.

39 Rocco, D., Moldoveanu, V.G., Feroci, M. et al. (2023). Electrochemical synthesis of carbon quantum dots. *ChemElectroChem* 10 (3): e202201104. https://doi.org/10.1002/celc.202201104.

40 Zhou, J. et al. (2007). An electrochemical avenue to blue luminescent nanocrystals from multiwalled carbon nanotubes (MWCNTs). *Journal of the American Chemical Society* 129 (4): 744–745. https://doi.org/10.1021/ja0669070.

41 Zhao, Q.-L., Zhang, Z.-L., Huang, B.-H. et al. (2008). Facile preparation of low cytotoxicity fluorescent carbon nanocrystals by electrooxidation of graphite. *Chemical Communications* 41: 5116–5118. https://doi.org/10.1039/b812420e.

42 Ming, H. et al. (2012). Large scale electrochemical synthesis of high quality carbon nanodots and their photocatalytic property. *Dalton Transactions* 41 (31): 9526–9531. https://doi.org/10.1039/C2DT30985H.

43 Kumar, R., Kumar, V.B., and Gedanken, A. (2020). Sonochemical synthesis of carbon dots, mechanism, effect of parameters, and catalytic, energy, biomedical and tissue engineering applications. *Ultrasonics Sonochemistry* 64: 105009. https://doi.org/10.1016/j.ultsonch.2020.105009.

44 Zhang, Y., Park, M., Kim, H.Y. et al. (2017). A facile ultrasonic-assisted fabrication of nitrogen-doped carbon dots/BiOBr up-conversion nanocomposites for visible light photocatalytic enhancements. *Scientific Reports* 7: 45086. https://doi.org/10.1038/srep45086.

45 Li, H., He, X., Liu, Y. et al. (2011). One-step ultrasonic synthesis of water-soluble carbon nanoparticles with excellent photoluminescent properties. *Carbon* 49 (2): 605–609. https://doi.org/10.1016/j.carbon.2010.10.004.

46 Duquenne, M., Prost, H., Schöpfel, J., and Dumeignil, F. (2020). Open bioeconomy—a bibliometric study on the accessibility of articles in the field of bioeconomy. *Publications* 8 (4): 55. https://doi.org/10.3390/publications8040055.

47 Li, W. et al. (2019). Kilogram-scale synthesis of carbon quantum dots for hydrogen evolution, sensing and bioimaging. *Chinese Chemical Letters* 30 (12): 2323–2327. https://doi.org/10.1016/j.cclet.2019.06.040.

48 Zhu, L., Yin, Y., Wang, C.-F., and Chen, S. (2013). Plant leaf-derived fluorescent carbon dots for sensing, patterning and coding. *Journal of Materials Chemistry C* 1 (32): 4925–4932. https://doi.org/10.1039/C3TC30701H.

49 Alhogail, S. et al. (2021). Simple and rapid peptide nanoprobe biosensor for the detection of Legionellaceae. *Analyst* 146 (11): 3568–3577.

50 Architha, N. et al. (2021). Microwave-assisted green synthesis of fluorescent carbon quantum dots from Mexican Mint extract for Fe3+ detection and bio-imaging applications. *Environmental Research* 199: 111263. https://doi.org/10.1016/j.envres.2021.111263.

51 Zhao, C., Li, X., Cheng, C., and Yang, Y. (2019). Green and microwave-assisted synthesis of carbon dots and application for visual detection of cobalt(II) ions and pH sensing. *Microchemical Journal* 147: 183–190. https://doi.org/10.1016/j.microc.2019.03.029.

52 Feng, X. and Zhang, Y. (2019). A simple and green synthesis of carbon quantum dots from coke for white light-emitting devices. *RSC Advances* 9 (58): 33789–33793. https://doi.org/10.1039/C9RA06946A.

53 Zhang, J., Yuan, Y., Liang, G., and Yu, S.-H. (2015). Scale-up synthesis of fragrant nitrogen-doped carbon dots from bee pollens for bioimaging and catalysis. *Advanced Science* 2 (4): 1500002. https://doi.org/10.1002/advs.201500002.

54 Tan, X.W., Romainor, A.N.B., Chin, S.F., and Ng, S.M. (2014). Carbon dots production via pyrolysis of sago waste as potential probe for metal ions sensing. *Journal of Analytical and Applied Pyrolysis* 105: 157–165. https://doi.org/10.1016/j.jaap.2013.11.001.

55 Liu, S. et al. (2012). Hydrothermal treatment of grass: a low-cost, green route to nitrogen-doped, carbon-rich, photoluminescent polymer nanodots as an effective fluorescent sensing platform for label-free detection of Cu(II) ions. *Advanced Materials* 24 (15): 2037–2041. https://doi.org/10.1002/adma.201200164.

56 Liu, Y., Zhao, Y., and Zhang, Y. (2014). One-step green synthesized fluorescent carbon nanodots from bamboo leaves for copper (II) ion detection. *Sensors and Actuators B: Chemical* 196: 647–652.

57 De, B. and Karak, N. (2013). A green and facile approach for the synthesis of water soluble fluorescent carbon dots from banana juice. *RSC Advances* 3 (22): 8286–8290. https://doi.org/10.1039/C3RA00088E.

58 Wang, N., Wang, Y., Guo, T. et al. (2016). Green preparation of carbon dots with papaya as carbon source for effective fluorescent sensing of Iron (III) and Escherichia coli. *Biosensors and Bioelectronics* 85: 68–75. https://doi.org/10.1016/j.bios.2016.04.089.

59 Lai, Z., Guo, X., Cheng, Z. et al. (2020). Green synthesis of fluorescent carbon dots from cherry tomatoes for highly effective detection of trifluralin herbicide in soil samples. *ChemistrySelect* 5 (6): 1956–1960. https://doi.org/10.1002/slct.201904517.

60 Yang, R., Guo, X., Jia, L. et al. (2017). Green preparation of carbon dots with mangosteen pulp for the selective detection of Fe3+ ions and cell imaging. *Applied Surface Science* 423: 426–432. https://doi.org/10.1016/j.apsusc.2017.05.252.

61 Adolfsson, K.H., Hassanzadeh, S., and Hakkarainen, M. (2015). Valorization of cellulose and waste paper to graphene oxide quantum dots. *RSC Advances* 5 (34): 26550–26558. https://doi.org/10.1039/C5RA01805F.

62 Jr, A., Afonso, G., Aquino, F.T. et al. (2022). Green synthesis of carbon nanodots from agro-industrial residues. *Carbon Letters* 32 (1): 131–141. https://doi.org/10.1007/s42823-021-00260-3.

63 Wu, D., Xu, H., and Hakkarainen, M. (2016). From starch to polylactide and nano-graphene oxide: fully starch derived high performance composites. *RSC Advances* 6 (59): 54336–54345. https://doi.org/10.1039/C6RA08194K.

64 Yao, J., Odelius, K., and Hakkarainen, M. (2020). Carbonized lignosulfonate-based porous nanocomposites for adsorption of environmental contaminants. *Functional Composite Materials* 1 (1): 5. https://doi.org/10.1186/s42252-020-00008-8.

65 Mewada, A. et al. (2013). Green synthesis of biocompatible carbon dots using aqueous extract of Trapa bispinosa peel. *Materials Science and Engineering: C* 33 (5): 2914–2917.

66 Xu, H., Xie, L., and Hakkarainen, M. (2017). Coffee-ground-derived quantum dots for aqueous processable nanoporous graphene membranes. *ACS Sustainable Chemistry & Engineering* 5 (6): 5360–5367. https://doi.org/10.1021/acssuschemeng.7b00663.

67 Erdal, N.B., Adolfsson, K.H., Pettersson, T., and Hakkarainen, M. (2018). Green strategy to reduced nanographene oxide through microwave assisted transformation of cellulose. *ACS Sustainable Chemistry & Engineering* 6 (1): 1246–1255. https://doi.org/10.1021/acssuschemeng.7b03566.

68 Hassanzadeh, S., Adolfsson, K.H., and Hakkarainen, M. (2015). Controlling the cooperative self-assembly of graphene oxide quantum dots in aqueous solutions. *RSC Advances* 5 (71): 57425–57432. https://doi.org/10.1039/C5RA09704E.

69 Klongklaw, K., Phiromkaew, B., Kiatsuksri, P. et al. (2023). Green one-step synthesis of mushroom-derived carbon dots as fluorescent sensors for Fe3+ detection. *RSC Advances* 13 (44): 30869–30875. https://doi.org/10.1039/D3RA06300C.

70 Yin, B. et al. (2013). Green synthesis of carbon dots with down- and up-conversion fluorescent properties for sensitive detection of hypochlorite with a dual-readout assay. *Analyst* 138 (21): 6551–6557. https://doi.org/10.1039/C3AN01003A.

71 Zhou, J., Sheng, Z., Han, H. et al. (2012). Facile synthesis of fluorescent carbon dots using watermelon peel as a carbon source. *Materials Letters* 66 (1): 222–224. https://doi.org/10.1016/j.matlet.2011.08.081.

72 Sun, D., Ban, R., Zhang, P.-H. et al. (2013). Hair fiber as a precursor for synthesizing of sulfur- and nitrogen-co-doped carbon dots with tunable luminescence properties. *Carbon* 64: 424–434. https://doi.org/10.1016/j.carbon.2013.07.095.

73 Zhao, C., Jiao, Y., Hu, F., and Yang, Y. (2018). Green synthesis of carbon dots from pork and application as nanosensors for uric acid detection. *Spectrochimica Acta Part A: Molecular and Biomolecular Spectroscopy* 190: 360–367. https://doi.org/10.1016/j.saa.2017.09.037.

74 Wang, S., Niu, H., He, S., and Cai, Y. (2016). One-step fabrication of high quantum yield sulfur- and nitrogen-doped carbon dots for sensitive and selective detection of Cr(VI). *RSC Advances* 6 (109): 107717–107722. https://doi.org/10.1039/C6RA21059G.

75 Liu, R. et al. (2014). A facile microwave-hydrothermal approach towards highly photoluminescent carbon dots from goose feathers. *RSC Advances* 5 (6): 4428–4433. https://doi.org/10.1039/C4RA12077A.

76 Dehvari, K., Liu, K.Y., Tseng, P.-J. et al. (2019). Sonochemical-assisted green synthesis of nitrogen-doped carbon dots from crab shell as targeted nanoprobes for cell imaging. *Journal of the Taiwan Institute of Chemical Engineers* 95: 495–503. https://doi.org/10.1016/j.jtice.2018.08.037.

77 Campalani, C. et al. (2021). Biobased carbon dots: from fish scales to photocatalysis. *Nanomaterials (Basel)* 11 (2): 524. https://doi.org/10.3390/nano11020524.

78 Torres, F.G., Gonzales, K.N., Troncoso, O.P., and Cañedo, V.S. (2023). Carbon quantum dots based on marine polysaccharides: types, synthesis, and applications. *Marine Drugs* 21 (6): https://doi.org/10.3390/md21060338.

79 Zhang, Y., Gao, Z., Yang, X. et al. (2019). Fish-scale-derived carbon dots as efficient fluorescent nanoprobes for detection of ferric ions. *RSC Advances* 9 (2): 940–949. https://doi.org/10.1039/C8RA09471C.

80 Gedda, G., Lee, C.-Y., Lin, Y.-C., and Wu, H. (2016). Green synthesis of carbon dots from prawn shells for highly selective and sensitive detection of copper ions. *Sensors and Actuators B: Chemical* 224: 396–403. https://doi.org/10.1016/j.snb.2015.09.065.

81 Elango, D., Packialakshmi, J.S., Manikandan, V., and Jayanthi, P. (2022). Sustainable synthesis of carbon quantum dots from shrimp shell and its emerging applications. *Materials Letters* 312: 131667. https://doi.org/10.1016/j.matlet.2022.131667.

82 Abe, R. (2010). Recent progress on photocatalytic and photoelectrochemical water splitting under visible light irradiation. *Journal of Photochemistry and Photobiology C: Photochemistry Reviews* 11 (4): 179–209. https://doi.org/10.1016/j.jphotochemrev.2011.02.003.

83 Kondarides, D.I., Daskalaki, V.M., Patsoura, A., and Verykios, X.E. (2008). Hydrogen production by photo-induced reforming of biomass components and derivatives at ambient conditions. *Catalysis Letters* 122 (1): 26–32. https://doi.org/10.1007/s10562-007-9330-3.

84 Dincer, I. and Acar, C. (2015). Review and evaluation of hydrogen production methods for better sustainability. *International Journal of Hydrogen Energy* 40 (34): 11094–11111. https://doi.org/10.1016/j.ijhydene.2014.12.035.

85 Sargin, I., Yanalak, G., Arslan, G., and Patir, I.H. (2019). Green synthesized carbon quantum dots as TiO2 sensitizers for photocatalytic hydrogen evolution. *International Journal of Hydrogen Energy* 44 (39): 21781–21789. https://doi.org/10.1016/j.ijhydene.2019.06.168.

86 Saravanan, K.R.A., Prabu, N., Sasidharan, M., and Maduraiveeran, G. (2019). Nitrogen-self doped activated carbon nanosheets derived from peanut shells for enhanced hydrogen evolution reaction. *Applied Surface Science* 489: 725–733. https://doi.org/10.1016/j.apsusc.2019.06.040.

87 Vyas, Y., Chundawat, P., Dharmendra, D. et al. (2021). Review on hydrogen production photocatalytically using carbon quantum dots: Future fuel. *International Journal of Hydrogen Energy* 46 (75): 37208–37241. https://doi.org/10.1016/j.ijhydene.2021.09.004.

88 Zhang, J., Kuang, M., Wang, J. et al. (2019). Fabrication of carbon quantum dots/TiO2/Fe2O3 composites and enhancement of photocatalytic activity under visible light. *Chemical Physics Letters* 730: 391–398. https://doi.org/10.1016/j.cplett.2019.06.011.

89 Zhou, P. et al. (2023). Solar-to-hydrogen efficiency of more than 9% in photocatalytic water splitting. *Nature* 613 (7942): 66–70. https://doi.org/10.1038/s41586-022-05399-1.

90 Goto, Y. et al. (2018). A particulate photocatalyst water-splitting panel for large-scale solar hydrogen generation. *Joule* 2 (3): 509–520. https://doi.org/10.1016/j.joule.2017.12.009.

91 Martindale, B.C.M., Hutton, G.A.M., Caputo, C.A., and Reisner, E. (2015). Solar hydrogen production using carbon quantum dots and a molecular nickel catalyst. *Journal of the American Chemical Society* 137 (18): 6018–6025. https://doi.org/10.1021/jacs.5b01650.

92 Mehta, A., Pooja, D., Thakur, A., and Basu, S. (2017). Enhanced photocatalytic water splitting by gold carbon dot core shell nanocatalyst under visible/sunlight. *New Journal of Chemistry* 41 (11): 4573–4581. https://doi.org/10.1039/C7NJ00933J.

93 Kang, Z.H. et al. (2016). Carbon dot and BiVO4 quantum dot composites for overall water splitting via a two-electron pathway. *Nanoscale* 8: https://doi.org/10.1039/C6NR05864G.

94 Wu, Z. et al. (2022). Bifunctional carbon dots as cocatalyst and reactor decorating an organic photocatalyst for H2 production from water-splitting in an emulsion. *Journal of Materials Chemistry A* 11 (1): 178–186. https://doi.org/10.1039/D2TA07562H.

95 Yang, P., Zhao, J., Wang, J. et al. (2015). Light-induced synthesis of photoluminescent carbon nanoparticles for Fe 3+ sensing and photocatalytic hydrogen evolution. *Journal of Materials Chemistry A* 3 (1): 136–138. https://doi.org/10.1039/C4TA05155F.

96 Wang, Z. et al. (2022). Solar hydrogen generation over carbon nitride photocatalyst promoted by water-soluble carbon dots. *ChemPhotoChem* 6 (12): e202200176. https://doi.org/10.1002/cptc.202200176.

97 Ma, D., Shi, J.-W., Zou, Y. et al. (2017). Highly efficient photocatalyst based on a CdS quantum Dots/ZnO nanosheets 0D/2D heterojunction for hydrogen evolution from water splitting. *ACS Applied Materials & Interfaces* 9 (30): 25377–25386. https://doi.org/10.1021/acsami.7b08407.

98 Z. Mo, J. Di, P.Yan, C. Lv, X. Zhu, D. Liu, Y. Song, C. Liu,Q. Yu, H. Li, Y. Lei, H. Xu, Q. Yan, An all-organic D-a system for visible-light-driven overall water splitting, *Small*, vol. 16, no. 48, pp. 2003914, 2020. https://doi.org/10.1002/smll.202003914.

99 Xu, J. et al. (2020). Vinylene-linked covalent organic frameworks (COFs) with symmetry-tuned polarity and photocatalytic activity. *Angewandte Chemie (International Ed. in English)* 59 (52): 23845–23853. https://doi.org/10.1002/anie.202011852.

100 Karim, M.R., Rahman, M.M., Asiri, A.M., and Hayami, S. (2020). Branched alkylamine–reduced graphene oxide hybrids as a dual proton–electron conductor and organic-only water-splitting photocatalyst. *ACS Applied Materials & Interfaces* 12 (9): 10829–10838. https://doi.org/10.1021/acsami.9b21200.

101 Carlsen, L., Bruggemann, R., and Sailaukhanuly, Y. (2013). Application of selected partial order tools to analyze fate and toxicity indicators of environmentally hazardous chemicals. *Ecological Indicators* 29: 191–202. https://doi.org/10.1016/j.ecolind.2012.12.028.

102 Rigodanza, F., Đorđević, L., Arcudi, F., and Prato, M. (2018). Customizing the electrochemical properties of carbon nanodots by using quinones in bottom-up synthesis. *Angewandte Chemie International Edition* 57 (18): 5062–5067.

103 Cao, S. and Yu, J. (2016). Carbon-based H_2-production photocatalytic materials. *Journal of Photochemistry and Photobiology C: Photochemistry Reviews* 27: 72–99. https://doi.org/10.1016/j.jphotochemrev.2016.04.002.

104 Sendão, R.M.S., Esteves da Silva, J.C.G., and Pinto da Silva, L. (2022). Photocatalytic removal of pharmaceutical water pollutants by TiO2—carbon dots nanocomposites: a review. *Chemosphere* 301: 134731. https://doi.org/10.1016/j.chemosphere.2022.134731.

105 Zhang, F., Wang, X., Liu, H. et al. (2019). Recent advances and applications of semiconductor photocatalytic technology. *Applied Sciences* 9 (12): 2489. https://doi.org/10.3390/app9122489.

106 V. Dutta, R. Verma, C. Gopalkrishnan,M. H. Yuan,K. M. Batoo, R. Jayavel, A. Chauhan, K.Y. A. Lin, R. Balasubramani,S. Ghotekar, Bio-inspired synthesis of carbon-based nanomaterials and their potential environmental applications: a state-of-the-art review, *Inorganics* Vol. 10 no. 10, pp. 1–33, 2022. https://doi.org/10.3390/inorganics10100169.

10

Carbon-Based Nanomaterials in Energy Storage Devices: Solar Cells

Seraj Ahmad[1], Manoj Kumar[1], Kahkashan Khatoon[1], Akram Ali[1], and Himanshu Arora[2]

[1] *CMP Degree College, University of Allahabad, Prayagraj, Uttar Pradesh, India*
[2] *Department of Chemistry, Faculty of Sciences, University of Allahabad, Prayagraj, Uttar Pradesh, India*

10.1 Introduction

Energy storage device advancements are greatly aided by carbon-based nanomaterials, especially when it comes to solar cells. The incorporation of carbon-based nanomaterials into solar cells, which are essential for using solar radiation to generate electricity, has several benefits. Photovoltaic (PV) cells, sometimes referred to as solar cells, are crucial for converting sunlight into electrical power. Researchers are progressively using carbon-based nanomaterials as essential components to improve their performance. These nanomaterials, which include graphene, CDs, and CNTs, improve solar cell technology in a number of ways [1].

The exceptional electrical conductivity of graphene and CNTs increases the mobility of electrons within the solar cells. Higher conversion efficiencies are the outcome of more effective charge separation and collection made possible by this greater mobility. The absorption of solar energy can be increased by engineering carbon-based nanomaterials to absorb particular light wavelengths. By functionalizing carbon nanomaterials with quantum dots or light-absorbing molecules, it is possible to increase the range of light absorption and increase the efficiency of solar cells in a variety of lighting scenarios. Because carbon nanomaterials are lightweight and flexible, thin, flexible, and transparent solar cells can be developed. These characteristics are particularly helpful for applications like portable electronics and building-integrated PVs [2].

Carbon-based Nanomaterials for Green Applications, First Edition. Edited by Upendra Kumar, Piyush Kumar Sonkar, and Suman Lata Tripathi.

Published 2025 by John Wiley & Sons, Inc.

The stability and corrosion resistance of carbon-based materials are well-known, and they contribute to the increased longevity and endurance of solar cells. This is essential for solar energy systems to be sustainable and cost-effective. More environmentally friendly energy solutions may result from the use of carbon-based compounds in solar cells. Certain materials are more environmentally friendly to produce and dispose of than some conventional materials, which makes them a desirable option for sustainable energy systems. Various solar cell types, such as silicon-based, organic, and perovskite-based cells, can incorporate carbon-based nanomaterials. Because of its adaptability, solar cell designs can be tailored and optimized to meet the needs of certain applications [3].

Solar cell technology is undergoing a revolution because of nanomaterials based on carbon. They are a viable option for increasing the effectiveness, adaptability, and sustainability of solar energy conversion due to their special qualities and versatility. Future advancements in solar cell performance and design are expected as long as this field of study is pursued, which will help ensure a future with more renewable and sustainable energy sources. Given the pressing need for sustainable energy sources and the growing global energy demand, solar cells have developed as a promising boon for efficient energy alteration. Solar cells, often known as PV cells, employ the PV effect to convert light from the sun into electrical power.

The qualities and characteristics of the materials used in the manufacture of solar cells are just one of the many variables that affect how well and efficiently they work. Recent years have seen a substantial increase in interest in carbon-based nanomaterials as prospective candidates for improving the effectiveness, stability, and cost-effectiveness of solar cells [1].

Carbon-based nanomaterials like graphene, CDs, and CNTs have special qualities that make them desirable for energy storage applications. These substances have outstanding mechanical strength, a wide surface area, strong electrical conductivity, and exceptional chemical stability, all of which are beneficial for enhancing the efficiency of solar cells [1]. Carbon-based nanomaterials like graphene, CDs, and CNTs have special qualities that make them desirable for energy storage applications. These substances have outstanding mechanical strength, a wide surface area, strong electrical conductivity, and exceptional chemical stability, all of which are beneficial for enhancing the efficiency of solar cells [1].

Furthermore accessible, non-toxic, and environmentally benign, carbon-based nanomaterials are perfect for usage in sustainable energy applications [2, 3].

In conclusion, carbon-based nanomaterials are transforming solar cell technology. They are a viable option for increasing the effectiveness, adaptability, and sustainability of solar energy conversion due to their special qualities and versatility. Future advancements in solar cell performance and design are expected as long as this field of study is pursued, which will help ensure a future with more renewable and sustainable energy sources. The objective of this long chapter is to present a comprehensive analysis of the role of carbon-based nanomaterials in energy storage systems, with particular emphasis on solar cells.

This chapter will cover the fundamental properties of carbon nanomaterials, their synthesis in detail, and their applications in various solar cell configurations. We'll also look at recent advancements and challenges in the use of carbon-based nanomaterials for efficient energy conversion and storage in solar cells.

10.2 Carbon Nanotubes

The tightly coiled graphene sheets give rise to the cylinder-shaped CNTs. They offer excellent heat conductivity, strong electrical conductivity, and a high aspect ratio, among other mechanical and electronic properties. These unique properties make CNTs highly desirable for various solar cell applications. CNTs have the ability to give or absorb electrons, which can help with efficient charge transfer and reduce internal energy losses in the device. Additionally, because of their one-dimensional structure, hybrid nanocomposites can be created, which enhances the efficiency of solar cells [4–7].

CNTs are desirable for solar cell applications because of their unique properties. They are highly electrically conductible, have a large surface area, and have strong mechanical strength [8]. These features facilitate efficient charge movement, reduce energy losses, and enhance the overall performance of solar cells. CNTs can act as materials that donate or take electrons, depending on how they are doped or functionalized. This helps with charge transport inside the device depicted in Figure 10.1 [10–14].

10.2.1 Synthesis Techniques Concerning Carbon Nanotubes

Many synthesis techniques have been developed as a result of the development of CNTs. These techniques include chemical vapor deposition (CVD), arc discharge, and laser ablation [15]. Among them, CVD is the most widely used method for creating high-quality CNTs with precise control over their size, chirality, and alignment [16]. When carbon-containing gases on a catalytic substrate are broken down at high temperatures, CNTs are created.

10.2.2 Carbon Nanotube Applications in Solar Cell Technology

Because of their exceptional qualities and the enormous advantages they could provide to PV technology, CNTs have attracted a lot of attention for their possible application in solar cells. Following are some important uses and benefits of CNTs in solar cells.

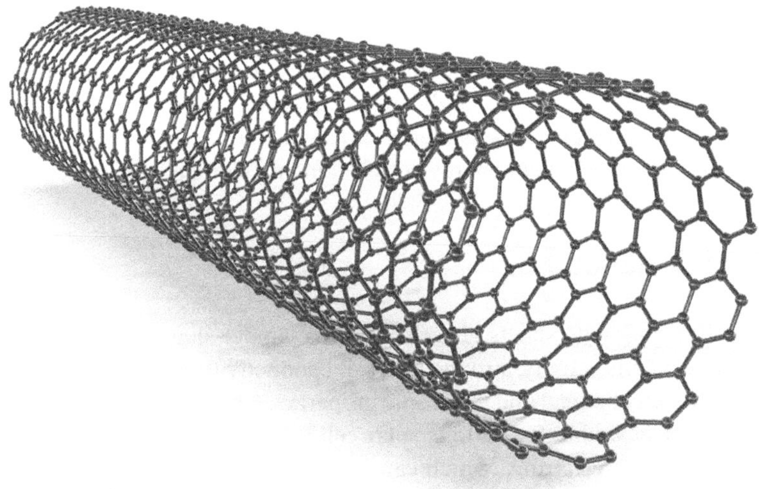

Figure 10.1 Carbon nanotubes (CNTs) composed of rolled-up graphene sheets [9]. *Source:* Dr_Microbe/Adobe stock.

10.2.2.1 Transparent Conductive Electrodes

As transparent conductive electrodes in solar cells, CNTs have been thoroughly investigated since they provide various advantages over more traditional materials like indium tin oxide (ITO). They have great electrical conductivity, good transparency, flexibility, and mechanical toughness [17]. CNT-based electrodes have been used in a variety of solar cell topologies, including perovskite solar cells, dye-sensitized solar cells (SSCs), and organic solar cells [18, 19].

10.2.2.2 Charge Transport Materials

In solar cells, CNTs can be used as effective charge transport components. As layers transport electrons or holes, they speed up the movement of charge carriers, lowering recombination losses and enhancing device performance [20]. CNTs have been demonstrated to be used as charge transport materials in perovskite, organic, and hybrid solar cells [21, 22].

10.2.2.3 Enhanced Electron Transport

CNTs exhibit excellent electrical conductivity, allowing for efficient transport of electrons within the solar cells. When used as an electron-transporting material, CNTs can aid in the transfer of electrons from the light-absorbing material to the electrical contacts, improving the overall cell efficiency.

10.2.2.4 Improved Charge Collection

In solar cells, CNTs can function as an efficient charge collector. The efficient collection of photogenerated charge carriers is made possible by their high surface area and electrical conductivity, which lowers recombination losses and enhances PV efficiency overall.

10.2.2.5 Transparency and Flexibility

For usage in transparent or semi-transparent solar cells, single-walled carbon nanotubes (SWCNTs) and some multi-walled carbon nanotubes (MWCNTs) are transparent in the visible spectrum. This property allows for innovative applications, such as integrating solar cells into windows, facades, and other transparent surfaces.

10.2.2.6 Lightweight and Flexible Design

CNTs are perfect for use in flexible and lightweight solar cell designs since they are lightweight and flexible. This flexibility can enable the creation of portable and wearable solar cells, as well as conformal solar panels for irregularly shaped surfaces [17].

10.2.2.7 Tunable Aspects of Optics

It is possible to tailor CNTs to absorb particular light wavelengths by varying their functionalization, chirality, and diameter. This tunability can be advantageous for optimizing light absorption and increasing the efficiency of solar cells in different lighting conditions.

10.2.2.8 Durability and Longevity

CNTs are highly stable and resistant to environmental factors, such as moisture and corrosion. This makes solar cells incorporating CNTs more durable and longer-lasting, reducing maintenance and replacement costs.

10.2.2.9 Compatibility with Other Materials

CNTs can be integrated with various other materials used in solar cells, including organic polymers, perovskite materials, and silicon, enabling the creation of hybrid solar cell structures that benefit from the unique properties of CNTs.

10.2.2.10 Scalability

The production of CNTs has become more scalable and cost-effective over time. This scalability, along with the potential for large-scale manufacturing, makes them a practical choice for solar cell applications.

While CNT use in solar cells holds great promise, researchers continue to work on addressing challenges such as improving the alignment and dispersion of CNTs in the active layers of the cells, as well as achieving higher manufacturing consistency. As these obstacles are removed, CNTs should become more significant in improving the performance and adaptability of solar cell technology, which will help expand the market for renewable energy sources.

10.2.3 Recent Advancements and Challenges

10.2.3.1 Recent Advancements

Recent studies have concentrated on developing fresh uses for CNTs in solar cells while also enhancing device designs and fabrication methods. CNT stability and scalability have been increased, which has improved the performance and efficiency of solar cells [14, 23]. CNTs have made notable advancements in solar cell technology, offering several benefits, but they also face some challenges that researchers are actively working to overcome. Following are some current developments and difficulties with CNT use in solar cells:

i) **Improved Effectiveness:** By adding CNTs, researchers have significantly improved the efficiency of solar cells. The total efficiency of solar cells has increased due to the improved charge separation and collection brought about by the use of CNTs as charge collectors or electron-transport materials.
ii) **Tandem Solar Cells:** CNTs have been integrated into tandem or multi-junction solar cell designs, where multiple layers of solar cells with varying band gaps are stacked to capture a broader spectrum of sunlight. CNTs can facilitate efficient charge transport between the different cell layers, improving energy conversion efficiency.
iii) **Perovskite Solar Cells:** CNTs have been used in perovskite solar cells as layers that transfer electrons. With improved stability and efficiency, these hybrid designs have demonstrated remarkable performance, offering a viable substitute for conventional silicon-based solar cells.
iv) **Transparent Solar Cells:** Transparent solar cells using CNTs have gained attention. These cells can be integrated into windows, building facades, and other transparent surfaces to generate electricity without obstructing light or visibility.
v) **Light Management:** Researchers have developed strategies to engineer CNTs for enhanced light absorption. Functionalizing CNTs with light-absorbing materials, like quantum dots, can broaden the absorption spectrum, allowing the capture of a broader range of sunlight wavelengths.

10.2.3.2 Challenges

Although there have been improvements, there are still problems with the incorporation of CNTs into solar cell systems. The main issues that need to be solved include large-scale synthesis, homogeneous dispersion of CNTs, contact resistance, and long-term stability [8]. To overcome these difficulties and fully realize the promise of CNTs in solar cell applications, more research and development are necessary.

i) **Scalability and Consistency:** Scaling up the production of high-quality CNTs for commercial applications remains a challenge. Achieving uniform

CNT properties and consistent performance in large-scale solar cell manufacturing is an ongoing hurdle.

ii) **Dispersion and Alignment:** Ensuring proper dispersion and alignment of CNTs within the active layers of solar cells is essential. Agglomerated or misaligned CNTs can lead to inefficiencies and reduce the device's overall performance.

iii) **Cost:** Although the cost of CNT production has decreased, it remains relatively high compared to some other materials. Reducing the cost of CNTs and developing cost-effective manufacturing processes is essential for broader adoption in the solar industry.

iv) **Environmental and Health Concerns:** The production and disposal of CNTs raise environmental and health concerns. Addressing these issues and ensuring the responsible use and disposal of CNT-based solar cells is crucial.

v) **Stability:** While CNTs are known for their stability, long-term stability in solar cell applications is still a concern. Ensuring that CNT-based solar cells can withstand environmental factors and maintain their performance over many years is an area of ongoing research.

vi) **Regulatory and Certification Challenges:** Establishing industry standards, certifications, and regulations specific to CNT-based solar cells is necessary to ensure their safety and reliability, but this process can be time-consuming.

In summary, CNTs have shown substantial promise in improving the efficiency, versatility, and sustainability of solar cells. However, addressing trials related to scalability, cost, dispersion, and long-term stability remains crucial for the broader adoption of CNT-based solar cell technologies. In order to overcome these obstacles and realize the full potential of CNTs in solar energy applications, research and development efforts are still underway.

10.3 Graphene

Two-dimensional substance called graphene has amazing thermal, optical, and electrical characteristics. It is composed of a single sheet of carbon atoms arranged in a lattice similar to a honeycomb. Graphene's remarkable transparency, broad absorption spectrum, and high carrier mobility make it an ideal material for transparent conductive electrodes in solar cells, as demonstrated in Figure 10.2. The carbon allotropes are typically preferred as nanofillers in advanced polymer nanocomposites (PNCs) over other nanofillers because of their exceptional properties, which include high aspect ratio, high surface area, higher tensile strength, good electrical and thermal conductivity, higher flexibility, and electromagnetic shielding properties. Graphene can be added to the active layers of solar cells as an

Figure 10.2 Graphene, a two-dimensional material composed of a single layer of carbon atoms arranged in a honeycomb lattice [24]. *Source:* Nobeastsofierce/Adobe Stock.

electron transport medium or charge transport bridge to improve charge collection and decrease recombination losses [25].

10.3.1 Synthesis Techniques

Several synthetic methods for producing graphene have been established, including CVD, mechanical exfoliation, epitaxial growth, and chemical reduction of graphene oxide [26]. Unlike CVD, which makes it possible to produce high-quality graphene in large quantities, mechanical exfoliation necessitates the removal of graphene layers from bulk graphite. Through a process known as epitaxial growth, graphene is formed on a substrate by means of techniques such as CVD or molecular beam epitaxy. Chemical reduction of graphene oxide allows oxidized graphene to be transformed into reduced graphene oxide with specified properties.

10.3.2 Utilizing Graphene in Solar Cell Applications

10.3.2.1 Transparent Conductive Electrodes

Transparent conductive electrodes based on graphene have been studied as possible substitutes for traditional materials such as ITO in solar cells [27]. Graphene is a strong candidate for improving electrical performance and light transmission due to its exceptional electrical conductivity and transparency [28]. Graphene-based electrodes have shown better performance and stability in a variety of solar cell types, including silicon-based solar cells, organic solar cells, and perovskite solar cells.

10.3.2.2 Charge Transport Layers

In solar cells, graphene can act as a useful charge transport layer that speeds up charge carrier collection and transmission [29]. It allows for effective charge extraction and lowers recombination losses when employed as an electron or hole transport layer. Solar cells have increased charge transport and improved device performance by adding graphene as a charge transport layer. A strong choice for this application is graphene due to its high carrier mobility, low sheet resistance, and compatibility with a variety of materials [30].

10.3.2.3 Light-Harvesting Enhancements

Materials based on graphene can also be used to expand solar cell ability to harvest light [31]. In dye-SSCs, functionalized graphene can operate as sensitizers or light-absorbing materials, extending the absorption spectrum and increasing overall device efficiency. Moreover, light absorption and power conversion efficiency in thin-film solar cells can be raised by utilizing the plasmonic properties of graphene to enhance light trapping and absorption [32].

10.3.3 Recent Advancements and Challenges

10.3.3.1 Recent Advancements

The advancement of graphene-based materials in solar cell technology has been the focus of recent research [33]. The creation of scalable synthesis methods, the investigation of innovative device topologies, and the incorporation of graphene into hybrid composites are a few of these developments [34]. The inclusion of materials based on graphene has resulted in significant advancements in solar cell efficiency, stability, and scalability [35].

The single layer of carbon atoms arranged in a hexagonal lattice, known as graphene, has attracted a lot of attention due to its possible uses in solar cells and other industries. Recent advancements in utilizing graphene in solar cells showcase its unique properties, but there are still challenges that researchers are actively addressing. Here are some of the recent advancements associated with graphene in solar cells:

1) **Enhanced Conductivity and Electron Mobility:** Graphene's high electrical conductivity facilitates efficient charge transport within the solar cell. This property is particularly valuable in improving electron mobility and reducing resistance, leading to enhanced overall cell performance.
2) **Transparent Conductive Films:** Graphene-based transparent conductive films are being developed as replacements for traditional ITO electrodes. These films are appropriate for use in transparent solar cells and other optoelectronic devices because they preserve high transparency while offering outstanding electrical conductivity.

3) **Hybrid Structures with Perovskite Solar Cells:** Graphene has been integrated into perovskite solar cells as a charge transport layer. In order to overcome issues with the deterioration of perovskite materials over time, this hybrid technique has demonstrated promise in improving the stability and performance of perovskite solar cells.
4) **Flexible and Lightweight Solar Cells:** Graphene's mechanical strength and flexibility make it ideal for usage in flexible and lightweight solar cell designs. Graphene-based materials can be incorporated into flexible substrates, enabling the development of bendable and lightweight solar panels.
5) **Plasmonic Effects for Light Absorption:** Graphene's plasmonic properties can be harnessed to enhance light absorption in solar cells. Researchers want to increase solar energy conversion efficiency and boost solar absorption by implementing graphene-based plasmonic devices.

10.3.3.2 Challenges

Despite the advances, integrating graphene-based materials into solar cells still faces a number of difficulties [36]. These include obtaining uniform and controlled deposition of graphene layers, producing high-quality graphene on a wide scale, and enhancing the long-term stability of graphene-based electronics [37]. Further issues that must be resolved include the high cost of large-scale graphene synthesis and the requirement for effective device integration. Here are some of the challenges associated with graphene in solar cells:

1) **Scalability and Production Cost:** Mass manufacturing of high-quality graphene at a reasonable cost remains a challenge. Techniques for scalable and cost-effective graphene production need to be developed to make graphene-based solar cells economically viable.
2) **Heterogeneity and Quality Control:** Ensuring the consistency and quality of graphene in large-scale manufacturing is challenging. Graphene produced through various methods can exhibit variations in properties, and maintaining uniformity is crucial for reliable solar cell performance.
3) **Integration with Existing Technologies:** Integrating graphene into existing solar cell technologies, such as silicon-based cells, poses challenges. Researchers are working on developing compatible interfaces and optimizing the integration of graphene in order to produce synergistic effects with other materials.
4) **Environmental Impact:** It is concerning how the manufacture of graphene affects disposal and environmental practices. Developing environmentally friendly production processes and recycling methods for graphene-based materials is essential for sustainable use.
5) **Stability and Longevity:** Graphene-based solar cells must demonstrate long-term stability and reliability under different environmental conditions.

Ensuring that graphene does not degrade over time and maintaining the structural integrity of the solar cells are ongoing research priorities.

6) **Regulatory Approval and Standardization:** Establishing regulatory frameworks and standards for graphene-based solar cells is important for commercial adoption. Certifications and guidelines need to be developed to ensure the safety and performance of these devices.

In summary, graphene has enormous potential to advance solar cell technology; but, widespread implementation will require resolving issues with scalability, affordability, integration, and environmental impact. In order to overcome these obstacles and realize graphene's full potential in the solar energy sector, research is still underway.

10.4 Carbon Dots

Small carbon nanoparticles called CDs have special quantum confinement characteristics that cause their optical and electrical properties to vary depending on their size. These nanoparticles have demonstrated potential as sensitizers in dye-SSCs, where they improve light absorption and efficiently separate charges. Charge transfer efficiency and device stability can be improved thanks to the surface chemical tunability of CDs. Additionally, they are well suited for mass manufacturing and commercialization due to their low cost and simplicity of synthesis [21]. Broad absorption spectra, adjustable photoluminescence, and high quantum yield are only a few of the distinctive optical characteristics of CDs [38]. Depending on their size, surface passivation, and chemical makeup, they can emit light at a variety of wavelengths, from ultraviolet to near-infrared. The excellent photostability, chemical stability, and photobleaching resistance of CDs also make them suitable for long-term applications in solar cells. These are classified by three important categories, i.e. carbon quantum dots, modified CDs, and composite CDs as shown in Figure 10.3.

10.4.1 Synthesis Techniques

Various production procedures have been devised to customize the optical characteristics, surface chemistry, and size of CDs [40]. Hydrothermal synthesis, solvothermal synthesis, microwave-assisted synthesis, and top-down strategies like chemical etching and laser ablation are examples of common synthesis techniques. During the manufacturing process, these methods enable precise control over the chemical and physical characteristics of CDs.

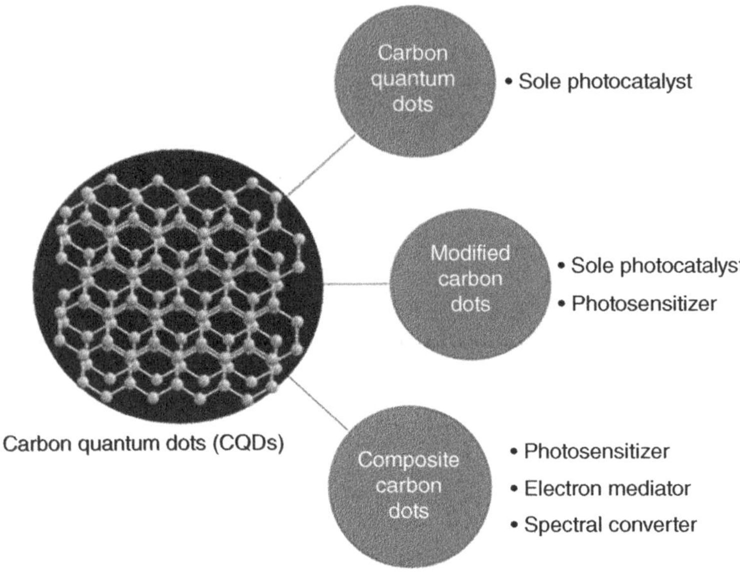

Figure 10.3 Carbon quantum dots (CQDs). *Source:* Ref. [39]/Frontiers Media S.A./ CC BY 4.0.

10.4.2 Applications of Solar Cell Carbon Dots

10.4.2.1 Light Harvesting and Sensitization

The promise of inexpensive, highly efficient PVs has drawn a lot of attention to SSCs in the search for carbon-free, renewable energy [41, 42]. CDs are able to enhance light absorption and optimize device performance in solar cells by acting as sensitizers or light-harvesting components. Due to their unique optical characteristics, which include broad absorption spectra and excellent photoluminescence quantum yield, they are beneficial for light harvesting in a range of solar cell topologies, including dye-SSCs and perovskite solar cells [43]. Surface engineering or functionalization of CDs can boost their ability to generate charges and absorb light, hence increasing the efficiency of solar cells.

10.4.2.2 Charge Separation and Transport of Electrons

In solar cells, CDs can act as materials for charge separation and electron transport. They enable effective charge transfer and minimize charge recombination due to their good electrical conductivity and electron-accepting characteristics, which improve device performance [44]. In solar cells, adding CDs as interfacial layers or within the active layer enhances electron transport, encourages charge separation, and reduces energy losses.

10.4.2.3 Energy Storage and Electrochemical Applications

In solar cells, CDs have demonstrated potential for electrochemical applications and energy storage. Due to their huge surface area, high conductivity, and superior stability, carbon materials are the family of electrode materials that dominate the market for commercial supercapacitors [45]. They can be used as additives or added to electrode materials to improve the electrochemical performance of energy storage devices such as lithium-ion batteries or supercapacitors. CDs' distinctive characteristics, such as their large surface area, strong electrochemical stability, and simple charge transfer kinetics, help them function more effectively electrochemically and store more energy.

10.4.3 Recent Advancements and Challenges

Novel synthesis techniques, higher photoluminescence characteristics, increased stability, and the investigation of innovative device topologies are some recent developments in the field of CDs for solar cell applications [46]. These developments have enhanced CD functionality, increased efficiency, and expanded CD uses in solar cells. However, issues including exact size and surface chemistry control, stability in challenging operating environments, and scalability of synthesis methods need to be addressed. Because of their special qualities, CDs, which are tiny carbon nanoparticles that are usually less than 10 nm, have drawn interest in the field of solar cells. These are some current developments and difficulties pertaining to the application of CDs in solar cells.

10.4.3.1 Recent Advancements

1) **Light Absorption and Photogenerated Charge Carrier Generation:** CDs exhibit tunable optical properties, allowing researchers to engineer their absorption spectrum. This tunability can be harnessed to enhance light absorption in solar cells, contributing to improved overall efficiency.
2) **Sensitization of Semiconductor Materials:** CDs can be used to sensitize semiconductor materials in solar cells, such as titanium dioxide (TiO_2) or other metal oxides. This sensitization enhances the separation and transfer of photogenerated charge carriers, leading to increased power conversion efficiencies.
3) **Hybrid Structures in Perovskite Solar Cells:** CDs have been integrated into perovskite solar cells to improve their stability and performance. The unique surface properties of CDs can passivate defects in perovskite materials, contributing to enhanced device stability and efficiency.

4) **Flexible and Transparent Solar Cells:** The size and structure of CDs make them suitable for use in flexible and transparent solar cells. Integration of CDs into flexible substrates allows for the development of lightweight and bendable solar panels.
5) **Surface Functionalization for Improved Electron Transfer:** Surface functionalization of CDs can be tailored to enhance their electron transfer properties. This modification facilitates efficient charge transfer within the solar cell, contributing to improved overall performance.

10.4.3.2 Challenges

1) **Synthesis and Reproducibility:** The synthesis of CDs with consistent properties, including size, surface chemistry, and optical characteristics, remains a challenge. Achieving reproducibility in large-scale manufacturing is crucial for commercial applications.
2) **Quantum Yield and Absorption Efficiency:** CDs often exhibit relatively low quantum yields and absorption efficiencies, which can limit their effectiveness in solar cells. Researchers are working on optimizing synthesis methods to improve these properties and maximize the utilization of incident light.
3) **Long-Term Stability:** Ensuring the long-term stability of CDs in solar cell applications is essential for their practical use. Stability concerns, including degradation over time and under different environmental conditions, need to be addressed for reliable and durable devices.
4) **Integration with Existing Technologies:** Integrating CDs into existing solar cell technologies, such as silicon-based cells, requires addressing compatibility issues and optimizing interfaces. Achieving effective integration is crucial for realizing the synergistic benefits of CDs in solar cells.
5) **Cost-Effectiveness:** The cost-effectiveness of CDs for large-scale solar cell production needs improvement. Developing cost-efficient synthesis methods and scalable manufacturing processes is a priority for making CD-based solar cells commercially viable.
6) **Regulatory Considerations:** Establishing regulatory guidelines and safety standards for CD-based solar cells is essential for their commercialization. Ensuring that these materials meet safety requirements and environmental standards is crucial for widespread adoption.

In conclusion, while CDs show promise in advancing solar cell technology, addressing challenges related to synthesis, reproducibility, stability, and cost is crucial for their successful integration into commercial PV applications. In order to overcome these obstacles and realize the full potential of CDs in solar energy conversion, research and development activities are still underway.

10.5 The Future of Carbon-Based Nanomaterials in Solar Cell Technology

The field of PVs could be greatly advanced by the use of carbon-based nanomaterials in solar cells, such as CNTs, graphene, and CDs. Because of these materials' unique properties and diverse range of applications, better energy conversion and storage are now possible. In this part, we discuss the possible applications of carbon-based nanomaterials in solar cells and emphasize key research and development areas.

10.5.1 Enhanced Light Harvesting and Absorption

The potential of carbon-based nanomaterials to improve light absorption and harvesting in solar cells is one of its main advantages. High surface areas and distinctive optical characteristics of CNTs, graphene, and CDs allow for effective light trapping and absorption over a broad wavelength range. In order to maximize light harvesting and improve light harvesting efficiency, future research will optimize the design and integration of these materials into solar cell architectures [20, 47].

10.5.2 Improved Charge Transport and Collection

The superior electrical conductivity of carbon-based nanomaterials makes them the perfect choice for enhancing charge transport and collecting in solar cells. Due to their high electron mobility, CNTs and graphene can efficiently move charges throughout a device, reducing losses and enhancing performance. CDs also help with effective charge collection and lower recombination losses by acting as electron mediators or charge transport materials. Future research should focus on creating cutting-edge methods to take use of the higher charge transport capabilities of carbon-based nanomaterials, which will increase solar cell efficiency [6].

10.5.3 Enhanced Stability and Durability

Stability and durability are crucial for the long-term proper operation of solar cells. Carbon-based nanomaterials are suitable for enhancing the durability of solar cell systems due to their high chemical inertness, intrinsic stability, and resistance to deterioration. Subsequent research endeavors will center on disentangling the processes via which carbon-based nanomaterials undergo degradation in diverse environmental contexts and formulating strategies to fortify and fortify them, thus guaranteeing the sustained functionality of solar cells [48].

10.5.4 Scalable Synthesis and Manufacturing

Scalable synthesis techniques and affordable manufacturing procedures must be created in order to fully utilize the promise of carbon-based nanomaterials in solar cells. High-quality CNTs, graphene, and CDs may now be produced on a lab scale using current synthesis methods such as CVD and solution-based processes. Scaling these methods up to industrial levels is still difficult, though. To enable large-scale manufacture of carbon-based nanomaterials for solar cell applications, future research should focus on improving synthesis procedures, investigating novel technologies, and developing scalable manufacturing processes [38].

10.5.5 Integration with New Advances in Solar Cell Technology

New advances in solar cell technology can be enhanced and given new capabilities by integrating carbon-based nanomaterials. For instance, integrating developing perovskite solar cells with CNTs or graphene can improve charge extraction and transport while addressing stability and scalability difficulties. Similarly, adding CDs to tandem solar cell topologies or organic solar cell designs can improve light harvesting and charge separation. To realize their full potential and attain better efficiencies, future research will investigate the synergistic integration of carbon-based nanomaterials with cutting-edge solar cell technologies [49].

10.5.6 Environmental Sustainability and Cost-Effectiveness

The advancement of carbon-based nanomaterials for solar cells must take into account both economic effectiveness and environmental sustainability. Researchers want to create sustainable synthesis techniques that use eco-friendly precursors and use less energy as the area develops. For carbon-based nanomaterials to be widely used in solar cell technologies, it will also be necessary to make efforts to maximize their cost-effectiveness through effective manufacturing processes, scalable production, and the use of abundant raw materials [50].

10.6 Conclusion

The efficiency and performance of solar cells are greatly improved by carbon-based nanomaterials like graphene, CNTs, and CDs because of their strong chemical stability, high electrical conductivity, and vast surface area. By enhancing light absorption, boosting charge transport, and lowering recombination losses, these materials create PV systems that are more resilient and effective. These nanomaterials have significant positive effects on the environment. The overall environmental impact can be minimized by using plentiful raw resources in an eco-friendly manner during production.

The promotion of green renewable and environmentally friendly energy sources is consistent with the larger objectives of green energy. Furthermore, the development of flexible and translucent solar cells made possible by carbon-based nanomaterials expands the uses of solar technology by allowing them to be incorporated into a range of surfaces including windows, structures, and portable electronics. These materials not only increase the efficiency of solar cells but are also essential for energy storage. CDs, for instance, can improve the performance of lithium-ion batteries and supercapacitors, leading to more effective energy storage systems. For renewable energy systems to be stable and reliable, energy generation and storage must work together harmoniously.

The creation of a more sustainable energy infrastructure is facilitated by the incorporation of carbon-based nanomaterials into solar cells and energy storage systems. These materials enable flexible and lightweight solar panels, which are especially advantageous for building-integrated PV systems and portable electronics. This increases the accessibility and usefulness of renewable energy for daily usage. Therefore, even if there are still issues with synthesis, cost, stability, and reproducibility, more research is being done to find solutions and help carbon-based nanomaterials reach their full potential. As these materials are developed further and used in commercial solar applications, substantial advancements in the field of efficient, flexible, and sustainable energy solutions are anticipated.

References

1 Smith, A.B., Johnson, C.D., and Thompson, E.F. (2018). Carbon nanotubes for energy storage applications. *Materials Today Chemistry* 8: 88–96.

2 Novoselov, K.S., Fal'ko, V.I., Colombo, L. et al. (2012). A roadmap for graphene. *Nature* 490 (7419): 192–200.

3 Li, X., Rui, M., Song, J. et al. (2015). Carbon and graphene quantum dots for optoelectronic and energy devices: a review. *Advanced Functional Materials* 25 (31): 4929–4947.

4 Gao, X., Jang, J., and Nagai, M. (2011). Hybrids of graphene and nanotubes: formation, properties, and applications. *Advanced Materials* 23 (33): 3686–3700.

5 Yang, S., Feng, X., Wang, X., and Müllen, K. (2013). Graphene-based carbon nitride nanosheets: efficient catalysts for electrochemical reduction of oxygen and hydrogen peroxide. *ACS Nano* 7 (4): 3576–3586.

6 Wang, H., Sun, X., Wu, J., and Bai, H. (2019). Carbon-based materials for perovskite solar cells: progress, challenges, and perspectives. *ACS Energy Letters* 4 (4): 799–818.

7 Yu, Y., Gao, J., Hummadi, K.K., and Wu, N. (2019). Recent advances in carbon dots for dye-sensitized solar cells. *Materials Today Chemistry* 13: 14–25.

8 Zheng, J., Zhang, L., Dong, Y. et al. (2020). Challenges and strategies for enhancing the stability of perovskite solar cells. *Advanced Materials Interfaces* 7 (5): 1901465.

9 https://www.newscientist.com/article/2093356-carbon-nanotubes-too-weak-to-get-a-space-elevator-off-the-ground/

10 Liu, Y., Zhang, X., and Li, Y. (2021). Carbon nanomaterials for efficient and stable perovskite solar cells: recent advances and perspectives. *Nano-Micro Letters* 13 (1): 38.

11 Kang, D. and Lee, J. (2021). Graphene-based counter electrodes for dye-sensitized solar cells. *Nanoscale* 13 (15): 6893–6915.

12 Jeon, Y. and Seo, J. (2021). Carbon-based nanomaterials for perovskite solar cells: recent advances and prospects. *Journal of Energy Chemistry* 59: 45–64.

13 Yuan, W., Liu, H., Xu, L., and Li, L. (2021). Carbon dots as light harvesters in perovskite solar cells. *Frontiers in Energy Research* 9: 661389.

14 Zhang, W., Liu, C., Li, G. et al. (2021). Recent advances in carbon-based materials for perovskite solar cells. *Nano Select* 2 (6): 793–807.

15 Liu, Y., Zhang, X., Chen, S. et al. (2021). Carbon nanomaterials for high-performance perovskite solar cells: advances and challenges. *Small Methods* 5 (4): 2001126.

16 Ma, J., Li, Z., Sun, J., and Sun, Y. (2021). Carbon-based materials for perovskite solar cells: a review. *Frontiers in Chemistry* 9: 666617.

17 Zhang, Q., Guo, X., Dong, Y., and Wang, C. (2019). Carbon-based nanomaterials for perovskite solar cells: a review. *Journal of Materials Chemistry A* 7 (36): 20673–20697.

18 Oviedo, C. and Mora-Seró, I. (2019). Perovskite solar cells: an overview of recent progress and perspectives. *Energy & Environmental Science* 12 (6): 1867–1888.

19 Sun, Y., Gao, S., Lei, F. et al. (2019). Recent progress in the stability of perovskite solar cells. *Journal of Materials Chemistry A* 7 (20): 11942–11960.

20 Chen, X., Li, Z., Zhu, J. et al. (2020). Carbon dots for perovskite solar cells: recent advances, challenges, and opportunities. *Journal of Materials Chemistry A* 8 (24): 12099–12121.

21 Shinde, D.V., Shinde, S.K., Mahadik, M.A. et al. (2020). Recent advances and challenges in the application of carbon dots in perovskite solar cells. *Journal of Energy Chemistry* 47: 119–135.

22 Wang, Z., Sun, H., Li, X. et al. (2018). Graphene-based materials for perovskite solar cells: a review of recent advances and prospects. *Nanoscale* 10 (3): 1055–1073.

23 Adi, K.K. and Wu, N. (2019). Recent advances in carbon dots for dye-sensitized solar cells. *Materials Today Chemistry* 13: 14–25.

24 https://newatlas.com/graphene-sheets-strength/27761/
25 (a) Khichar, K.K., Dangi, S.B., Dhayal, V. et al. (2020). Structural, optical, and surface morphological studies of ethyl cellulose/graphene oxide nanocomposites. *Polymer Composites* 41 (7): 2792–2802.
26 Novoselov, K.S., Geim, A.K., Morozov, S.V. et al. (2004). Electric field effect in atomically thin carbon films. *Science* 306 (5696): 666–669.
27 Li, X., Magnuson, C.W., Venugopal, A. et al. (2009). Large-area graphene single crystals grown by low-pressure chemical vapor deposition of methane on copper. *Journal of the American Chemical Society* 131 (43): 15939–15944.
28 Bae, S., Kim, H., Lee, Y. et al. (2010). Roll-to-roll production of 30-inch graphene films for transparent electrodes. *Nature Nanotechnology* 5 (8): 574–578.
29 Li, X., Cai, W., An, J. et al. (2008). Large-area synthesis of high-quality and uniform graphene films on copper foils. *Science* 324 (5932): 1312–1314.
30 Li, X., Zhu, Y., Cai, W. et al. (2009). Transfer of large-area graphene films for high-performance transparent conductive electrodes. *Nano Letters* 9 (12): 4359–4363.
31 Zhang, Y., Guo, L., Wei, S. et al. (2010). Direct imprinting of microcircuits on graphene oxides film by femtosecond laser reduction. *Nano Today* 5 (1): 15–20.
32 Karim, N.A., Alias, M.S., and Yang, H. (2021). Recent developments for the application of 3D structured material nickel foam and graphene foam in direct liquid fuel cells and electrolyzers. *Catalysts 11* (2): 279.
33 Gao, L., Ren, W., Xu, H. et al. (2010). Repeated growth and bubbling transfer of graphene with millimetre-size single-crystal grains using platinum. *Nature Communications* 3 (1): 1–7.
34 Chen, J.H., Jang, C., Adam, S. et al. (2008). Charged-impurity scattering in graphene. *Nature Physics* 4 (5): 377–381.
35 Souma, S. and Ogawa, M. (2020). Impact of electron–phonon scattering on the strain-induced current-blocking effect in graphene field-effect transistors. *Journal of Applied Physics* 127 (9).
36 Schedin, F., Geim, A.K., Morozov, S.V. et al. (2007). Detection of individual gas molecules adsorbed on graphene. *Nature Materials* 6 (9): 652–655.
37 Tan, J., Ji, Y.N., Deng, W.S., and Su, Y.F. (2021). Process intensification in gas/liquid/solid reaction in trickle bed reactors: a review. *Petroleum Science* 18 (4): 1203–1218.
38 Hu, C., Li, M., Qiu, J., and Sun, Y.P. (2019). Design and fabrication of carbon dots for energy conversion and storage. *Chemical Society Reviews* 48 (8): 2315–2337.
39 Jung, H., Sapner, V.S., Adhikari, A. et al. (2022). Recent progress on carbon quantum dots based photocatalysis. *Frontiers in Chemistry* 10: 881495.
40 Zhou, W., Zhuang, J., Li, W. et al. (2017). Towards efficient dual-emissive carbon dots through sulfur and nitrogen co-doped. *Journal of Materials Chemistry C* 5 (32): 8014–8021.

41 Kamat, P.V. (2008). Quantum dot solar cells. Semiconductor nanocrystals as light harvesters. *The Journal of Physical Chemistry C* 112 (48): 18737–18753.

42 Kamat, P.V. (2010). Photosensitization of SnO_2 and other oxides. *Dye-sensitized Solar Cells* 165–182.

43 Emdadi, B., Carbon Dots Dye-Sensitized Solar Cells Application. 2022.

44 Dave, K. and Gomes, V.G. (2019). Carbon quantum dot-based composites for energy storage and electrocatalysis: mechanism, applications and future prospects. *Nano Energy* 66: 104093.

45 Majumder, M. and Choudhary, R.B. (2011). Gravimetric and volumetric capacitive performance of. *Science* 334: 917–918.

46 Yang, S.T., Cao, L., Luo, P.G. et al. (2013). Carbon dots for optical imaging in vivo. *Journal of the American Chemical Society* 131 (32): 11308–11309.

47 Di, J., Yong, Z., Zheng, X. et al. (2013). Aligned carbon nanotubes for high-efficiency Schottky solar cells. *Small* 9 (8): 1367–1372.

48 Zhai, Y., Zhang, B., Shi, R. et al. (2022). Carbon dots as new building blocks for electrochemical energy storage and electrocatalysis. *Advanced Energy Materials 12* (6): 2103426.

49 Xu, B., Qi, S., Jin, M. et al. (2019). 2020 roadmap on two-dimensional materials for energy storage and conversion. *Chinese Chemical Letters 30* (12): 2053–2064.

50 Gao, N., Huang, L., Li, T. et al. (2020). Application of carbon dots in dye-sensitized solar cells: a review. *Journal of Applied Polymer Science 137* (10): 48443.

11

Carbon-Based Nanomaterials in Energy Storage Devices: Fuel Cells and Biofuel Cells

Ponnusamy Thillai Arasu[1], Arumugam Murugan[2], G. Kanthimathi[3], A. Malar Retna[4], S. Daphne Rebekal[5], Natarajan Raman[6], Robin Kumar Samuel[7], and Tola Jebssa Masho[1]

[1] *Department of Chemistry, College of Natural and Computational Sciences, Wollega University, Nekemte, Ethiopia*
[2] *Department of Chemistry, North Eastern Regional Institute of Science and Technology, Nirjuli, Arunachal Pradesh, India*
[3] *Department of Chemistry, Ramco Institute of Technology, Rajapalayam, Virudhunagar, Tamil Nadu, India*
[4] *Department of Chemistry, Scott Christian College (Autonomous), Nagercoil, Tamil Nadu, India*
[5] *Department of Chemistry, Sarah Tucker College, Tirunelveli, Tamil Nadu, India*
[6] *Department of Chemistry, VHNSN College (Autonomous), Virudhunagar, Tamil Nadu, India*
[7] *Department of Chemistry, Good Shepherd College of Engineering and Technology, Kanyakumari, Tamil Nadu, India*

11.1 Introduction

Researchers who have been studying the world's energy issues have recently concentrated on biofuel cells (BFCs), which are currently thought to be an explorable technology. BFCs have various uses, including self-powered biosensors, worn and implanted biomedical equipment, and sewage treatment. Nevertheless, BFCs continue to pose challenges in terms of their limited power generation capacity and their ability to maintain stable operation over extended periods of time. The majority of BFCs that have been built thus far have the capacity to provide bioelectronics devices with temporary power [1], which requires efficient production of bioelectrodes in order to use enzymatic BFCs (EBFCs) in real applications. Graphene possesses unique physiochemical features that render it very appropriate as a construction material for electrochemical fuel cells. Despite its promising characteristics, graphene, a type of all of the carbon at the nanoscale size, is not as widely utilized in EBFCs as carbon nanotubes (CNTs). This review's main focus is on the state of research on graphene-based electrodes at the moment, including modified electrodes constructed of graphene derivatives and composites and

Carbon-based Nanomaterials for Green Applications, First Edition. Edited by Upendra Kumar, Piyush Kumar Sonkar, and Suman Lata Tripathi.

Published 2025 by John Wiley & Sons, Inc.

free-standing graphene electrodes. The bioelectrochemical applications highlight the particular attributes of graphene-based electrodes, including their extensive porosity, excellent electrical conductivity, and mechanical malleability [2]. The catalyst employed in a fuel cell determines the characteristics of a BFC, a device that directly produces power from biological materials. Many benign chemical fuels with specific components to oxide fuels can be used in BFCs, including glucose, lactate, urate, starch, fructose, alcohol, and amines. Their potential for the production of renewable energy, as remote device power sources, etc., is demonstrated by their readily accessible fuel sources and comfortable operating conditions. Instead of using precious metals as catalysts to oxidize fuel, enzymatic BFCs use enzymes. The EBFC system is characterized by several limitations, such as inadequate fuel oxidation, reduced power density, unstable operation, diminished voltage output, decreased energy concentration, insufficient resilience, and unpredictability in long-term usage. The creation of non-enzymatic BFCs is therefore required. The review study delves deeper into the different production pathways and catalytic properties of NEBFCs. Nanotechnologies are being developed to improve the mechanical adaptability, biocatalytic properties, ability to be implanted, compatibility with living organisms, and ability to degrade naturally of BFCs [3]. Bacterial cells and nanomaterials cooperated to enhance the electrical energy output of microbial biological fuel elements (MBFE). It has been demonstrated that oxidizing ethanol with *Gluconobacter* bacteria membrane fractions shared with thermo-expanded graphite can transfer electrons directly and generate energy [4]. Multi-walled CNTs can be added to fuel cell electrodes to enhance the power of microbial fuel cells. Without the usage of risky redox mediators, electron movement to the electrode's outermost layer and charge collection across the electrode were made possible by the co-gels' higher electrical conductivity and nanoscale structure. Enzymatic BFCs are reusable and may be loaded again due to the long-lasting co-gels and our meticulously crafted loading process. The current research demonstrated the system's ability to perform and identified components that may be improved to enhance EBFC performance [5]. The primary biomass components, such as cellulose, alginate, and starch, are used by direct biofuel-fed fuel cells to produce energy. These fuel cells can only be utilized in conjunction with wastewater treatment operations to harvest energy as biomass breaks down into organic wastes, though, because of their low power density. Furthermore, in addition to the progress in breakdown technology, there were other efforts to develop fuel cells that derive energy from biomass or supplementary biofuels such as monosaccharides, bioethanol, and biohydrogen. Although the electrical power intensity achieved by biofuel-fed fuel cells has so far proved adequate to meet the needs of portable devices, the main hurdles still remain in the cost and lifetime of fuel cell manufacturing [6]. Figure 11.1 provides a concise overview of the primary uses of various types of fuel cells.

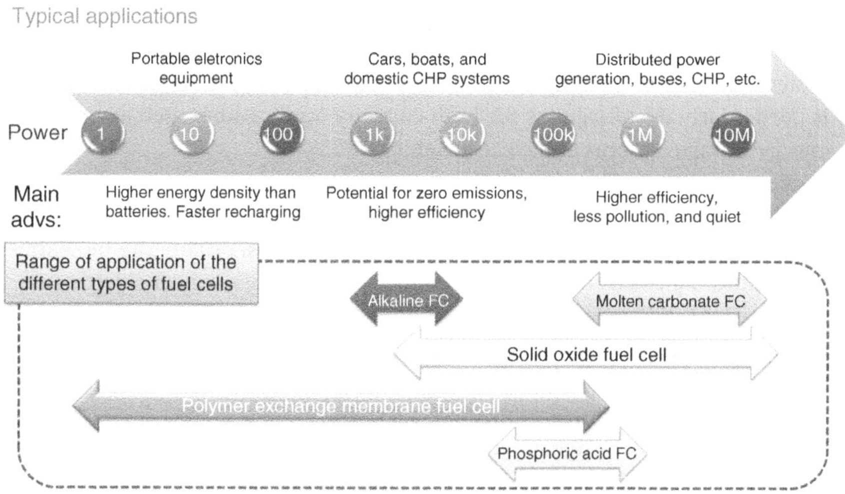

Figure 11.1 A comprehensive summary of the primary uses and benefits of various types of fuel cells. The power unit is measured in watts, and CHP refers to combined heat and power systems. *Source:* Ref. [6]/MDPI/CC BY 4.0.

11.2 Carbon-Based Nanomaterials' Function in Energy Storage

Researchers have produced and demonstrated the promise of numerous materials in energy storage applications, including graphene, conducting polymers, multi- and single-walled CNTs, and various metal oxides.

CNTs, which are one-dimensional nanostructures, are increasingly being used for energy storage due to their exceptional electrical, mechanical, and thermal characteristics, as well as their enormous surface area, excellent surface-to-weight ratio, and excellent ability to store electricity [7]. An extensive examination of the literature has shown numerous studies focused on utilizing CNT-based materials for various forms of energy storage.

11.3 Carbon Nanotube-Based Materials for Use in Batteries

The science of energy storage has undergone a revolution thanks to the remarkable capabilities and adaptability of CNTs. Constructed from hexagonally organized carbon atoms, these cylindrical nanostructures have created interesting opportunities to improve battery performance in a variety of applications.

1) **Enhanced Conductivity:** The exceptional electrical conductivity of CNTs is well known. They provide faster electron movement when integrated into battery electrodes, leading to faster rates of charge and discharge. This benefit is particularly advantageous for high-capacity applications such as grid-scale energy storage systems and electric vehicles (EVs).
2) **Increased Energy Density:** CNTs may store more energy in a given volume or mass because of their huge surface-to-weight ratio and high surface area. This characteristic enhances the energy density of batteries, prolonging the battery life of portable electronics and augmenting the driving range of electric cars.
3) **Mechanical Robustness:** The exceptional mechanical strength of CNTs prolongs the life and resilience of battery components. This longevity is particularly useful in applications like wearable technology and aerospace where batteries may be subjected to mechanical stress.
4) **Thermal control:** Longevity and safety of batteries depend on efficient thermal control. Because of their remarkable thermal conductivity, CNTs help dissipate heat produced during the charge and discharge cycles, lowering the possibility of thermal runaway and improving battery performance overall.
5) **Versatility:** CNTs can be designed and implemented to fulfill particular battery specifications. They can function as active ingredients in lithium-ion, sodium-ion, and supercapacitor batteries, among other battery types.
6) **Future Innovations:** New uses for CNTs in energy storage are still being investigated. This includes researching CNT-based anodes and cathodes with higher capacity and cycle life, as well as developing flexible and stretchable batteries for wearable devices.

Even though CNTs have a lot of potential to improve batteries, there are still issues that need to be resolved, including large-scale production, cost-effectiveness, and environmental concerns. However, the substantial study and advancement in this area show the ongoing dedication to utilizing the exceptional qualities of CNTs to produce more effective and long-lasting battery technologies. CNTs are likely to play a critical role in shaping the progress of technologies for storing energy, as long as there continue to be developments in materials science and engineering.

11.4 Carbon Nanotube Varieties

CNTs exist in several forms, primarily characterized by their architectural composition. There are two primary types of CNTs: SWCNTs and MWCNTs.

These groups were further separated according to their chiralities and other structural characteristics. Here is a quick synopsis of each kind.

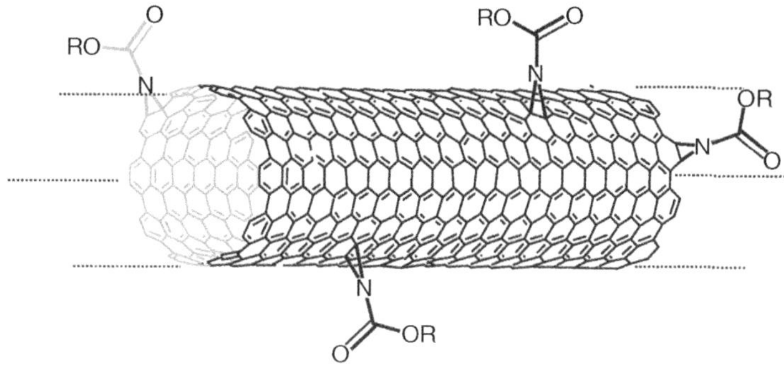

Figure 11.2 Single-walled carbon nanotube replication. *Source:* Ref. [9]/American Chemical Society.

11.4.1 Single-Walled Carbon Nanotubes (SWCNTs)

SWCNTs consist of a monolayer of carbon atoms arranged in a cylindrical structure that resembles a rolled-up graphene sheet (Figure 11.2). The particular arrangement of carbon atoms determines their chirality, which can either make them metallic or semiconducting. High tensile strength, exceptional electrical conductivity, and distinctive electronic characteristics are well-known attributes of SWCNTs [8, 9].

11.4.2 Multi-Walled Carbon Nanotubes (MWCNTs)

MWCNTs consist of many layers of CNTs stacked one on top of the other, giving the appearance of nested tubes. Because they are made of more layers than SWCNTs, they have more mechanical strength. Depending on how the layers are arranged and interact, MWCNTs can also have a range of electrical characteristics.

11.4.2.1 Chirality

The particular configuration of carbon atoms within the cylindrical structure is known as the chirality of CNTs. A nanotube's chirality defines its metallic or semiconducting nature. Chirality comes in three primary forms:

1) **Armchair (n, n):** Armchair nanotubes exhibit a metallic behavior because the indices (n, n) are equivalent. Their electrical conductivity is really good.
2) **Zigzag ($n, 0$):** These metallic nanotubes have one index that is equal to zero ($n, 0$).
3) **Chiral (n, m):** The values of the two indices (n, m) differ for chiral nanotubes. They can be semiconducting or metallic, depending on the precise values of n and m. There are three varieties of CNTs as shown in Figure 11.3.

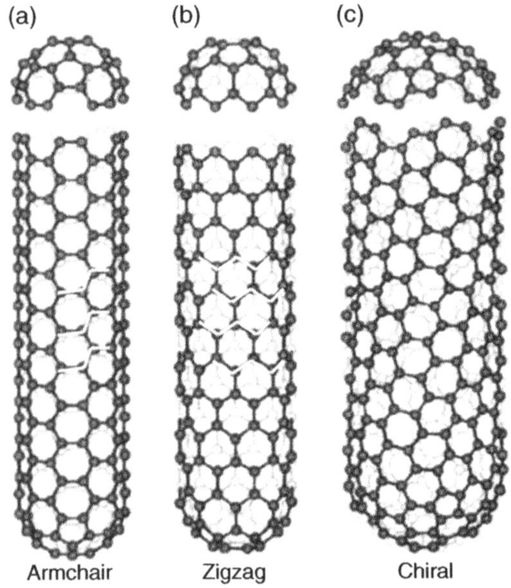

Figure 11.3 There are three types of carbon nanotubes (CNTs): (a) armchair with the indices $(n, m) = (5, 5)$; (b) zigzag with the indices $(n, m) = (9, 0)$; and (c) chiral with the indices $(n, m) = (10, 5)$. *Source:* The above image is a reproduction from an open-access journal, in accordance with the provisions of a Creative Commons Attribution License. Ref. [10]/ MDPI/CC BY 4.0. The content is subject to copyright and is owned by 2020qadi.

a) **Diameter and qualities associated with diameter:** Another important parameter influencing a CNT's qualities is its diameter. While larger-diameter nanotubes can be metallic, smaller-diameter nanotubes are usually semiconducting.
b) **Length:** The length of a CNT can range from a few nanometers to several centimeters. In some applications, like as reinforcing elements in composites, longer nanotubes are preferable.
c) **Functionalization:** By introducing chemical groups or molecules, researchers can alter the surface of CNTs. Their features can be customized for certain uses through this modification.

CNTs are highly versatile materials that can be utilized in various applications, including composites, electronic devices, energy storage, and sensors. The desired qualities and intended use in a given application determine the type and structure of CNT to be used.

Due to their meticulous manufacturing, CNTs are extensively utilized in various industries including energy management, electronics, and nanomaterials research, including chemical processing. Their notable attributes encompass remarkable robustness, as well as exceptional conductivity in both heat and electricity, rendering them a subject of significant interest in various fields, particularly in the realm of energy storage devices, signifying a significant advancement in the realm of science and technology [11]. CNTs possess a hexagonal carbon lattice that is crucial for efficiently generating electrical charges within the tube. Due to this particular attribute, CNTs are valuable in uses for energy storage such as electrochemical supercapacitors and lithium-ion batteries [12]. CNTs possess not only excellent electrical conductivity but also a substantial surface area and electrochemical stability. These characteristics render them a suitable material for enhancing energy storage and conversion devices. Moreover, CNTs possess a diverse array of applications across several scientific and technical fields due to their unique properties and structural characteristics [13]. CNTs have applications in power storage systems, including lithium-ion batteries [14], supercapacitors [15], and fuel cells [16]. CNTs are included in the anode and cathode materials of supercapacitors and fuel cells to enhance their electrical conductive properties [17].

11.5 Carbon Nanoparticles

Carbon nanoparticles have demonstrated potential in a range of energy storage applications, especially in the development of batteries and supercapacitors. They are highly sought-after competitors for batteries to store energy because of their exceptional features, including their large surface area, superb conductivity of electricity and chemical stability. These are the ways in which carbon nanoparticles are applied.

11.5.1 Supercapacitors

Carbon nanoparticles, such as graphene and CNTs, can be used as materials for electrodes in supercapacitors, especially those with well-defined structures [14, 15]. High capacitance is made possible by the huge surface area that these materials provide for ion adsorption. Rapid energy storage and release are made possible by the high electrical conductivity of carbon nanoparticles, which guarantees effective charge/discharge operations. Compared to conventional batteries, carbon-based supercapacitors have a longer life cycle since they don't go through chemical reactions as they charge or discharge.

11.5.2 Batteries

Carbon nanoparticles can be used as parts of different battery types as well as lithium-ion batteries. In the anode materials, they are frequently utilized [18]. By enhancing the conductivity of the electrode and offering stable structures for lithium-ion insertion/extraction, CNTs and graphene can enhance the performance of lithium-ion batteries. To further enhance battery performance, carbon nanoparticles can be employed as conductive additives and as supports for different types of electrode materials.

11.5.3 Fuel Cells

Carbon nanoparticles are used in fuel cell electrodes due to their large surface area and excellent ability to conduct electricity [19]. They aid in the efficient catalysis of reactions, such as the fuel cell's oxygen reduction reaction (ORR), which is essential for producing energy.

11.5.4 Hybrid Energy Storage Systems

Hybrid systems for storing energy can include carbon nanoparticles to combine the benefits of batteries and supercapacitors. This approach is appropriate for applications that require both a significant amount of energy and rapid energy delivery. It aims to attain both an elevated energy density and a substantial amount of power at the same time [20]. It is important to remember that even though carbon nanoparticles have a lot of potential, putting them into practice usually requires overcoming obstacles like cost and scalability, and making sure their production and disposal methods don't harm the environment.

The primary feature of quantum dots, which are semiconductor particles with nanoscale dimensions, is their exceptional optical and electrical properties. They are frequently utilized in sensors, photovoltaic devices, and display technologies. But using them as energy storage devices, like supercapacitors or batteries, is not a common or extensively researched function. The probable reasons are:

Limited capacity for storing energy: The main applications of quantum dots are in size-dependent electronics, including programmable bandgaps and quantum confinement effects. Although these characteristics are useful for optoelectronic and electrical applications, they do not naturally provide a large amount of energy storage, which is essential for batteries [21].

Material limitations: Semiconducting materials such as lead sulfide (PbS) or cadmium selenide (CdSe) are commonly used to create quantum dots. There may be toxicity issues with certain materials, which could be detrimental for widely used energy storage applications. Although there is a lot of work being done to create environmentally friendly quantum dots, their capacity to store energy is still quite low.

Challenges with charge storage: The diminutive dimensions and enormous surface area of quantum dots may be advantageous in some situations, but they can also provide problems with charge storage. It can be difficult to control ion and electron movement at the nanoscale, which could make them less suitable for large-scale energy storage.

Research status: Research on the application of quantum dots in storing energy is ongoing, although it is not as developed or established compared to the utilization of different substances like CNTs, graphene, or metal oxides. These materials are more commonly used in batteries and supercapacitors.

In conclusion, although quantum dots were being investigated for a different types of usage and present intriguing opportunities in the field of nanotechnology, their use as the main component of that store energy like battery packs or supercapacitors is not as common. Future developments might reveal creative ways to use quantum dots for energy storage since researchers are always looking for new materials and methods to improve energy storage technology. However, optoelectronics and nanoscale sensors continue to be their principal applications at the moment.

11.5.5 Quantum Dots

Traditionally, quantum dots are not frequently used as storage for energy devices like supercapacitors or battery packs. Quantum dots, which are semiconductor nanoparticles, are commonly used in LEDs, photovoltaic systems, and displays due to their unique electrical and optical characteristics [22]. Nonetheless, several recent studies indicate that quantum dots might indirectly contribute to energy storage or be used in particular specialized applications:

Photovoltaic: Research has looked into using quantum dots in solar cells. Quantum dots can be engineered to absorb light at specific wavelengths, allowing for better spectral matching with the solar spectrum. Although the main purpose of solar cells is to convert sunlight into electricity, they can indirectly aid in energy storage by producing electrical energy that can be instantly consumed or stored in batteries [23].

Quantum Dot Batteries (Speculative): Although this is still a relatively new topic of study, some research has looked into using quantum dots in experimental battery designs. Despite being in the initial phases of advancement and encountering many technical challenges, quantum dots have the potential to be utilized in distinct ways, such as enhancing charge and discharge rates or extending the longevity of batteries [24–26].

Quantum Dots for Capacitors: According to some research, quantum dots could be used as a material to boost supercapacitors' efficiency [27–29]. Their large surface area and compact size may increase the supercapacitor electrodes' capacitance. Although there is promise in this notion, research is still needed before it can be put into practice.

It is important to remember that energy storage does not usually involve the usage of quantum dots alone. Rather, they are more frequently employed in solar cell and other energy conversion technologies. However, since scientists are always looking into new materials and technologies, it is likely that as knowledge advances and engineering methods advance, quantum dots will play a bigger part in energy storage applications. However, the field is still in its early stages. Before quantum dots are used as a viable component of energy storage systems, there are numerous technical issues that need to be resolved.

11.6 Carbon Nanosheets

Carbon nanosheets (CNSs) are a type of nanomaterial composed of carbon atoms organized in a 2D shape resembling a sheet. These structures can be graphene, CNTs, or graphene oxide (GO) sheets, among other forms. Because of their remarkable qualities, CNSs are currently receiving a lot of attention for their potential to utilize technologies for storing energy, such as supercapacitors and battery packs [30].

High Surface Area: The extensive surface area of CNSs is advantageous for energy storage applications. More electrochemically active locations where charge storage can take place are made possible by this large surface area.

Good Electrical Conductivity: CNSs have good electrical conductivity, which is necessary for energy storage devices to charge and discharge quickly.

Chemical Stability: CNSs have a prolonged lifespan in energy storage devices due to the inherent chemical stability of carbon-based materials.

Tunable Properties: To optimize the efficiency of CNSs for certain energy storage purposes, the manufacturing process can be modified by altering the layer count, functionalization, or doping.

CNSs are commonly utilized in energy storage for two main purposes:

1) CNSs possess the capacity to serve as material for electrodes for supercapacitors. Supercapacitors are devices for storing energy that utilize the electrostatic dispersion of ions at the interface between the electrolyte and the electrode to store and discharge energy. CNSs possess a substantial surface area and exhibit conductivity to electricity, enabling them to store a significant amount of energy and for rapid charging and discharging. As a result, they are well-suited for applications such as EVs and renewable energy systems that require higher power densities and rapid charging [31].
2) CNSs can also be used in lithium-ion batteries, which are widely used in a variety of portable electronic gadgets. These batteries can use them as anode materials since they have a large capacity and better cycling stability. The two-dimensional (2D) structure of CNSs can support the electrode material's expansion and contraction during cycles of charge and discharge, enhancing the battery's overall longevity.

CNSs possess significant promise for energy storage technologies. There are still many unanswered questions in this area of study, including those related to scalability, cost-effectiveness, and the creation of novel synthesis methods. However, their distinct qualities make them an attractive option for upcoming developments in energy storage technology [32].

11.7 Biofuels

It is acknowledged that biofuels are a practical energy substitute. Biofuel is produced by thermochemical or biochemical conversion and is often more productive and efficient [33]. The term biofuel is a type of fuel that is produced from the biomass of plants, which may contain leftovers from recently deceased organisms [34]. Advanced biofuels, also known as second-generation biofuels, are derived from lignocellulosic material, which includes forestry crops and waste from farming. On the other hand, traditional biofuels, or first-generation biofuels, are produced from cellulose, glucose, or plant-based oils [35]. For USA, renewable energy is a crucial component of policy. In the economy, the biofuel sector serves a dual purpose by providing both energy and basic chemicals [36]. Enhancing the production of biomass resources, testing biofuels, energy compatibility, and emission modeling are some of the aspects of biofuel research. Using thermal or microbiological methods to generate new starch and cellulose compounds is also beneficial [37].

11.7.1 Biofuel Classification

The four primary categories of biofuels are solid, gaseous, bioethanol, and biobutanol. Bioethanol and biodiesel are two biorenewable liquid transportation fuels with the capacity to substitute diesel and petrol. Ethanol from biomass is a feasible alternative to fuel because it is predominantly obtained from used for food. Biodiesel has recently become popular due to its environmental benefits. Liquid biofuels are crucial for the future as they serve as a substitute for gasoline and diesel. Biofuels are typically seen as providing multiple benefits, such as supply chain safety, social structure, agriculture, regional growth, sustainability, and lowering greenhouse gas emissions. The cost of producing biofuel varies depending on the feedstock, conversion method, production volume, and geographic location. When comparing sulfur, aromatic content, and flash point, biodiesel performs better than regular diesel. Bioethanol is derived from the sugarcane plant sweet beets, and maize, and can also be obtained from tapioca and other sources, cellulose-containing plants (trees and grasses), and crop waste [38]. Production of biobutanol is mostly constant. It is commonly mislabeled as a new fuel and utilized as a basic chemical and solvent. Solid biofuels are utilized. The primary fuel used in homes was firewood. Before the eighteenth century, it served as the primary source of fuel worldwide for both heating and cooking needs [39].

11.7.1.1 Biogas

A mixture of several gases called biogas is created when biological material (cow dung) breaks down anerobically without the presence of free oxygen. CH_4 (50–75%), CO_2 (25–50%), N_2 (0–10%), and H_2 (0.1–1%) make up biogas. Anerobic bacteria produce biogas, which is continually collected in large quantities throughout the process [40].

11.7.2 Background of Biofuel

In 2011, biofuels accounted for three percent of the total energy utilized for vehicle travel worldwide. In 2011, the country of America, which is the leading global producer of biofuels, derived four percent of its transportation energy from biofuels. In 2009, biofuels constituted 23% of Brazilian transportation energy. Large-scale global production of biofuel can decrease emissions associated with transportation [41].

The history of biofuels begins with biomass, which is a very flexible material that has long been created from cotton fiber biomass and is largely utilized as fuel from firewood mass. Prior to now, fuels made from different biomasses might have been added to fuels based on gasoline. The current energy goal and climate change encourage the production and consumption of renewable fuels, such as bioenergy. The annual biomass energy derived by native plants is three to four times greater than the world's current energy consumption. For internal combustion engines, biofuels come in three varieties. Explosive fuel is used in the first internal combustion engine.

Since it takes millions of years for fossil fuels to convert into biofuels, they cannot be regarded as renewable. Sustainable farming practices can help them form year after year. Renewable refers to the ability of the source to form and not run out. The only energy that doesn't harm the environment, cause acid rain, or contribute to global warming is green energy. Examples of renewable energy sources that aren't necessarily environmentally friendly include biofuels, as they contribute significantly to greenhouse gas emissions [42].

Many countries have taken steps to ensure energy security, especially in the wake of the 1970s oil crisis that put an end to cheap and plentiful fuel. Biofuels are viewed as an alternative from an economic, social, and environmental perspective. Liquid, gas, and solid fuels that are mostly made of biomass are referred to as biofuels. A variety of biomass, including agricultural plants, lignocellulosic materials, and organic wastes, are utilized to produce biofuel. The history of biofuels and bioenergy parallels that of civilization. Since the discovery of fire by man, solid biofuels like wood, dung, and charcoal have been utilized. In many nations, these are still in use today for heating and cooking. At least since the early ages, liquid biofuels such as olive and whale oil have been utilized. In the middle of the eighteenth and early nineteenth centuries, whale oil was the fuel of choice for home lighting. Because of this, whale oil grew more expensive, and before whales went extinct in the middle of the 1800s, humans began utilizing kerosene made from fossil fuels, which was cheaper. Fossil fuels might have provided the whales with protection during this time. Wood fuels were mostly supplanted by fossil fuels after 1900. This made it easier to reduce the then-unsustainable usage of timber resources, whose price had skyrocketed during the previous century. The United States patented the first internal combustion engine in 1826. It was intended to operate on an ethanol and turpentine blend [43].

11.8 Morphological and Evolutionary Characteristics of Enzyme-Based Biofuels

Topics covered include the morphological and evolutionary characteristics of enzyme-based biofuels, as well as the electron transfer mechanism involved in enzyme-based biofuels.

Electron transfer mechanisms are frequently used to analyze the structure and function of the chain of electron transportation in electroactive bacterial cells and to understand the connection between the design and function of oxidoreductases. Additionally, it is the theoretical foundation for the design of the bioelectrocatalytic reactor, electrode, and reaction process. The exchange of electrons between the electrocatalysts and the electrodes connects the electrochemical and biological processes. There are two techniques to couple the electrode process and enzyme reaction.

i) Mediated electron transfer (MET)
ii) Direct electron transfer (DET)

11.8.1 Mediated Electron Transfer

Electrons are moved from the enzyme to the electrode surface during the mediated transfer of electrons, facilitated by electron transfer mediators, which are essential redox chemicals. These intermediaries are chemical compounds that may readily engage in the redox process and transfer electrons artificially, causing the transfer of electrons to occur quickly [33, 34]. Redox protein molecules combined with chemical and inorganic substances can be employed as mediators.

Let's first examine what makes an ideal mediator before delving into the mechanics. Ideal mediators are individuals who demonstrate the capacity to promptly respond within the diminished enzyme or the coenzyme, exhibit bidirectional heterogeneity kinetics, and have the capability to rebuild an oxidized mediator. These perfect mediators should also have low and pH-independent characteristics, stable reduced and oxidized forms, and the inability to react with oxygen. The mediator primarily interacts with the reduced enzyme or coenzyme during the catalytic process. In order to facilitate rapid electron transport, it is secondarily diffused to the electrode exterior area. For the mediator, in order to effectively immobilize the enzyme-containing membrane upon administration to the BFC, it must be incorporated into the enzyme. Methyl violet, methylene blue, alizarin yellow, Prussian blue, phenazines, azure A, azure C, thionine, and toluidine blue are some of the mediators that are employed in BFCs [35, 36]. For example, ferricyanide has been shown to be used extremely infrequently due to the decreased stability and dependence of pH in organic dyes. Reductively mediated electron transfer (RMET) and oxidative mediated electron transfer (OMET) are the two forms of MET.

To enable spontaneous MET in OMET, the reduction potential of the mediator must exceed that of the redox cofactor of the bioelectrocatalyst. The mediator's reduction potential must exceed that of the redox cofactor [37] for RMET. Research is being done on artificial redox mediators to improve the efficiency of electron transfer between bioelectrocatalyst and electrodes. A few examples of synthetic electron transfer mediators are anthraquinone 2,6-disulfonate [39] and methyl viologen [38] and red that is neutral [40]. Microbial cells have the ability to create metabolites as a natural electron mediator [41]. As an illustration, S. oneidensis MR-1 functions as a DET and a MET. S. oneidensis MR-1 secretes flavin molecules, which enhance the capability of its outer membrane to transfer electrons as redox cofactors/mediators [42, 43]. As different oxidoreductases have different options for electron mediators based on the architectures of their active sites, their ability to establish electrochemical communication with the active region of the enzyme is assessed after considering favorable electrical characteristics. For example, a specific type of glucose-oxidizing enzyme can use naphthoquinone derivatives for the effective performance of MET. However, it has been experimentally demonstrated quite clearly that other species sharing the same redox cofactor did not exhibit expected performance. Once again, this outcome emphasizes the significance of the accessibility of the electron mediator's cofactor.

11.8.1.1 NAD^+-Dependent Enzymes

MET produces nicotinamide coenzymes, which are particularly significant because NAD(H)/NADP(H) is a coenzyme required by 90% of known oxidoreductases. Nicotinamide coenzyme regeneration is particularly important for industrial applications due to its increasing cost and wide range of applications. Ferreodoxin

NAD(P)$^+$ reductase (FNR), lipoamide dehydrogenase, diaphorase, and AMAPORS (artificial mediator accepting oxidoreductase) are enzymes involved in the MET-based nicotinamide-coenzyme regeneration system. These enzymes are appropriate for facilitating electron transfer between the electrode and nicotinamide coenzyme, as well as for regenerating nicotinamide coenzymes. They serve as intermediaries between one-electron mediators (such as cytochromes, FeS clusters, quinones, and flavins) and two-electron mediators. A few synthetic electron mediators can also be used for this, including flavins, viologens, ferrocenes, and cobalt sepulchrate.

11.8.2 Direct Electron Transfer

Direct electron transfer (DET) is the process by which oxidoreductases can transport electrons directly to an electrode without the need for an electron mediator. Enzymatic tests conducted in the 1980s have demonstrated that several enzymes can interact directly with an electrode while catalyzing substrate oxidation. There is no need for any extra mediators because DET can create catalytic currents and reduce the reaction overvoltage on its own [44]. The redox enzymes (oxidoreductases) work as electrocatalysts during DET when there is no mediator present. They speed up the rate of electron transfer between the electrode and the substrate molecule [45, 46]. For DET to happen, there must be less than 20 Å of space between the electrode's surface and the oxidoreductase's redox cofactor [47]. Because the active centers responsible for the reduction of oxidoreductases are typically located deep inside the protein structure, redox enzymes can be used for bioelectric fuels of the DET type. Using a specific docking motif that securely positions the protein in the desired orientation is a common way to shorten the distance between the electrode and the redox cofactor of oxidoreductase. Lee and colleagues linked a gold-binding peptide (GBP) with glucose dehydrogenase (GDH). To meet the needs of DET, the GBP sped up the alignment of GDH on the gold electrode surface and greatly cut down on the space between the electrode and the FAD cofactor of GDH [48]. Another method that is frequently used to immobilize enzymes on various electrode surfaces is cross-linked hydrogels, which also have the ability to stabilize proteins through electrostatic contact. Liu and his colleagues used an agarose hydrogel to stick three hemoglobin (Hb) proteins to edge-plane pyrolytic graphite electrodes. The proteins are myoglobin (Mb), horseradish peroxidase (HRP), and Hb. This immobilization technique facilitated the rapid detection of each protein using DET [49]. Several investigations have employed pyrene to directly affix proteins to a carbon electrode [50]. Hickey and colleagues have developed a technique that uses linear poly(ethylenimine) (LPEI) chemically modified with pyrene groups (pyrene-LPEI) to facilitate direct electrochemical communication with redox-active proteins. Cross-linking immobilizes a range of oxidoreductases on carbon electrodes in the presence of multi-walled CNTs

(MWCNTs). This immobilization allows for the application of pyrene-LPEI to facilitate DET. We have effectively used this approach to promote the direct bioelectrocatalytic reduction of O_2.

Three oxidoreductase modification techniques to boost electron transport:

i) **Oxidoreductase truncation:** The acceleration of electron transport can be achieved by eliminating the haem binding domain, which does not participate in the electron transfer pathway.
ii) **Surface modification:** Glucose oxidase deglycosylation reduces the distance between the electrode and active site, hence enhancing the electron flow rate.
iii) **Active site mutation:** This involves genetically modifying glucose oxidase to introduce a free thiol group near its active site. This modification enables the attachment of a maleimide-modified gold nanoparticle to a specific location on the enzyme, allowing for a direct electrical connection. As a result, Au-Np gold nanoparticles are formed.

11.9 Immobilization of Enzymes

The biological catalysts that live cells create are called enzymes. They have numerous uses in a variety of industries, including biochemical research, agriculture, medicine, and others. They catalyze all biochemical activities in the living body, which are essential to human life. Because the enzymes are thermo-labile, they face the challenge of maintaining their viability after extended use. Consequently, since Nelson and Griffin [51] discovered that an enzyme in a form that is insoluble in water demonstrates catalytic activity, research toward that goal has been conducted. Enzyme immobilization refers to the process of confining protein molecules to a solid framework or supporting that is different from the place where substrates or end products can be accessed. Enzyme immobilization can be accomplished by affixing the enzymes onto or within a select few appropriate support materials. It is crucial to acknowledge that both the substrate molecules and the resulting products must possess the ability to readily enter and exit the phase in which the enzymes are confined. Typically, enzymes can be immobilized by utilizing inorganic materials and inert polymers as a matrix or support system. The appropriate carrier matrix must exhibit features such as cost-effectiveness, inertness, stability, physical strength, the capacity to enhance the specificity of the enzyme or action, regenerability, minimum product inhibition, prevention of non-specific adsorption and bacterial contamination, and survivability [52]. The several techniques for immobilizing enzymes are categorized into groups, such as membrane confinement, entrapment, covalent bonding/cross-linking, and adsorption/carrier-binding.

11.9.1 Adsorption/Carrier-Binding Method

This process entails the formation of distinct ionic bonds per protein molecules, along with hydrophobic interactions, which facilitate the adhesion of enzyme molecules to the transport matrix's surface. The process involves either dehydrating the molecules of the enzyme on the surfaces of electrodes or dispersing the enzyme in a matrix. The most commonly used matrix materials are water-soluble transporters, such as carbohydrate substitutes, plastic or glass, and synthetic polymers [53–55]. Using this approach, the enzyme and the support matrix will form a robust link. The addition of substrate and physiological circumstances like high pH or temperature might weaken the binding. Because they interact hydrophobically with surfaces, adsorbed enzymes are typically resistant to proteolysis and aggregation [56]. Because of its enhanced cation exchange policy and capacity to store water, coconut fibers are frequently employed as a matrix material. After acetylation, kaolin, another environmentally favorable matrix, exhibits great enzyme retainability. Additional matrix materials that have been employed by different researchers include micro-/mesoporous materials with a wide surface area and functionalized thiol groups, microcrystalline cellulose with an irreversible binding capacity, etc. [57–60]. Several performances for immobilizing enzymes and cells are shown in Figure 11.4.

The molecular sieves are silanized by the presence of silanols, which, in turn, facilitates enzyme immobilization by hydrogen bonding [62]. Additional intriguing chemical alterations of the current support matrices provide opportunities for enhanced immobilization. The lipase enzyme was absorbed into hydrophobic granules composed of polypropylene (Accurel EP-100), and the profiles of lipase water activity were analyzed. The reduction of Accurel's particle size was discovered to have a substantial impact on both the enantiomeric ratio and reaction speeds in biocatalysis [63]. Experimental assessment demonstrates that immobilizing Yarrowia lipolytica lipase on octyl-agarose and octadecyl-sepa beads with the aid of physical adsorption leads to higher production and stability, enhanced process regulation, and a cost-efficient solution in comparison to using lipase in its free form. The primary factor behind this was the hydrophobic nature of the octadecyl-sepa beads, which enhanced the affinity between the enzyme and the support. After being adsorbed onto poly(3-hydroxybutyrate-co-hydroxyvalerate), Candida rugosa lipase maintained 94% of its activity even after being treated at 50 °C for four hours and reused for 12 cycles [64]. Mishra and Pithawala [65] discovered that by attaching to activated byssus threads treated with 1,4-butanediol diglycidyl ether, the urease enzyme exhibited improved pH stability and the capacity to maintain 50% activity even in dry conditions. The investigations showed that the support matrix materials, which are environmentally safe, are compatible with living organisms, can be broken down naturally, are affordable, have a long lifespan, and are efficient.

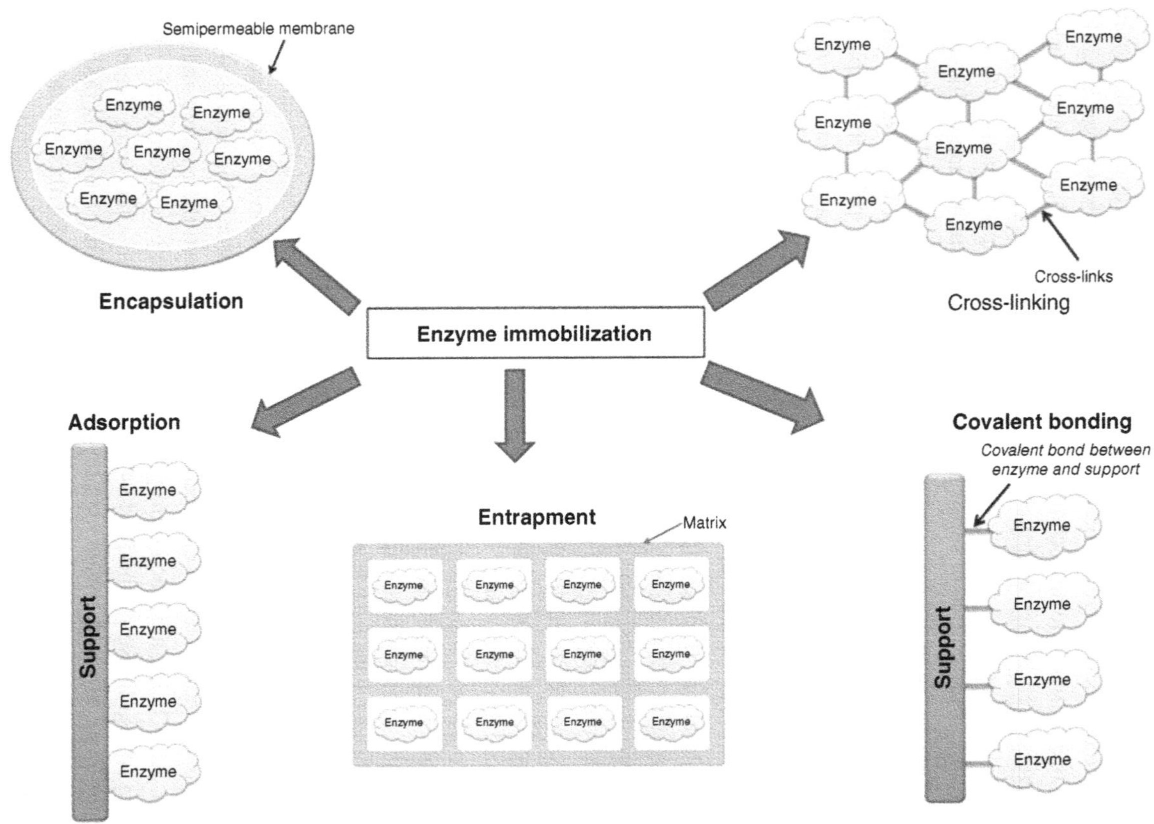

Figure 11.4 Various techniques for immobilizing enzymes and cells. *Source:* Ref. [61] with permission of Elsevier.

11.9.2 Covalent Bonding

The primary prerequisite for this form of covalent binding is the establishment of a covalent connection between the enzyme and the support material. Aspartic acid, histidine, arginine, and other amino acids with side chains serve as the main sources of covalent bonds that are established between the matrix and the enzyme. The inclusion of diverse functional groups, such as carboxyl, amino, indole, phenolic, sulfhydryl, thiol, imidazole, and hydroxyl groups, enhances reactivity. To achieve optimal enzyme activity, it is necessary to prevent the deactivation of the residues of amino acids located in the active location. Commonly employed defense tactics include:

i) Covalent bonding between the enzyme and the matrix when a competing inhibitor or substrate is present.
ii) Through the reversible covalent enzymatic–inhibitor complex's synthesis.
iii) By adding additional residues, a chemically altered soluble enzyme can be covalently linked to the matrix.
iv) A precursor of zymogen. Moreover, utilizing peptide-modified surfaces for enzyme linkages can lead to enhanced specific activity and stability, while also controlling protein orientation [66]. When the enzyme is chemically bonded to cyanogen bromide (CNBr)-agarose and CNBr-sepharose, using a carbohydrate component and glutaraldehyde as a connecting arm, it demonstrates exceptional resistance to heat [67]. Moreover, it has been shown that chemically bonding enzymes to modified silica gel carriers, which are produced by removing unreacted aldehyde groups using SBA-15 supports, enhance enzyme stability and transform the enzymes into highly active biocatalysts.

Furthermore, it was shown that the covalent bonding between enzymes and mesoporous silica and chitosan enhanced the enzymes' thermal stability and half-life. The field has experienced a significant change with the use of nanodiametric supports for immobilizing enzymes. Studies have indicated that the residual activity of cross-linking enzymes can be improved by combining them with electrospun nanofibers, which have greater porosity and surface area [68, 69]. Researchers in this field have used attapulgite nanofibers as a support material for covalent binding with the alcohol dehydrogenase enzyme [70] due to their availability in different nanosizes and their ability to tolerate high temperatures. The immobilization of enzymes on magnetic nanoclusters through covalent attachment in various orientations has proven to be highly beneficial in the pharmaceutical business due to its improved operational stability, longevity, and capacity to be reused. Preserving the structural and functional properties of immobilized enzymes is crucial, and this can be achieved by utilizing a cross-linking agent. Glutaraldehyde is a commonly used cross-linking agent that acts as a bifunctional cross-linker. It is soluble in water-based solvents and has the capability to create strong covalent bonds inside and between subunits.

11.9.3 Affinity Immobilization

Affinity binding refers to the process of immobilizing an enzyme onto a support matrix through particular interactions. The enzyme's specificity is utilized through affinity immobilization to maintain its stability under various physiological circumstances. There are two potential approaches that could be employed to accomplish this goal. The initial approach involves preconjugating the matrix with an affinity ligand specific to the target enzyme. An alternate approach entails conjugating or altering the enzyme to a distinct molecule that exhibits a strong attraction to the matrix [71]. Enzyme refining has also utilized affinity adsorbents [55, 72]. Enzymes exhibit greater concentrations and enhanced stability and effectiveness when they are attached to a complex support matrix with affinity, such as agarose-linked multi-layered concanavalin A and alkali-stable chitosan-coated porous silica beads. Implementing the bioaffinity-stacking method has the potential to greatly enhance the enzyme's ability to bind and be reused. The primary factor responsible for this phenomenon is the existence of non-covalent forces, including hydrogen bonding, van der Waals forces, and columbic interactions.

11.9.4 Entanglement

The entrapment approach entails immobilizing the enzyme within a polymeric network by either covalent or non-covalent bonds. This allows the substrate and product to pass through while yet keeping the enzyme in place. In contrast to the previously stated methods, the enzyme in this approach is not linked to the support matrix. Enzyme caging can be achieved through three methods: (1) enclosing the enzyme within a highly interconnected polymer matrix; (2) dissolving the enzyme in a phase that is not water-based; or (3) employing a selectively permeable microcapsule to isolate the enzyme from a large solution. Enzyme entrapment can be accomplished by many methods, such as microencapsulation, gel entrapment, and fiber entrapment. An encapsulating technique utilizing a hybrid carrier composed of alginate, gelatin, and calcium has been found to be successful in enhancing mechanical stability and preventing enzyme leakage. The lipase enzyme in C. rugosa demonstrated enhanced enzymatic activity and entrapment efficiency when encapsulated in chitosan. Furthermore, it ceased the process of leaching and the tendency to crumble. Recent reports indicate that the co-entrapment of lipase and magnetic nanoparticles within biomimetic silica, in the presence of different silane additives, has resulted in a considerable enhancement of enzyme activity. The main reason for this is the biocompatibility and non-toxicity of the support matrix, which renders it susceptible to chemical changes. Furthermore, due to its hydrophilic properties, it exhibits a strong attraction to proteins. Furthermore, studies have demonstrated that lipases exhibit exceptional thermal stability and exhibit a high level of resistance to organic solvents when encapsulated with κ-carrageenan. The utilization of nanostructured supports, such as electrospun nanofibers and pristine materials, for the immobilization of enzymes, has been found to have applications in various industries, including biomedicine, biofuel, chemistry, and biosensors.

11.9.5 Ionic Binding

The ionic binding approach relies on the interaction between the enzyme protein and water-insoluble carriers that include ion-exchange residues. By adjusting the ionic strength, polarity, and temperature, this kind of non-covalent immobilization can be reversed. The concepts of protein-ligand interaction found in chromatography bear resemblance to this principle. For this kind of enzyme immobilization, support materials like polysaccharides and artificial polymers with ion-exchange centers are typically utilized. There are benefits to this approach, such as less complicated enzyme-carrier binding and softer conditions than covalent binding. Furthermore, ionic interaction results in minute changes to the enzyme's active site and conformation, which almost always results in enhanced enzyme activity. On the other hand, using a substrate solution with high ionic strength or a solution with varying pH values may cause enzyme leakage from the carrier. The source of this issue lies in the weak affinity between the enzyme proteins and the carrier molecules. The affinity between the enzyme and the carrier in the case of ionic binding is significantly greater than that of physical adsorption. Ionic bonding, on the other hand, is comparatively weaker than covalent bonding.

11.9.6 Immobilization Associated with Metals

This process involves the precipitation of metal salts on the surface of the support matrix. These metals possess the ability to bind to the nucleophilic groups of the carrier. Heating can cause the metal ion to precipitate on the support matrix. This approach of immobilizing the enzyme results in a comparatively higher 30–80% enzyme activity [73]. The technique is straightforward, simple, and reversible. It is possible to separate the enzyme and carrier molecule by lowering the pH of the solution. Furthermore, this technique might replenish the matrix and the enzyme.

11.10 Graphene and CNT Applications in Fuel Cells

Graphene and CNTs are utilized as the counter-electrode in dye-sensitized solar cells (DSSCs) due to their efficient catalytic activity and excellent conductivity. CNTs and graphene have the potential to serve as catalysts for a wider range of electrochemical reactions, including those in fuel cells, in addition to their current application in catalyzing electrochemical reactions in DSSCs. Fuel cells have the ability to directly transform chemical energy into electrical energy. This energy transfer is facilitated by the electrochemical reactions that take place on the anode and cathode of the fuel cell. The electrochemical process of ORR takes place on the anode of a fuel cell, while fuel oxidation happens on the cathode. Fuel cells necessitate electrodes with robust catalytic activity for their operation. Fuel cells typically derive their electricity from hydrogen, methanol, ethanol, and formic acid. Although platinum is widely used as a catalyst for both the fuel oxidation reaction and ORR, its excessively high cost poses a major barrier to the widespread adoption of fuel cell technology in the industry. In order to do this, it is imperative to decrease the Pt loading in the catalyst or, alternatively, create catalysts that do not contain Pt.

Carbon nanomaterials are suitable for supporting nanoscale catalysts because they have a larger surface area. Consequently, a reduced amount of catalyst may be necessary for fuel cells. Further investigations have been conducted on CNTs and graphene as catalyst carriers, specifically for metal nanoparticles. There are three main ways for depositing metal nanoparticles onto carbon nanomaterials.

One of the ways involves the electrochemical deposition of metal nanoparticles, such as Pt, Pd, or their alloy nanoparticles, onto carbon nanomaterials [74]. Graphene was found to have metal nanoparticles deposited on it using the process of electroless deposition, as reported in reference [75]. Another approach entails utilizing a reducing agent to concurrently decrease GO and the precursor of the metal catalyst in a solution. The metal precursor undergoes reduction on rGO, resulting in the formation of metal nanoparticles. Similarly, GO undergoes reduction to become rGO. A study was conducted on the methanol oxidation reaction (MOR) using metal nanoparticles on graphene as a catalyst for fuel oxidation. In the study, Pt nanoparticles or alloy nanoparticles were employed to cover reduced graphene oxide (rGO) as an electrocatalyst for the MOR [76].

11.10.1 Comparative Performance Analysis of Existing Fuel Cell

Every type of fuel cell possesses distinct characteristics that render them appropriate for particular applications and operational circumstances. The choice of fuel cell technology depends on the intended application, operating environment, fuel availability, and cost considerations as tabulated in Table 11.1.

Table 11.1 Comparative performance analysis of existing fuel cell.

Fuel cell type	Efficiency	Operating temperature (°C)	Fuel	Applications	Advantages	Limitations
Proton exchange membrane fuel cells (PEMFCs)	40–60% (up to 85% CHP)	50–100	Hydrogen	Transportation (vehicles), portable power, stationary power generation	High power density, quick start-up/shut-down, low operating temperature	Sensitive to fuel impurities, expensive catalyst materials, limited durability and lifespan
Solid oxide fuel cells (SOFCs)	45–65% (up to 85% CHP)	600–1000	Hydrogen, natural gas, biogas, hydrocarbons	Stationary power generation, industrial power systems, auxiliary power units	Fuel flexibility, high efficiency, solid electrolyte reduces corrosion issues	High operating temperature, long start-up times, material degradation, high costs
Alkaline fuel cells (AFCs)	60–70%	60–90	Hydrogen	Space missions, backup power systems, submarines	High efficiency, simple design, relatively low cost	Sensitive to CO_2, requires high-purity hydrogen and oxygen, limited commercial use
Phosphoric acid fuel cells (PAFCs)	37–42% (up to 85% CHP)	150–200	Hydrogen	Stationary power generation, combined heat and power systems	Tolerant to fuel impurities, established commercial use, can use impure hydrogen	Moderate efficiency, corrosive electrolyte, larger and heavier
Molten carbonate fuel cells (MCFCs)	45–55% (up to 85% CHP)	600–700	Hydrogen, natural gas, biogas	Stationary power generation, combined heat and power systems	Fuel flexibility, high efficiency, can operate on CO_2-containing fuels	High operating temperature, corrosive electrolyte, durability and longevity issues

Notes:
Efficiency is expressed as a range which varies depending on various operating conditions and configurations. Higher efficiencies can be achieved in CHP systems.
Operating temperature plays a crucial role in determining the appropriateness of a fuel cell for various applications and has an impact on start-up times, thermal management, and the durability of materials.
Fuel flexibility is a significant benefit of certain types of fuel cells is their ability to be adaptable to many types of fuel, including a range of hydrocarbons and biofuels.
Applications emphasize the main purposes of each type of fuel cell, showcasing their advantages and operational traits.
Advantages and limitations refer to the primary benefits and obstacles that come with each fuel cell technology, which have a significant impact on their adoption and progress.

11.11 Conclusion

This chapter covers the following subjects: the definition, categorization, history, structure, composition (based on enzymes), electron transfer method, and enzyme immobilization of biofuels. Carbon-based nanomaterials have created a new opportunity for the advancement of enzyme-based biofuel cells (EFCs) with exceptional efficiency. Electrodes for electrochemical fuel cells are composed of various materials such as graphene, nanosheets, nanodots, nanotubes, and their composites with conducting polymers. These materials facilitate the transmission of electric charge between biocatalysts and the electrodes. Regarding the study, a diverse range of carbon nanomaterials can be utilized as effective conducting supports to improve the performance of EFC. Carbon nanomaterials and their composites exhibit excellent performance in catalyzing reactions at the interface between the substrate and the electrode, as demonstrated in this chapter.

11.12 Expected Future Application of Fuel Cells and Biofuel Cells

The exploration of carbon-based nanomaterials for the development of EFCs represents a significant advancement in biofuel technology. The comprehensive coverage of topics in this chapter, including the definition, categorization, history, structure, composition based on enzymes, electron transfer mechanism, and enzyme immobilization, highlights the potential of these materials to revolutionize EFC performance.

11.13 Future Applications

The future of biofuel technology is poised for transformative developments with the integration of carbon-based nanomaterials. These materials, including graphene, nanosheets, nanodots, nanotubes, and their composites with conducting polymers, enable highly efficient charge transfer between biocatalysts and electrodes. As research progresses, a diverse array of carbon nanomaterials is anticipated to serve as effective conducting supports, significantly enhancing the performance of EFCs.

1) **Next-Generation Biofuel Cells:** The implementation of advanced carbon nanomaterials will lead to the creation of next-generation biofuel cells with superior efficiency and power density. These enhanced biofuel cells are expected to outperform traditional fuel cells, making them suitable for a broader range

of applications, including portable electronics, wearable devices, and microelectromechanical systems (MEMS).

2) **Sustainable Energy Solutions:** With their ability to improve the efficiency of enzymatic reactions at the electrode interface, carbon nanomaterials will contribute to more sustainable and eco-friendly energy solutions. This will bolster international endeavors to diminish dependence on fossil fuels and mitigate the release of greenhouse gases, in accordance with objectives of environmental preservation.
3) **Biomedical Applications:** The high efficiency and biocompatibility of EFCs enhanced with carbon nanomaterials could pave the way for innovative biomedical applications. These include implantable medical devices that harness bioenergy from bodily fluids, providing a continuous and reliable power source for critical health monitoring and therapeutic devices.
4) **Industrial Bioprocesses:** In industrial settings, the enhanced performance of EFCs could revolutionize bioprocesses, making them more energy-efficient and cost-effective. This could lead to advancements in biomanufacturing, waste treatment, and bio-based production systems, thereby driving the growth of the bioeconomy.
5) **Environmental Monitoring:** The sensitivity and specificity of carbon nanomaterial-enhanced biofuel cells make them ideal for environmental monitoring applications. They can be utilized in sensors to detect pollutants and toxins, offering real-time data for environmental protection and management.

In conclusion, the integration of carbon nanomaterials into EFCs holds immense potential for future applications. By improving the efficiency and functionality of EFCs, these advancements will facilitate the development of sustainable, high-performance energy systems, significantly impacting various sectors including energy, healthcare, industry, and environmental management.

References

1 Haque, S.U., Nasar, A., Duteanu, N. et al. (2023). Carbon based-nanomaterials used in biofuel cells—a review. *Fuel* 331: 125634. https://doi.org/10.1016/j.fuel.2022.125634.

2 Tang, J., Yan, X., Engelbrekt, C. et al. (2020). Development of graphene-based enzymatic biofuel cells: a minireview. *Bioeletrochemistry* 134: 107537. https://doi.org/10.1016/j.bioelechem.2020.107537.

3 Geetha, M., Sadasivuni, K.K., Al-Ejji, M. et al. (2022). Review of progress and prospects in research on enzymatic and non-enzymatic biofuel cells; specific emphasis on 2D nanomaterials. *Current Biotechnology* 11: 212–229. https://doi.org/10.2174/2211550112666221201152211.

4 Verona, E., Kolesov, V.V., and Reshetilov, A.N. (2018). Biofuel elements on the basis of carbon nanomaterials for microenergy. *AIP Conference Proceedings* 1990: 020006. https://doi.org/10.1063/1.5047760.
5 Campbell, A.S., Jeong, Y.J., Geier, S.M. et al. (2015). Membrane/mediator-free rechargeable enzymatic biofuel cell utilizing graphene/single-wall carbon nanotube Co-gel electrodes. *ACS Applied Materials & Interfaces* 7: 4056–4065. https://doi.org/10.1021/am507801x.
6 Hoa, L.Q., Vestergaard, M.C., and Tamiya, E. (2017). Carbon-based nanomaterials in biomass- based fuel-fed fuel cells. *Sensors* 17: 2587. https://doi.org/10.3390/s17112587.
7 Zhang, H., Cao, G., Wang, Z. et al. (2008). Growth of manganese oxide nanoflowers on vertically-aligned carbon nanotube arrays for high-rate electrochemical capacitive energy storage. *Nano Letters* 8: 2664–2668. https://doi.org/10.1021/nl800925j.
8 Harris, P.J.F. (2004). Carbon nanotube composites. *International Materials Review* 49: 31–43. https://doi.org/10.1179/095066004225010505.
9 Holzinger, M., Abraham, J., Whelan, P. et al. (2003). Functionalization of single-walled carbon nanotubes with (R-) oxycarbonyl nitrenes. *Journal of the American Chemical Society* 125: 8566–8580. https://doi.org/10.1021/ja029931w.
10 Qasir, A., Pinke, P., and Dusza, J. (2020). Silicon nitride-based composites with the addition of CNTs—a review of recent progress, challenges, and future prospects. *Materials* 13: 2799. https://doi.org/10.3390/ma13122799.
11 Yogeswari, B., Khan, I., Satish Kumar, M. et al. (2022). Role of carbon-based nanomaterials in enhancing the performance of energy storage devices: design small and store big. *Journal of Nanomaterials* 2022: 4949916. 10.1155/2022/4949916.
12 Sun, L., Wang, X., Wang, Y., and Zhang, Q. (2017). Roles of carbon nanotubes in novel energy storage devices. *Carbon* 122: 462–474. https://doi.org/10.1016/J.CARBON.2017.07.006.
13 Bernholc, J., Roland, C., and Yakobson, B.I. (1997b). Nanotubes. *Current Opinion in Solid State & Materials Science* 2 (6): 706–715. https://doi.org/10.1016/s1359-0286(97)80014-9.
14 Ding, R., Huang, Y., Li, G. et al. (2020). Carbon anode materials for rechargeable alkali metal ion batteries and in-situ characterization techniques. *Frontiers in Chemistry* 8: 607504. https://doi.org/10.3389/fchem.2020.607504.
15 Pan, H., Li, J., and Feng, Y. (2010). Carbon nanotubes for supercapacitor. *Nanoscale Research Letters* 5: 654–668. https://doi.org/10.1007/s11671-009-9508-2.
16 Akbari, E. and Buntat, Z. (2017). Benefits of using carbon nanotubes in fuel cells: a review. *International Journal of Energy Research* 41: 92–102. https://doi.org/10.1002/er.3600.
17 Amin, R., Kumar, P.R., and Belharouak, I. (2021). *Carbon Nanotubes: Applications to Energy Storage Devices*. IntechOpen eBooks https://doi.org/10.5772/intechopen.94155.

18 Nzereogu, P., Omah, A., Ezema, F. et al. (2022). Anode materials for lithium-ion batteries: a review. *Applied Surface Science Advances 9*: 100233. https://doi.org/10.1016/j.apsadv.2022.100233.

19 Jaleh, B., Nasrollahzadeh, M., Eslamipanah, M., Nasri, A., Shabanlou, E., Manwar, N. R., Zboril, R., Fornasiero, P., & Gawande, M. B. (2022). The role of carbon-based materials for fuel cells performance. *Carbon*, *198*, 301–352. https://doi.org/10.1016/j.carbon.2022.07.023

20 Tie, D., Huang, S., Wang, J. et al. (2019). Hybrid energy storage devices: advanced electrode materials and matching principles. *Energy Storage Materials* 21: 22–40. https://doi.org/10.1016/j.ensm.2018.12.018.

21 Weddel, A.S., Magno, M., Merrett, G.V., et al. (2013). A survey of multi-source energy harvesting systems. *2013 Design, Automation & Test in Europe Conference & Exhibition (DATE)*, Grenoble, France, 905–908. https://doi.org/10.7873/DATE.2013.190.

22 Xu, Q., Niu, Y., Li, J. et al. (2022). Recent progress of quantum dots for energy storage applications. *Carbon Neutrality* 1: 13. https://doi.org/10.1007/s43979-022-00002-y.

23 Toyoda, T. and Shen, Q. (2012). Quantum dot-sensitized solar cells: effect of nanostructured TiO_2 morphologies on photovoltaic properties. *Journal of Physical Chemistry Letters* 3: 1885–1893. https://doi.org/10.1021/jz3004602.

24 Wild, M., O'Neill, L., Zhang, T. et al. (2015). Lithium sulphur batteries, a mechanistic review. *Energy & Environmental Science* 8: 3477–3494. https://doi.org/10.1039/C5EE01388G.

25 Pang, Q., Liang, X., Kwok, C.Y., and Nazar, L.F. (2016). Advances in lithium–sulphur batteries based on multifunctional cathodes and electrolytes. *Nature Energy* 1: 16132. https://doi.org/10.1038/nenergy.2016.132.

26 Xu, Z.-L., Lin, S., Onofrio, N. et al. (2018). Exceptional catalytic effects of black phosphorus quantum dots in shuttling-free lithium sulphur batteries. *Nature Communications* 9: 4164. https://doi.org/10.1038/s41467-018-06629-9.

27 Jia, H., Cai, Y., Lin, J. et al. (2018). Heterostructural graphene quantum dot/MnO_2 nanosheets toward high-potential window electrodes for high-performance supercapacitors. *Advancement of Science* 5: 1700887. https://doi.org/10.1002/advs.201700887.

28 Zahir, N., Magri, P., Luo, W. et al. (2021). Recent advances on graphene quantum dots for electrochemical energy storage devices. *Energy and Environmental Materials* 5: 201–214. https://doi.org/10.1002/eem2.12167.

29 Noori, A., El-Kady, M.F., Rahmanifar, M.S. et al. (2019). Towards establishing standard performance metrics for batteries, supercapacitors and beyond. *Chemical Society Reviews* 48: 1272–1341. https://doi.org/10.1039/C8CS00581H.

30 Cott, D.J., Verheijen, M., Richard, O. et al. (2013). Synthesis of large area carbon nanosheets for energy storage applications. *Carbon* 58: 59–65. https://doi.org/10.1016/j.carbon.2013.02.030.

31 Zhao, X., Tian, H., Zhu, M. et al. (2009). Carbon nanosheets as the electrode material in supercapacitors. *Journal of Power Sources* 194: 1208–1212. https://doi.org/10.1016/j.jpowsour.2009.06.004.

32 Song, R., Song, H., Zhou, J. et al. (2012). Hierarchical porous carbon nanosheets and their favorable high-rate performance in lithium ion batteries. *Journal of Materials Chemistry* 24: 12369–12374. https://doi.org/10.1039/C2JM31910A.

33 Dahman, Y., Syed, K., Begum, S. et al. (2019. In Woodhead Publishing Series in Composites Science and Engineering). Biofuels: their characteristics and analysis. In: *Biomass, Biopolymer-Base Materials, and Bioenergy* (ed. D. Verma, E. Fortunati, S. Jain, and X. Zhang), 277–325. Duxford, UK: Woodhead Publising 10.1016/B978-0-08-102426-3.00014-X.

34 Beilke, M.C., Klotzbach, T.L., Treu, B.L. et al. (2009). Enzymatic biofuel cells. Micro fuel cells. In: *Micro Fuel Cells: Principles and Applications* (ed. T.S. Zhao), 180–241. Burlington, MA: Academic Press https://doi.org/10.1016/B978-0-12-374713-6.00005-6.

35 Chaubey, A. and Malhotra, B.D. (2002). Mediated biosensors. *Biosensors and Bioelectronics* 17: 441–546. https://doi.org/10.1016/s0956-5663(01)00313-x.

36 Ruzgas, T., Csoregi, E., Emneus, J. et al. (1996). Peroxides – modified electrodes: fundamentals and applications. *Analytica Chimica Acta* 330: 123–138. https://doi.org/10.1016/0003-2670(96)00169-9.

37 Gorton, L. (1995). Carbon paste electrodes modified with enzymes, tissues and cells. *Electroanalysis* 7: 23–45. https://doi.org/10.1002/elan.1140070104.

38 Milton, R.D. and Minteer, S.D. (2017). Direct enzymatic bioelectrocatalysis: differentiating myth and reality. *Journal of Royal Society Interface* 14: 20170253. https://doi.org/10.1098/rsif.2017.0253.

39 Steinbusch, K.J.J., Hamelers, H.V.M., Schaap, J.D. et al. (2010). Bioelectrochemical ethanol production through mediated acetate reduction by mixed cultures. *Environmental Science & Technology* 44: 513–517. https://doi.org/10.1021/es902371e.

40 Hatch, J.L. and Finneran, K.V. (2008). Influence of reduced electron shuttling compounds on biological hydrogen production in the fermentative pure culture *Clostridium beijerinckii*. *Current Microbiology* 56: 268–273. https://doi.org/10.1007/s00284-007-9073-9.

41 Park, D.H. and Zeikus, J.G. (1999). Utilization of electrically reduced red by acitobacillus succinogenes: physiological function of neutral red in membrane driven fumarate reduction and energy conservation. *Journal of Bacteriology* 181: 2403–2410. https://doi.org/10.1128/jb.181.8.2403-2410.1999.

42 Schroder, U. (2007). Anodic electron transfer mechanisms in microbial fuel cells and their energy efficiency. *Physical Chemistry Chemical Physics* 9: 2619–2629. https://doi.org/10.1039/B703627M.

43 Okamoto, A., Hashimoto, K., Nealson, K.H., and Nakamura, R. (2013). Rate enhancement of bacterial extracellular electron transport involves bound flavin semiquinones. *Biological Sciences* 110: 7856–7861. https://doi.org/10.1073/pnas.1220823110.

44 Jiang, X., Hu, J., Fitzgerald, L.A. et al. (2010). Probing electron transfer mechanisms in Shewanella oneidensis MR-1 using a nanoelectrode platform and single cell imaging. *Proceedings of the National Academy of Sciences of the United States of America* 107: 16806–16810. https://doi.org/10.1073/pnas.1011699107.

45 Ghindilis, A.L., Atanasov, P., and Wilkins, E. (1997). Enzyme catalyzed direct electron transfer: fundamentals and analytical applications. *Electroanalysis* 9: 661–674. https://doi.org/10.1002/elan.1140090902.

46 Gorton, L., Csöregi, E., Dominguez, E. et al. (1991). Selective detection in flow analysis based on the combination of immobilized enzymes and chemically modified electrodes. *Analytica Chimica Acta* 250: 203–248. https://doi.org/10.1016/0003- 2670(91)85072-Z.

47 Marcus, R.A. and Sutin, N. (1985). Electron transfers in chemistry and biology. *Biochimica et Biophysica Acta* 811: 265–322. https://doi.org/10.1016/0304-4173(85)90014-X.

48 Lee, Y.S., Baek, S., Lee, H. et al. (2018). Construction of uniform monolayer and orientation tunable enzyme electrode by a synthetic glucose dehrogenase without electron transfer subunit via optimized site specific gold binding peptide capable of direct electron tranfer. *ACS Applied Materials & Interfaces* 10: 28615–28626. https://doi.org/10.1021/acsami.8b08876.

49 Liu, H.H., Tian, Z.-Q., Lu, Z.-X. et al. (2004). Direct electrochemistry and electrocatalysis of heme protiens entrapped in agarose hydrogel films. *Biosensors and Biolelectronics* 20: 294–304. https://doi.org/10.1016/j.bios.2004.01.015.

50 Mazurenko, L., Monslave, K., Infossi, P. et al. (2017). Impact of substrate difusion and enzyme distribution in 3D porous electrodes, a combined electrochemical and modelling study of a thermostable hydrogen/ oxygen enzymatic fuel cell. *Energy & Environmental Science* 10: 1966–1982. https://doi.org/10.1039/C7EE01830D.

51 Nelson, J.M. and Griffin, E.G. (1916). Adsorption invertase. *Journal of the American Chemical Society* 38: 1109–1115. https://doi.org/10.1021/ja02262a018.

52 Szymańska, K., Bryjak, J., and Jarzębski, A.B. (2009). Immobilization of invertase on mesoporous silicas to obtain hyper active biocatalyst. *Topics in Catalysis* 52: 1030–1036. https://doi.org/10.1007/s11244-009-9261-x.

53 Al-Adhami, A.J.H., Bryjak, J., Greb-Markiewicz, B., and Peczynska-Czoch, W. (2002). Immobilization of wood rotting fungi laccases on modified cellulose and acrylic carriers. *Process Biochemistry* 37: 1387–1394. https://doi.org/10.1016/S0032- 9592(02)00023-7.

54 Rosa, C.C., Cruz, H.J., Vidal, M., and Oliva, A.G. (2002). Optical biosensor based on nitrite reductase immobilised in controlled pore glass. *Biosensors and Bioelectronics* 17: 45–52. https://doi.org/10.1016/S0956-5663(01)00263-9.

55 Wu, S.-C. and Lia, Y.-K. (2008). Application of bacterial cellulose pelletsin enzyme immobilization. *Journal of Molecular Catalysis B: Enzymatic* 54: 103–108. https://doi.org/10.1016/j.molcatb.2007.12.021.

56 Spahn, C. and Minteer, S.D. (2008). Enzyme immobilization in biotechnology. *Recent Patents on Engineering* 2: 195–200. https://doi.org/10.2174/187221208786306333.

57 Brigida, A.I.S., Calado, V.M.A., Goncalves, L.R.B., and Coelho, M.A.Z. (2010). Effects of chemical treatments on properties of green coconut fiber. *Carbohydrate Polymers* 79: 832–838. https://doi.org/10.1016/j.carbpol.2009.10.005.

58 Dey, G., Nagpal, V., and Banerjee, R. (2002). Immobilization of α-amylase from *Bacillus circulans* GRS 313 on coconut fiber. *Applied Biochemistry and Biotechnology* 102–103: 303–313. https://doi.org/10.1385/abab:102-103:1-6:303.

59 Karagulyan, H.K., Gasparyan, V.K., and Decker, S.R. (2007). Immobilization of fungal β- glucosidase on silica gel and kaolin carriers. In: *Biotechnology for Fuels and Chemicals, ABAB Symposium* (ed. W.S. Adney, J.D. McMillan, J. Mielenz, and K.T. Klasson), 159–167. Humana Press https://doi.org/10.1007/978-1-60327-526-2_18.

60 Diaz, J.F. and Balkus, K.J. (1996). Enzyme immobilization in MCM-41 molecular sieve. *Journal of Molecular Catalysis B: Enzymatic* 2: 115–126. https://doi.org/10.1016/S1381-1177(96)00017-3.

61 Sirisha, V.L., Jain, A., and Jain, A. (2016). Enzyme immobilization: an overview on methods, support material, and applications of immobilized enzymes. In: *Advances in Food and Nutrition Research Research*, vol. 79 (ed. S.-K. Kim and F. Toldrá), 179–211. Amsterdam, The Netherlands: Academic Press https://doi.org/10.1016/bs.afnr.2016.07.004.

62 Sabbani, S., Hedenström, E., and Nordin, O. (2006). The enantioselectivity of *Candida rugosa* lipase is influenced by the particle size of the immobilising support material Accurel. *Journal of Molecular Catalysis B: Enzymatic* 42: 1–9. https://doi.org/10.1016/j.molcatb.2006.05.003.

63 Cabrera-Padilla, R.Y., Lisbosa, M.C., Fricks, A.T. et al. (2012). Immobilization of *Candida rugosa* lipase on poly(3-hydroxybutyrate-co-hydroxyvalerate): a new eco-friendly support. *Journal of Industrial Microbiology & Biotechnology* 39: 289–298. https://doi.org/10.1007/s10295-011-1027-3.

64 Mishra, N., Pithawala, K., and Bahadur, A. (2011). Byssus thread: a novel support material for urease immobilization. *Applied Biochemistry and Biotechnology* 165: 1568–1576. https://doi.org/10.1007/s12010-011-9376-y.

65 Fu, J., Reinhold, J., and Woodbury, N.W. (2011). Peptide modified surfaces for enzyme Immobilization. *PLoS One* 6: e18692. 10.1371/journal.pone.0018692.

66 Hsieh, H.-J., Liu, P.-C., and Liao, W.-J. (2000). Immobilization of invertase via carbohydrate moiety on chitosin to enhance its thermal stability. *Biotechnology Letters* 22: 1459–1464. https://doi.org/10.1023/A:1005602812037.

67 Kim, J., Jia, H., and Wang, P. (2006). Challenges in biocatalysis for enzyme based biofuel cells. *Biotechnology Advances* 24: 296–308. https://doi.org/10.1016/j.biotechadv.2005.11.006.

68 Xing, R., Liu, N., Liu, Y. et al. (2007). Novel solid acid catalysts: sulfonic acid group-functionalized mesostructured polymers. *Advanced Functional Materials* 17: 2455–3461. https://doi.org/10.1002/adfm.200600784.

69 Yusdy, Patel, S.R., Yap, M.G.S., and Wang, D.I.C. (2009). Immobilization of l-lactate dehrogenase for chiral synthesis of pharmaceutical compounds. *Biochemical Engineering Journal* 48: 13–21. https://doi.org/10.1016/j.bej.2009.07.017.

70 Nisha, S., Karthick, S.A., and Gobi, N. (2012). A review on methods, application and properties of immobilized enzyme. *Chemical Science Review and Letters 1* (3): 148–155.

71 Ho, L.-F., Li, S.-Y., Lin, S.-C., and Hsu, W.-H. (2004). Integrated enzyme purification and immobilization processes with immobilzed metal affinity adsorbnts. *Process Biochemistry* 39: 1573–1581. https://doi.org/10.1016/S0032-9592(03)00288-7.

72 Szymańska, K., Bryjak, J., and Jarzębski, A.B. (2009). Immobilization of invertase on mesoporous silicas to obtain hyperactive biocatalysts. *Topics in Catalysis* 52: 1039–1036. https://doi.org/10.1007/s11244-009-9261-x.

73 Yasin, G., Arif, M., Nizam, M.N. et al. (2018). Effect of surfactant concentration in electrolyte on the fabrication and properties of nickel-graphene nanocomposite coating synthesized by electrochemical co-deposition. *RSC Advances* 8: 20039–20047. https://doi.org/10.1039/C7RA13651J.

74 Kurkowska, M., Awietjan, S., Kozera, R. et al. (2018). Application of electroless deposition for surface modification of the multiwall carbon nanotubes. *Chemical Physics Letters* 702: 38–43. https://doi.org/10.1016/j.cplett.2018.04.056.

75 Goswami, A., Rathi, A.K., Aparicio, C. et al. (2017). In situ generation of Pd–Pt core-shell nanoparticles on reduced graphene oxide (Pd@Pt.rGO) using microwaves: applications in dehalogenation reactions and reduction of olefins. *ACS Applied Materials Inferfaces* 9: 2815–2824. https://doi.org/10.1021/acsami.6b13138.

76 Liu, Z., Abdelhafiz, A.A., Jiang, Y. et al. (2019). Pt/graphene with intercalated carbon nanotube spacers introduced by electrostatic self-assembly for fuel cells. *Materials Chemistry and Physics* 225: 371–378. https://doi.org/10.1016/j.matchemphys.2018.12.100.

12

Carbon-Based Nanomaterials in Energy Storage Devices: Supercapacitors

Shikha Chander[1], *Veerabathuni Jaya Usha Praveena*[2], *and Meenu Mangal*[3]

[1] *Department of Chemistry, St. Francis College for Women, Hyderabad, Telangana, India*
[2] *Department of Physics, St. Francis College for Women, Hyderabad, Telangana, India*
[3] *Department of Chemistry, Poddar International College, Jaipur, Rajasthan, India*

12.1 Introduction

A two-dimensional material known as graphene is made up of carbon atoms in a methodic hexagonal lattice. Strong mechanical properties, excellent electrical conductivity, and thermal stability are just a few of its remarkable attributes. These qualities make graphene a great choice for supercapacitors, which are electrical devices that store energy by accumulating electric charges on the surface of the electrode. Supercapacitors have superior cycle life, charging and discharging speeds, and environmental friendliness to batteries. Their low energy density makes them unsuitable for a variety of applications, as shown in Figure 12.1.

Recent advances in graphene materials have aimed to address this issue by increasing the electrodes' specific surface area, porosity, and capacitance. Scientists have developed techniques for synthesizing graphene aerogels, which are porous structures with a large surface area and low density that can be used to make high-density supercapacitors. Alternatively, graphene can be combined with other materials, such as metal oxides, polymers, or carbon nanotubes (CNTs), to create hybrid structures that improve the electrochemical performance of supercapacitors. These hybrid materials may also introduce novel functionalities, such as pseudocapacitance and the role of redox reactions in charge storage.

Even though graphene materials have shown significant potential for supercapacitor applications, there are still issues to address, including scalability, stability,

Carbon-based Nanomaterials for Green Applications, First Edition. Edited by Upendra Kumar, Piyush Kumar Sonkar, and Suman Lata Tripathi.

Published 2025 by John Wiley & Sons, Inc.

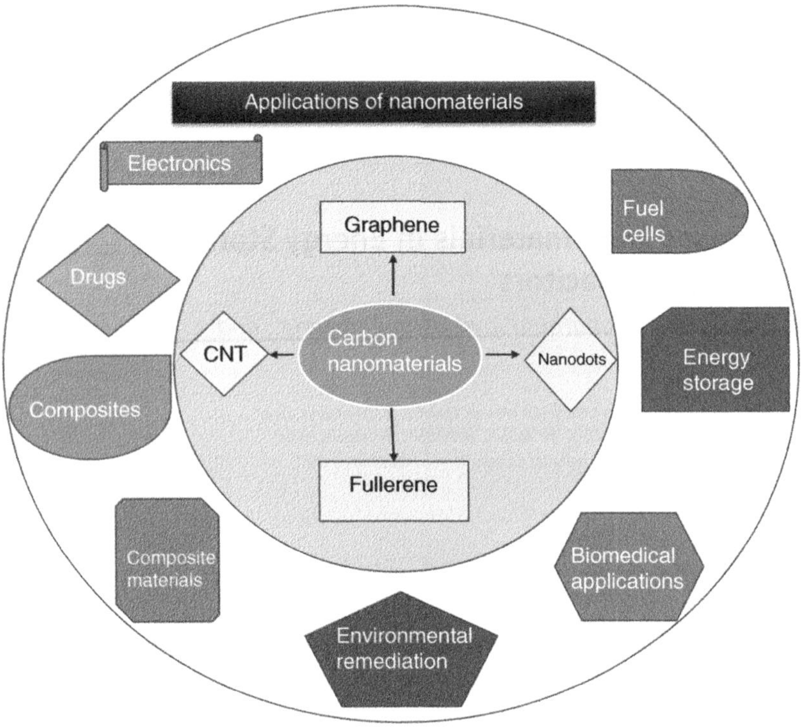

Figure 12.1 Flow chart of carbon nanomaterials and their applications.

and cost. Materials are produced from graphene for supercapacitors [1]. Supercapacitors' better power density and cycle life allow them to ultimately meet or outperform batteries in terms of energy density; nevertheless, further research and development are needed to enhance their synthesis, characterization, and integration. With exceptional qualities high electrical conductivity mechanical strength and thermal stability make graphene an excellent choice for supercapacitor applications.

12.1.1 Graphene Supercapacitors as Energy Storage Devices

Supercapacitors are energy storage devices that work by accumulating electric charge on electrode surfaces. In terms of charging and discharging speeds, cycle life, and environmental friendliness, they outperform batteries [2]. Their low energy density, however, limits their use in many applications.

Graphene can benefit supercapacitors in several ways:

By increasing specific surface area of electrodes, because graphene has a very high surface area of about 2630 $m^2 g^{-1}$, it can provide more space for charge storage. This can increase the capacitance of the supercapacitors, which is the amount of charge stored per unit voltage. Graphene can form porous structures with low density and high porosity, such as graphene aerogels, by increasing the porosity of the electrodes. This might make it simpler for the electrolyte to access the electrode surface, facilitating diffusion and charge transfer processes. This has the potential to increase supercapacitors' power density or the amount of energy that can be delivered or received in a given amount of time.

Graphene may be mixed with other materials, such as metal oxides, polymers, and CNTs, to build hybrid structures that increase the electrochemical performance of supercapacitors [3]. These hybrid materials may have novel functionalities, such as pseudocapacitance, which allows charge storage via a redox process. The energy density, or the amount of energy held per unit volume or mass, can increase in supercapacitors. Supercapacitors store energy by building an electric charge on electrodes. Supercapacitors have various advantages over batteries, including the ability to charge or discharge completely in seconds or minutes, whereas batteries might take hours or days.

Unlike batteries, which lose capacity after hundreds or thousands of cycles, supercapacitors may sustain millions of charge–discharge cycles without substantial deterioration.

Supercapacitors contain no toxic or hazardous elements, whereas batteries can harm the environment and human health. However, supercapacitors have a big disadvantage: they have a poor energy density. This implies they ably store less energy per unit of space or mass than batteries [4]. As a result, they are unsuitable for many applications that demand a lot of energy storage.

Graphene is a substance that can increase energy storage by expanding the performance of supercapacitors. Supercapacitors are electrical energy capacity devices that collect charges between two anodes separated by an electrolyte. Compared to conventional batteries, they have a higher control thickness, quicker charging and discharging, a longer cycle life, and are more naturally friendly. In any case, they have various disadvantages, including low energy density and high cost. Graphene can expel these limits while at the same time expanding the energy thickness and proficiency of supercapacitors. Graphene is a two-dimensional substance made up of carbon atoms in a methodic hexagonal cross-section. A hexagonal arrangement is a grid of standard hexagons connected by edges. The connections that connect one carbon atom to the other three carbon particles make a flat sheet of hexagons in graphene. This structure gives graphene its special properties, such as a tremendous surface range, amazing conductivity, and high quality. One of the critical properties of graphene is its huge surface area, which permits extra charge capacity sites. Another eminent highlight of graphene is its high conductivity, which supports particle and electron movement. Graphene's morphologies and functionalizations may, moreover, be adjusted, influencing the material's capacitance, porosity, and stability.

Researchers have created supercapacitors with higher capacitance, higher voltage, lower resistance, and longer cycle life by utilizing graphene as the electrode material. Electrolyte types that can be utilized with graphene-based supercapacitors incorporate fluid, organic, ion-liquid, and solid-state electrolytes [5]. These electrolytes may affect the supercapacitors' security, particle diffusion, and working voltage. Graphene has incredible potential as a material to progress energy storage through supercapacitor performance. Supercapacitors based on graphene have a lot of promise for creating high-performing, sensibly estimated, and ecologically inviting energy storage devices for a range of uses.

Graphene can overcome these confinements and increase the energy density and effectiveness of supercapacitors. Carbon molecules are organized hexagonally to frame the two-dimensional material known as graphene. Graphene's structure is associated with that of a honeycomb, with six carbon molecules shaping a standard hexagon in each cell. In graphene, each carbon particle is covalently bound to three other carbon molecules to shape a flat sheet of hexagons. The solid bonds shaping the carbon molecules in graphene are capable of their high mechanical quality and adaptability. The carbon molecules, moreover, have one free electron each, which can move openly along the sheet, giving graphene its high electrical conductivity and thermal conductivity. One of the key properties of graphene is its high surface range, which can provide more sites for charge capacity. High conductivity, which can offer assistance with particle and electron transport, is another critical characteristic of graphene. Graphene influences capacitance in two ways: by expanding the electrostatic capacitance and by expanding the pseudocapacitance. Electrostatic capacitance is the capacity of a cathode to store charge by creating an electric field between the anode and the electrolyte. Electrostatic capacitance depends on the surface zone and the separation between the electrode and the electrolyte. Because of its expansive surface range, graphene can hold more charge per unit mass or volume. Due to its amazingly thin structure, graphene can increase the quality of the electric field by decreasing the distance between the electrode and the electrolyte. The capacity of an electrode to store charge through reversible chemical reactions with the electrolyte is known as pseudocapacitance. Pseudocapacitance is contingent upon the electrolyte type and surface chemistry. The chemical reactivity and pseudocapacitance of graphene can be increased by incorporating dopants, other materials, or various functional groups.

Graphene influences capacitance in two ways: by expanding the electrostatic capacitance and by expanding the pseudocapacitance. Electrostatic capacitance is the capacity of an anode to accumulate charge by creating an electric field between the electrode and the electrolyte. Electrostatic capacitance depends on the surface area and the distance between the electrode and the electrolyte. Graphene has an exceptionally high surface region, which implies it can store more charge per unit mass or volume. Since graphene has an exceptionally thin structure, it can diminish the separation between the electrode and the electrolyte, expanding the strength of the electric field.

An electrode's capacity to hold onto charge through reversible chemical reactions with the electrolyte surface chemistry and electrolyte type affects pseudocapacitance. Another qualification between pseudocapacitance and electrostatic capacitance is that the former is connected to an increase in the oxidation state of the electrode or electrolyte, permitting it to store a more prominent amount of charge per unit potential difference for any kind of refinement. The two sorts of pseudocapacitance are redox and intercalation pseudocapacitance. Redox reactions including the exchange of electrons between the electrode and the electrolyte, such as oxidation and decrease, are the source of redox pseudocapacitance. Particles from the electrolyte are added to or expelled from the electrode's cross-section structure due to changes in interatomic separations, which causes intercalation pseudocapacitance.

To increase graphene's chemical reactivity and pseudocapacitance, you can include dopants, other materials, or distinctive functional groups. To functionalize graphene, for illustration, oxygen-holding groups like hydroxyl, carboxyl, or epoxy groups can be included. This increases graphene's redox pseudocapacitance by expanding the number of dynamic sites accessible for electron transfer. Besides, doping graphene with heteroatoms such as nitrogen or boron increases its intercalation pseudocapacitance. Because of these imperfections, the lattice structure will hold more ions.

Scientists have made supercapacitors with more prominent capacitance, higher voltage, lower resistance, and longer cycle life by utilizing graphene as the electrode material. An extension of electrolytes, such as organic, ionic liquid, solid-state, and aqueous electrolytes, can be used with graphene-based supercapacitors. These electrolytes might affect a supercapacitor's working voltage, particle diffusion, and safety. Researchers were able to make devices with higher capacitance, voltage, resistance, and cycle life by utilizing graphene as the electrode material in supercapacitors. Graphene-based supercapacitors can utilize electrolytes based on organic materials, ionic liquids, solid-state, and aqueous solutions. These electrolytes may have an effect on the safety, working voltage, and ion diffusion of supercapacitors.

12.2 Carbon Nanotube

Functionalization of CNTs is one-dimensional nanostructures that exhibit exceptional chemical and physical characteristics, including great mechanical strength, electrical and thermal conductivity, and a huge surface area. Unfortunately, their low solubility and poor dispersion in many solvents and matrices limit their applicability [6, 7]. To address this issue, the functionalization of CNTs is a common approach for changing their surface chemistry and adding new functionalities. Carbon functionalization can be classified as either covalent or non-covalent. Covalent functionalization is the process by which CNTs undergo electrical and other changes. CNTs' capacity to preserve inherent traits including hydrogen bonding, electrostatic interactions, van der Waals forces, and π-π stacking is dictated by their physical interactions with functional molecules. Both kinds of functionalization can increase CNT solubility, biocompatibility, and reactivity while also facilitating their integration with other materials and systems. This will be beneficial to an extensive range of applications in energy, environmental engineering, biotechnology, and nanotechnology. The following diagrams show the various functionalization and composite forms:

Figures 12.2 and 12.3 depict the functionalization of CNTs with a variety of chemicals, both covalent and non-covalent. Copolymer grafts, radical addition,

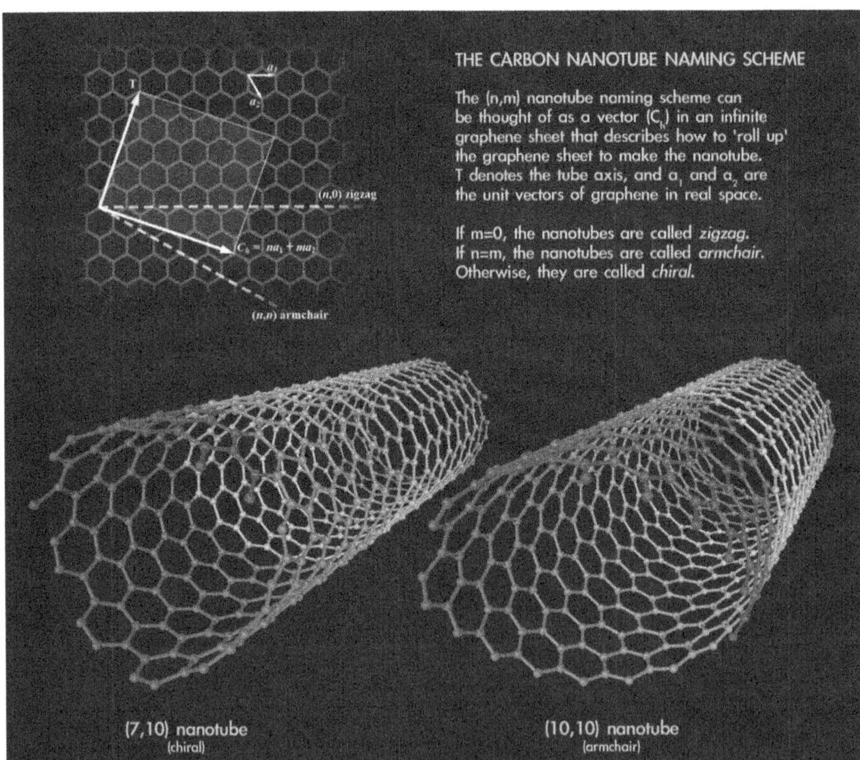

Figure 12.2 Covalent functionalization of carbon nanotubes. *Source:* Mstroeck/ Wikipedia/CC BY 3.0.

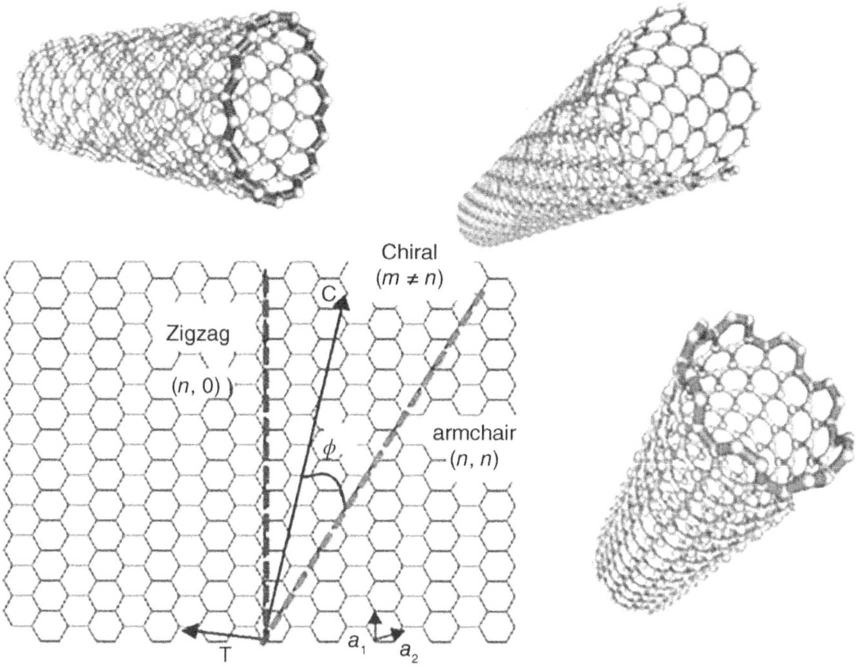

Figure 12.3 Carbon nanotube biosensors. *Source:* Tilmaciu and Morris [8]/Frontiers Media S.A./CC BY 4.

cycloaddition, and oxidation are all examples of covalent functionalization processes. Non-covalent functionalization of CNTs is possible via encasing, adsorbing, or wrapping them in polymers, surfactants, biomolecules, or nanoparticles. The required uses and attributes of functionalized CNTs determine the appropriate functionalization procedure.

The figure below shows the covalent and non-covalent functionalization of CNTs with various chemicals. Several strategies for covalent functionalization exist, including cycloaddition, radical addition, polymer grafting, and oxidation. Encasing, wrapping, or adsorbing CNTs with polymers, surfactants, nanoparticles, or biomolecules can result in non-covalent functionalization of nanotubes. The optimal functionalization procedure is determined by the desired uses and attributes of functionalized CNTs. One use for CNT-based sensors is the very sensitive and precise detection of gases, biomolecules, and chemical agents. CNTs work as transducers, converting chemical interactions involving molecules, surface functional groups, and analytes into electrical signals. CNTs work as transducers, turning chemical interactions involving analytes and molecules or surface functional groups into electrical impulses. CNTs functionalized with metal CNTs work as sensors, converting chemical interactions involving molecules, surface functional groups, and analytes into electrical signals. CNTs work as sensors, turning chemical interactions involving analytes and molecules or surface functional groups into electrical impulses. CNTs functionalized with metal nanoparticles, for example, may detect hydrogen gas by detecting the change in resistance generated by hydrogen molecules dissociating on the metal surface. Functionalized CNTs with antibodies can recognize certain antigens by establishing a complex between the antigen and the CNT surface antibody, resulting in capacitance changes.

Chemical agents can be identified by the change in conductance brought about by the analytes' selective binding to the polymer cavities on the surface of CNTs functionalized with molecularly imprinted polymers.

12.2.1 Functioning of the CNT Detectors

Specific electrical characteristics of CNTs are determined by their diameter, functionalization, chirality, and structure. CNT-based sensors can be powered by these characteristics. CNTs can be metallic or semiconducting, owing to their packing in a hexagonal cross-section along the tube. Semiconducting CNTs have changing conductances due to extreme conditions such as temperature, strain, electric field, and chemical environment, while CNTs have steady conductance. Unlike CNTs, which have a consistent conductance, semiconducting CNTs have a conductance that is modulated by external variables such as temperature, strain, electrical field, and chemical environment. CNTs can be functionalized with chemicals or groups to alter their electrical properties and react to particular analytes [9]. The separation of hydrogen gas causes charge transfer between the metal and CNT. The arrangement of antibody–antigen complexes adjusts CNT capacitance, which in turn modifies the surface dielectric constant of CNTs. When analytes bind to the polymer cavities in functionalized CNTs, which are made up of molecularly engraved polymers, the conductance changes, altering the charge thickness of the CNT surface. By connecting functionalized CNTs to electrodes or circuits, electrical signals indicating the presence or concentration of analytes can be identified, measured, and correlated. CNTs are classified as metallic or semiconducting based on their electronic band structure, which directs the flow of electrons along the tube. The degree to which carbon molecules bend around the tube axis, known as chirality, determines electronic band structure of CNTs. Two integers represent the number of 60° revolutions of a unit vector along the tube axis required to connect two adjacent carbon atoms on the hexagonal lattice. These numbers represent the CNTs' chirality. Because metallic CNTs do not have a band gap, electrons can always find an energy level to possess. As a result, metallic CNTs have a constant conductancc that is free of external components. A limited band hole in the band structure of semiconducting CNTs means that electrons cannot participate in any energy levels within a specific energy range. The conductance of semiconducting CNTs changes when external components such as temperature, strain, electric field, or chemical environment modify the CNTs' energy levels or electron check.

Mechanical strain, temperature variations, chemical conditions, and external electric fields may all alter the conductivity of semiconducting CNTs. The number of electrons that can cross the band gap is determined by their energy levels inside CNTs. A greater electric field can enhance semiconductor conductance by either producing new energy levels inside the band gap or narrowing it. An electric field can diminish the conductance of semiconducting CNTs by expanding the band gap or removing pre-existing energy levels. Temperature changes can affect both energy levels and number of electrons in CNTs.

Temperature changes can affect both the energy levels and the number of electrons in CNTs. Raising the temperature of semiconducting CNTs raises the thermal energy of the electrons, which may induce the electrons to jump the band gap or occupy higher energy levels, improving the conductivity of the CNTs. Lower temperatures may cause semiconducting CNTs to remain at lower energy levels or below the band gap, lowering electron thermal energy and conductance [10, 11]. Mechanical strain alters the nanostructure of CNTs and can affect their electrical structure and characteristics.

Tensile tension can increase the conductivity characteristics of semiconducting CNTs, perhaps lowering their band gap or turning them metallic by making them smaller and more chiral. Converting them to metal. Under compressive pressure, a semiconducting CNT can lose its chirality and diameter, transforming into an insulator. However, CNTs' surface potential and charge can alter in response to their chemical environment, influencing how they interact with other molecules or groups. Semiconducting CNTs may be converted into insulators by compressing their diameter and chirality. However, when exposed to a chemical environment, CNTs can change their surface potential and charge, influencing how they interact with other molecules or groups. Semiconducting CNTs can become insulators by reducing their diameter and chirality under compressive pressure; nevertheless, when exposed to chemical conductance, CNTs can impact the interactions of other molecules or groups by altering their surface charge and potential. CNTs can change the way other molecules or groups interact with them by altering their surface charge and potential. Semiconducting CNTs may be converted into insulators by compressing their diameter and chirality. However, when exposed to a chemical environment, CNTs can change their surface potential and charge, influencing how they interact with other molecules or groups. Conductance can be lowered by boosting Fermi levels and lowering electron density in a chemical environment that absorbs electrons from CNT particles.

Tensile power modifies both the geometry and electrical structure of CNTs, thereby changing the material's conductance. Under tensile strain, a material will lengthen or stretch along its length. Tensile strain causes a decrease in diameter and an increase in length in CNTs by altering their aspect ratio and chirality. A CNT's aspect ratio is its length-to-diameter ratio, and its chirality is the angle at which its carbon atoms are arranged around its axis. Electron mobility along a tube is governed by the electronic structure of CNTs, which is shaped by their geometry [12]. The distribution of electron energy levels in a solid, or band structure, is what determines the electronic structure of CNTs. To join two nearby carbon atoms on the hexagonal lattice, a unit vector along the tube axis must be rotated by 60° several times, represented by the pair of integers (n, m) that represent the CNT chirality. The number of times a unit vector along the tube axis is rotated by 60° to join two nearby carbon atoms on the hexagonal lattice is represented by the pair of integers (n, m) that represent the CNT chirality. Depending on whether $(n - m)$ is divisible by three, CNTs can have a metallic or semiconducting band structure. Because there is no band gap in metallic CNTs, electrons can always find an available energy level to occupy. Finite band gaps in conducting CNTs indicate that there are no available energy levels for electrons to occupy within a given energy range. Tensile strain can decrease the band gap of a CNT or turn it metallic by increasing the material's chirality by widening the gap between n and m. Increased conductance can be achieved by increasing the number of electrons that can cross the band gap and occupy any energy level. Tensile strain can also alter the curvature of CNTs, affecting orbital hybridization and overlap. The process of combining atomic orbitals to produce new hybrid orbitals with unique energies and forms is known as orbital hybridization. Orbital overlap is the quantity of space that two orbitals share and utilize for communication. By making CNTs more curved by reducing their diameter, tensile strain can enhance carbon atom overlap and sp_2 hybridization. By enhancing electron mobility and delocalization, conductance in the tube can be increased.

12.3 Functionalization of Carbon Nanotubes

CNTs are functionalized by altering their surface or structure with different chemical groups or molecules to tailor their properties to specific applications [13, 14]. Functionalized CNTs are more compatible with a wide range of materials, have higher reactivity, and are more soluble. Here are some common techniques and applications for functionalizing carbon.

Functionalization Techniques: Covalent functionalization of carbon nanotubes (CNTs) is the process of attaching chemical moieties to the CNT tubular structure by forming covalent bonds that share atleast one pair of electrons with the CNT and the inserted chemical moiety. The process of making something functional **Chemical Structure of Diazo**: This process attaches functional groups to CNTs via the interaction of diazonium salts.

Chemical Vapor Deposition (CVD) is a technique that deposits functional groups on CNT surfaces using gas-phase functionalization and reactive precursors. Diagonal Cycloaddition 1, 3: Azide-alkyne cycloaddition, also known as "click chemistry," is a method of attaching functional compounds to CNTs.

Surfactant Functionalization: Non-Covalent Wrapping: This technique involves dispersing CNTs in a solution without altering their covalent structure.

12.3.1 Applications of Functionalized CNTs

In the realm of materials science, FCNTs are widely used to produce excellent nanocomposites. The use of off-CNTs for the creation of sophisticated polymeric composites offers significant benefits over the use of non-functionalized CNTs by improving the dispersion of CNTs in polymeric matrices. Unlike unfunctionalized CNTs, which have reduced dispersion and interfacial bonding, the functional moiety of CNTs increases interfacial bonding, hence reducing fiber pullout under mechanical stress or entanglement.

Functionalized CNTs: Production, characteristics, and uses in medication delivery, material and biological sciences, and water purification. Functionalized CNT synthesis properties and applications in drug delivery, water purification, materials science, and biological sciences are accessible because they function as support materials, functionalized CNTs have a broad range of applications in catalysis. To improve CNT characteristics for use as catalyst support, researchers studied a variety of surface fictionalization approaches. They also reviewed available methods for loading catalyst nanoparticles on F-CNTs. By treating CNTs with acid, oxygen functional groups were created on their surface that served as anchors for nanoparticles of various metals. At 350–380 °C, the functional materials that were synthesized had an improved NO_x removal effectiveness of more than 90%. Further research has been done to show that significantly higher catalytic activity is provided by metal nanoparticles supported by CNTs.

It's important to note that the choice of functionalization method, as shown in Figure 12.4, depends on the desired application and the specific properties needed for that application. Additionally, careful consideration must be given to the potential impact of functionalization on the overall properties of CNTs.

To eliminate amorphous carbon, the CNTs were oxidized at 350 °C in the muffle furnace, followed by a 31 wt% HCl treatment for 18 hours. The Fe catalysts were removed from the CNTs using an HCl treatment. These CNTs were then vacuum-filtered and washed several times with DI water to achieve a pH of 7. The resulting CNT powder was refluxed in 10 M HNO_3 for four hours at 180 °C to produce functional groups (OH and COOH) on the CNT surface. After achieving a pH of 7 through repeated vacuum filtrations with DI water, the functionalized CNT mixture was dried in a vacuum for an entire night at 90 °C.

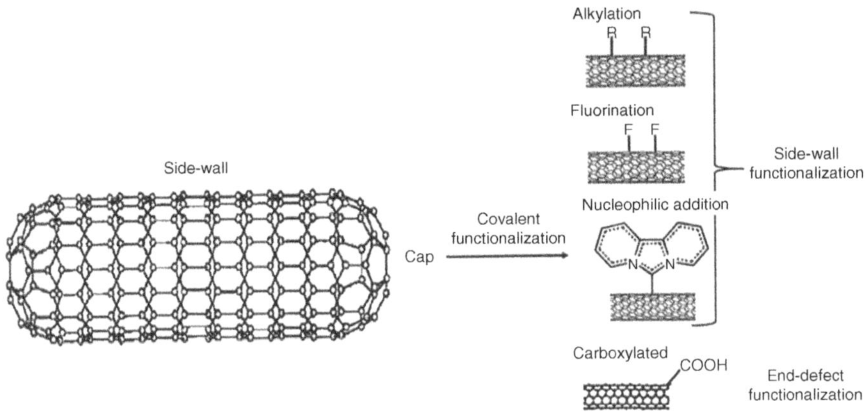

Figure 12.4 Covalent functionalization.

12.3.2 Precursor Features

Natural graphite flakes (purity 99.9%, mesh size −250 mm), sodium nitrate ($NaNO_3$) (FINAR Reagents), extra-pure potassium permanganate ($KMnO_4$) (SRL), 30% hydrogen peroxide (H_2O_2) (EMPARTA), hydrochloric acid (35.4%) (HCl) (SDFCL), and hydrazine hydrate (N_2H_4).

12.4 Reduced Graphene Oxide (rGO) Synthesis

As explained below, a modified Hummers technique was used to create the rGO, with a few minor adjustments made in the pre-oxidation phase. During the pre-oxidation step, 1 g of $NaNO_3$ and 2 g of natural graphite flakes were dissolved in concentrated H_2SO_4 and agitated for 18 hours. There was now a visibly thick, black solution. After that, the beaker was kept at 10 °C by being placed in an ice-cube-filled container. 3 g of $KMnO_4$ were gradually added to the liquid while stirring slowly and steadily, all while keeping the temperature at 10 °C. After that, the ice cube container was taken out, and the beaker was stirred continuously for three hours, or until it reached room temperature.

At this point, a thin greenish slurry was noticed. The next step, which involved adding a total of 92 ml of DI water, was done over 30 minutes with continual stirring. The temperature was kept at 40 °C. The next stage was to heat the mixture to 60 °C and stir for half an hour. To rapidly and efficiently boil the mixture, it was heated to 90 °C and stirred for half an hour. The liquid began to transform from a greenish color to a thick, brownish paste consistency. To halt the reaction, a solution of 292 ml DI water and 9 ml 30% H_2O_2 was added. Graphene oxide shows a light yellowish hue, indicating that it is being created. It was laid aside overnight for consideration. The mixture was then acidified to a pH of 7 with 10 wt% HCl and washed with DI water via a vacuum filtration system to recover the pH.

To create a thick, homogeneous brown solution, 300 mg of GO was dispersed in 300 ml of DI water and sonicated for 30 minutes. In the preceding experiment, a tiny quantity of ammonia liquid was added to the solution to give it a pH of 10. The aforementioned solution was heated under stirring at 80 °C for two hours with 0.5 ml of N_2H_4. Then, the solution was chilled and soaked in DI water several times before being neutralized with lime powder and finally rinsed with water. The name "reduced graphene oxide" (rGO) was used to obtain the product.

12.4.1 Synthesis of rGO-FCNT Hybrid

Twenty-five mg of rGO was added to 100 ml of ethanol, and the mixture was sonicated for 30 minutes. After stirring the aforementioned solution for 30 minutes, 50 mg of FCNT was added, and it was then sonicated for an additional 30 minutes. The remaining ingredients were added after the solution had been shaken for 15 minutes [15]. The solution was cleaned by bathing it in DI water, rinsing it several times, and then letting it dry in a vacuum for an entire night at 900 °C. The powder (rGO-FCNT) turned light brown when it was combined with a tiny quantity of black substance after it had dried.

12.5 Characterization

The characterization methodologies and sample preparation processes employed in this work are detailed below. Using copper K radiation and a Bruker Advance D8 Powder X-ray Diffractometer, the detailed structural data was acquired at a scanning rate of 1.20/min (step size of 0.02° for one second). The sample powder was used in this investigation after being finely ground.

A Wintec Alpha 300 with a 532 nm laser was utilized for the Raman analysis. The sample was a fine powder for the analysis.

Utilizing an electron scanning microscope (FESEM) from ZEISS, the sample's morphology was ascertained. Finely ground powder made up the analytical sample.

After that, a 300-mesh carbon-coated grid was submerged in two to three drops of analysis solution and allowed to dry overnight.

Finally, BioLogic Science Instruments was used to conduct an electrochemical material analysis. Here are comprehensive instructions on how to prepare electrodes and produce electrochemical cells, analyze input electrochemical data, and compute different supercapacitor parameters using analytical formulas. The list is as follows:

12.5.1 Preparation of Electrodes and Cells

Carbon black as a conductive ingredient and polyvinylidene difluoride (PVDF) are the active ingredients. The pestle and mortar had been ground to a fine powder. A small number of goats were introduced to test the slurry with *N*-methyl 2-pyrrolidone (NMP). Then, 350 m of thick graphite foil was placed on top of the slurry. Overnight, the foil was vacuum-dried at 80 °C. A Swagelok cell was used to build the two-electrode cell. The cell is made up of two electrodes spaced apart by a Whatman separator that has been wet with an electrolyte solution of 3 M KOH.

12.5.2 Input Parameters for Analysis

A 1 V input (potential window) ramp with variable scanning rates of 5, 10, 20, 500, 100, 200, and 500 mV s^{-1} is positioned between the potentials (*x*-axis) and the current (*y*-axis). Cyclic stability was determined using the same method. PEIS was performed using an input sine wave with a V_{rms} amplitude of 5 mV, a frequency range of 0.1 Hz to 10 kHz, and a starting voltage equal to the open-circuit voltage of the electrochemical cell just before this study. PEIS is used to draw the Nyquist plot (−Im(*Z*) vs Re(*Z*)). Evaluating key performance parameters is given below.

Using Cyclic Voltammetry (CV), specific capacitance is expressed as follows:

$$C_S = \frac{Idv \cdot m}{V_S}; \text{ units: F g}^{-1} \tag{12.1}$$

where

m denotes the mass of active material in the electrochemical cell
V_S stands for scan rate
dV for potential window.

$$\text{Energy density: in}1000/3600\text{: } E_d = \frac{C_S \cdot dV}{2}; \text{ units: W kg}^{-1} \tag{12.2}$$

where

dV is the applied voltage
C_S is the specific capacitance derived from Eq. (12.1)

$$\text{Power density: } P_d = \frac{C_S \cdot dV \cdot V_S}{2}; \text{ units: W kg}^{-1} \tag{12.3}$$

where

V is the potential window
C_S is the specific capacitance derived from Eq. (12.1)
V_S is the scan rate derived from galvanostatic charge–discharge.

$$\text{Specific capacitance: } C_s = \frac{4I \cdot dt \cdot m}{dV}\text{; units: F g}^{-1} \quad (12.4)$$

where

I is the cathodic current applied
dt is the discharging time
dV is the discharging voltage
m is the mass of the total active materials of the electrochemical cell.

The energy density I_S calculated with a similar equation is given in Eq. (12.2), except that C_S is the value obtained from Eq. (12.5).

$$\text{Power density: } P_d = \frac{E_d \cdot 3600}{t}\text{; units: W kg}^{-1} \quad (12.5)$$

where

E_d is energy density obtained from the above discussion
t is discharging time.

12.6 Results and Discussion

12.6.1 Raman Analysis

Raman spectroscopy is a spectroscopic method for monitoring when different spectroscopic and other modes of rotation are in the system. It is based on inelastic dispersion, or scattering Raman, of monochromatic light, generally in the visible, close laser infrared range, or near UV range. The lamp interacts with or without phonon system excitations that change the energy of the laser photons. The energy shift offers information on the vibrational modes of this system. The Raman spectrum was measured by this Raman microscope in environmental settings, utilizing a 514 nm laser.

The Raman spectrum is the most defining characteristic of FCNTs. Generally speaking, the structure, disorder, and imperfections of a crystal are described using Raman spectroscopy on graphs. The Raman spectrum of FCNTs and rGO is displayed in Figure 12.5. The D peak is at 1347 and 1339 cm^{-1} for rGO and FCNT, respectively, considering the defects are responsible for the D peaks in the GO and rGO. There is no amorphous carbon content; rGO-FCNT peaks are, respectively, found at 1572 and 1576 cm^{-1} due to the crystalline nature of the graphite structure. The D may be detected from the Raman FCNT spectrum G peaks of 1339 and 1584 cm^{-1}, the second-order observed mode attributed to the first D mode overtone between 2450 and 2650 cm^{-1}, a combination mode of D and G modes between 2775 and 2950 cm^{-1}.

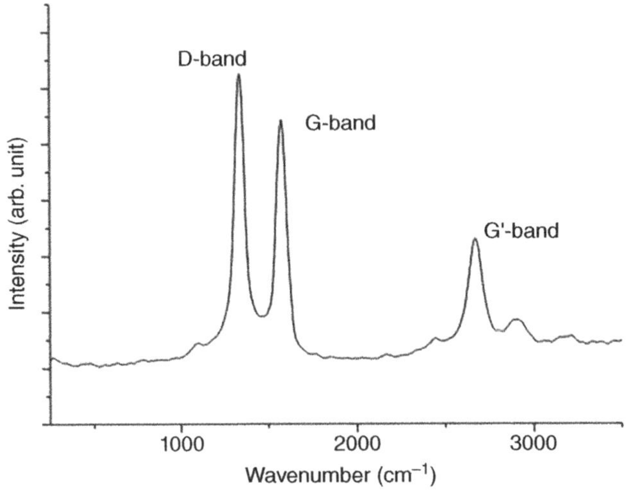

Figure 12.5 FCNT-rGO Raman spectroscopic analysis.

12.6.2 Powder X-Ray Diffraction (XRD)

To get a wide variety of crystallographic information, X-ray diffraction (XRD) benefits from the coherent scattering of X-rays by crystalline materials. Each disperser retrieves a tiny amount of its intensity as an incoming spherical wave. Only in directions where the wavelength multiple μ equals the path length difference, 2*d* sin – Wintec, do scoop waves synchronize when a symmetrically organized dispersion with a separation of *d* is involved. Scattering waves only synchronize in directions where the wavelength multiple μ equals the path length difference, 2*d* sin – Wintec when a symmetrically organized dispersion with a separation of d is involved. With a separation of *d*, sphere waves synchronize in a symmetrically organized dispersion, with the wavelength multiple μ equal to the difference in path length, 2*d* sin – Wintec. In this example, a point of reflection in the diffraction pattern is created by deforming a portion of the incoming beam at an angle of 2. The samples in this thesis were obtained with a D8 Advance Bruker Powder X-Ray Diffractometer using Cu $K\alpha$ radiation, a wavelength of 1.54 Å, and a scan rate of 1.20/min (step size of 0.02° for one second). XRD was used to characterize the crystalline detail of the samples. Powder from the sample that had been finely ground was used for this analysis.

Figure 12.6 displays the XRD graphite powder, rGO, and FCNT patterns. At a maximum of 25.8, an interlayer distance of 3.34 Å is reached. In the rGO XRD pattern, the interlayer distance increases and the peak for adding oxygen functional groups to the graphite is moved to 11.5. But when oxygen levels drop,

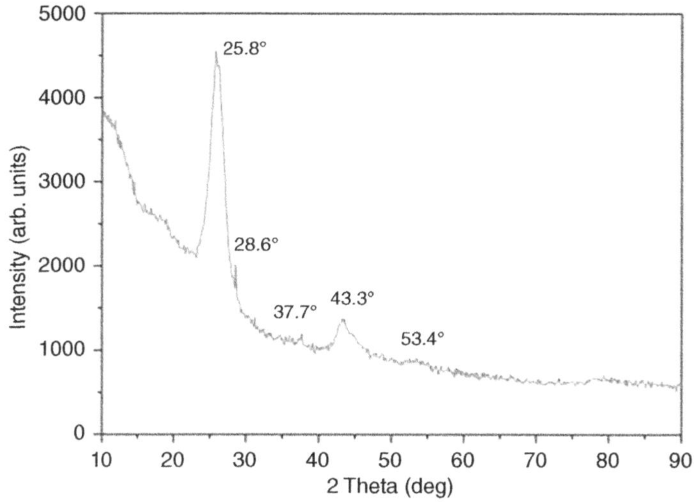

Figure 12.6 FCNT-rGO X-ray diffraction pattern.

the interlayer disappears. The maximum is shifting to 25.8, and the spacing is decreasing again. FCNT graphite carbon comes from 26.1 and is at peak diffraction.

12.6.3 FTIR Analysis

It is commonly known that functional chemical groups absorb specific light frequencies. It is well known that certain light frequencies are absorbed by functional chemical groups. It is commonly known that functional chemical groups absorb specific light frequencies. The fact that light is absorbed at specific frequencies by functional chemical groups is commonly known. Specific frequency ranges. Consequently, the chemical structure may be the characteristic that has been noted and identified. FT-IR spectra were recorded for this thesis on an FT-IR variant 3100 equipped with KBr pellet technology.

FT-IR of rGO-FCNTs is shown in Figure 12.7, with various treatments ranging from 500 to 4000 cm^{-1}. The wide peak in the FTIR spectra of virgin FCNTs is 3435 cm^{-1} and corresponds to the hydroxyl group's O—H stretch, which could be due to carboxyl group oscillation. During manufacturer purification, partial oxidation can lead to the formation of carboxyl groups on the surfaces of pure FCNTs. The O—H of carboxylic groups is responsible for the extending (O=H) of the FCNT-COOH spectra, which also showed peaks at 2917 cm^{-1} and 2845 cm^{-1}, which are associated with symmetric and asymmetric C—H stretching. C=O might

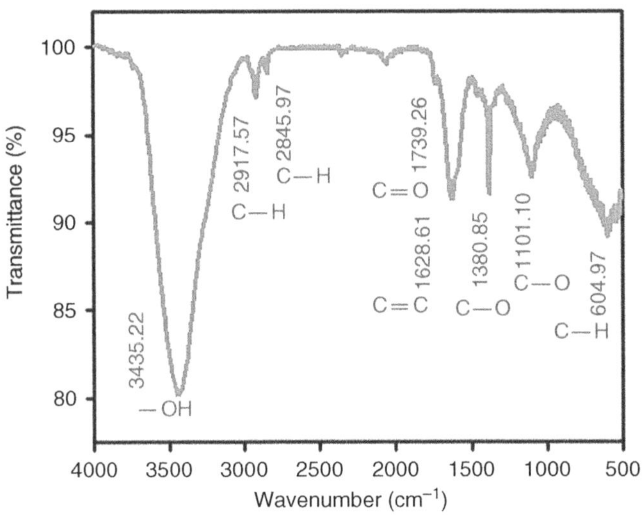

Figure 12.7 FTIR analysis of FCNT (functionalized CNT).

explain the peak at 1738 cm^{-1}, while the CNT backbone stretching could explain the peak at 1628 cm^{-1}. The C—O bending deformation mode of the carboxylic acid group hydrogen-bonded COOH is associated with the peak at 1380 cm^{-1} for acid-functionalized CNTs, while the signal at 1101 cm^{-1} may be associated with C—O stretching in the same functionalities. The peak at 1101 cm^{-1} is caused by the C=C stretching of the CNTs, whereas the peak at 605 cm^{-1} is caused by the C—H bond stretching.

12.6.4 Scanning Electron Microscopy Analysis

A scanning electron microscope (SEM) is a type of electron microscope that uses a concentrated stream of electrons to scan a sample and make images of it. When electrons collide with atoms in a sample, secondary and backscattered electrons are produced, which can be measured and contain information about the surface topography and composition of the material. In the present chapter, a Wintec field-emission SEM (FESEM, JEOL, JSM-6701F) was used to measure SEM images at 5 kV.

Figure 12.8 depicts the morphology of the FCNT, whereas the morphology of CNTs varies greatly from low to high. The FCNTs have a visible, massive layer structure with a smooth fracture surface. The fracture surface displayed a ductile fracture with numerous rubs and tearing edges as the concentration of CNTs increased. There is no significant zone of CNT aggregation when distinct spectroscopic nanotubes are evenly distributed throughout the matrix, as demonstrated

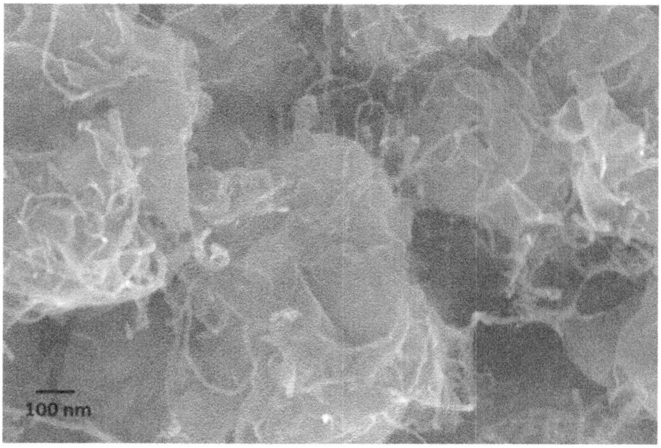

Figure 12.8 FESEM image of FCNT-rGO.

in the image. The results of the XRD investigation corroborate this. The unique characteristics of the CNTs implanted into the matrix enhance the mechanical characteristics of sintered bulk samples.

Although scanning electron microscopy (SEM) studies offer significant insights into the microscopic and structural properties of CNTs, they do not fully capture the ultimate nature of these materials. Only by using FESEMs can CNTs and nanofibers be distinguished. As a result, the TEM inspection of the samples reveals more information about the CNTs obtained.

12.6.5 Transmission Electron Microscopy Analysis

Microscopy of the transmission electron (TEM) is a method of microscopy using electron beams delivered via an ultra-thin probe that interacts with the example as it passes. The interaction of the picture is created by electrons conveyed by the material, magnifying and concentrating images on an imaging instrument. An electron microscope JEOL-2100 was running, and the internal structure of the acceleration voltage was shown with 200 kV materials.

Figure 12.9 shows the TEM images of the operating CNT-rGO. The colors in various parts of the image, as well as the apparent gray tone shift, demonstrate the mixing of the nanotubes. FCNTs have an average diameter of 5 nm. According to the literature, the MWCNT wall distance is approximately 0.34 nm, implying that the nanotubes have 4–10 layers. Figure 12.8 depicts a multi-wall CNT with about four walls with a diameter of 3 nm, confirming the earlier calculation and the assumption of wall distance. With a greater number of functional groups, nanoparticles tend to congregate on the surface and at the crossings of nanotubes.

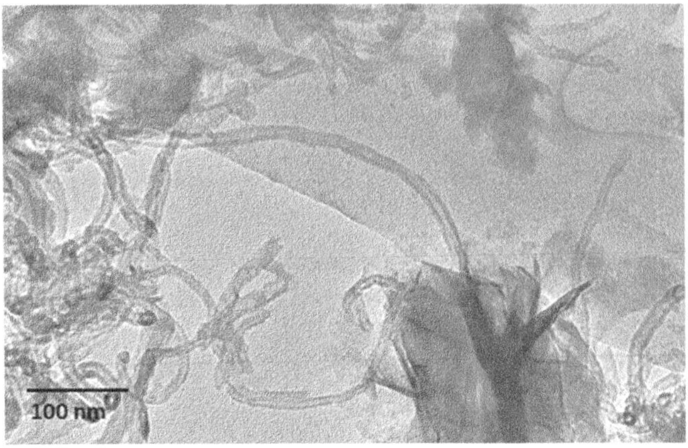

Figure 12.9 TEM image of CNT-rGO.

This aggregation tendency is linked to rGO nanoparticles with high surface energies that are not bonded to the nanotube.

12.7 Applied Electrochemistry

Electrochemical data from galvanostatic charge–discharge methods and CV were used to study the performance of rGO-FCNT composites as supercapacitor electrodes. The performance of rGO-FCNT composites as supercapacitor electrodes was studied utilizing galvanostatic charge–discharge and electrochemical data from CV. The performance of rGO-FCNT composites as supercapacitor electrodes was investigated utilizing electrochemical data obtained from galvanostatic charge–discharge methods and CV.

12.7.1 Galvanostatic Charge Discharge

The charge storage phenomena in the current system have been thoroughly examined [16–17].

The specific capacitances of rGO, rGO-FCNT, and electrodes at 0.5 A g^{-1} are 62.4, 112.75, and F g^{-1}, respectively, as shown in Figure 12.10a. The charge–discharge profile on the GCD curve was asymmetric and triangle-shaped, indicating that the material has excellent reversibility and capacitive properties. The material had a smaller potential drop of 0.019 V at 0.5 A g^{-1}, indicating that it provides the electrolyte with more faradaic centers, reducing usable voltage loss. Figure 12.10b depicts the specific capacitance behavior as the current density increases. Capacitance remained high at a current density of 10 A g^{-1},

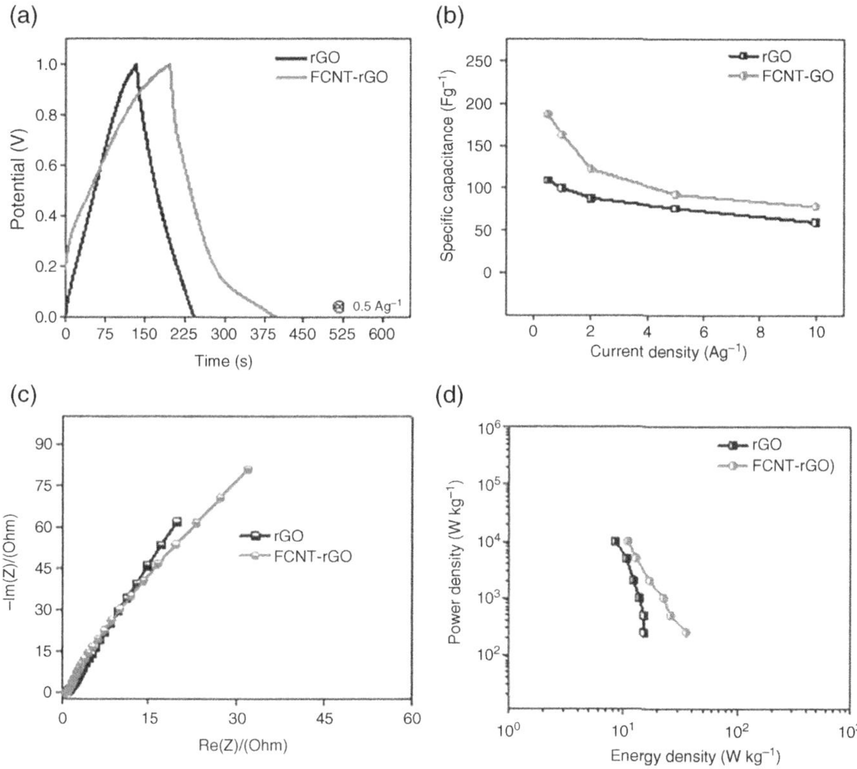

Figure 12.10 (a) Galvanostatic charge/discharge (GCD) curves at 0.5 A g^{-1} (b) rGO and rGO-FCNT performance curves at 0.5, 1, 2, 5, and 10 A g^{-1} (c) Nyquist plot with an inset illustrating the plot at higher frequencies (d) The Ragone plot (energy density vs. power density).

corresponding to an 85% retention of capacitance as the current density increased from 0.5 to 10 A g^{-1}, implying exceptional rate capability.

12.7.2 Electrochemical Impedance Spectroscopy (EIS)

Figure 12.10c depicts the Nyquist plots of the rGO and rGO-FCNT electrodes. Almost all curves are composed of a semicircle arc at high frequency and a nearly straight line at low frequency. The intercept between the Nyquist plots and the x-axis in the high-frequency region is quite small. In this test, the electrodes demonstrated an extremely low electrolyte resistance (R_e) system. Additionally, certain electrode materials have a low charge transfer resistance, which is reflected in the semicircle's small diameter. These Nyquist plots contain an arc in the region of high frequency. In Nyquist plots, the line in the low-frequency region shows the system's Warburg impedance. The electrode in the test system has a high slope, indicating superior capacitive behavior. Figure 12.10c shows that rGO-FCNT has a lower Warburg resistance than the other electrodes, indicating that it has superior capacitive and ion diffusion kinetics properties.

The "Ragone plot" in Figure 12.10d depicts the material's specific energy and specific power properties.

12.8 Conclusions

The goal of this study was to suggest a synthesis strategy that would improve the electrochemical performance of FCNT-reduced graphene oxide-based electrodes for use in high-energy supercapacitor applications. In this context, bulk rGO powder was irradiated with various sources to examine the impact on its physicochemical and electrochemical characteristics. For supercapacitor applications, we created a nanostructured electrode material called FCNT-rGO that performed better electrochemically than CNT-rGO. Of these, FCNT-rGO demonstrated an 84% capacitance retention over a cyclic period and a high specific capacitance of 112.75 F g^{-1} at a scan rate of 0.5 A g^{-1}.

12.9 Future Scope

Functionalization is necessary because it provides several active sites for successful faradic redox processes as well as increases the electrochemical performance of the electrode. It also serves as a support layer for nanoparticles, preventing them from dissolving and aggregating when combined with rGO. These findings support the idea that FCNT-rGO, with its great cycle stability and increased supercapacitance, is a viable electrode material for high-capacity energy storage.

References

1 Yan, J., Wei, T., Shao, B. et al. (2010). Electrochemical properties of graphene nanosheet/carbon black composites as electrodes for supercapacitors. *Carbon* 48: 1731–1737.

2 Lei, Z., Christov, N., Zhao, X.S. et al. (2011). Intercalation of mesoporous carbon spheres between reduced graphene oxide sheets for preparing high-rate supercapacitor electrodes. *Energy Environmental Sciences* 4: 1866–1873.

3 Qiu, L., Yang, X., Gou, X. et al. (2010). Dispersing carbon nanotubes with graphene oxide in water and synergistic effects between graphene derivatives. *Chemistry A European Journal* 16: 10653–10658.

4 Yang, Z., Liu, M., Zhang, C. et al. (2013). Carbon nanotubes bridged with graphene nanoribbons and their use in high-efficiency dye-sensitized solar cells. *Angewandte Chemie International Edition* 52: 3996–3999.
5 Yu, D. and Dai, L. (2010). Self-assembled graphene/carbon nanotube hybrid films for supercapacitors. *Journal of Physical Chemistry Letters* 1: 467–470.
6 Sun, H., You, X., Deng, J. et al. (2014). Novel graphene/carbon nanotube composite fibers for efficient wire-shaped miniature energy devices. *Advanced Materials* 26: 2868–2873.
7 Yu, D., Goh, K., Wang, H. et al. (2014). Scalable synthesis of hierarchically structured carbon nanotube-graphene fibers for capacitive energy storage. *Nature Nanotechnology* 9: 555–562.
8 Tilmaciu, C.M. and Morris, M.C. (2015). Carbon nanotube biosensors. *Frontiers in Chemistry* 3: 59.
9 You, B., Wang, L., Yao, L. et al. (2013). Three-dimensional N-doped graphene-CNT networks for supercapacitors. *Chemical Communications* 49: 5016–5018.
10 Wu, Y., Liu, S., Zhao, K. et al. (2015). Facile synthesis of 3D graphene hydrogel/carbon nanofibers composites for supercapacitor electrodes. *ECS Solid-State Letters* 4: M23–M25.
11 Feng, X., Liang, Y., Zhi, L. et al. (2009). Synthesis of microporous carbon nanofibers and nanotubes from conjugated polymer networks and their evaluation in an electrochemical capacitor. *Advanced Functional Materials* 19: 2125–2129.
12 Sun, G., Zhang, X., Lin, R. et al. (2015). Hybrid fibers made of molybdenum disulfide, reduced graphene oxide, and multi-walled carbon nanotubes for solid-state, flexible, asymmetric supercapacitors. *Angewandte Chemie International Edition* 54: 4651–4656.
13 Sun, G., Liu, J., Zhang, X. et al. (2014). Fabrication of ultralong hybrid microfibers from nanosheets of reduced graphene oxide and transition-metal dichalcogenides and their application as supercapacitors. *Angewandte Chemie International Edition* 53: 12576–12580.
14 Cao, X., Zheng, B., Shi, W. et al. (2015). Reduced graphene oxide-wrapped MoO_3 composites prepared by using metal–organic frameworks as precursors for all-solid-state flexible supercapacitors. *Advanced Materials* 27: 4695–4701.
15 Saifuddin, N., Raziah, A.Z., and Junizah, A.R. (2013). Carbon nanotubes: a review on structure and their interaction with proteins. *Journal of Chemistry* 2013: 676815.
16 Chander, S. and Mangal, M. (2020). rGonanostructure electrode material for high-performance super-capacitor application. *To Chemistry Journal* 6: 52–62.
17 Labulo, A.H., Martincigh, B.S., Omondi, B., and Nyamori, V.O. (2017). Advances in carbon nanotubes as efficacious supports for palladium-catalyzed carbon–carbon cross-coupling reactions. *Journal of Materials Science* 52 (16): 9225–9248.

13

A Review of Effective Biomass, Chemical, Recycling and Storage Processes for Electrical Energy Generations

Suman Lata Tripathi[1], *Krishan Arora*[1], *and Celestine Lwendi*[2]

[1] *School of Electronics and Electrical Engineering, Lovely Professional University, Phagwara, Punjab, India*
[2] *School of Creative Technologies, University of Bolton, Bolton, United Kingdom*

13.1 Introduction

Energy consumption is rising as a result of rapid industrialization and population increase, with a predicted 30% growth by 2040 within domestic and industrial sectors. Coal and oil are used to generate electricity, warm the area, enzymes, chemicals, and energies in the global market. Oil and coal make for 81% of the electricity production, nuclear technology accounts for 5%, and wind and solar power contribute 14% (out of which the biomass contribution is around 70%). The biochemical sector accounts for 11 and 8% of the total global production demands for natural gas and oil correspondingly; although, about 50% of the market's energy requirements are being used as raw material for products [1]. Clean energy production is seen as a critical and necessary defensive measure to ensure depletion. The term "biomass energy" refers to hard, liquid, or gaseous oils that are made primarily with biorenewable raw ingredients. Biomass energy has garnered more attention among emerging renewable technologies than other sustainable energy resources such as solar power, WE, geothermal power, ocean energy and hydroelectric [2]. Biofuel is a word used to describe physiological sources derived from animals and plants [3, 4]. This is the most important renewable source of energy for generating power, hot air use, and bioenergy manufacturing, to mention a few applications [5]. Energy consumption by different sectors is shown in Figure 13.1.

Carbon-based Nanomaterials for Green Applications, First Edition. Edited by Upendra Kumar, Piyush Kumar Sonkar, and Suman Lata Tripathi.

Published 2025 by John Wiley & Sons, Inc.

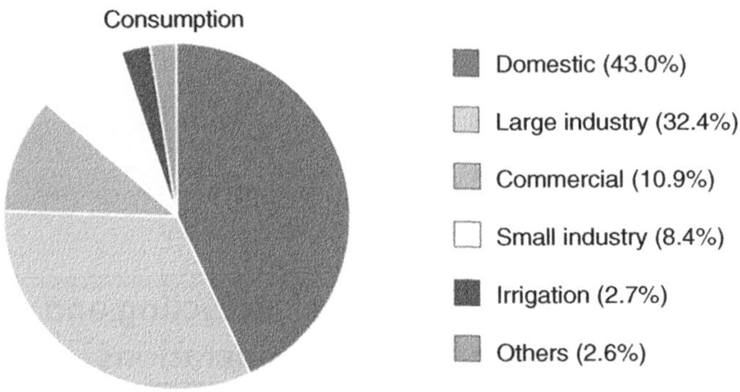

Figure 13.1 Energy consumption in different sectors.

Bioenergy, as a fall substitute for conventional coal and oil, has the potential to significantly reduce carbon dioxide (CO_2) emissions from the transportation industry in the future. Bioenergy may be especially inspiring for use in aircraft, maritime, and heavy trucks, where options such as battery systems might not have been appropriate or where hydrogen technology is lacking. Although it has easy availability, upgrading lignocellulose biomass into biofuels necessitates substantial treatment. Because bioenergy has a short lifespan and a substantial oxygen content, one strategy for producing bioenergy involves converting the grain in two steps. Through the distillation method, bioenergy is transformed into an adaptive biocrude near the supplier of the biomass in the very first step. Biomass energy has become the fourth largest primary source of energy [6].

Despite the wind and solar power that are only often used to generate electricity, biofuel is generally utilized to make bioethanol that may be used for combined heat and power but also transport. Bioethanol is produced from the converted form of organic, which would be created by a photosynthetic process using water and CO_2. Vegetation is recognized as a global process for synthesizing bioenergy in which sunlight energy is absorbed from the standpoint of electricity storage [7]. Bioenergy obtained from plant sources had already developed during the last few years, from first-generation bioenergy derived from agricultural harvests to second-generation bioenergy derived from indigestible lignocelluloses to third-generation bioenergy derived from algal biomass but also finally to fourth-generation biofuels derived from genetically modified microalgae biomass [8, 9]. Biomaterials can also be burnt to produce both heat and electricity output. Non-conventional energies play a significant role in the electrical sector's "decarbonizing" cycle, which refers to decreasing the concentration of CO_2 in the atmosphere, and emissions created by the combustion of coal. Sustainable sources of clean energy, windy, hydropower, and geothermal heat, as well as energy from nuclear fission, are carbon-free within the energy business [10]. The transportation industry consumes 3% of the total overall energy consumption and emits 23% of worldwide emissions of CO_2 based on energy usage. Around 96% of global transportation energy requirements are satisfied by oil products, with the remainder being covered by electrical and bioenergy; as a result, the growing consumption of biofuels in the transportation industry is a top priority in decarbonizing this sector. Bioenergy and some other alternative energies, such as wind and solar power, can help partially reduce carbon emissions in the transportation sector [11]. Nutrition-based biofuels have already been capped in the Eurozone United States. Although the bioenergy business uses only 2% of the worldwide area for feedstocks, the "fueled vs. feed" discussion suggests that bioenergy utilized for industrial applications is a delicate topic in nations. Several agricultural residues and bioenergy products, like wood chips and crop residues, are underutilized. A range of proven processes, including burning, pyrolysis, and compression, can be used to transform lignocellulose biomass energy into electrical power [12].

A number of transformation techniques or biomass energy conversion techniques like thermo-chemicals, chemical conversion, and biochemical and bioelectrochemical conversion techniques have been proposed to expand their value in business and many other areas. Bioenergy is now predominantly used mostly for feedstock, followed by nutrition, and thereafter power, lubricants, and petrochemical material generation. It is responsible for 13% of the overall global ultimate energy use.

Here a comprehensive literature review is done on where various authors have used several biomass conversion techniques for energy generation through effective biomass chemical, recycling, and storage processes. Various raw materials along with their utilization specifically in power generation and other useful biomass products have been discussed.

13.2 Bio-Raw Materials and Utility

Bioenergy can be twisted by scorching or gasifying dry matter or taking biogas done controllable anaerobic fermentation (AF). Co-firing bioenergy and fossil fuel extraction (typically coal) in fossil fuel power plants is just a small way to decrease carbon emissions, recover cost-effectiveness, and reduce toxic emissions. Rapid burning of wood chips, forest residues, and other kinds of woody matter is usually used to generate heating (cogeneration) at the scale of separate structures [13, 14]. Integrated solar-thermal (CHP) plants are the most cost-effective way to utilize bioenergy. Such plants collect excess heat and/or heat from bioenergy plants and transport it to neighboring areas for warming or refrigeration systems for freezing [15].

A variety of biofuels can be developed from organic matter, reducing demand for petrol products and upgrading the transportation market's global emission characteristics. The current market for biofuels is dominated by ethyl alcohol from starch and sugar, and biodiesel from canola, oilseed rape, and palm oil. However, a lot of corporations are working hard to improve and take actions for biofuels made from quasi-feedstocks like solid waste, micro-organisms, evergreen vegetation, and woody biomass [16]. Lignocellulose alcohol, biobutanol, ethanol, and a variety of synthesized fuel counterparts are among such biofuels. Bioenergy will have remained only an easily accessible clean energy source, and sustainable transportation fuel unless we are capable of mass-producing a considerable number of electric vehicles that run on non-conventional sources. Bioenergy may be used to replace oil and coal in the generation of energy, and it can also be used to replace countless commercial materials and components obtained from petrochemical feedstock biobased polyurethane foam, polymers, fertilizers, grease, and toxic products are just a few examples [17].

Bioenergy can also be developed wherever living things can survive, and a wide diversity of biomass materials is accessible. Most organic fertilizers can also be converted into petroleum products, steam, electricity, and/or organic goods. Bioenergy is a popular fuel throughout many regions, particularly in developing nations for cookery and warmth [8]. Many advanced economies are boosting their use of cellulosic biofuels for vehicles and power production as a way to reduce greenhouse emissions through coal consumption. Bioenergy is made up of heat energy that has been stored either from the sun. Photosynthesis is a process by which plants make biomass [18]. Organic matter can be produced by burning for heating or processed into sustainable gasoline using several techniques.

Wood and wood processing wastes: Timber and woodworking mill dust and garbage, as well as black liquid from papermaking mills, are all sources of firewood, wood products, and charcoal [19].

Agricultural crops and waste materials: Food crop-manufacturing leftovers, as well as maize, beans, sugar beets, swap grasses, broadleaf weeds, and algae.

Biogenic materials in municipal solid waste: Meals, garden, and timber scraps, as well as sheet, textile, and woolen goods.

13.3 Biomass Energy Conversion Techniques

There seem to be four main types of energy technologies accessible right now that could lead in particular energy efficiency and non-conventional goods. The different types of Biomass Energy Conversion Techniques are shown in Figure 13.2.

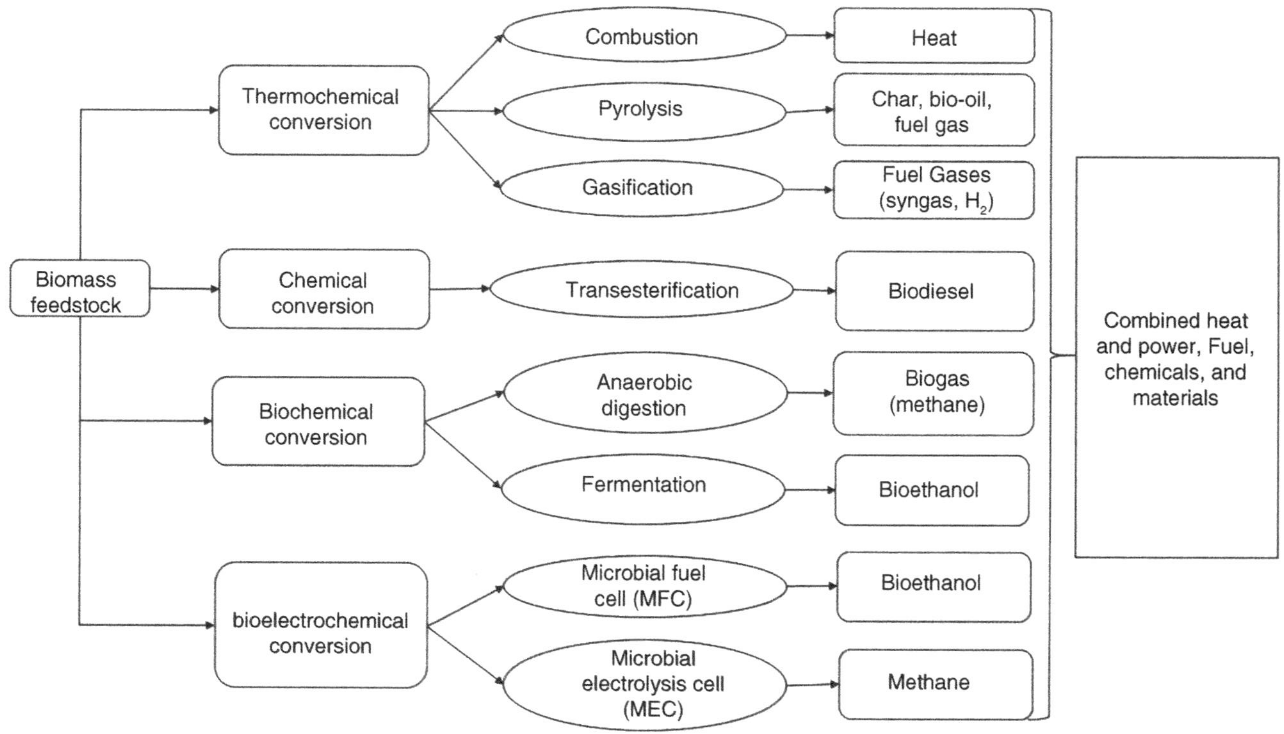

Figure 13.2 Bioenergy production routes.

13.3.1 Thermochemical Conversion

One of the most prevalent methods for biomass conversion to usable energy is direct burning. All bioenergy may be produced by burning to warm residences and beverages, as well as to provide warmth for industries and generate power in turbine generators. All use of steam, even without atmospheric oxygen, to transform cellulose into some other types of energy and goods is called thermal transition. Burning, pyrolysis and gasification are examples [20].

13.3.1.1 Combustion

The burning of material in the atmospheric oxygen is known as combustion shown in Figure 13.3. Excess heat is being used to supply heated air, steam, or electrical energy using a heat recovery boiler to power a combustion engine. Bioenergy may be used in conjunction with traditional non-renewable energy power plants [21].

13.3.1.2 Biomass Pyrolysis

Biomass pyrolysis converts biological byproducts into gasoline, petroleum, and ash at a constant condition and without O_2. Gases and petroleum could be utilized to drive an engine, as well as some techniques can also convert the gases into gasoline and compounds. Pyrolysis is a process of heat breakdown in the lack of air [22]. Biogarbage is warmed whether in the inert atmosphere (i.e. passively) or by partial oxidation of most of the garbage in a constrained excess air supplied even during catalytic pyrolysis. This causes the garbage to decompose thermally, yielding a mixture of coal, gas, and fluid petroleum as shown in Figure 13.4. It can be used as a liquefied fuel or improved and converted into valuation chemicals [23].

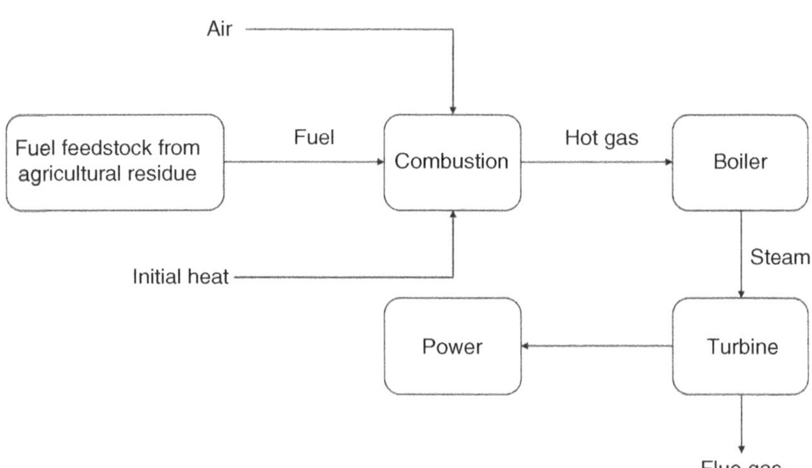

Figure 13.3 Flow diagram of combustion process.

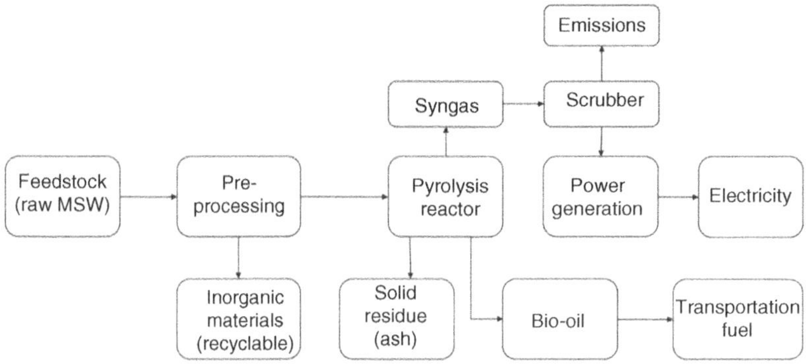

Figure 13.4 Pyrolysis process flow diagram.

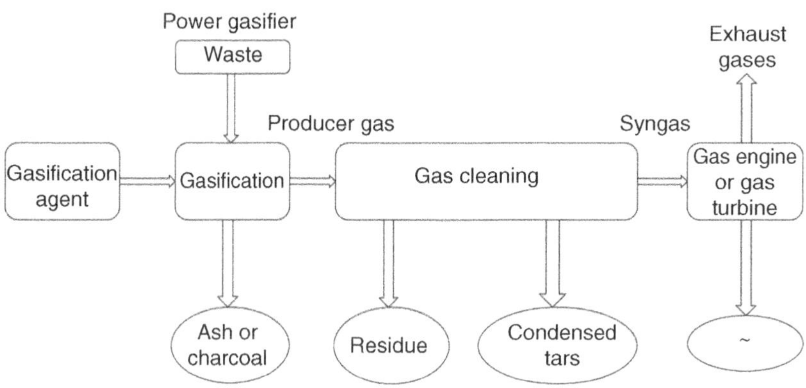

Figure 13.5 Gasification process.

13.3.1.3 Gasification

Gasification is a robust technical route that converts organic matter to energy as well as other chemicals avoiding burning through a steady state using heating, vapor, and air shown in Figure 13.5. Thermal decomposition is another term for gasing [10]. Hot temperatures are used in regulated partial oxidation to create a gaseous fuel and char, which is then chemically reduced. The use of bioenergy for agricultural byproducts in gas turbine engines is a common application. Petroleum, diesel fuel, and pharmaceuticals are only a few of the advanced applications [24].

13.3.2 Chemicval Conversion

Biochemical transition is the process of converting waste into petroleum products, the first most common of which is methanol. The papermaking company always used biofuel hydrogenation processes [25]. Transesterification is a biochemical

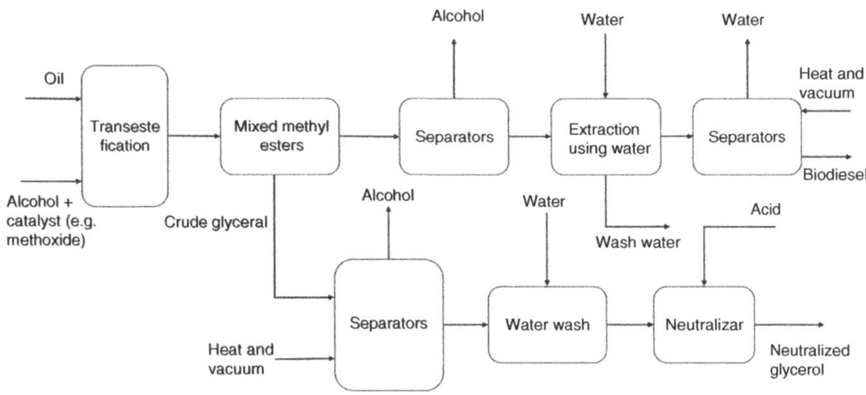

Figure 13.6 Biodiesel process using transesterification.

method of transforming vegetable oils, red meats, and lubricates into monoglycerides, that are being used to produce ethanol [26].

13.3.2.1 Transesterification

Transesterification is described as the reaction that occurs when quasi-oil is permitted to interact with ethanol shown in Figure 13.6. Compounds seem to be the most widely used distilled liquors in this process because of their low price and efficiency. This process is commonly used to lower the stiffness of quasi-oil and convert glycerol to esters. Catalysis transesterification and quasi-transesterification are the two methods for transesterification [27]. Catalysis transesterification is well known to have two issues. The fundamental issue is that the procedures take a very long time and need extraction of the biofuels from oil, ethanol, catalysts, and saponified contaminants' combination. Detoxification of ethanol is significantly easier because there is no need for a catalyst. Although during the supercharged transesterification method, it prevents hydrolysis reaction and sulfonation. The extreme heat and pressure of a supercharged methanol transesterification method, on the other hand, lead to high expenditure on the equipment [28].

13.3.3 Biochemical Conversion

Biological transformation, which comprises anaerobic digestion (AD) and fermenting, also comprises all use of catalysts, bacteria, or other micro-organisms to decompose waste into fluids and gaseous fuel. Electricity, vehicle fuel, and sustainable compounds can all be made from this substrate. Via AD, fermentation, or composting, it utilizes the catalysts of a bacterium or other microbes to decompose waste [13, 29].

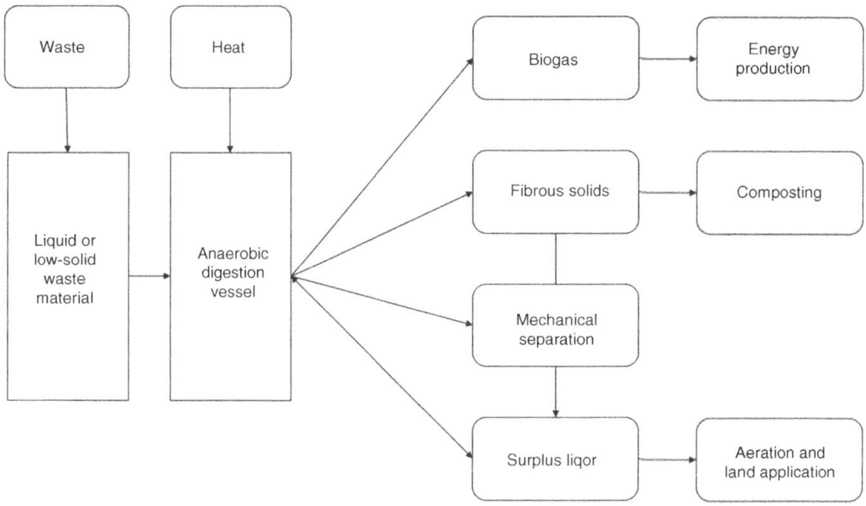

Figure 13.7 Process diagram of anaerobic digestion.

13.3.3.1 Anaerobic Digestion

Microscopic organisms decompose organic matter in the absence of air, which is known as AD as shown in Figure 13.7. Cow dung, forage, food manufacturing, and lignocellulose might all be used as biofuel in AF [30]. Micro-organisms produce approximately 90% of the energy output of cellulose into bioenergy, which involves between 50 and 70% methanol. Biofuel could be utilized directly in reignite gas turbine engines, as well as processed for pollutant removal and improved performance, or dry reformed to make compounds. AD aids in the reduction of biomass waste as well as the recovery of fuel [31, 32]. In a nutshell, the organic part of MSW can be used as fuel for AD. The contribution of MSW is normally degraded, and the methane is used for CHP or as a transportation fuel; nonetheless, the quasi-proportion of MSW is either burnt or gasified [33]. Biogas is created under aerobic circumstances. Propane (a high-temperature gas) can be turned into alcohol with ease.

13.3.3.2 Fermentation

Micro-organisms digest vegetable glucose and generate methanol, propanol, and other products through fermentation [34]. Classical methanol fermentation utilizes sugarcane plants; however, other resources including grain plants or cellulosic materials feedstock can also be employed as shown in Figure 13.8. Fermentation is attributed to the presence of bacteria at air pressure and air temperature on a continuous basis. During fermentation, ethyl acetate concentrations range from 10 to 18% by proportion [35]. In the lack of air, the biological material is

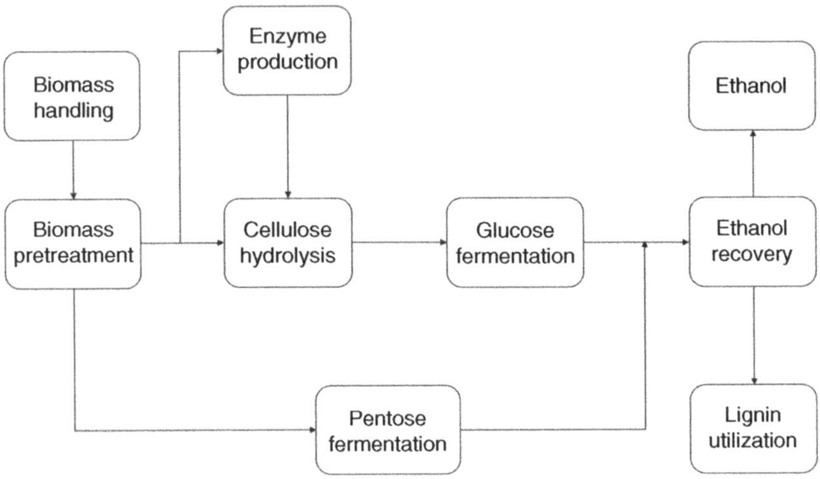

Figure 13.8 Fermentation process.

turned into acids or alcohol (ethylene, lactic, etc.), leaving nutrition behind. Biofuel, a sustainable vehicle fuel, may be made from a liquid culture of certain fungal isolates [36]. The planet warms throughout the digestion with this chemical portion, reaching as much as 65 °C, but begins to reduce within two months. Fermentation, on either hand, takes a significant amount of time and produces a variety of gases, particularly minimal amounts of carbon monoxide and H_2S [37].

13.3.4 Bioelectrochemical Conversion

By microbicide transition, a fast-emerging bioelectrochemical process can leverage the mechanism of electrical and chemical bioconversion of matter to create useable byproducts, lubricants, and bioelectricity [38]. BES has a number of setups that can turn biochemical energy from garbage (oxidizable materials) further into electrical energy or other useful goods. BES research has mostly emphasized its excellent power transition efficacy, microbe-anode interface, and resizing [39]. The efforts to control plant matter even at fairly low biological O_2 requirements, concentrations inside the liquid waste as well as at lower doses of high volatility fatty amines are two particular applications of bioelectrochemical engineering over commercially accessible traditional AD mechanisms. An extremely quick bioelectrochemical technique called anti-mycobacterial transformation can take advantage of the technique called electrical and chemical biotransformation of materials to produce useful lignocellulose, fluids, and biomass energy [40, 41]. BES consists of various configurations that can convert biological energy from waste (oxidizable organic compounds) into electric power or other useful products [42]. BES studies have largely centered on its overwhelming power transformation accuracy, microbial population interfacing, and scaling [43]. Two specific applications of bioelectrochemical sciences over widely viable classical AF techniques are still the attempts to stop crops at even relatively low biological O_2 requirements, accumulations inside the wastewaters, and reduced dosage of extremely volatile fatty acids [44].

13.3.4.1 Microbial Fuel Cells

Sometimes cellulose is broken down into carbohydrates, such as glycogen [45]. The glucose is converted into quickly saturated fats, ethyl alcohol, H_2, and CO_2 throughout the fermentation process shown in Figure 13.9. Ultimately, electrical genesis occurs, resulting in the production of power and the transport of CO_2. The photocatalytic micro-organisms in electrical evolution transform organic power into electricity [46]. Its anode is anaerobiosis, with bacteria and raw material in the negative electrode. Micro-organisms oxidize the fuel, producing CO_2, electrons, and protons in the process [47]. The aquarobics chamber is the positive electrode, and then a proton swapping layer divides the two compartments, allowing only hydrogen ions to cross, similar to other fuel cell technology.

Microbial fuel cells (MFCs) come in two varieties: mediator and mediator-less. First, the mediator type contains a mediator, which is a substance that transmits ions from its germs in the cells toward the negative electrode [48]. This was first developed during the early twentieth century. Thionine, methyl-viologen, methyl-blue, tartaric acid, and neutral red are many of these compounds [49]. These substances are both costly and hazardous. MFCs without a mediator were a much more recent innovation, dating back to the mid-1960s. On its outer surface, these units possess electrochemical oxidation redox proteins such as cytochromes that can transmit electrons directly to the anode [50, 51].

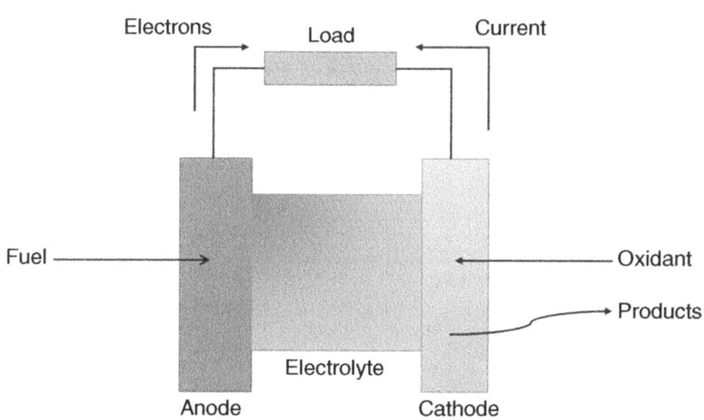

Figure 13.9 Microbial fuel cell (MFC).

13.3.4.2 Microbial Electrolysis Cells (MECs)

All of those are slightly altered microbiological fuel cells that do not use a mediator (MFC) as shown in Figure 13.10. Although MFCs create power out from the bacterial degradation of organic materials in freshwater, MECs substantially invert the method to create H_2 or CH_4 by delivering a voltage to microbes to augment the voltage provided first by microbial degradation of organics (often from garbage) [52]. Another second PEM power source or ignition system can be employed to generate energy from the H_2 or CH_4 emitted. (Hydrogen-powered vehicles, for example.) MECs hold great potential for the future of the whole power sector, and they are still in the initial phases of development [53]. The MEC is composed of two modifications to an MFC: introducing a considerable amount of charge (>0.2V) toward the negative electrode voltage caused by micro-organisms, but not using O_2 at the positive electrode [54]. The inclusion of voltage allows pure H_2 to be produced at the positive terminal. As a result, the MEC/BEAMR (bioelectrochemically assisted microbial reactors) technology is used as an anaerobic digester. The voltage required can be generated using either MFC power or hydrogen gas produced by the MEC in a traditional H_2 fuel cell [55]. The theory behind the whole technology would be that the gene's protons and ions can indeed be rearranged as H_2 gas at the positive electrode, a practice known as the H_2 evolution reaction (HER) [56]. Theoretically, 0.41 V is required to create H_2 from acetate, but bacteria only generate 0.2–0.3 V. In the MEC/BEAMR, we simply have to add approximately 0.2 V and more to generate H_2 gas. Such voltage is substantially lower than that required for water electrolytic, which is typically around 1.8 V. Separation of water requires energy, but once O_2 is employed at the positive electrode, "splitting" apart organic materials by microbes is a reversible chemical

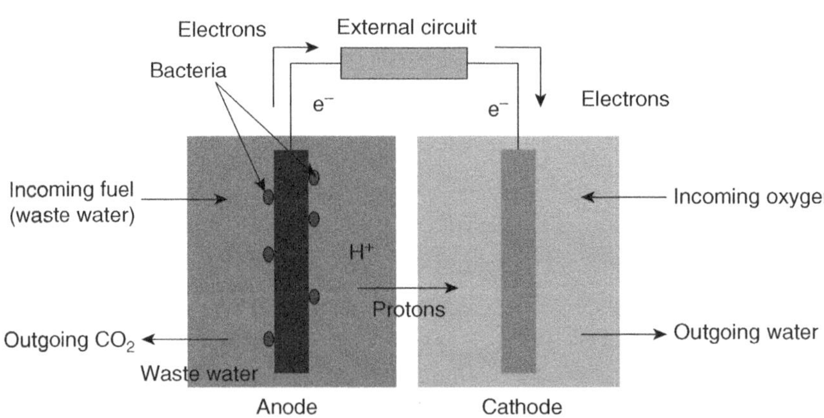

Figure 13.10 Microbial electrolysis cell (MEC).

process [57]. There is really no O_2 present in the MEC method. Hence, H_2 creation is not really spontaneous unless a little increase in voltage is provided to the electricity generated by the bacterium. As a result, the MEC treatment is more akin to "organic material electrolysis" [58].

13.4 Application Areas of Biomass Energy

Biomass energy is fruitful in every aspect from small domestic purposes to large-scale industry needs, specifically domestic space heating and electricity generation for small and medium industries. Since biomass energy is mainly generated from biowaste, therefore, it supports waste management, the concept of clean water, and the overuse of water in thermal and hydroelectric plants. Utilization of agricultural waste for biomass energy generation controls air pollution through controlled emission of carbon and nitrogen in air during generation processes. An energy-secured world is possible with the best utilization, generation, and storage of biofuels. The biofuel conversion efficiency is a major challenge in the effective utilization of biomass that can be improved by applying suitable optimization algorithms at a different level of biomass conversion processes [59, 60].

13.5 Comparative Analysis of Modern Biomass Energy Conversion Techniques

A comparative analysis of different biomass resources with different conversion techniques and their utilization are illustrated in Table 13.1. Here in this table, various types/sources of biomaterial are compared with their conversion techniques, their operating temperatures, the percentage of energy conversion in the form of electricity and other useful things, and their application areas. Various biomass materials have already been discussed earlier. Now only their operating conditions and application areas are discussed here in this table from the literature of various papers [87–90].

13.6 Optimization Techniques for Effective Biomass Conversion and Supply Chain Management

Modeling and optimization techniques need to be added at different stages of biomass conversion processes, recycling, and process supply chain management. Uncertainties and barriers of bioenergy market lead to the rigorous analysis through

Table 13.1 Comparison of existing biomass conversion processes and applications.

Reference	Type/source of biomaterial	Conversion techniques	Operating conditions (°C)	% Energy conversion	Application area
[9]	Biomass powered	Biomass pyrolysis	500–800	70%	Electricity, biofuel, heat, biogas, oil
[12]	Biomass powered	Biomass liquefaction	250–375	47%	Heat, biogas, oil
[33]	Biomass fired	Direct combustion	800–1000	61% for 5% moisture content	Exothermic steam, heat, electricity
[33]	Biogas fired	Anaerobic digestion	25–35	40% for electricity and 50% for heat, biogas and oil	Endothermic biosyngas, electricity, heat, and oil
[33]	Biomass powered	Thermal gasification	1400–1700	16–30% for electricity and 27–51% for methanol	Electricity, methanol
[61]	Methanol	Methanol ICV	250–280	14%	Automobile
[61]	Methanol	Methanol fired	250–300	26%	Electricity
[61]	Biomass	Ethanol fermentation	32–35	90%	Ethanol
[62]	Biomass	Biomass reforming	800–900	79%	Hydrogen
[62]	Methanol	Methanol reforming	800–900	83%	Ethanol
[63]	Biomass powered	Supercritical water gasification	400–600	29% for electricity and 62% for methanol	Electricity and methanol
[64]	Sulfur and acetate	Microbial fuel cells	900–1000	90%	Electricity, bio-hydrogen, wastewater treatment, biosensor development.
[65]	Decomposition of organic compounds	Microbial electrolytic cells	30–31	60–70%	Electricity, methane, and hydrogen

biomass supply chain optimization that is based on design and management processes at different upstream, midstream, and downstream levels [66, 67]. Upstream includes biomass production, harvest, collection, and pre-treatment processes while midstream includes biomass energy conversion processes further ended with downstream energy storage.

Reference	Optimization levels	Process steps or processes	Decision variables	Techniques or models
[68–70]	Upstream	Biomass production, harvest, collection, pre-treatment	Location, capacity, technology, sourcing Inventory planning, risk management, and fleet management	Linear programming, integer programming, mixed integer linear programming, and strategic decision making
[71–76]	Midstream	Energy conversions	Biomass particle size, shape, particle density, specific heat, thermal conductivity	Bioethanol logistics network, computational fluid dynamics (CFD) simulations
[77–79]	Downstream	Biomass storage	Location of biomass storage	Integer programming, multi-objective model

13.7 Government Policies and Marketing Strategies

A global survey of 2030 states about the countries' policies toward sustainable goals with a multi-dimensional approach to integrate social, economic, climate change, resource efficiency, etc. [70]. The US government started regulating the release of greenhouse gas (GHG) emissions from power plants, factories, and cars to achieve a projected goal of reducing global warming [68]. The UN sustainable teamwork on sustainable developments by forming a regional collaborative platform (RGP) that identifies the challenges based on specific regions and facilitates expertise and resources. A framework is proposed for COVID-19 recovery by facilitating better education, health, social protection, and job opportunities along with protective measures against environmental changes. The Indian sustainable development goals (SDG), mainly Swachh Bharat Mission and Smart Cities and Namami Gange Mission also focus toward waste management and planning along with financial growth and healthy environment perspectives.

13.8 Applications of Biomass Energy and Biomass Products

1) **Biomass: A Renewable Energy Resource:** Biomass is a flexible source of non-conventional energy [80, 81]. It may be transformed into liquid fuels, such as diesel, jet fuel, and gasoline, that are comparable to fossil fuels. With the use of biofuel techniques, carbon from waste streams and biomass can be recycled to create renewable energy, bioproducts, and fuels with fewer emissions for automobiles, trucks, aeroplanes, and ships.
2) **Biofuels: Energy for Transportation:** Another type of renewable feedstock that can be transformed into biofuels, or liquid fuels, for transportation, is biomass. Cellulosic ethanol, biodiesel, and renewable hydrocarbon "drop-in" fuels are examples of biofuels. Ethanol and biodiesel are the two types of biofuels that are used most frequently nowadays. Biofuel can be used in most on-road vehicles as well as in aeroplanes.
3) **Biopower: Energy for Heat and Electricity:** Utilizing techniques similar to those utilized with fossil resources, biopower technologies transform renewable biofuels into electric power and heat. Burning, bacterial decomposition and conversion into a gas or a liquid fuel are the three processes that can extract the energy from biomass to create biopower [82]. Both the flexibility and dependability of the electric grid can be improved by using biopower.
4) **Miscellaneous:** Some other applications include food additives and nutritional supplements, detergents, cleaning products, and plastics and other materials [83]. Bioenergy crops are used to maintain CO_2 levels in the atmosphere by supporting climate mitigation [84]. Biomass energy has the potential to support green energy or renewable energy [85, 86] and will be an addition to the clean energy concept for a sustainable environment.

13.9 Conclusions

Huge availability of biowaste from domestic, industrial and agricultural sectors led researchers to develop efficient eco-friendly techniques for biomass energy production that can meet the requirements of small-level to large-level industry needs. Advanced fuel cells, supercapacitors, and batteries proved as an effective means toward the best utilization of biochar or biomass. Government plans and policies are playing key roles in efficient and effective utilization of energy generation and storage of biowaste toward achieving the global SDGs by 2030. Sulfur and acetate-based microbiofuel cells are promising means for electricity generation, wastewater treatment, and biosensor development with a suitable operating temperature of 900–1000 °C. Biomass through ethanol fermentation is found to be a suitable process in low operating temperature range of 32–35 °C with 90% conversion efficiency. Biomass pyrolysis technique is preferred for biofuel conversion in a medium temperature of 500–800 °C range with 70% conversion efficiency. A wide variety of biomass-based applications are explored in different areas like biomass energy, biofuel cells, biomass crops, and bioproducts through efficient and effective use of biowaste along with mitigation of climate changes.

References

1 REN21. (2019) Renewables 2018, global status report; renewable energy policy network for the 21st century. 1–324. https://www.ren21.net/wp-content/uploads/2019/08/Full-Report-2018.pdf (accessed August 2019).

2 Manzano-Agugliaro, F., Alcayde, A., Montoya, F.G. et al. (2013). Scientific production of renewable energies worldwide: an overview. *Renewable and Sustainable Energy Reviews* 18: 134–143.

3 Pappu, A., Pickering, K.L., and Thakur, V.K. (2019). Manufacturing and characterization of sustainable hybrid composites using sisal and hemp fibres as reinforcement of poly (lactic acid) via injection moulding. *Industrial Crops and Products* 137: 260–269.

4 Sharma, B., Thakur, S., Mamba, G. et al. (2021). Titania modified gum tragacanth based hydrogel nanocomposite for water remediation. *Journal of Environmental Chemical Engineering* 9 (1): 104608. 10.1016/j.jece.2020.104608.

5 Ma, H., Guo, Y., Qin, Y., and Li, Y.-Y. (2018). Nutrient recovery technologies integrated with energy recovery by waste biomass anaerobic digestion. *Bioresource Technology* 269: 520–531.

6 Bui, H.-H., Tran, K.-Q., and Chen, W.-H. (2015). Pyrolysis of microalgae residues—a kinetic study. *Bioresource Technology* 199: 362–366.

7 Chen, W.-H., Lin, B.-J., Lin, Y.-Y. et al. (2021). Progress in biomass torrefaction: principles, applications and challenges. *Progress in Energy and Combustion Science* 82: 100887.

8 Havlík, P., Schneider, U.A., Schmid, E. et al. (2011). Global land-use implications of first and second generation biofuel targets. *Energy Policy* 39: 5690–5702.

9 Bach, Q.-V. and Chen, W.-H. (2017). Pyrolysis characteristics and kinetics of microalgae via thermogravimetric analysis (TGA): a state-of-the-art review. *Bioresource Technology* 246: 88–100.

10 Abdoulmoumine, N., Adhikari, S., Kulkarni, A., and Chattanathan, S. (2015). A review on biomass gasification syngas cleanup. *Applied Energy* 155: 294–307.

11 Simionescu, M., Albu, L.-L., Raileanu Szeles, M., and Bilan, Y. (2017). The impact of biofuels utilisation in transport on the sustainable development in the European Union. *Technological and Economic Development of Economy* 23 (4): 667–686.

12 Patel, M., Zhang, X., and Kumar, A. (2016). Techno-economic and life cycle assessment on lignocellulosic biomass thermochemical conversion technologies: a review. *Renewable and Sustainable Energy Reviews* 53: 1486–1499.

13 Eurostat. (2019). Supply, transformation and consumption of renewable energies-annual data. Eurostat, Statistical Office of the European Union Luxembourg.

14 Vízvári, B. and Bacsi, Z. (2002). Technological development and the stability of technology in crop production. *Journal of Central European Agriculture* 3 (1): 63–72.

15 Lam, S.S., Mahari, W.A.W., Ok, Y.S. et al. (2019). Microwave vacuum pyrolysis of waste plastic and used cooking oil for simultaneous waste reduction and sustainable energy conversion: recovery of cleaner liquid fuel and techno-economic analysis. *Renewable and Sustainable Energy Reviews* 115: 109359.

16 CEFIC (2020). *Facts & Figures of the European Chemical Industry*, 1–78. Brussels, Belgium: European Chemical Industry Council (CEFIC).

17 van der Stelt, M.J.C., Gerhauser, H., Kiel, J.H.A., and Ptasinski, K.J. (2011). Biomass upgrading by torrefaction for the production of biofuels: a review. *Biomass and Bioenergy* 35: 3748–3762.

18 Nunes, L.J.R., Matias, J.C.O., and Catalão, J.P.S. (2014). A review on torrefied biomass pellets as a sustainable alternative to coal in power generation. *Renewable and Sustainable Energy Reviews* 40: 153–160.

19 Naik, S.N., Goud, V.V., Rout, P.K., and Dalai, A.K. (2010). Production of first and second generation biofuels: a comprehensive review. *Renewable and Sustainable Energy Reviews* 14: 578–597.

20 Toor, S.S., Rosendahl, L., and Rudolf, A. (2011). Hydrothermal liquefaction of biomass: a review of subcritical water technologies. *Energy* 36: 2328–2342.

21 Sami, M., Annamalai, K., and Wooldridge, M. (2001). Co-firing of coal and biomass fuel blends. *Progress in Energy and Combustion Science* 27: 171–214.

22 Mehmood, S., Reddy, B.V., and Rosen, M.A. (2012). Energy analysis of a biomass co-firing based pulverized coal power generation system. *Sustainability* 4: 462.

23 Kumar, H., Mohapatra, S., and Singh, R.I. (2017). Review on CFD modelling of fluidized bed combustion systems based on biomass and co-firing. *Journal of the Institution of Engineers (India)* 99: 449.

24 Ud Din, Z. and Zainal, Z.A. (2016). Biomass integrated gasification–SOFC systems: technology overview. *Renewable and Sustainable Energy Reviews* 53: 1356–1376.

25 Liu, Z. and Zhang, F.-S. (2008). Effects of various solvents on the liquefaction of biomass to produce fuels and chemical feedstocks. *Energy Conversion and Management* 49 (34): 98–504.

26 Pham, T.P.T., Kaushik, R., Parshetti, G.K. et al. (2015). Food waste-to-energy conversion technologies: current status and future directions. *Waste Management* 38: 399–408.

27 Dabe, S.J., Prasad, P.J., Vaidya, A.N., and Purohit, H.J. (2019). Technological pathways for bioenergy generation from municipal solid waste: renewable energy option. *Environmental Progress & Sustainable Energy* 38: 654–671.

28 Arora, K., Joshi, G.P., Ragab, M. et al. (2022). Bilateral contract for load frequency and renewable energy sources using advanced controller. *Computers, Materials & Continua* 73 (2): 3165–3180. https://doi.org/10.32604/cmc.2022.026966.

29 Kircher, M. (2019). Bioeconomy: markets, implications, and investment opportunities. *Economies* 7 (3): 73.

30 Pasangulapati, V., Ramachandriya, K.D., Kumar, A. et al. (2012). Effects of cellulose, hemicellulose and lignin on thermochemical conversion characteristics of the selected biomass. *Bioresource Technology* 114: 663–669.

31 Searcy, E. and Flynn, P. (2009). The impact of biomass availability and processing cost on optimum size and processing technology selection. *Applied Biochemistry and Biotechnology* 154 (1–3): 92–107.

32 Malik, A. and Mohapatra, S.K. (2013). Biomass-based gasifiers for internal combustion (IC) engines—a review. *Sadhana* 38 (3): 461–476.

33 van de Kaa, G., Kamp, L., and Rezaei, J. (2017). Selection of biomass thermochemical conversion technology in the Netherlands: a best worst method approach. *Journal of Cleaner Production* 166: 32–39.

34 Chen, H.Z. (1992). Advances in solid-state fermentation. *Research and Application of Microbiology (China)* 3: 7–10.

35 Lina, C.-W., Chih-Hung, W., Tran, D.-T. et al. (2011). Mixed culture fermentation from lignocellulosic materials using thermophilic lignocellulose-degrading anaerobes. *Process Biochemistry* 46: 489–493.

36 Ghorbani, F., Younesi, H., Sari, A.E., and Najafpour, G. (2011). Cane molasses fermentation for continuous ethanol production in an immobilized cells reactor by *Saccharomyces cerevisiae*. *Renewable Energy* 36: 503–509.

37 Ganesh, S., Fazlullah Khan, A.K., Suresh, M., and Senthil, N. (1995). Character associated for ethanol yield in sweet sorghum. *The Madras Agricultural Journal* 82 (5): 361–363.

38 Arora, K., Kumar, A., Kamboj, V.K. et al. (2021). Impact of renewable energy sources into multi area multi-source load frequency control of interrelated power system. *Mathematics* 2021 9 (2): 186. https://doi.org/10.3390/math9020186.

39 Rabaey, K., Butzer, S., Brown, S. et al. (2010). High current generation coupled to caustic production using a lamellar bioelectrochemical system. *Environmental Science and Technology* 44 (11): 4315–4321.

40 Steinbusch, K.J.J., Hamelers, H.V.M., and Buisman, C.J.N. (2008). Alcohol production through volatile fatty acids reduction with hydrogen as electron donor by mixed cultures. *Water Research* 42 (15): 4059–4066.

41 Fultz, M.L. and Durst, R.A. (1982). Mediator compounds for the electrochemical study of biological redox systems—a compilation. *Analytica Chimica Acta* 140 (1): 1–18.

42 (2009). Selective inhibition of methanogenesis to enhance ethanol and n-butyrate production through acetate reduction in mixed culture fermentation. *Bioresource Technology* 100 (13): 3261–3267.

43 Hawkes, F.R., Kim, J.R., and Premier, G. (2010). Bioelectrochemical systems: from extracellular electron transfer to biotechnological application. In: *Integrated Environmental Technology Series* (ed. K. Rabaey, L.T. Angenent, U. Schroder, and J. Keller). London, New York: IWA Publishing.

44 Li, M., Zhao, M., Li, F. et al. (2017). A niobium and tantalum co-doped perovskite cathode for solid oxide fuel cells operating below 500 °C. *Nature Communications* 8: 13990–13997.

45 Liu, H., Ramnarayanan, R., and Logan, B.E. (2004). Production of electricity during wastewater treatment using a single chamber microbial fuel cell. *Environmental Science & Technology* 38: 2281–2285.

46 Arora, K., Kumar, A., Kamboj, V.K. et al. (2020). Optimization methodologies and testing on standard benchmark functions of load frequency control for interconnected multi area power system in smart grids. *Mathematics* 8 (6): 980. https://doi.org/10.3390/math8060980.

47 Logan, B.E. et al. (2006). Microbial fuel cells: methodology and technology. *Environmental Science & Technology* 40: 5181–5192.

48 Rabaey, K., Rabaey, K., Lissens, G. et al. (2003). A microbial fuel cell capable of converting glucose to electricity at high rate and efficiency. *Biotechnology Letters* 25: 1531–1535.

49 Davis, J.B. and Yarbrough, H.F. (1962). Preliminary experiments on a microbial fuel cell. *Science* 137: 615–616.

50 Min, B., Cheng, S., and Logan, B.E. (2005). Electricity generation using membrane and salt bridge microbial fuel cells. *Water Research* 39: 1675–1686.

51 Cheng, S., Liu, H., and Logan, B.E. (2006). Increased power generation in a continuous flow MFC with advective flow through the porous anode and reduced electrode spacing. *Environmental Science & Technology* 40: 2426–2432.

52 Lu, L., Xing, D., and Ren, N. (2012). Pyrosequencing reveals highly diverse microbial communities in microbial electrolysis cells involved in enhanced H_2 production from waste activated sludge. *Water Research* 46: 2425–2434.

53 Kleijn, J.M., Lhuillier, Q., and Jeremiasse, A.W. (2010). Monitoring the development of a microbial electrolysis cell bioanode using an electrochemical quartz crystal microbalance. *Bioelectrochemistry* 79: 272–275.

54 Logan, B.E., Call, D., Cheng, S. et al. (2008). Microbial electrolysis cells for high yield hydrogen gas production from organic matter. *Environmental Science & Technology* 42: 8630–8640.

55 Escapa, A., Manuel, M.F., Moran, A. et al. (2009). Hydrogen production from glycerol in a membraneless microbial electrolysis cell. *Energy and Fuels* 23: 4612–4618.

56 Kundu, A., Sahu, J.N., Redzwan, G., and Hashim, M. (2013). An overview of cathode material and catalysts suitable for generating hydrogen in microbial electrolysis cell. *International Journal of Hydrogen Energy* 38: 1745–1757.

57 Liu, H., Hu, H., Chignell, J., and Fan, Y. (2009). Microbial electrolysis: novel technology for hydrogen production from biomass. *Biofuels* 1: 129–142.

58 Foley, J.M., Rozendal, R.A., Hertle, C.K. et al. (2010). Life cycle assessment of high-rate anaerobic treatment, microbial fuel cells, and microbial electrolysis cells. *Environmental Science & Technology* 44: 3629–3637.

59 Anajemba, J.H., Yue, T., Iwendi, C. et al. (2020). Mittal optimal cooperative offloading scheme for energy efficient multi-access edge computation. *IEEE Access* 8: 53931–53941. https://doi.org/10.1109/ACCESS.2020.2980196.

60 Ngabo, D., Dong, W., Ibeke, E. et al. (2021). Tackling pandemics in smart cities using machine learning architecture. *Mathematical Biosciences and Engineering* 18 (6): 8444–8461. 10.3934/mbe.2021418.

61 Yoshidaa, Y., Dowakib, K., Matsumurac, Y. et al. (2003). Comprehensive comparison of efficiency and CO_2 emissions between biomass energy conversion technologies—position of supercritical water gasification in biomass technologies. *Biomass & Bioenergy* 25 (3): 257–272.

62 Czernik, S., French, R., Feik, C., and Chornet, E. (2002). Hydrogen by catalytic steam reforming of liquid byproducts from biomass thermoconversion processes. *Industrial & Engineering Chemistry Research* 41 (17): 4209–4215.

63 Kruse, A. and Karlsruhe, F. (2008). Supercritical water gasification. *Biofuels, Bioproducts & Biorefining* 2: 415–437.

64 Goswami, R. and Mishra, V.K. (2017). A review of design, operational conditions and applications of microbial fuel cells. *Biofuels* https://doi.org/10.1080/17597269.2017.1302682.

65 Gernaey, K.V., Huusom, J.K. and Gani, R. (Eds.) (2015). A comparative study between neural networks (NN)-based and adaptive-PID controllers for the optimal bio-hydrogen gas production in microbial electrolysis cell reactor, *12th International Symposium on Process Systems Engineering and 25th European Symposium on Computer Aided Process Engineering*, ScienceDirect, pp. 1529–1534.

66 DeMeyer, A., Cattrysse, D., Rasinmäki, J., and Van Orshoven, J. (2014). Methods to optimise the design and management of biomass-for-bioenergy supply chains: a review. *Renewable and Sustainable Energy Reviews* 31: 657–670.

67 An, H., Wilhelm, W.E., and Searcy, S.W. (2011). Biofuel and petroleum-based fuel supply chain research: a literature review. *Biomass and Bioenergy* 35 (9): 3763–3774. https://doi.org/10.1016/j.biombioe.2011.06.021.

68 Awudu, I. and Zhang, J. (2012). Uncertainties and sustainability concepts in biofuel supply chain management: a review. *Renewable and Sustainable Energy Reviews* 16 (2): 1359–1368. https://doi.org/10.1016/j.rser.2011.10.016.

69 Ba, B.H., Prins, C., and Prodhon, C. (2016). Models for optimization and performance evaluation of biomass supply chains: an operations research perspective. *Renewable Energy* 87 (2): 977–989. https://doi.org/10.1016/j.renene.2015.07.045.

70 Gaur, J., Amini, M., and Rao, A.K. (2017). Closed-loop supply chain configuration for new and reconditioned products: an integrated optimization model. *Omega* 66 (B): 212–223. https://doi.org/10.1016/j.omega.2015.11.008.

71 Rentizelas, A.A., Tatsiopoulos, I.P., and Tolis, A. (2009). An optimization model for multi-biomass tri-generation energy supply. *Biomass and Bioenergy* 33 (2): 223–233. https://doi.org/10.1016/j.biombioe.2008.05.008.

72 Azadeh, A., Babazadeh, R., and Asadzadeh, S.M. (2013). Optimum estimation and forecasting of renewable energy consumption by artificial neural networks. *Renewable and Sustainable Energy Reviews* 27: 605–612. https://doi.org/10.1016/j.rser.2013.07.007.

73 Arce, M.E. (2015). The influence of parameters variability on biomass selection for energy use. *Energy, Sustainability and Society* 5: 13. https://doi.org/10.1186/s13705-015-0042-z.

74 Ebadian, M., Sowlati, T., Sokhansanj, S. et al. (2013). Modeling and analysing storage systems in agricultural biomass supply chain for cellulosic ethanol production. *Applied Energy* 102: 840–849. https://doi.org/10.1016/j.apenergy.2012.08.049.

75 Mehrabian, R., Zahirovic, S., Scharler, R. et al. (2012). A CFD model for thermal conversion of thermally thick biomass particles. *Fuel Processing Technology* 95: 96–108. https://doi.org/10.1016/j.fuproc.2011.11.021.

76 Luo, H., Wang, X., Liu, X. et al. (2022). A review on CFD simulation of biomass pyrolysis in fluidized bed reactors with emphasis on particle-scale models. *Journal of Analytical and Applied Pyrolysis* 162: 105433. https://doi.org/10.1016/j.jaap.2022.105433.

77 Martinez-Valencia, L., Garcia-Perez, M., and Wolcott, M.P. (2021). Supply chain configuration of sustainable aviation fuel: review, challenges, and pathways for including environmental and social benefits. *Renewable and Sustainable Energy Reviews* 152: 111680. https://doi.org/10.1016/j.rser.2021.111680.

78 Kim, S., Kim, S., and Kiniry, J.R. (2018). Two-phase simulation-based location-allocation optimization of biomass storage distribution. *Simulation Modelling Practice and Theory* 86: 155–168. https://doi.org/10.1016/j.simpat.2018.05.006.

79 Shang, X., Yang, K., Jia, B. et al. (2021). Heuristic algorithms for the bi-objective hierarchical multimodal hub location problem in cargo delivery systems. *Applied Mathematical Modelling* 91: 412–437. https://doi.org/10.1016/j.apm.2020.09.057.

80 Perea-Moreno, M.-A., Samerón-Manzano, E., and Perea-Moreno, A.-J. (2019). Biomass as renewable energy: worldwide research trends. *Sustainability* 11: 863. https://doi.org/10.3390/su11030863.

81 Reid, W.V., Ali, M.K., and Field, C.B. (2020). The future of bioenergy. *Global Change Biology* 26 (1): 274–286. https://doi.org/10.1111/gcb.14883.

82 Jones, M.B. and Albanito, F. (2020). Can biomass supply meet the demands of bioenergy with carbon capture and storage (BECCS)? *Global Change Biology* 26 (10): 5358–5364. https://doi.org/10.1111/gcb.15296.

83 Sertolli, A., Gabnai, Z., Lengyel, P., and Bai, A. (2022). Biomass potential and utilization in worldwide research trends—a bibliometric analysis. *Sustainability* 14: 5515. https://doi.org/10.3390/su14095515.

84 Li, W., Ciais, P., Han, M. et al. (2021). Bioenergy crops for low warming targets require half of the present agricultural fertilizer use. *Environmental Science & Technology* 55 (15): 10654–10661. https://doi.org/10.1021/acs.est.1c02238.

85 Tripathi S. L., Padmanabhan S Green Energy: Fundamentals, Concepts, and Applications, Newark, NJ: Scrivener Publishing, Wiley 2020. doi:https://doi.org/10.1002/9781119760801

86 Tripathi, S.L., DubeyM, K., Rishiwal, V., and Padmanabhan, S. (2021). *Introduction to AI Techniques for Renewable Energy System*. Boca Raton, FL: Taylor & Francis https://doi.org/10.1201/9781003104445.

87 Liu, H., Cheng, S., and Logan, B.E. (2005). Production of electricity from acetate or butyrate in a single chamber microbial fuel cell. *Environmental Science & Technology* 39: 658–662.

88 Suman Lata Tripathi, Souvik Ganguli, Abhishek Kumar, Tengiz Magradze "Intelligent Green Technologies for Sustainable Smart Cities" Hoboken, NJ: Wiley, 2022, https://doi.org/10.1002/9781119816096 https://onlinelibrary.wiley.com/doi/book/10.1002/9781119816096

89 Arora, K., Tripathi, S.L., and Padmanaban, S. (2022). *Smart Electrical Grid System: Design Principle, Modernization, and Techniques*, 1e. Boca Raton, FL: CRC Press https://doi.org/10.1201/9781003242277.

90 Arora, K., Kumar, V., Prashar, D., and Tripathi, S.L. (2023). *Modern Computational Techniques for Engineering Applications*, 1e. Boca Raton, FL: CRC Press 10.1201/9781003407409.

14

Carbon-Based Nanomaterials for Pollutants' Treatment

Gaganpreet[1] and Y. Pathania[2]

[1] *Department of Physics, Post Graduate Government College for Girls, Sector 11, Chandigarh, India*
[2] *Department of Physics, DAV Post Graduate College, Sector 10, Chandigarh, India*

14.1 Introduction

Accelerated surge in urbanization, industrialization, and population growth poses a great threat to the environment globally by releasing pollutants at an alarming rate. Any undesired, harmful entities/chemicals that have an adverse effect on living organisms and environment are termed pollutants [1]. According to WHO, about seven million premature deaths per year from ischemic heart disease, stroke, lung cancer, chronic obstructive pulmonary disease, and acute lower respiratory infections are because of air pollution [2]. The adverse impact of polluted air on public health and the economy is significant, as it leads to an increase in global health expenses, accounting for 6.1% of the global gross domestic product (GDP) [3]. The presence of an excessive amount of air pollution is also accountable for the occurrence of acid rain, which consequently disrupts the quality of water and soil by contaminating them with toxic pollutants and disturbing the soil pH. Contaminants of water have many detrimental health effects like nephritic syndrome, kidney damage, anemia, and miscarriages [3]. According to reports, in 2022, 1.7 billion individuals were exposed to drinking water that was contaminated with fecal matter, thereby posing a significant threat to their health and well-being [4].

Furthermore, soil pollution also affects human health and global food security. The use of fertilizers, pesticides, and industrial activity affects the soil quality and growth of organic food [5–9]. This hampers the realization of sustainable

Carbon-based Nanomaterials for Green Applications, First Edition. Edited by Upendra Kumar, Piyush Kumar Sonkar, and Suman Lata Tripathi.

Published 2025 by John Wiley & Sons, Inc.

development goals (SDGs), encompassing the eradication of poverty, the elimination of hunger, and the assurance of healthy lives, reversing the loss of land degradation and biodiversity, and making cities safe and resilient [10]. Sustaining and improving the quality of water, soil, and air is a great challenge in the present global scenario. However, recent advancements in the field of nanotechnology are playing a crucial role in designing new materials and addressing these global issues [1, 5–7, 11]. The most abundant element carbon has several allotropes, commonly diamond, graphite, and fullerenes. Coal, charcoal, lamp-black, coke, and carbon black are examples of amorphous carbon allotropes that have not been observed in their pure state. Unlike crystalline forms of carbon, their structures cannot be manipulated, that is why they have not garnered much interest. Materials derived from elemental carbon known as carbonaceous materials, i.e. fullerenes, activated carbon (AC), graphene, CNTs, etc., are extensively used for the detection, removal, and treatment of pollutants. These materials exhibit excellent photocatalytic, adsorption, electrocatalysis, and high surface area properties owing to their bulk counterparts because of which they are used for pollutant removal [11]. Another class is carbon-based hybrid nanomaterials which formed as a combination of pure carbonaceous materials and other functional materials also emerged as good candidates for pollutant treatment [12]. The addition of metal oxides, polymers, and nanoparticles to these carbon-based structures creates a distinctive hybrid material. High aspect ratios, catalytic reactive, tunable band gaps, and different morphologies of these materials make them promising contenders for their usage in remediating the pollutants in the environment's soil, water, and air [13]. Based on the dimensions, the nanomaterials are classified as

1) Zero dimensional (0-D), when all three dimensions of a structure are below 100 nm. Example: quantum dots [14].
2) One dimension (1-D), which has two dimensions smaller and one dimension greater than 100 nm. Examples: CNTs [15], nanowires (NW) [16].
3) Two dimensions (2-D), which have two dimensions greater than 100 nm. Examples: graphene, phosphorene [17], etc.
4) Three dimensions (3-D), which do not have any confined dimension to nanoscale in any dimensions. Example: fullerenes [18].

Early detection of pollutants and efficient techniques for their removal is one of the important environmental concerns. There are different approaches involved in nanoremediation such as photodegradation, adsorption, nanobioremediation, electronanobioremediation, and heterogeneous catalysis, through which contaminants can be removed or immobilized. There are various adsorbent materials such as agro-industrial waste, zeolites, AC, clays, alumina, and hybrid nanoporous solids, i.e. metal-organic frameworks (MOFs). But carbon-based

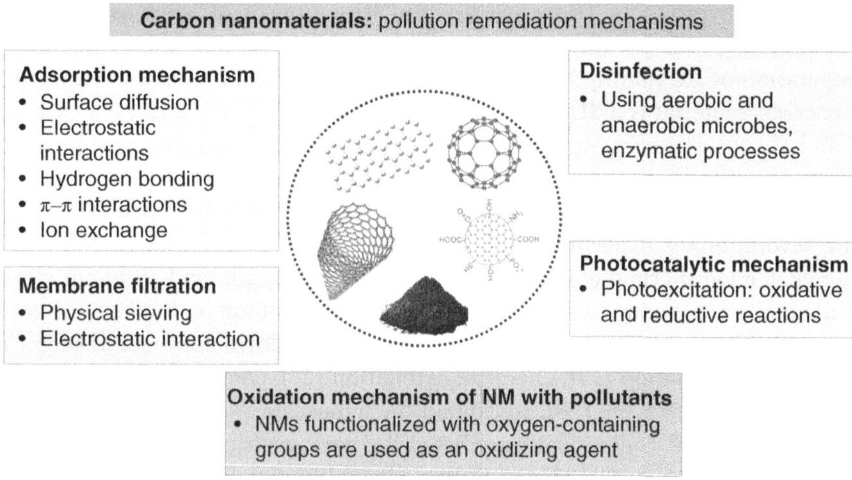

Figure 14.1 Different treatment mechanisms involved in mitigating pollutants from the environment.

materials occupy a special place because of their large porosity, good adsorption capacities and surface area [5]. Air pollution is mainly tackled through adsorption mechanisms, while water pollution is remediated using physical and biological methods, oxidation, filtration, and distillation. Soil pollutants, on the other hand, are treated through microbial actions and nanosorbents. The selection of carbon-based nanomaterials plays a crucial role in removing pollutants by activating specific mechanisms in a flexible manner. The mechanisms depicted in Figure 14.1 are further elucidated in the subsections of the chapter, correlating them with various types of pollutants and carbon nanomaterials.

In the next sections, various allotropes of carbon and hybrid carbonaceous materials are discussed and their applications in the treatment of environmental pollutants.

14.2 Allotropic Forms of Carbonaceous Nanomaterials

Out of the numerous existing carbonaceous materials, highly porous materials, i.e. ACs are first discussed. It is a good adsorbent material with a large surface area ranging, charge-holding capacity, and high porosity [19], which helps in the elimination of various contaminants from the gaseous and aqueous media. AC can be obtained from wood, bituminous coal, anthracite, and biomass agricultural byproducts. There are different structures of AC such as powdery, granular, and fibrous physical forms, which are used for water cleaning. Preparation of AC mainly involves physical and chemical processes. Next, carbonaceous material is a 2D graphene extracted from graphite and each carbon atom has sp^2 hybridization [20]. A layer of graphene was first extracted in 2004 out of graphite by peeling off the layer of carbon successfully [21]. In 2010, Konstantin Novoselov and Andre Geim were awarded the Nobel Prize in Physics "for revolutionary experiments on graphene" [22]. It has unique physical, chemical, good thermal conductivity, high tensile strength, and excellent electronic properties owing to its tunable band gap. Depending on the layer of graphene, thickness, and nature, there are different top-down approaches such as mechanical exfoliation [23], chemical exfoliation [24], chemical synthesis [25], and mechanical cleaving [26]. To explain the nanoscale and edge effects of graphene, Fujita et al. theoretically predicted the narrow strips of graphene called graphene nanoribbon (GNR) [27]. GNR has a width up to 50 nm and depending on the synthesis method, its length ranges up to several micrometers [28]. Synthesis of GNRs is mainly by bottom-up synthesis from polycyclic molecules [29], photolithography and electron beam lithography [30], and unwrapping of CNTs [31]. Other carbon structures like CNTs and fullerenes can be obtained from graphene.

A single layer of graphene and a multi-layer of graphene on rolling up give SWCNTs and MWCNTs, respectively, which exhibit unique physicochemical properties [32]. Their diameters are of the order of several nanometers with large specific areas and have mesopores, which help in heavy metal ion removal via adsorption [33]. There are different methods to synthesize various types of CNTs: laser ablation [34], carbon arc-discharge technique [34], electrolysis [35], chemical vapor deposition (CVD) [34], sonochemical or hydrothermal methods [36]. Functionalization of the CNT surface by adding the various functional groups enhances the adsorption capacities through nonmagnetic metal oxide coating, oxidation, and grafting of iron oxide processes [37]. The adsorption process of contaminants on CNTs is contingent upon the surface properties of the CNTs. This process can occur through either physisorption or chemisorption mechanisms, which are determined by the size of ions and the hydration energy of the contaminants. They find widespread application in sensing devices as it has good response time, selectivity, sensitivity, and cost-effectiveness [38]. The functionalization of CNTs with organic molecules improves its adsorption capabilities [39]. There are other 2-D materials that have been extensively explored in recent years like phosphorene, MoS_2, and WSe_2, which have a finite band gap, are stable, and have high tensile strength because of which they find many applications in the area of gas sensing [17], bio-applications, and removing the salt ions from water [40, 41].

Nanodiamonds are other carbonaceous structures whose core is made of diamond and outer layers are of amorphous carbon. Depending on the surface, shape, and core, they exhibit unique physicochemical properties. Compared to other nanomaterials, they possess superior hardness, chemical stability, electrical resistivity, high thermal conductivity, and good mechanical strength [42]. Synthesis of nanodiamonds mainly involves three main techniques: CVD [42], detonation synthesis [43], and milling of high-pressure, high-temperature (HPHT) [44] micro-sized diamonds.

There are various allotropic forms of carbon such as graphite, diamond, and fullerene. C_{60} fullerene is also known as Buckminsterfullerene, consisting of polygons, pentagons and hexagons [18]. The formula C_{20+k} is utilized, with k representing an integer value. It has the ability to manifest in diverse structures, such as hollow spheres, or CNTs. C_{60} is the most common fullerene composed of 60C atoms and they are synthesized by CVD, arc discharge, and laser ablation. Figure 14.2 shows different synthesis methods for the carbonaceous materials. There are other fullerenes also reported such as C_{20}, C_{70}, C_{120}, C_{185}, C_{540}, and C_{720}. Their notable characteristic lies in their high strength and low density. Moreover, they display the ability to partake in various reactions, including oxidation, reduction, and nucleophilic reactions [45, 46]. Additionally, their commendable conductivity renders them highly suitable for a wide range of applications in different fields. Laser vaporization of carbon in an inert atmosphere is used for the production of small amounts of fullerene. Arc heating and laser irradiation methods are used to synthesize C_{60} at a large scale.

14.3 Synergistic Approaches for Carbonaceous Materials

Certain shortcomings of pristine carbon materials, such as low conductivity, poor mechanical strength, and less chemical reactivity [47] lead to synergistic approaches. In this approach, more than two materials are combined to create a hybrid material/structure with improved properties essentially required for the pollutants' removal.

Carbon-based hybrid nanomaterials are gaining significant attention as a potential solution for pollutant treatment. Their distinct characteristics, including a high surface area, enhanced reactivity, and exceptional adsorption capability, make them particularly suitable for this purpose [47–49]. These materials can be easily tailored by combining various carbon nanomaterials, such as metal nanoparticles, metal oxides, and polymers. Magnetic nanoparticles-carbon nanotubes (MNP-CNT) and hybrid materials [48] have unique properties of MNPs and CNTs, allowing for improved pollutant removal efficiency. The removal of organic pollutants from all environments can be achieved by utilizing either adsorption or

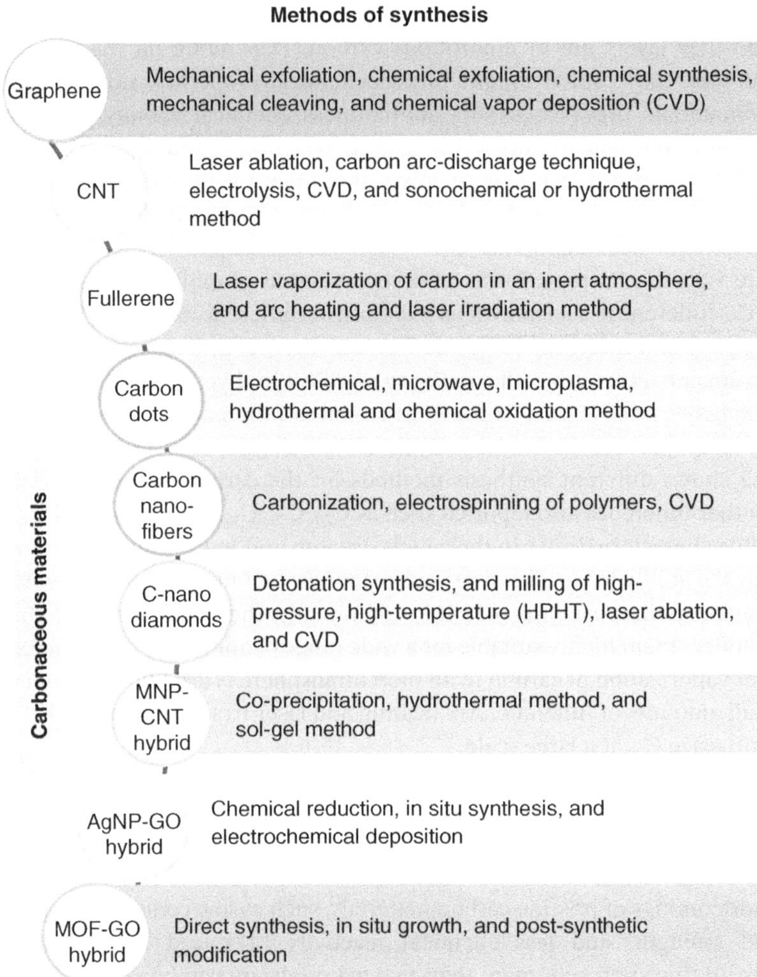

Figure 14.2 Commonly used synthesis methods for carbon-based materials. Abbreviations used: C-Nanodiamonds–carbon nanodiamonds; MNP-CNT–metallic nanoparticle-carbon nanotube; hybrid AgNP-GO–hybrid silver nanoparticle-graphene oxide; MOF-GO–MOF-graphene oxide.

magnetic separation techniques. Magnetic properties of MNPs make it convenient to extract and reclaim the hybrid materials from the contaminated medium, making them highly convenient for pollutant removal. The adsorption capacity of MNPs is significantly improved by the addition of CNTs. This improvement is primarily due to the exceptional adsorption properties and large surface area of

CNTs. Interaction between the organic pollutants and MNP-CNT hybrid materials occurs through various mechanisms, including physical adsorption, chemical adsorption, and π–π interactions [49]. The primary mechanism involved in physical adsorption is the Van der Waals forces existing between the pollutants and the surface of the hybrid material. In contrast, the hybrid material enables the chemical adsorption of functional groups through various distinct interactions, including electrostatic interactions, and hydrogen bonding. Furthermore, the π–π interactions between the conjugated structure of the organic pollutants and the graphitic structure of CNTs contribute to the enhanced adsorption capacities, thereby improving their removal efficiency of MNP-CNT hybrid materials. The MNP-CNT hybrid materials can be easily fabricated using hydrothermal method, sol–gel method, and co-precipitation [50]. Synthesis process allows controlled size, morphology, and composition of the hybrid materials to optimize their performance for specific pollutant removal applications. The silver nanoparticle-graphene oxide (AgNPs-GO) hybrid material is a unique mixture of the remarkable adsorption capacity of graphene oxide and the exceptional antibacterial characteristics of silver nanoparticles [51, 52]. AgNPs can release silver ions, which have cytotoxic effects on bacteria, disrupting their cellular membranes and inhibiting their growth. The extensive adsorption sites for bacteria are further enhanced by the porous configuration and the large surface area of GO, resulting in increased antimicrobial activity. The incorporation of GO in the hybrid material also proves to be beneficial in the extraction of heavy metals through adsorption mechanisms [53]. The functional groups like OH, COOH, and epoxy, on the GO surface, can effectively bind with heavy metal ions through complexation, electrostatic interactions, and ion exchange processes. This adsorption capacity allows the AgNPs-GO hybrid material to extract heavy metal pollutants from water and other environmental media. Synthesis of AgNPs-GO hybrid materials can be achieved through chemical reduction, in situ synthesis, and electrochemical deposition [54]. By utilizing these techniques, it becomes possible to regulate the dimensions, configuration, and arrangement of AgNPs on the GO substrate, thereby optimizing the performance of the hybrid material. Another, combination of MOFs and graphene oxide results in a MOF-GO hybrid material with high adsorption capacity and surface area, as well as remarkable reactivity [55]. This approach is employed to extract organic pollutants and heavy metals from water sources [56]. MOFs are constructed by coordinating metal ions or clusters with organic ligands, forming their unique structural composition. This coordination process gives rise to a crystalline material that exhibits a porous structure with a large surface area. The high porosity and specific surface area of MOFs enable them to adsorb and trap organic molecules effectively. The incorporation of MOFs into the MOF-GO hybrid enhances the stability and adsorption of the material. Synthesis of MOF-GO hybrid materials can be accomplished through several methods, including direct synthesis, in situ growth, and post-synthetic modification [57]. The selection of suitable MOFs and their integration with GO allows for tailoring the hybrid material's properties according to the specific application requirements. The exceptional adaptability of MOF-GO hybrids allows for the efficient removal of a wide array of organic pollutants, ranging from dyes, aromatics, pesticides, and pharmaceuticals, to VOCs. This is attributed to the advantageous physicochemical properties of GO, a two-dimensional carbon-based material, which includes a large surface area, excellent dispersal, and strong adsorption capability. The functional groups present on the GO surface, such as hydroxyl, carboxyl, and epoxy groups, facilitate interactions with organic pollutants through various mechanisms, such as hydrogen bonding, π–π stacking, and electrostatic interactions.

As discussed in the above section, interactions of contaminants with carbonaceous materials in water, air, and soil through physical and chemical phenomena can significantly affect the environmental fate of contaminants and their remediation rates. So, in the next section, application or potential of carbonaceous materials for specific contaminants' removal is discussed.

14.4 Role of Carbonaceous Materials in Environmental Remediation

14.4.1 Removal of Air Pollutants

Utmost need for clean air and the survival of life demands improved techniques to purify air. Release of harmful gases from industries and vehicles such as NO_2, CO_x, and CH_4 is the source of air pollutants that have dangerous effects on the atmosphere and human health. In addition to these toxic gases, numerous VOCs, such as toluene, ethanol, formaldehyde, and acetone, can cause harm to human health. Hence, there is a need for sensors that are highly sensitive, flexible, and cost-effective for the detection of toxic gases and treatment is a very important parameter for a clean environment. Gas sensors can be chemiresistive [58], field-effect transistor [59] and micromechanical (MEMS) [60], and are based on different mechanisms. Sensors based on carbon nanostructure materials like CNTs and graphene are sensitive even at low concentrations of explosive and greenhouse gases [61]. Numerous studies have been conducted utilizing first-principle density functional theory (DFT) to report on the detection of toxic gases NH_3 and NO_2 over pristine and doped phosphorene [17, 61]. By monitoring the change in conductance of adsorbed materials/devices, gases can be effectively detected [17]. Functionalized CNT strongly and selectively interacts with the air pollutants. The detection of gases like CO and CO_2 was achieved by analyzing the impedance changes using CNTs as chemical sensors [62]. Lee et al. reported disposable NO_2 gas sensors using carbon adhesive tape (CAT) as a gas sensing element. The carbon black material in CAT is known for its remarkable ability to adsorb NO_2 gas molecules [63]. Research findings indicate that functionalized CNTs exhibit a greater response and extended recovery time due to the robust chemical bonding between gas molecules and the functional groups of CNTs [64]. Additionally, there have been reports on the detection of NH_3 using the tetrapodal ZnO network (ZnO-T) that has been decorated with MWCNTs [64]. For the detection and monitoring of pollutants, carbon nanodots have been used to design high-performance sensors [65]. The adsorption and photocatalytic degradation of air pollutants, specifically VOCs, have been effectively addressed through the development of a hybrid material composed of zinc oxide nanoparticles and multi-walled CNTs [66]. This material has exhibited high adsorption capacity and excellent photocatalytic activity toward toluene. The incorporation of nanoparticles and organic materials in GO production has led to the emergence of new NO_2 sensors and clean air purifiers. Wang et al. [67] have employed Pd nanoparticles and tin dioxide nanoparticles on reduced GO to fabricate Pd-Tin dioxide-reduced GO hybrids for NO_2 gas detection, while Zhang et al. [68] have developed a reduced GO/Gold nanocomposite-based NO_2 sensor. There is increasing interest in using air filtration to clean areas such as factories and hospitals, as it can provide an environmentally friendly and cost-effective way to manage breathed-in air. Ag/CNT hybrids have been tested for antimicrobial air filtration [69]. A hybrid catalyst that can effectively remove formaldehyde gas has been validated by Miyawaki et al. [70]. The frequently employed carbonaceous adsorbents for different pollutants are listed in Table 14.1.

14.4.2 Removal of Water Pollutants

Ensuring the accessibility of clean water to cater to the demands of the public is also a major societal challenge. Water is polluted by the release of heavy metal ions such as arsenic, chromium, cadmium, nickel, zinc, and mercury from industries, bacteria, and viruses, which pose a great risk to living beings and the environment. These pollutants in the water are responsible for many deadly diseases such as kidney damage, hepatitis, miscarriages, and anemia [87, 88]. There are various methods available to remove metal ions like reverse osmosis, coagulation, biosorption, extraction, and precipitation filtration [89]. The best method for the heavy metal ions removal is adsorption because it is highly efficient, and cost effective. Contaminants from the water are removed by the adsorption processes using materials like silica, clays, zeolites, composites, and activated charcoal [4, 90]. For the treatment of water depending on the interaction of the pollutants and the adsorbents various organic and inorganic

Table 14.1 Various carbon-based adsorbents for the removal of air pollutants.

S. No.	Air pollutants	Carbon-based adsorbents
1.	NH_3	AC [71], graphene [72], graphene nanocomposites [73], ZnO-T decorated with MWCNTs [64], 3D MoS_2/RGO sensor [74], Biochar [75]
2.	Gasoline	AC [71]
3.	CO_2	Biochar [76], CNTs [62], C_{60} fullerene [77], Biochar [75, 78], graphene [72], graphene nanocomposites [73]
4.	Hg	AC [79], MnO_x/graphene and Ag/graphene composites [80] MnO_x—CeO_2-modified activated coke [81] CoO_x—CeO_2-modified activated coke [82] Cu—Ce-modified bio-chars [83]
5.	SO_2	Graphene oxide (GO) [84]
6.	CO	CNTs [62], C_{60} fullerene [77], 3D MoS_2/RGO sensor [74]
7.	NO_2	CAT [63], graphene [72], graphene nanocomposites [73], Pd—Tin dioxide-reduced GO hybrids [65], reduced GO/Gold nanocomposite [68], 3D MoS_2/RGO sensor [74], GO flakes [85]
8.	N_2O	C_{60} fullerene [77]
9.	NO	C_{60} fullerene [77]
10.	C_6H_5OH	H_2S-impregnated activated carbons as adsorbents [86]
11.	CH_2O	Carbon hybrid catalyst [70]
12.	C_2H_5OH, H_2, HCHO	3D MoS_2/RGO sensor [74]
13.	Volatile organic compounds (VOC) gas molecules	Graphene [72]

Abbreviations used: AC–Activated carbon; ZnO-T–Tetrapodal ZnO network; MWCNTs–multi-walled carbon nanotubes; MnO_x–manganese oxide; 3D MoS_2/RGO–molybdenum disulfide (MoS_2)/reduced graphene oxide; MnO_x—CeO_2–manganese–cerium oxide; CoO_x—CeO_2–cobalt cerium oxide; Cu—Ce–Copper–cerium, CNTs–carbon nanotubes; cMWCNT–carboxyl-multi-walled carbon nanotube; GO–graphene oxide; CAT–carbon adhesive tape; H_2S–hydrogen sulfide.

water pollutants can be effectively removed [91]. Carbon nanomaterials such as graphene, CNTs, and fullerenes are widely used for the water treatment and removal of contaminants and pollutants [92]. C_{60} fullerenes exhibit excellent absorption properties for organic contaminants and their mixture with other contaminants, i.e. triclosan, diuron, or venlafaxine affects the toxicity of organic pollutants in river biofilms [93]. Scientific investigations have demonstrated that the application of functionalized fullerenes can effectively eradicate pathogenic micro-organisms present in water by means of a photocatalytic process [94]. Researchers have developed water-soluble fullerene compounds and employed them as sensitizers to generate reactive oxygen species (ROS) in water when exposed to visible and ultraviolet radiation. These ROS have the ability to degrade organic pollutants in water, while the water-soluble fullerenes also serve as antioxidants [95]. The high porosity and ease of functionalization of single-walled CNTs make them a popular choice for pollution and water treatment applications. Nanofilter synthesized using CNTs by a continuous spray pyrolysis method by using n-hexane as a carbon source is used for bacteriophage MS2 virus removal from water [96]. Molecular dynamics studies also show how the bare single-walled CNT and functionalization with different groups affect the adsorption of heavy metal ions from aqueous media. Functionalized SWCNT with COOH have larger adsorption capacities of heavy metal ions [97]. Membranes composed of SWCNTs-polysulfone nanocomposites are employed for heavy metal separation purposes as well [98, 99]. A novel approach involving the use of microwave-assisted solvothermal synthesis has been employed to create stable aqueous carboxyl and amine co-functionalized graphitic carbon nitride quantum dots (gCN QDs). These quantum dots have been successfully bio-conjugated with functionalized agarose, enabling their application in the efficient removal of heavy metal ions in a micro-cartridge format [100]. Salt and other ions are removed using the membrane technology in the reverse osmosis process [41]. Owing to excellent photoluminescence and photocatalytic properties, carbon nanodots are a new game-changing approach for solar energy-driven wastewater pollutant degradation [101, 102]. Hybrid material composed of AC and iron oxide has been used for the extraction of organic dyes from water [103]. The material showed high adsorption capacity and rapid adsorption kinetics toward methylene blue (MB), with a maximum adsorption capacity of 208 mg/g. The MWCNT-starch-iron oxide composite has been found to be more effective than MWCNT-iron oxide in removing both anionic dye methyl orange (MO) and cationic dye MB from aqueous solutions [104].

Additionally, AgNP-GO hybrid materials can be immobilized on different substrates, such as membranes, filters, or beads, to facilitate their practical application in water treatment systems [108, 109]. The immobilization ensures easy separation of the hybrid material from treated water, enabling its reusability and reducing the risk of nanoparticle release into the environment. The material showed excellent adsorption capacity and selectivity toward lead ions, with a maximum adsorption capacity of 178 mg/g [109]. Furthermore, the immobilization of MOF-GO hybrids on various substrates, such as membranes, filters, or beads, can enhance their practical application in water treatment systems. The immobilization allows easy separation of the hybrid material from treated water, facilitating its reuse and preventing the release of nanoparticles into the environment. Composite carbon-iron consisting of nanoscale zero-valent iron (nZVI) clusters on AC colloids is used for the removal of contaminants in groundwater remediation processes [110]. Removal of various water pollutants is summarized in Table 14.2. Pollutants in the water can flow to the soil.

Table 14.2 Various carbon-based adsorbents for the removal of water pollutants.

S. No.	Water pollutants	Carbon-based adsorbents
1.	Heavy metals, i.e. V, Co, Cu, Zn, Ti, Fe, As, Mo, Ni, Cd, Ag, Sn, Pt, Au, Hg, Pb, Cr, Mn	Graphene [4, 105]
2.	Phenolics and naphthol	Graphene [105]
3.	Viruses, bacteria, and organic pollutants	CNTs [106]
4.	Triclosan, diuron, or venlafaxine	C60 fullerenes [91]
5.	Hg	SWCNT-based nanocomposites [96]
6.	MS2 virus	Nanofilter synthesized using CNTs [96], graphitic carbon nitride quantum dots (gCN QDs) [100]
7.	4-Aminophenol, 4-chlorophenol, 4-nitrophenol	cMWCNT/cyclodextrin edge-functionalized graphene composite [86]
8.	Pb	CNTs purified by HNO_3 [99]
9.	Cd^{2+}, Cu^{2+}, Pb^{2+}, and Hg^{2+}	SWCNT and functionalization with different groups, SWCNTs-polysulfone nanocomposite [97, 98]
10.	Anionic dye methyl orange (MO) and cationic dye methylene blue (MB)	MWCNT-starch-iron oxide [104], oxidized MWCNTs [107]
11.	*Escherichia coli*, *Staphylococcus aureus*, and *Pseudomonas aeruginosa*, and heavy metals including Pb, Cd, Hg, Cu, Cd, Hg, and Cu	AgNPs-GO hybrid material [108]

Abbreviations used: MWCNTs–multiwalled carbon nanotubes; cMWCNT–carboxyl-multi-walled carbon nanotube SWCNT–Single-walled carbon nanotubes; AgNPs-GO–Silver nanoparticles-graphene oxide.

14.4.3 Soil Remediation

The enhancement of soil quality has been extensively researched through the utilization of carbon-based materials. Discharge of contaminated water with excessive heavy metal (and metalloids) to the agricultural lands is degrading the land [6]. Most of the metals do not undergo chemical or microbial degradation but organic contaminants get oxidized by microbial action. Accumulation of these elements in agricultural products such as rice grains through irrigated water results in a threat to the health of living beings and the ecosystem [7, 8]. Biochar, formed of carbonaceous materials, such as organic waste, which in turn helps in waste management, has gained attention because of distinctive physio-chemical characteristics and varied applications [111]. This material is widely utilized for soil and water treatment due to its affordability, effectively eliminating pollutants like heavy metal ions, pesticides, dyes, VOCs, pharmaceuticals, and aromatic hydrocarbons [112]. Liu et al. [113] investigated the potential effects of synthesized novel graphene-like biochar-supported nanoscale zero-valent iron (nZVI) for the Cd(II) and As(III) removal under aerobic conditions. Nanoporous carbon materials such as biochar and graphene are extensively used as potential substrates for the contaminant. Furthermore, research has demonstrated that carbonaceous materials, such as biochar and graphene and its oxide, exhibit promising qualities as substrate candidates, as they have the ability to significantly augment contaminant removal capacities [114, 115]. Wang et al. [114] synthesized biochar-supported nZVI by precipitating the nanoparticles on carbon surfaces for the Arsenate (As(V)) removal. Widely tributyltin (TBT) used organotin biocide as a wood preservative, antifouling paints on ships, and wood and agriculture fungicides [116] exhibit toxicity toward aquatic life and human beings. Excellent adsorption mechanism of Biochar via electrostatic interaction, ion exchange, and precipitation is used for the adsorption of organic contaminants. Functionalized carbon-based materials are employed as highly effective and ideal amendments for the remediation of agricultural soil [117]. The combination of CNTs with TiO_2 nanoparticles has shown effective removal of organic contaminants in soil through adsorption and photocatalytic degradation [118]. CNTs enhance the adsorption capacity, while the TiO_2 nanoparticles provide photocatalytic activity under light irradiation. CNTs have also been utilized in combination with other substances to create composites, which have proven effective in extracting heavy metals from polluted soils. CNTs and nZVI particle nanocomposites showed high adsorption efficiency to remove hexavalent chromium, selenium, and cobalt from aqueous solutions [119]. These hybrid nanomaterials can be applied as amendments to the soil, improving its adsorption capacity and reducing the mobility of heavy metals. A few of the soil pollutants are listed in Table 14.3.

Table 14.3 Various carbon-based adsorbents for the removal of soil pollutants.

S. No.	Soil pollutants	Carbon-based adsorbents
1.	Heavy metal ions	Biochar [120], biochar-supported nanohydroxyapatite [121], biochar-supported nZVI [2, 122], CNTs and nano, Nzvi [119], Graphene [5]
2.	Fluoride	Amorphous Al_2O_3 supported on CNT [6]
3.	Cd	Biochar-supported nanoiron-phosphate particles [120, 123] biochar-supported nZVI [2], GO [3]
4.	Pesticides, dyes, volatile organic compounds, pharmaceuticals, and aromatic hydrocarbons	Biochar [120]
5.	Tributyltin (TBT)	AC [124]

Abbreviations used: nZVI–nanoscale zero-valent iron; CNTs–carbon nanotubes.

14.5 Environmental Impact of Carbon-Based Nanomaterials

The manufacturing process of carbon-based nanomaterials involves energy-intensive methods that often demand high temperatures and toxic chemicals. Inadequate management of these processes can result in the release of greenhouse gasses and pollution. To illustrate, the extraction and purification of CNTs may require the use of harmful solvents and generate waste that necessitates proper disposal [120]. Such practices can raise concerns about the potential long-term consequences on ecosystems and human health. Additionally, workers involved in wastewater treatment may face exposure to chemicals such as chlorine, sodium hydroxide, and sulfuric acid, which can cause skin and eye irritation, burns, and respiratory problems [125]. Research has indicated that specific carbon-based nanomaterials, particularly those with a high aspect ratio such as CNTs, can pose a threat to living organisms [126]. These hazards encompass cellular and organ damage, inflammation, and the potential for carcinogenicity. Upon entering the environment, these nanomaterials can interact with organisms and ecosystems. Due to their minute size and extensive surface area, they have the potential to be readily absorbed by aquatic organisms and soil micro-organisms, potentially leading to detrimental effects. The influence of these nanomaterials on the ecological equilibrium is currently under investigation. Establishing effective management and recycling practices for carbon-based nanomaterials is essential in order to prevent their uncontrolled release into the environment, whether they ultimately end up in electronic waste (e-waste) or other products upon reaching the culmination of their life cycle [127]. To mitigate the environmental impacts of carbon-based nanomaterials, ongoing research is focused on developing sustainable production methods, efficient waste treatment procedures, and safe disposal techniques. Furthermore, regulations and guidelines need to be implemented to ensure the responsible use and management of these materials to minimize their potential environmental harm. ISO, the International Organization for Standardization, has formulated a set of standards specifically addressing nanotechnology, which encompasses ISO/TS 80004-1:2015., and provides definitions and descriptions of nanomaterials. ISO 19007:2019 provides guidelines for the development and implementation of safety practices for nanomaterials [124].

14.6 Conclusions: Technological Challenges and Future Prospects

The particular mechanisms for removing pollutants from the environment, including adsorption, filtration, photocatalysis, disinfection, and oxidation, are contingent upon factors such as the nature and concentration of pollutants, treatment process design and optimization, operating conditions, and the necessity for pre-treatment or a combination of multiple treatment steps. In practice, a blend of different removal mechanisms is frequently utilized to attain the targeted level of pollutant elimination. For instance, the removal of suspended solids and particulate matter typically involves sedimentation, filtration, and coagulation. Heavy metals are eliminated through ion exchange and adsorption mechanisms. Pathogens are rendered inactive through disinfection and are then eliminated via filtration and sedimentation. The typical removal efficiency for pathogens ranges from 90 to 99.999% [128] in air and from 90 to 99% for heavy metals in wetlands [129]. Additionally, nanomaterials remove air pollutants through catalytic processes, achieving removal rates ranging from 55 to 100% for gases like CO, NO, and others [130].

Fabrication of carbon nanomaterials usually involves intricate synthesis techniques that require costly starting materials and high temperatures [34]. However, the challenge lies in scaling up production while keeping the costs low. Moreover, even slight variations in production conditions can impact the properties and performance of these nanomaterials, necessitating strict control over the manufacturing processes. The existing synthesis methods for carbon nanomaterials are often time-consuming and impractical for large-scale production. In order to facilitate commercialization, developing scalable synthesis techniques is of paramount significance that can maintain the quality and performance of the products. Furthermore, the process of functionalizing these nanomaterials for specific applications may present challenges in terms of maintaining product quality and scalability. It is worth noting that carbon nanomaterials possess unique properties that may pose health and safety concerns during production and handling processes. Therefore, it is crucial to establish robust protocols for the safe handling, storage, and characterization of large quantities of carbon nanomaterials to ensure the workers' safety and compliance with regulations. Continuous research and development efforts should be directed toward scalable synthesis techniques, purification methods, and safe handling protocols in order to successfully commercialize and widely adopt carbon nanomaterials. Population growth, demographic changes, urbanization, and climate change are a great threat as it is difficult to meet the needs of drinking water, clean air, and soil free from contaminants. Usage of carbonaceous materials is increasing at a considerable rate as they are effective in removing and treating organic and inorganic environmental pollutants and lead to clean and healthy environments.

Conflicts of Interest

The authors declare no conflict of interest.

Authors' Contributions

Both the authors contributed equally to this chapter and have read and approved the final version.

References

1 Wu, Y., Pang, H., Liu, Y. et al. (2019). Environmental remediation of heavy metal ions by novel nanomaterials: a review. *Environmental Pollution* 24: 6608–6620. https://doi.org/10.1016/j.envpol.2018.12.076.
2 WHO Global Air Quality Guidelines. Particulate Matter (PM2.5 and PM10), Ozone, Nitrogen Dioxide, Sulfur Dioxide and Carbon Monoxide. Geneva: World Health Organization 2021. http://iris.who.int/bitstream/handle/10665/345329/9789240034228-eng.pdf?sequence=1.
3 World Bank (2022). *The Global Health Cost of PM2.5 Air Pollution: A Case for Action Beyond 2021*. World Bank Publications https://doi.org/10.1596/978-1-4648-1816-5.

4 UN-Water. (2021). Summary progress update 2021: SDG 6—water and sanitation for all. https://www.unwater.org/sites/default/files/app/uploads/2021/12/SDG-6-Summary-Progress-Update-2021_Version-July-2021a.pdf.
5 Fu, T., Zhang, B., Gao, X. et al. (2023). Recent progresses challenges, and opportunities of carbon-based materials applied in heavy metal polluted soil remediation. *Science of the Total Environment* 856: 158810. https://doi.org/10.1016/j.scitotenv.2022.158810.
6 Yang, Q., Li, Z., Lu, X. et al. (2018). A review of soil heavy metal pollution from industrial and agricultural regions in China: pollution and risk assessment. *Science of the Total Environment* 642: 690–700. https://doi.org/10.1016/j. scitotenv.2018.06.068.
7 Qureshi, A.S., Hussain, M.I., Ismail, S., and Khan, Q.M. (2016). Evaluating heavy metal accumulation and potential health risks in vegetables irrigated with treated wastewater. *Chemosphere* 163: 54–61. https://doi.org/10.1016/j.chemosphere.2016.07.073.
8 Williams, P.N., Ming, L., Guoxin, S. et al. (2009). Occurrence and partitioning of cadmium, arsenic and Lead in mine impacted Paddy Rice: Hunan, China. *Environmental Science & Technology* 43: 637–642. https://doi.org/10.1021/es802412r.
9 Chibuike, G.U. and Obiora, S.C. (2014). Heavy metal polluted soils: effect on plants and bioremediation methods. *Applied and Environmental Soil Science* 2014: 1–12. 10.1155/2014/752708 FAO and UNEP 2021.
10 FAO and UNEP (2021). *Global Assessment of Soil Pollution: Report*. Rome, Italy: FAO and UNEP https://doi.org/10.4060/cb4894en.
11 Liu, D., Gu, W., Zhou, L. et al. (2022). Recent advances in MOF-derived carbon-based nanomaterials for environmental applications in adsorption and catalytic degradation. *Chemical Engineering Journal* 427: 131503. https://doi.org/10.1021/acscatal.6b01222.
12 Karri, R.R., Mubarak, N.M., Koduru, J.R. et al. (2022). Role of hybrid nanomaterials for a sustainable environment. In: *Hybrid Nanomaterials for Sustainable Applications*, 1–24. San Diego, CA: Elsevier https://doi.org/10.1016/B978-0-323-98371-6.00017-3.
13 Bartke, S., Hagemann, N., Harries, N. et al. (2018). Market potential of nanoremediation in Europe—market drivers and interventions identified in a deliberative scenario approach. *Science of the Total Environment* 619–620: 1040–1048. https://doi.org/10.1016/j.scitotenv.2017.11.215.
14 Cui, L., Ren, X., Sun, M. et al. (2021). Carbon dots: synthesis, properties and applications. *Nanomaterials* 11: 3419. https://doi.org/10.3390/nano11123419.
15 Dai, H. (2002). Carbon nanotubes: synthesis, integration, and properties. *Accounts of Chemical Research* 35 (12): 1035–1044. https://doi.org/10.1021/ar0101640.

16 Huang, Y., Zhu, G., Zou, K. et al. (2021). Highly efficient removal of organic pollutants from wastewater using a recyclable graphene oxide membrane intercalated with g-C_3N_4@TiO_2-nanowires. *Journal of Molecular Liquids* 337: 116461. https://doi.org/10.1016/j.molliq.2021.116461.

17 Gaganpreet, "Enhanced sensitivity of band gap engineered phosphorene towards NH_3 and NO_2 toxic gases", *Applied Surface Science* 2019, 507, 144967, https://doi.org/10.1016/j.apsusc.2019.144967

18 Kroto, H. (1992). C_{60}: buckminsterfullerene, the celestial sphere that fell to earth. *Angewandte Chemie* 31 (2): 111–129. https://doi.org/10.1002/anie.199201113.

19 Oubagaranadin, J. and Murthy, Z.V.P. (2011). *Activated Carbons: Classifications, Properties and Applications*. New York: Elsevier.

20 Zhen, Z. and Zhu, H. (2018). Structure and properties of graphene. In: *Graphene*, 1–12. New York: Academic Press.

21 Novoselov, K.S., Jiang, D., Schedin, F. et al. (2005). Two-dimensional atomic crystals. *Proceedings of the National Academy of Sciences of the United States of America* 102 (30): 10451–10453. https://doi.org/10.1073/pnas.0502848102.

22 Kumar, S., Prakash, H., Verma, J. et al. (2023). A review on properties and environmental applications of graphene and its derivative-based composites. *Catalysts* 13: 111. https://doi.org/10.3390/catal13010111.

23 Liang, X., Allan, S.P.C., Zhang, Y. et al. (2009). Electrostatic force assisted exfoliation of prepatterned few-layer graphenes into device sites. *Nano Letters* 9 (1): 467–472. https://doi.org/10.1021/nl803512z.

24 Viculis, L.M., Mack, J.J., and Kaner, R.B. (2003). A chemical route to carbon nanoscrolls. *Science* 299 (5611): 1361–1362. https://doi.org/10.1126/science.1078842. PMID: 12610297.

25 Park, S., Ruoff, R. S., "Chemical methods for the production of graphenes", *Nature Nanotechnology* 2009, 4, 4, 217–224, https://doi.org/10.1038/nnano.2009.58. Erratum in: Nature Nanotechnology 2010, 5, 4, 309. PMID: 19350030.

26 Li, X., Cai, W., An, J. et al. (2009). Large-area synthesis of high-quality and uniform graphene films on copper foils. *Science* 324 (5932): 1312–1314. https://doi.org/10.1126/science.1171245.

27 Fujita, M., Wakabayashi, K., Nakada, K., and Kusakabe, K. (1920). Peculiar localized state at zigzag graphite edge. *Jornal of Physical Society of Japan* 65 (7): 1996. https://doi.org/10.1143/JPSJ.65.1920.

28 Hernandez, Y., Pang, S., Feng, X., and Müllen, K. (2012). 8.16—graphene and its synthesis. In: *Polymer Science: A Comprehensive Reference* (ed. K. Matyjaszewski and M. Möller), 415–438. Amsterdam, The Netherlands: Elsevier.

29 Chen, L., Hernandez, Y., Feng, X., and Müllen, K. (2012). From nanographene and graphene nanoribbons to graphene sheet chemical synthesis. *Angewandte*

Chemie, International Edition 51: 7640–7654. https://doi.org/10.1002/anie.201201084.
30 Wang, X. and Dai, H. (2010). Etching and narrowing of graphene from the edges. *Nature Chemistry* 2: 661–665. https://doi.org/10.1038/nchem.719.
31 Kosynkin, D.V., Lu, W., Sinitskii, A. et al. (2011). Highly conductive graphene nanoribbons by longitudinal splitting of carbon nanotubes using potassium vapor. *ACS Nano* 5: 968–974. https://doi.org/10.1021/nn102326c.
32 Solorio-Rodriguez, S.A., Williams, A., Poulsen, S.S. et al. (2023). Single-walled vs. multi-walled carbon nanotubes: influence of physico-chemical properties on toxicogenomics responses in mouse lungs. *Nanomaterials* 13 (6): 1059. 10.3390/nano13061059.
33 Burakov, A.E., Galuni, E.V., Burakova, I.V. et al. (2018). Adsorption of heavy metals on conventional and nanostructured materials for wastewater treatment purposes: a review. *Ecotoxicology and Environmental Safety* 148: 702–712. https://doi.org/10.1016/j.ecoenv.2017.11.034.
34 Szabó, A., Perri, C., Csató, A. et al. (2010). Synthesis methods of carbon nanotubes and related materials. *Materials* 3 (5): 3092–3140. https://doi.org/10.3390/ma3053092.
35 Novoselova, I.A. (2008). Electrolytic synthesis of carbon nanotubes from carbon dioxide in molten salts and their characterization. *Physica E Low-Dimensional Systems and Nanostructures* 40 (7): 2231–2237. https://doi.org/10.1016/j.physe.2007.10.069.
36 Jeong, S.H., Ko, J.H., Park, J.B., and Park, W.A. (2004). Sonochemical route to single-walled carbon nanotubes under ambient conditions. *Journal of the American Chemical Society* 126 (49): 15982–15983. https://doi.org/10.1021/ja0451867. PMID: 15584730.
37 Gulati, S., Lingam, B.H.N., Kumar, S. et al. (2022). Improving the air quality with functionalized carbon nanotubes: sensing and remediation applications in the real world. *Chemosphere* 299: 134468. https://doi.org/10.1016/j.chemosphere.2022.134468.
38 Yamada, K., Kim, C.T., Kim, J.H. et al. (2014). Single walled carbon nanotube-based junction biosensor for detection of Escherichia coli. *PLoS One* 9 (9): e105767. https://doi.org/10.1371/journal.pone.0105767.
39 Gupta, V., Tyagi, I., and Sadegh, H. (2015). Nanoparticles as adsorbent; a positive approach for removal of noxious metal ions: a review. *Environmental Science, Chemistry, Materials Science* https://doi.org/10.3923/std.2015.195.214.
40 Gaganpreet, Pathania, Y., "Charged nanoporous phosphorene as a water desalination membrane: insights from molecular dynamics", *Nanotechnology* 2023, 19. https://doi.org/10.1088/1361-6528/ad0504.
41 Pathania, Y. and Gaganpreet (2021). Self-passivated nanoporous phosphorene as a membrane for water desalination. *Desalination* 497: 114777.

42 Mochalin, V.N., Shenderova, O., Ho, D., and Gogotsi, Y. (2011). The properties and applications of nanodiamonds. *Nature Nanotechnology* 7 (1): 11–23. https://doi.org/10.1038/nnano.2011.209. PMID: 22179567.

43 Osswald, S., Yushin, G., Mochalin, V.N. et al. (2006). Control of sp2/sp3 carbon ratio and surface chemistry of nanodiamond powders by selective oxidation in air. *Journal of the American Chemical Society* 128 (35): 11635–11642. https://doi.org/10.1021/JA063303N.

44 Auciello, O., Renou, S., Kang, K. et al. (2022). A biocompatible ultrananocrystalline diamond (UNCD) coating for a new generation of dental implants. *Nanomaterials (Basel, Switzerland)* 12 (5): 782. https://doi.org/10.3390/nano12050782.

45 Shevlin, P.B. (2000). Lecture notes on fullerene chemistry: a handbook for chemists by Roger Taylor (University of Sussex). Imperial College Press: London 1999, xiv + 268 pp. *Journal of the American Chemical Society* 122 (19): 4846. https://doi.org/10.1021/ja9957635.

46 Krusic, P. J., Wasserman, E., Keizer, P.N., Morton, J. R, Preston, K F. "Radical reactions of C_{60}", *Science* 1991, 254, 5305, 1183–1185, https://doi.org/10.1126/science.254.5035.1183

47 Laurila, T., Sainio, S., and Miguel, A. (2017). Hybrid carbon based nanomaterials for electrochemical detection of biomolecules. *Progress in Materials Science* 88: 499–594. https://doi.org/10.1016/j.pmatsci.2017.04.012.

48 Ricardo, A.I.C., Bernardo, F.J.G., Zougagh, M. et al. (2017). Magnetic nanoparticles—carbon nanotubes hybrid composites for selective solid-phase extraction of polycyclic aromatic hydrocarbons and determination by ultra-high performance liquid chromatography. *Analytical and Bioanalytical Chemistry* 409: 5125–5132. https://doi.org/10.1007/s00216-017-0459-7.

49 Pan, B. and Xing, B. (2008). Adsorption mechanisms of organic chemicals on carbon nanotubes. *Environmental Science & Technology* 42 (24): 9005–9013. https://doi.org/10.1021/es801777n.

50 Singh, R.K., Patel, K.D., Kim, J. et al. (2014). Multifunctional hybrid nanocarrier: magnetic CNTs ensheathed with mesoporous silica for drug delivery and imaging system. *ACS Applied Materials & Interfaces* 6 (4): 2201–2208. 10.1021/am4056936.

51 Cobos, M., De-La-Pinta, I., Quindós, G. et al. (2020). Graphene oxide–silver nanoparticle nanohybrids: synthesis, characterization, and antimicrobial properties. *Nanomaterials* 10 (2): 376. https://doi.org/10.3390/nano10020376.

52 De Medeiros, A.M.Z., Khan, L.U., Da Silva, G.H. et al. (2021). T graphene oxide-silver nanoparticle hybrid material: an integrated nanosafety study in zebrafish embryos. *Ecotoxicology and Environmental Safety* 209: 111776. https://doi.org/10.1016/j.ecoenv.2020.111776.

53 Ahmad, S.Z.N., Salleh, W.N.W., Ismail, A.F. et al. (2020). Adsorptive removal of heavy metal ions using graphene-based nanomaterials: toxicity, roles of functional groups and mechanisms. *Chemosphere* 248: 126008. https://doi.org/10.1016/j.chemosphere.2020.126008.

54 Kumari, S., Sharma, P., Yadav, S. et al. (2020). A novel synthesis of the graphene oxide-silver (GO-AG) nanocomposite for unique physiochemical applications. *ACS Omega* 5 (10): 5041–5047. https://doi.org/10.1021/acsomega.9b03976.

55 Choudhary, P., Kumari Ola, S., Chopra, I. et al. (2022). Metal–organic framework (MOF)/graphene–oxide (GO) nanocomposites materials: a potential formulation for anti-corrosive coatings- a review. *Materials Today Proceedings* 79: 172–178. https://doi.org/10.1016/j.matpr.2022.09.603.

56 Yan, C., Jin, J., Wang, J. et al. (2022). Metal–organic frameworks (MOFs) for the efficient removal of contaminants from water: underlying mechanisms, recent advances, challenges, and future prospects. *Coordination Chemistry Reviews* 468: 214595. https://doi.org/10.1016/j.ccr.2022.214595.

57 Gao, M., Liu, G., Gao, Y. et al. (2021). Recent advances in metal-organic frameworks/membranes for adsorption and removal of metal ions. *Trends in Analytical Chemistry* 137: 116226. https://doi.org/10.1016/j.trac.2021.116226.

58 Dariyal, P., Sharma, S., Chauhan, G.S. et al. (2021). Recent trends in gas sensing via carbon nanomaterials: outlook and challenges. *Nanoscale Advances* 3 (23): 6514–6544. https://doi.org/10.1039/d1na00707f.

59 Sacco, L., Forel, S., Florea, I., and Cojocaru, C.S. (2020). Ultra-sensitive NO_2 gas sensors based on single-wall carbon nanotube field effect transistors: monitoring from ppm to ppb level. *Carbon* 157: 61–639. https://doi.org/10.1016/j.carbon.2019.10.073.

60 Hou, Z., Wu, J., Zhou, W. et al. (2007). A MEMS-based ionization gas sensor using carbon nanotubes. *IEEE Transactions on Electron Devices* 54: 1545–1548.

61 Chadha, G. and Priyanka, C. (2018). Enhanced CO_2 adsorption on doped Au_{32} gold nanocages: a density functional approach. *Materials Research Express* 5: 065038. https://doi.org/10.1088/2053-1591/aaca82.

62 Varghese, O.K., Kichambre, P.D., Gong, D. et al. (2001). Gas sensing characteristics of multi-wall carbon nanotubes. *Sensors and Actuators B: Chemical* 81 (1): 32–41. https://doi.org/10.1016/S0925-4005(01)00923-6.

63 Lee, S.W., Lee, W., Lee, D. et al. (2018). A simple and disposable carbon adhesive tape-based NO_2 gas sensor. *Sensors and Actuators B: Chemical* 266: 485–492. https://doi.org/10.1016/j.snb.2018.03.161.

64 Schütt, F., Postica, V., Adelung, R., and Lupan, O. (2017). Single and networked ZnO-CNT hybrid Tetrapods for selective room-temperature high-performance ammonia sensors. *ACS Applied Materials & Interfaces* 9 (27): 23107–23118. https://doi.org/10.1021/acsami.7b03702.

65 Sawalha, S., Moulaee, K., Nocito, G. et al. (2021). Carbon-dots conductometric sensor for high performance gas sensing. *Carbon Trends* 5: 100105. https://doi.org/10.1016/j.cartre.2021.100105.

66 Saleem, H., Zaidi, S.J., Ismail, A.F., and Goh, P.S. (2021). Advances of nanomaterials for air pollution remediation and their impacts on the environment. *Chemosphere* 287: 132083. https://doi.org/10.1016/j.chemosphere.2021.132083.

67 Wang, Z., Zhang, T., Zhao, C. et al. (2018). Anchoring ultrafine Pd nanoparticles and SnO_2 nanoparticles on reduced graphene oxide for high-performance room temperature NO_2 sensing. *Journal of Colloid and Interface Science* 514: 599–608. https://doi.org/10.1016/j.jcis.2017.12.075.

68 Zhang, H., Li, Q., Huang, J. et al. (2016). Reduced graphene oxide/au nanocomposite for NO_2 sensing at low operating temperature. *Sensors* 16 (7): 1152. https://doi.org/10.3390/s16071152.

69 Jung, J.H., Hwang, G.B., Lee, J.E., and Bae, G.N. (2011). Preparation of airborne ag/CNT hybrid nanoparticles using an aerosol process and their application to antimicrobial air filtration. *Langmuir* 27 (16): 10256–10264. https://doi.org/10.1021/la201851r.

70 Miyawaki, J., Lee, G., Yeh, J. et al. (2012). Development of carbon-supported hybrid catalyst for clean removal of formaldehyde indoors. *Catalysis Today* 185 (1): 278–283. https://doi.org/10.1016/j.cattod.2011.09.036.

71 Njewa, J.B., Vunain, E., and Biswick, T. (2022). Synthesis and characterization of activated carbons prepared from agro-wastes by chemical activation. *Journal of Chemistry* 9975444. https://doi.org/10.1155/2022/9975444.

72 Cui, Y., Liu, Q., He, Z. et al. (2020). Epitaxial graphene gas sensors on SiC substrate with high sensitivity. *Journal of Semiconductors* 41: 032101. https://doi.org/10.1088/1674-4926/41/3/032101.

73 Karmakar, N.S., Fernandes, R.P., Jain, S. et al. (2017). Room temperature NO_2 gas sensing properties of p-toluenesulfonic acid doped silver-polypyrrole nanocomposite. *Sensors and Actuators B: Chemical* 242: 118–126. https://doi.org/10.1016/j.snb.2016.11.039.

74 Chen, T., Yan, W., Xu, J. et al. (2019). Highly sensitive and selective NO_2 sensor based on 3D MoS_2/rGO composites prepared by a low temperature self-assembly method. *Journal of Alloys and Compounds* 793: 541–551. https://doi.org/10.1016/j.jallcom.2019.04.126.

75 Krounbi, L., Enders, A., Anderton, C.R. et al. (2020). Sequential ammonia and carbon dioxide adsorption on pyrolyzed biomass to recover waste stream nutrients. *ACS Sustainable Chemistry & Engineering* 8 (18): 7121–7131. https://doi.org/10.1021/acssuschemeng.0c01427.

76 Kamran, U., Heo, Y.J., Lee, J.W., and Park, S.J. (2019). Functionalized carbon materials for electronic devices: a review. *Micromachines* 10 (4): 234. https://doi.org/10.3390/mi10040234C21.

77 Luo, M., Liang, Z., Peera, S.G. et al. (2020). Theoretical study on the adsorption and predictive catalysis of MnN_4 embedded in carbon substrate for gas molecules.

Applied Surface Science 525: 146480. https://doi.org/10.1016/j.apsusc.2020.146480.

78 Sethupathi, S., Zhang, M., Rajapaksha, A.U. et al. (2017). Biochars as potential adsorbers of CH_4, CO_2 and H_2S. *Sustainability* 9 (109): 121–131.

79 Fan, L., Ling, L., Wang, B., and Zhang, R. (2020). The adsorption of mercury species and catalytic oxidation of hg° on the metal-loaded activated carbon. *Applied Catalysis A: General* 520: 13–23. https://doi.org/10.1016/j.apcata.2016.03.036.

80 Xu, H., Qu, Z., Huang, W. et al. (2015). Regenerable ag/graphene sorbent for elemental mercury capture at ambient temperature. *Colloid Surface A* 476: 83–89. https://doi.org/10.1016/j.colsurfa.2015.03.027.

81 Xie, Y., Li, C., Zhao, L. et al. (2015). Experimental study on Hg^0 removal from flue gas over columnar MnO_x—CeO_2 activated coke. *Applied Surface Science* 333: 59–67. https://doi.org/10.1016/j.apsusc.2015.01.234.

82 Wu, H., Li, C., Zhao, L. et al. (2015). Removal of gaseous elemental mercury by cylindrical activated coke loaded with CoO_x—CeO_2 from simulated coal combustion flue gas. *Energy and Fuels* 29: 6747–6757. https://doi.org/10.1021/acs.energyfuels.5b00871.s001.

83 Xu, W., Adewuyi, Y.G., and Liu, Y.X. (2018). Removal of elemental mercury from flue gas using CuO_x and CeO_2 modified rice straw chars enhanced by ultrasound. *Fuel Processing Technology* 170: 21–31. https://doi.org/10.1016/j.fuproc.2017.10.017.

84 Zhang, Y., Yuan, S., Feng, X. et al. (2016). Preparation of nanofibrous metal–organic framework filters for efficient air pollution control. *Journal of the American Chemical Society* 138: 5785–5788. https://doi.org/10.1021/jacs.6b02553.

85 Prezioso, S., Perrozzi, F., Giancaterini, L. et al. (2013). Graphene oxide as a practical solution to high sensitivity gas sensing. *Journal of Physical Chemistry C* 117 (20): 10683–10690. https://doi.org/10.1021/jp3085759.

86 Gao, J., Liu, M., Song, H. et al. (2016). Highly-sensitive electrocatalytic determination for toxic phenols based on coupled cMWCNT/cyclodextrin edge-functionalized graphene composite. *Journal of Hazardous Materials* 318: 99–108. https://doi.org/10.1016/j.jhazmat.2016.06.051.

87 Yang, X., Zhou, T., Ren, B. et al. (2018). Removal of Mn (II) by sodium alginate/graphene oxide composite double-network hydrogel beads from aqueous solutions. *Scientific Reports* 8 (110): 717. https://doi.org/10.1038/s41598-018-29133-y.

88 Bali, M. and Tlili, H. (2018). Removal of heavy metals from wastewater using infiltration percolation process and adsorption on activated carbon. *International Journal of Environmental Science and Technology* 16 (1): 249–258. https://doi.org/10.1007/s13762-018-1663-5.

89 Fu, F. and Wang, Q. (2011). Removal of heavy metal ions from wastewaters: a review. *Journal of Environmental Management* 92 (3): 407–418. https://doi.org/10.1016/j.jenvman.2010.11.011.

90 Younas, F., Mustafa, A., Farooqi, Z.U.R. et al. (2021). Current and emerging adsorbent technologies for wastewater treatment: trends, limitations, and environmental implications. *Water* 13 (2): 215. https://doi.org/10.3390/w13020215.

91 Sabzehmeidani, M.M., Mahnaee, S., Ghaedi, M. et al. (2021). Carbon based materials: a review of adsorbents for inorganic and organic compounds. *Materials Advances* 2 (2): 598–627. https://doi.org/10.1039/d0ma00087f.

92 Tang, W., Zeng, G., Gong, J. et al. (2014). Impact of humic/fulvic acid on the removal of heavy metals from aqueous solutions using nanomaterials: a review. *Science of the Total Environment* 468–469: 1014–1027. https://doi.org/10.1016/j.scitotenv.2013.09.044.

93 Freixa, A., Acuña, V., Gutierrez, M. et al. (2018). Fullerenes influence the toxicity of organic micro-contaminants to river biofilms. *Frontiers in Microbiology* 9: 1426. https://doi.org/10.3389/fmicb.2018.01426.

94 Brunet, L., Lyon, D.Y., Hotze, E.M. et al. (2009). Comparative photoactivity and antibacterial properties of C60 fullerenes and titanium dioxide nanoparticles. *Environmental Science & Technology* 43 (12): 4355–4360. https://doi.org/10.1021/es803093t.

95 Pickering, K.D. (2005). Photochemistry and environmental applications of water soluble fullerene compounds. In: *Civil and Environmental Engineering*, 113. Houston, TX: Rice University.

96 Mostafavi, S., Mehrnia, M., and Rashidi, A. (2009). Preparation of nanofilter from carbon nanotubes for application in virus removal from water. *Desalination* 238 (1–3): 271–280. https://doi.org/10.1016/j.desal.2008.02.018.

97 Anitha, K., Namsani, S., and Singh, J.K. (2015). Removal of heavy metal ions using a functionalized single-walled carbon nanotube: a molecular dynamics study. *The Journal of Physical Chemistry. A* 119 (30): 8349–8358. https://doi.org/10.1021/acs.jpca.5b03352.

98 Gupta, V.K., Agarwal, S., and Saleh, T.A. (2011). Synthesis and characterization of alumina-coated carbon nanotubes and their application for lead removal. *Journal of Hazardous Materials* 185 (1): 17–23. https://doi.org/10.1016/j.jhazmat.2010.08.053.

99 Li, Y.H., Wang, S., Wei, J. et al. (2002). Lead adsorption on carbon nanotubes. *Chemical Physics Letters* 357 (3): 263–266. https://doi.org/10.1016/S0009-2614(02)00502-X.

100 Shorie, M., Kaur, H., Chadha, G. et al. (2019). Graphitic carbon nitride QDs impregnated biocompatible agarose cartridge for removal of heavy metals from contaminated water samples. *Journal of Hazardous Materials* 367: 629–638. https://doi.org/10.1016/j.jhazmat.2018.12.115.

101 Bhunia, S., Ghorai, N., Burai, S. et al. (2021). Unraveling the carrier dynamics and photocatalytic pathway in carbon dots and pollutants of wastewater system. *Journal of Physical Chemistry C* 125 (49): 27252–27259. https://doi.org/10.1021/acs.jpcc.1c06135.

102 Akbar, K., Moretti, E., and Vomiero, A. (2021). Carbon dots for photocatalytic degradation of aqueous pollutants: recent advancements. *Advanced Optical Materials* 9 (17): 2100532. https://doi.org/10.1002/adom.202100532.

103 Tuna, A.Ö.A., Özdemir, E., Şimşek, E.B., and Beker, Ü. (2013). Removal of As(V) from aqueous solution by activated carbon-based hybrid adsorbents: impact of experimental conditions. *Journal of Chemical Engineering* 223: 116–128. https://doi.org/10.1016/j.cej.2013.02.096.

104 Gamage, A., Punniamoorthy, T., and Madhujith, T. (2022). *Starch-Based Hybrid Nanomaterials for Environmental Remediation*. London, UK: IntechOpen https://doi.org/10.5772/intechopen.101697.

105 Ersan, G., Apul, O.G., Perreault, F., and Karanfil, T. (2017). Adsorption of organic contaminants by graphene nanosheets: a review. *Water Research* 126: 385–398. https://doi.org/10.1016/j.watres.2017.08.01.

106 Hammer, T., Sachinidou, P., He, X. et al. (2020). Advanced filtration and lung deposition models of airborne multi-walled carbon nanotubes for inhalation exposure assessment. *NanoImpact* 19: 100240. https://doi.org/10.1016/j.impact.2020.100240.

107 Aslam, M.M., Kuo, H.W., Den, W. et al. (2021). Functionalized carbon nanotubes (CNTs) for water and wastewater treatment: preparation to application. *Sustainability* 13 (10): https://doi.org/10.3390/su13105717.

108 Liu, Q., Ren, J., Lu, Y. et al. (2021). A review of the current in-situ fouling control strategies in MBR: biological versus physicochemical. *Journal of Industrial and Engineering Chemistry* 98: 42–59. https://doi.org/10.1016/j.jiec.2021.03.042.

109 Ibrahim, Y., Wadi, V.S., Ouda, M. et al. (2022). Highly selective heavy metal ions membranes combining sulfonated polyethersulfone and self-assembled manganese oxide nanosheets on positively functionalized graphene oxide nanosheets. *Journal of Chemical Engineering* 428: 131267. https://doi.org/10.1016/j.cej.2021.131267.

110 Bleyl, S., Kopinke, F., and Mackenzie, K. (2012). Carbo-iron®—synthesis and stabilization of Fe(0)-doped colloidal activated carbon for in situ groundwater treatment. *Journal of Chemical Engineering* 191: 588–595. https://doi.org/10.1016/j.cej.2012.03.021.

111 Premarathna, K., Rajapaksha, A.U., Sarkar, B. et al. (2019). Biochar-based engineered composites for sorptive decontamination of water: a review.

Chemical Engineering Journal 372: 536–550. https://doi.org/10.1016/j.cej.2019.04.097.

112 El-Naggar, A., Ahmed, N., Mosa, A. et al. (2021). Nickel in soil and water: sources, biogeochemistry, and remediation using biochar. *Journal of Hazardous Materials* 419: 126421. https://doi.org/10.1016/j.jhazmat.2021.126421.

113 Liu, K., Li, F., Cui, J. et al. (2020). Simultaneous removal of cd(II) and as(III) by graphene-like biochar-supported zero-valent iron from irrigation waters under aerobic conditions: synergistic effects and mechanisms. *Journal of Hazardous Materials* 395: 122623. https://doi.org/10.1016/j.jhazmat.2020.122623.

114 Wang, S., Gao, B., Li, Y. et al. (2017). Adsorptive removal of arsenate from aqueous solutions by biochar supported zero-valent iron nanocomposite: batch and continuous flow tests. *Journal of Hazardous Materials* 322: 172–181. https://doi.org/10.1016/j.jhazmat.2016.01.052.

115 Fang, L., Borggaard, O.K., Marcussen, H. et al. (2010). The pH-dependent adsorption of tributyltin to charcoals and soot. *Environmental Pollution* 158: 3642–3649. https://doi.org/10.1016/j.envpol.2010.08.003.

116 Hoch, M. (2001). Organotin compounds in the environment—an overview. *Applied Geochemistry* 16 (7–8): 719–743. https://doi.org/10.1016/S0883-2927(00)00067-6.

117 Qiu, M., Liu, L., Ling, Q. et al. (2022). Biochar for the removal of contaminants from soil and water: a review. *Biochar* 4 (1): 19. https://doi.org/10.1007/s42773-022-00146-1.

118 Abega, A.V., Ngomo, H.M., Nongwe, I. et al. (2019). Easy and convenient synthesis of CNT/TiO_2 Nanohybrid by in-surface oxidation of Ti^{3+} ions and application in the photocatalytic degradation of organic contaminants in water. *Synthetic Metals* 251: 1–14. https://doi.org/10.1016/j.synthmet.2019.03.012.

119 Vilardi, G., Mpouras, T., Dermatas, D. et al. (2018). Nanomaterials application for heavy metals recovery from polluted water: the combination of Nano zero-valent iron and carbon nanotubes. Competitive adsorption non-linear Modeling. *Chemosphere* 201: 716–729. https://doi.org/10.1016/j.chemosphere.2018.03.032.

120 Akinpelu, A.A., Ali, M.E., Johan, M.R. et al. (2019). Polycyclic aromatic hydrocarbons extraction and removal from wastewater by carbon nanotubes: a review of the current technologies, challenges and prospects. *Process Safety and Environment Protection* 122: 68–82. https://doi.org/10.1016/j.psep.2018.11.006.

121 Xiong, T., Yuan, X., Wang, H. et al. (2018). Implication of graphene oxide in cd-contaminated soil: a case study of bacterial communities. *Journal of Environmental Management* 205: 99–106. https://doi.org/10.1016/j.jenvman.2017.09.067.

122 Hankins, N.P., Lu, N., and Hilal, N. (2006). Enhanced removal of heavy metal ions bound to humic acid by polyelectrolyte flocculation. *Separation and Purification Technology* 51 (1): 48–56. https://doi.org/10.1016/j.seppur.2005.12.022.

123 Zhu, S., Ho, S.-H., Huang, X. et al. (2017). Magnetic nanoscale zerovalent iron assisted biochar: interfacial chemical behaviors and heavy metals remediation performance. *ACS Sustainable Chemistry & Engineering* 5 (11): 9673–9682. https://doi.org/10.1021/acssuschemeng.7b00542.

124 Clifford, C.A., Stinz, M., Hodoroaba, V.-D. et al. (2020). International standards in nanotechnologies. In: *Characterization of Nanoparticles: Measurement Processes for Nanoparticles*, 1e (ed. V.-D. Hodoroaba, W.E.S. Unger, and A.G. Shard), 511–526. Amsterdam, The Netherlands: Elsevier https://doi.org/10.1016/B978-0-12-814182-3.00026-2.

125 Singh, A.K. and Chandra, R. (2019). Pollutants released from the pulp paper industry: aquatic toxicity and their health hazards. *Aquatic Toxicology* 211: 202–216. https://doi.org/10.1016/j.aquatox.2019.04.007.

126 Madannejad, R., Shoaie, N., Jahanpeyma, F. et al. (2019). Toxicity of carbon-based nanomaterials: reviewing recent reports in medical and biological systems. *Chemico-Biological Interactions* 307: 206–222. https://doi.org/10.1016/j.cbi.2019.04.036.

127 Cardoso, C.E., Almeida, J.C., Lopes, C.B. et al. (2019). Recovery of rare earth elements by carbon-based nanomaterials—a review. *Nanomaterials* 9 (6): 814. https://doi.org/10.3390/nano9060814.

128 Jangra, R., Ahlawat, K., Dixit, A., and Prakash, R. (2023). Efficient deactivation of aerosolized pathogens using a dielectric barrier discharge based cold-plasma detergent in environment device for good indoor air quality. *Scientific Reports* 13. https://doi.org/10.1038/s41598-023-37014-2.

129 Fu, G., Zhou, S., Zhao, Y. et al. (2022). Removal efficiency of heavy metals such as lead and cadmium by different substrates in constructed wetlands. *Processes* 10 (12): 2502. https://doi.org/10.3390/pr10122502.

130 Manimegalai, S., Sundaram, V.V., Santhana, R.D. et al. (2023). Carbon-based nanomaterial intervention and efficient removal of various contaminants from effluents a review. *Chemosphere* 312: 137319. https://doi.org/10.1016/j.chemosphere.2022.137319.

15

Carbon Nanomaterials for Detection and Degradation of Wastewater Inorganic Pollutants: Present Status and Future Prospects

Prem Rajak[1], Ruchika Agarwal[1], Sohini Goswami[1], Satadal Adhikary[2], Suchandra Bhattacharya[3], Abhratanu Ganguly[1], and Sayantani Nanda[1]

[1] *Department of Animal Science, Kazi Nazrul University, Asansol, West Bengal, India*
[2] *Post Graduate Department of Zoology, A. B. N. Seal College, Cooch Behar, West Bengal, India*
[3] *Department of Chemistry, A. B. N. Seal College, Cooch Behar, West Bengal, India*

15.1 Introduction

In recent years, there has been a notable surge in consumer demand for commercially available dyes and associated products. To color the products, industries like paper, textile, and leather manufacturers use various organic dyes, both natural and synthetic, and then release the waste into water sources [1]. Pharmaceuticals, dioxins, chemical pesticides, herbicides, phenols, microbes, and textile dyes have been found in effluents from wastewater discharge. Consequently, the issue of water pollution has emerged as a pervasive worldwide concern, which significantly hampers the prospects of achieving sustainable development. Numerous traditional approaches have been devised for the treatment of wastewater, including, but not limited to, sedimentation, ultrafiltration (UF), reverse osmosis (RO), ion exchange, flocculation, adsorption, oxidation, and advanced oxidation processes (AOPs) [2]. Membrane-based filtration techniques are widely used to treat industrial and municipal wastewater. Membrane-based filtration techniques can be categorized as microfiltration (MF), UF, nanofiltration (NF), and RO based on the size of the membrane pores. These membranes are typically employed to eliminate contaminants of varying sizes. MF membranes are effective for removing particles with a size range of 100–1000 nm, UF membranes are suitable for

Carbon-based Nanomaterials for Green Applications, First Edition. Edited by Upendra Kumar, Piyush Kumar Sonkar, and Suman Lata Tripathi.

Published 2025 by John Wiley & Sons, Inc.

particles sized between 5 and 100 nm, and NF membranes are designed for particles in size range of 1–5 nm. The application of polymer-supported ultrafiltration (PSU) has also been used for the treatment of wastewater containing heavy metals. Use of different adsorbents can enhance the efficacy of heavy metal removal in membrane-based filtration techniques. Biosorption is also effective in eliminating heavy metals, by binding them to inactive microbial biomass through adsorption and chelation mechanisms. Membrane bioreactors (MBRs) are another advanced approach that utilizes the advantages of both membrane filtration and activated sludge processes (ASPs) to improve the quality of treated effluent. This is accomplished by integrating biological approach with membrane separation. Bioadsorbents such as rice husk, coconut shells, native algae, native fruit peels, leaves, roots, bark of native plants, and recycled paper sludge have also emerged as highly effective substitutes for eliminating inorganic pollutants from wastewater (Figure 15.1). They are being used to remove heavy metals and dyes from water bodies. AOPs are chemical treatments that use hydroxyl radicals ($OH^{\cdot}$) to effectively oxidize various pollutants. $OH^{\cdot}$ exhibits a notable oxidation potential and with significantly faster reaction rate compared to commonly used oxidants such

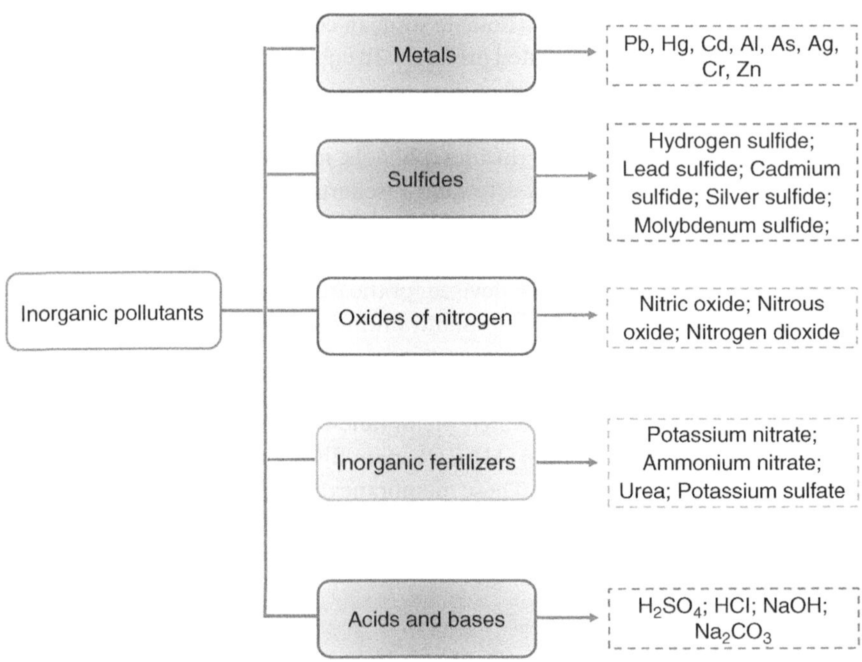

Figure 15.1 Inorganic pollutants and their categories.

as chlorine, oxygen, ozone, H_2O_2, or potassium permanganate. Therefore, the interaction between $OH^{\cdot}$ and both inorganic and organic solutes can produce effective results. AOPs utilize various oxidants to break down numerous pollutants by transforming them from their reduced states into their ultimately harmless oxidized states. While this system employs various oxidants, they all generally generate the same radicals. These techniques have a significant capacity to cleanse water from pollutants that are difficult to eliminate using biological means. All these techniques are effective. However, they have their own drawbacks such as cost inefficiency, time consuming, and frequent maintenance.

AOPs, including the Fenton reaction and photocatalytic processes, are among the conventional techniques that have drawn significant attention due to their numerous benefits. The advantages encompass high efficacy, low operating expenses, straightforward design, ambient temperature and pressure operation, and sustainability. The Fenton process involves the facile decomposition of H_2O_2, with the sole requirement of a renewable energy source for the purpose of photocatalysis. Fenton's oxidation is a frequently employed, resilient, and practical AOP that finds extensive application in the breakdown of organic contaminants. Application of this technology can serve two purposes: reducing the harmful effects of wastewater or facilitating the degradation of primary contaminants, enabling their entry into the sewer system. In the context of the photocatalytic technique, the use of solid substances, also known as photocatalysts, expedites the pace of reaction, hence facilitating the removal of specific contaminants from water, in conjunction with the utilization of UV-visible light sources. To achieve optimal performance, photocatalysts must possess favorable photon-capturing characteristics while also exhibiting a reduced rate of recombination for the photogenerated charge carriers. The following are the primary prerequisites for oxidation-reduction reactions to effectively facilitate the breakdown of typical organic contaminants, specifically dyes. The need for a potential Fenton catalyst in broad-spectrum commercial treatment for the removal of aqueous contaminants has prompted the examination of carbon-based materials (CBMs). CBMs have gained attention due to their cost-effectiveness, environmentally benign after-effects, and sustainability. The distinctive characteristics of CBMs render them highly sought-after materials [3]. The physical characteristics of substances made from carbon vary depending upon the pattern of the arrangement of carbon atoms. CBMs have made substantial contributions to the field of photocatalysis due to their notable characteristics, including a large surface area, superior conductivity, exceptional stability, extraordinary mechanical strength, and finally, their environmentally benign nature and global accessibility. Given the significance of carbon materials, their antecedents must possess qualities such as cost-effectiveness, efficacy, natural origin, and recyclability. Bio-based raw materials have the potential to be utilized for all the activities mentioned above due to their extensive range of potential compositions and porosity scales [4]. Carbon is found in several forms with different dimensionalities, ranging from zero-dimensional to three-dimensional nanostructures. Certain forms like graphene and its various derivatives have gained considerable interest due to their emerging features, which are harnessed to develop tailored functional CBMs for suitable applications. Although the Fenton and photocatalytic reactions include lower catalytic activity of carbon, CBMs with various morphologies have been found to display enhanced catalytic effects. Classical photocatalysts like ZnO, TiO2, and ZnS have been broadly explored, but due to their large band gap (~3.2 eV), they are now being substituted by other types of photocatalysts like the metal-free catalysts such as carbon-based nanomaterials. With a small band gap, carbon-based nanomaterials show distinctive morphologies, and are non-toxic [5].

As stated earlier, despite the low catalytic activity of carbon in Fenton and photocatalytic reactions, CBMs with various morphologies have been shown to enhance catalytic effects. Carbon-based photocatalysts have acquired significant traction due to their capacity to assemble into diverse dimensions and topologies [6]. Hence, carbon and carbon-based compounds find widespread application in the purification of water through the process of photocatalysis. Several technological advancements have been developed to mitigate the issue of limited efficacy in targeted pollutant elimination. Consequently, numerous scholarly investigations have prioritized the integration of the principles of nanotechnology with the physical and chemical surface modification of CBMs in order to generate customized CBMs capable of effectively addressing various obstacles encountered during pollution remediation. Given the substantial progress in the application of CBMs for the eradication of pollutants, particularly organic ones in aquatic ecosystems, it is crucial to do a thorough investigation to bring together the current understandings in this field of research. Furthermore, a comprehensive evaluation encompassing AOP-type reactions, like the Fenton and photocatalysis, employing CBMs for the elimination of aqueous contaminants is currently lacking. In the context of heterogeneous Fenton systems, various oxidants, including peroxydisulfate (PDS), H_2O_2, and peroxymonosulfate (PMS), have been employed to initiate the activation of graphitic carbon nitride (g-C_3N_4)-based compounds for the purpose of eliminating pollutants in wastewater [7]. The current article primarily concentrates on the discussion and summarization of (1) various carbon-based nanomaterials in zero-dimensional, one-dimensional, and two-dimensional forms; (2) integration of CBMs with Fenton-like reactions utilizing the oxidant H_2O_2, as well as photocatalytic reactions, for the treatment of wastewaters from different sources contaminated by dyes and pharmaceutical products; and (3) the prevailing challenges and further opportunities associated to the utilization of CBMs.

15.2 Properties of Carbon Nanomaterials

Nanomaterials refer to materials that possess a minimum of one external dimension falling within the range of 1–100 nm. They have several applications in biology and medical sciences. Carbon is commonly regarded as one of the most versatile elements in the periodic table, owing to its inherent strength and ability to form bonds with various other elements. At the atomic scale, carbon has a high propensity for polymerization, resulting in the formation of an exceptionally elongated carbon-based chain. Due to the presence of four electrons in its outer electron shell, carbon has the ability to engage in the formation of covalent bonds, which can manifest as single, double, or triple bonds. These bonds can be established with other elements, as well as with other carbon atoms [8]. Carbon-based nanomaterials (CMBs) are notable to be one of the very effective nanomaterials owing to their vast accessibility and lesser production cost paralleled to other nanomaterials. CBMs show variable morphologies, as a result of arrangement of the carbon atoms. Specifically, CBMs have fundamentally contributed toward photograph catalysis in light of their high surface area, extraordinary conductivity, magnificent synthetic security, and exceptional mechanical strength, as well as their ecological agreeableness and boundless accessibility [9]. Carbon exhibits a diverse range of allotropic forms, spanning from zero-dimensional to three-dimensional nanostructures. Notably, certain structures such as graphene and its derivatives are gaining increasing attention, particularly when novel features are discovered and harnessed for the development of specific functional CBMs [10]. Reusability of these carbon nanomaterials is essential because of the expense involved in their production. It's important to know that a nanoparticle with good dispersion is made up of very small particles. Subsequently, when carbon nanomaterials are utilized, an extra partition procedure holds these carbon nanomaterials. Layer filtration is one such conceivable partition technique in which the immobilization of nanomaterials on the film channels facilitates the division cycle. Different types of pollutants that may contaminate water can be detrimental to the health of organisms [11–13]. There have been a plethora of studies that demonstrate that certain nanomaterials, for example, CNTs, graphite, fullerenes, etc., can be reused a few times for water treatment which, in the long run, turn out to be cost-effective [14].

15.3 Common Types of Carbon Nanomaterials

Carbon nanomaterials are classified into different groups based on the arrangement of carbon atoms (Figure 15.2).

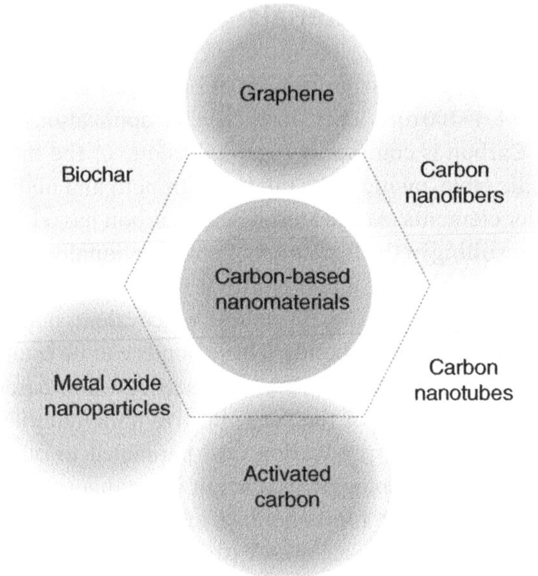

Figure 15.2 Various types of carbon nanomaterials.

15.3.1 Carbon Nanotubes

Carbon nanotubes (CNTs) are chambers formed by macromolecules that span as little as a couple of nanometers and as extensive as 20 cm long. The fundamental building block of the cylinder walls is carbon atoms arranged in a pattern of hexagonal grid, which is identical to the graphite's nuclear planes. CNTs can be single-walled (SWCNTs) or multi walled CNTs (MWCNTs); where SWCNTs are comprised of a solitary layer of carbon sheets while MWCNTs comprise up to many concentric cylinders [15]. Similarly, CNTs comprise of empty and layered structure alongside a huge explicit surface region, which prompts a notable application in removal of heavy metals from the treated water by using the adsorption technique [16]. The prospect of CNTs in the adsorption method for the removal of water contaminants is currently under investigation by researchers. Most natural pollutants are adsorbed onto the external surface and inward parts of open-ended CNTs. The morphology of CNTs is supposed to significantly affect the adsorption of natural pollutants, explicitly for those stacked as the π–π interactive force. Because of the nanoscale curvature effect, the capacities of adsorption of tetracene and phenanthrene on the surface of the tube of CNTs differed significantly [17]. CNTs have drawn extensive interest as building blocks. Polynuclear aromatic hydrocarbons (PAHs), Dioxin, polybrominated diphenylethers (PBDEs), chlorobenzenes and chlorophenols, phthalate esters, trihalomethane, dyes, imidacloprid, thiamethoxam, and acetamiprid, as well as sulfuron derivatives, atrazine, dicamba, nonylphenol, and bisphenol A are the most common organic compounds [18]. All these contaminants like PAHs, pesticides, etc., can pose serious threats to biodiversity. The performance of CNTs is often suboptimal due to impurities generated during their manufacture. The impurities would be carbonaceous products and build-ups from the metal used for the readiness of CNTs. Therefore, CNTs are subjected to an alkaline or acidic solution to get rid of the impurities in a way so that the properties remain unchanged. The charged practical gathering owing to the adsorption of ionic molecules in the CNT surface inhibits nanotube aggregation by the coulombic repulsion forces between modified CNTs. Also, other CBMs have been successfully coated onto CNTs to improve their ability to remove water contaminants. CNTs have effectively ended up being a remarkable adsorbent for the expulsion of pollutants from water which prompts a healthy climate [19].

15.3.2 Carbon Nanofibers

Due to their comprehensive properties of ultra-high specific surface area, chemical resistance, electrical conductivity, high stiffness, and thermal stability, carbon nanofibers (CNFs) have received profound attention in recent times. CNFs are characterized as round and hollow nanostructures with graphene layers built looking like cones, plates, or cups, with a typical measurement of 70–200 nm and a typical size of 50–200 μm [20]. CNFs are typically collected in the form of a nonwoven mat in a random orientation. Other than such two-layered (2D) gathering, 1D and 3D congregations of CNFs would generally expand the spectrum of usage. The most straightforward method for acquiring 3D gathering is to build the assortment time, i.e. increment of thickness of the film. A different but effective method is to scatter electrospun CNFs and afterward acquire aerogels through basic or freeze drying. The 3D designs can be broadly utilized in energy and other applications. CNFs comprise a remarkable adsorption framework, where nonpolar particles can be really eliminated from fluids like water [21]. CNFs exhibit considerable potential as viable options for the development and production of innovative photocatalysts, owing to their unique combination of nanoscale properties and fibrous architectures [22]. Furthermore, the porosity of the structure and content of CNFs can be readily manipulated through the application of suitable synthetic techniques. The electrospinning method has gained prominence as an efficient method for producing CNFs due to its straightforward operation, cost-effectiveness, and compatibility with various carbon precursors. As a result, an extensive amount of study has been dedicated to exploring the application of electrospun CNFs in the domain of photocatalysis [23]. For ZnO-CNFs biosynthesis, herbal concentrates of *Stachys pilifera* Benth (SB), and *Thymus daenensis* (TD) were utilized as both green-lessening specialists and watery carbon-rich sources. Photocatalytic activity of the combined ZnO-CNFs was concentrated in the photodegradation of antibiotics and anti-microbial deposits in contaminated water.

15.3.3 Graphitic Carbon Nitride

A promising technology is membrane separation, which can effectively remove an array of prevailing pollutants from water while using little energy and leaving a minimal carbon footprint. For water and wastewater treatment, graphitic carbon nitride (g-C_3N_4) has received a lot of attention especially for the removal of heavy metals, phenolic compounds, organic dyes, oils, pharmaceuticals etc. Due to its distinctive structure and outstanding catalytic activity, g-C_3N_4 has emerged as a promising membrane material. Polymeric g-C_3N_4 having a structure resembling graphene is getting increasing attention in environmental remediation due to its excellent physicochemical stability, easy and inexpensive synthesis, and unique catalytic activities. g-C_3N_4 is a polymeric-based material that comprises for the most part of C and N and is efficiently connected by very much arranged tris-triazine-based designs [24]. g-C_3N_4 has a 2D structure of layers, with weak van der Waals interactions holding the adjacent layers together [25]. g-C_3N_4 is a 2D carbon-based metal-free polymer (conjugated) that has six nitrogen lone-pair electrons in its structure. The aforementioned characteristic renders it a favorable medium for the immobilization and dispersion of metals, hence attracting significant interest from researchers worldwide [26]. The oxygenated functional groups found in the one-of-a-kind structure of g-C_3N_4 aid in the reduction of H_2O_2 into the highly reactive $HO^{\cdot}$ radicals. The six nitrogen solitary sets of electrons that are accessible in g-C_3N_4 go about as electron donor, and the adjoining carbon atom goes about as the electron acceptor. Sadly, the Fenton-like response of immaculate g-C_3N_4 for wastewater treatment is restricted in view of its feeble redox capacity, unfortunate electron transport limit, and low surface area. Consequently, g-C_3N_4's chemical properties are enhanced and its application is encouraged by metal doping. Additionally, the self-Fenton system, which is applied to get rid of the refractory pollutants, enables g-C_3N_4 to produce high amounts of H_2O_2 on its own in the presence of sufficient oxygen [27]. Furthermore, due to its unique blend of organic and inorganic properties, g-C_3N_4 is often regarded as an ideal matter for the development of photocatalytic membranes. Under the visible light, photocatalytic membranes that have been enhanced with g-C_3N_4-based nanomaterials, including g-C_3N_4 nanoparticles and nanocomposites, exhibit remarkable efficacy in the degradation of pharmaceutical compounds present in wastewater [28].

15.3.4 Activated Carbon

Activated carbon (AC) is a classic and extensively employed adsorbent for the elimination of inorganic and organic contaminants from water and wastewater. The material is commonly employed due to its superior internal microporosity, significantly large surface area (500–1500 $m^2 g^{-1}$), and diverse array of functional groups on the surface. However, there is increasing scientific attention to the field of ACs, which stems from the requirement to enhance the affinity of ACs toward certain contaminants, hence facilitating their removal from various water sources [29]. The adsorption process's use of AC is largely determined by porous carbons' surface chemistry and pore structure. Oxygen, nitrogen, sulfur, and hydrogen are by and large present in AC as useful gatherings or potential atoms, chemically clung to the construction. Carbonyl, carboxyl, lactones, quinones, and phenols are among the main functional groups in their structure that are thought to be accountable for the uptake of pollutants. These functional groups may have a major role in the particular adsorption properties of AC [30]. It has been widely reported that ACs work well as adsorbents for a variety of pollutants [31]. It is common knowledge that AC is more effective than metals and other inorganic pollutants at removing organic compounds. By utilizing various chemicals or the appropriate treatment methods, efforts are ongoing to significantly enhance the carbon surface's ability to remove specific contaminants from the aqueous phase [32]. Carbon's chemical and physical structure can be altered through a variety of means, including activation conditions (like temperature, different agents, and duration of the process), presence of other additives, among many other things [33].

15.3.5 Nanoporous Carbons

As environmentally friendly and non-toxic functional materials, nanoporous carbons (NPCs) hold great promise for addressing the environmental crisis and energy scarcity. Lately, the creation of adsorbent in light of biomaterials, for example, farming waste has been expanding because of the plentiful asset of the horticultural waste. Along with their abundance, procurement of these agricultural items is inexpensive, and the expenditure of producing an adsorbent by utilizing agricultural waste is low. Then again, these biomaterials are profoundly viable at low metal focuses and it is effectively reusable. Subsequently, these biomaterials will undergo compression and amalgamation to form a carbon-based adsorbent, which can be employed in the purification method for water. The carbon-based adsorbent developed from biomaterials like sawdust, crab shells, waste from olive trees, and AC materials [34]. These adsorbents, based on carbon showed a decent adsorption limit on the expulsion of impurities water. Yet the effectiveness was not sufficiently satisfactory. Consequently, the adsorption limit of this actuated vehicle can be expanded by expanding the particular surface area or decreasing the molecule size. Therefore, this enacted carbon can change into nanoporous actuated carbon by incorporating fictionalizations to introduce greater efficacy of expulsion of pollutants from target water. NPCs were made by heating zeolitic imidazolate frameworks based on zinc and cobalt (ZIF-8/ZIF-67) and their bimetallic ZIFs. They are used for energy storage, environmental remediation, and conversion. Pure ZIFs had high phenol adsorption capacities for adsorption-based wastewater treatment. Orchestrated Co/NPCs from ZIF-67 are showing phenomenal adsorption of dyes in wastewater. Pollutant removal via adsorption or photocatalysis was also investigated for carbons and carbon-ZnO hybrids from ZIF-8 [35]. According to reports, AC Fibers (ACFs) were employed for the purpose of disinfecting pathogens such as *E.coli* in drinking water. The adhesion of *E.coli* onto ACF was found to be more efficient compared to other carbon materials or granular AC.

15.3.6 Graphene in Wastewater Treatment

As a result of its novel 2D structure and tremendous thermal, electrical, and mechanical properties, graphene was also used in the study to demonstrate its ability to adsorb contaminants from water. Graphene oxide (GO) nanosheets which were synthesized from graphite using the adjusted Hummers strategy were utilized as the adsorbent for expulsion of Cd^{2+} and Co^{2+} from huge volumes of fluid arrangements. Graphene, GO, reduced GO, and peeled GO called graphene subordinates have emerged as promising candidates for the treatment of wastewater due to their exceptional characteristics [36]. As a result of its greater electrical conductivity, large surface area, and cost-effectiveness during mass production, graphene has been thoroughly investigated as a potential photocatalyst. Additionally, because the valence band (VB) and conduction band (CB) of graphene contain bonding and anti-bonding (*) orbitals that touch the corners of the Brillouin zone, graphene monolayer is a zero-bandgap semiconductor [37]. Co-catalysts in composites, standalone photocatalysts, and supporting materials for semiconductor catalysts are all used for graphene and its various derivatives. For water treatment, graphene and its derivatives work well when combined with conventional photocatalysts made of metals and metal or nonmetal oxides, phosphides, or sulfides [38]. By facilitating electron transport in redox reactions, this electronic transport property aids catalytic activity. Parallel and tertiary composite frameworks, which consist of metal oxides with graphene derivatives, have garnered significant attention as promising graphene-based composites for remediation purposes. This is primarily due to their optical properties, layered electronic band structures, and high apparent light absorbance. Exfoliated graphite likewise can be utilized for the expulsion of dyes and motor oil from wastewater. The exfoliated graphite created from the mixtures of regular graphite with sulfuric acid has an adsorption potential for methylene blue and methyl orange [39].

15.4 Elimination of Inorganic Contaminants from Wastewater

Considerable focus is currently directed toward the utilization of carbon-based nanomaterials for the purpose of eliminating inorganic contaminants from wastewater. This attention stems from their conspicuous properties, like greater adsorption capabilities, large surface area, and the ability to modify the surface chemistry (Figure 15.3).

15.4.1 Adsorption

The process of adsorption involves the accumulation of gases and liquids (adsorbate) on the surface of a solid (adsorbent). The technique is regarded as a highly efficient, easy-to-use, and low-cost solution for the removal of inorganic as well as organic pollutants from wastewater [40, 41]. Adsorbents are absorbed and deabsorbed from particle pores by external intraparticle mass transfer and internal diffusion. Chemisorption and physisorption are two methods of

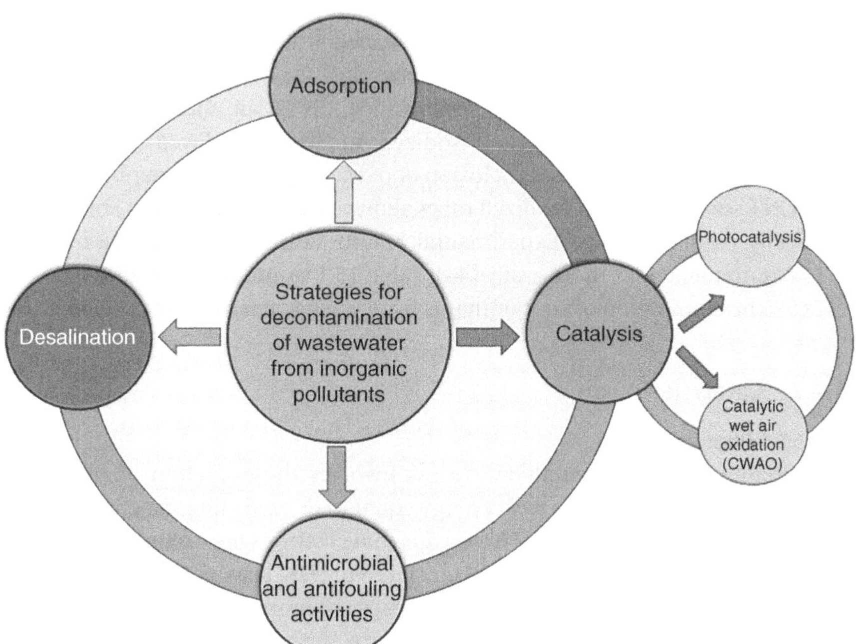

Figure 15.3 Figure illustrates the strategies used to decontaminate wastewater using nanomaterials derived from carbon to remove inorganic pollutants from the wastewater.

adsorption, which depend on the interactions between adsorbate and adsorbent. Weak van der Waals forces work between the adsorbate and the adsorbent, which results in the process of physical adsorption or physisorption. In terms of thermodynamics, physical adsorption is considered a reversible exothermic reaction. A multilayer of adsorbate forms on the adsorbent during this process [42]. The adsorbent forms a monolayer of adsorbate during chemisorption. A variety of factors affect chemisorption and physisorption, such as pressure, temperature, adsorbent surface area, and adsorbate nature.

ACs are one of the numerous materials that may be utilized as adsorbents, along with zeolites and resins. This is because they have several benefits, including broad-spectrum removal, pollutant strength, thermal stability, and chemical inertia. In spite of this, ACs are often lacking in certain parameters, such as kinetics, poor adsorption, and regeneration problems. To counter these challenges, the second generation of carbonaceous adsorbents has been introduced as ACFs. By opening directly above the carbon matrix, ACFs reduce the distance between adsorption sites and diffusion sites. In ACFs, this increases adsorption kinetics compared to ACs. There seems to be no difference between CNTs with a one-dimensional structure versus ACFs with a three-dimensional structure, where all adsorption sites exist on the surface of the inside and outside layers. In addition to their layered and hollow architecture and flexible surface chemistry, CNTs may form a convincing third generation of carbonaceous adsorbents [43].

CNTs were compared with traditional materials for their adsorption potential for heavy metals. Plasma-oxidized multiwalled CNTs, an alternative to multiwalled CNTs, which are chemically oxidized, might be employed as adsorbents for certain metals. Owing to the inclusion of oxygen-containing functional groups in the CNT surface, plasma-oxidized tubes showed a better adsorption ability than chemically oxidized tubes. Experimental results indicate that plasma-oxidized CNTs desorb metal ions more easily [44]. Table 15.1 summarizes the utilization of CNTs for the elimination of contaminants from wastewater by adsorption method.

15.4.2 Catalysis

15.4.2.1 Photocatalysis

The initiation of a photocatalytic process involves the excitation of photogenerated pairs of electron holes with energy equal to or exceeding the energy of the band gap of the semiconductor. This means that electrons are excited in the VB of the semiconductor, followed by their transfer to the CB, leaving holes behind (h^+). Hence, photocatalysts with narrower band gaps are more likely to capture visible-light photons. Further, electrons and holes are separated from photogenerated electrons. In contrast to this, the bulk charge carriers experience phonon or heat production that culminates into a reduction in excited carriers. On the surface,

Table 15.1 An overview of studies that have been conducted on the use of CNTs for the removal of contaminants from wastewater.

Adsorbents	Target pollutant	pH	Adsorbent dosage (g)	Adsorbent capacity ($mg\,g^{-1}$)	Contact time (min)	Adsorption mechanism	Reference
Zero-valent iron/MWCNT	Arsenate	8	4	250	60	Complex mechanism	[45]
CNT-sediments	Cd(II)	10	10	1.482	300	Physisorption	[46]
Iron-imprinted polymers/MWCNT	Pb(II)	6	0.02	83.20	15	Chemisorption	[47]
MWCNT	Cu(II)	6	0.3	24.49	—	Physisorption	[48]
Oxidized-MWCNT	Cu(II)	5.5	0.02	14	120	Chemisorption	[49]
Chitin/magnetite/MWCNT	Cr(VI)	2	0.05	11.3	45	Chemisorption	[50]
Zero-valent iron/MWCNT	Arsenite	7	4	200	90	Complex mechanism	[45]
MWCNT	Pb(II)	6	0.3	97.08	—	Physisorption	[48]
MWCNT	Cd(II)	11	0.3	10.86	—	Physisorption	[48]

electrons and holes can take part in multiple chemical reactions, and the two charge carriers can also be combined. One of the most important features of these is the photogenerated electrons, which are thought to be a reductant for directly reducing a number of heavy metals. The separated holes have a notable tendency for oxidation, leading to their ability to react with hydroxyl ions (OH^-) or water molecules, resulting in the formation of $^{\bullet}OH$. Additionally, these separated holes can directly participate in oxidative degradation, serving as the key process to the synthesis of $^{\bullet}OH$. In addition, the electrons that are separated have the ability to react with oxygen molecules in water, resulting in the formation of superoxide radicals ($^{\bullet}O_2^-$). These radicals can subsequently undergo decomposition, leading to the generation of $^{\bullet}OH$. Once these contaminants adhere to the surface of the catalytic material, they augment the mobility of charges, hence further amplifying the redox capability of the catalytic material. The catalyst generates active species, and those species undergo a series of chemical reactions. Photo-Fenton reactions are also characterized by a similar photocatalytic process. Through two further processes, including (1) the Fe^{3+} to Fe^{2+} photoreduction and (2) the peroxide photolysis by shorter wavelengths, more $^{\bullet}OH$ are produced from the Fenton reagents, i.e. Fe^{2+} and H_2O_2 under UV-Vis having wavelength (λ) of 600 nm [51].

Reinforced photocatalytic composites were created by integrating CNTs with semiconductors, including TiO_2, ZnO, and other similar compounds [52]. This approach yields increased surface area, enhanced photocatalytic efficiency, and improved electrical conductivity [53].

The treatment of chromium in water has been achieved through photocatalysis. Heavy metal ions have been removed using TiO_2—ZrO_2, which exhibits excellent sorption characteristics and high chemical stability. This material showed excellent efficiency in degrading Cr(VI) and Cu(II) in one step, showing 91% and 100% removal rates of Cu(II) and Cr(VI) after four cycles, respectively [54].

The removal of Pb(II) from water solutions can be accomplished with ease using heterogeneous photocatalysis [55]. A superparamagnetic nanohybrid of $NiFe_2O_4$ and Pd has been successfully synthesized [56]. In comparison to bare $NiFe_2O_4$, $NiFe_2O_4$-Pd removed Cd^{2+} and Pb^{2+} ions with an efficiency of 97% and 98%, respectively, showing a higher level of photocatalytic activity. Hg(II) in aqueous solutions was photocatalytically abated by Au-decorated TiO_2 nanotubes. Among other examples, mesoporous CuO/g—C_3N_4 heterostructures demonstrated an impressive rate of photoreduction of Hg(II) of 628.74 mol g^{-1} h^{-1} [57].

15.4.2.2 Catalytic Wet Air Oxidation

The process of wet air oxidation involves the utilization of atmospheric oxygen to oxidize organic materials present in water, both in suspended and dissolved forms. This technique has been extensively employed for approximately six decades as an effective means of managing wastewater. However, its application in the procedure of treatment of wastewater from industries is not practical due to the significant expenses associated with its operation and the harsh reaction conditions it requires. Additionally, it has been observed that the incorporation of catalysts, typically noble metals, into this procedure enhances the efficacy of oxidation and renders the method economically viable [43]. As part of a research project, several researchers used CNTs to treat biological and toxic wastewater with catalytic wet air oxidation (CWAO) [58–62]. CNTs are designed with a mesoporous structure, as opposed to AC, allowing the diffusion of contaminants onto the surface of the catalyst much more easily compared to other catalysts. Additionally, CNTs have been used by many scientists in conjunction with catalysts such as Ru, Pt, and Pd to improve catalytic performance and reduce the cost of these catalysts [59, 60]. Functionalized phenol multiwalled CNTs (MWCNTs) is a batch reactor, which was used to test MWCNT catalysts during CWAO of phenol with an initial concentration of 1000 $mg\,l^{-1}$ for phenol and a temperature of 160 °C along with the magnitude of oxygen pressure of 2 MPa. In the absence of a catalyst, approximately 11.6% of phenol is eliminated during a period of 120 minutes. After 120 minutes, MWCNTs-A remove 10% total organic carbon (TOC) and 22% phenol. After 120 minutes of CWAO of phenol using MWCNTs-B as a catalyst, 76% TOC is reduced and almost all the phenol is removed. MWNTs-B have good CWAO activity. As a result of adsorbed liquid oxygen, oxygen atoms were dissociated on graphite sheets of MWCNTs. As a result, MWNTs are promising materials for application in CWAO not only as a support but also as a catalyst [62].

15.4.3 Antimicrobial and Antibiofouling Activities

The utilization of carbon nanomaterials has become popular for application in coating ships and other submerged structures and equipment due to their remarkable mechanical capabilities, in addition to their antibacterial and antifouling (AF) characteristics [63–65]. Biofouling occurs when micro-organisms adhere to surfaces, multiply rapidly, and form a biofilm that supports the establishment of macro-organisms [66]. As a result of this biological process, the efficiency of water-filtering devices and membrane separation processes is compromised and marine surfaces are deteriorated [67–70]. Carbon nanomaterials are becoming increasingly important as fillers for polymer composites [71]. This nanomaterial can augment the mechanical strength and biofouling resistance of the composite [72, 73].

CNTs and GP have both displayed notable antimicrobial properties against both Gram-negative as well as Gram-positive bacteria, along with antibacterial activity [74, 75]. With respect to CNTs, single-walled CNTs (SWCNTs) have demonstrated considerably greater antibacterial activity than MWCNTs toward Gram-positive bacteria [76]. Carbon-based nanomaterials are characterized by their complex architecture and the multitude of elements that can potentially influence their ability to kill bacteria. However, their complexity and the extent of our understanding regarding these materials are still limited. Besides composition and geometry, bacteria type, morphology as well as growth state affect their effectiveness [77, 78]. The antibacterial capabilities of these nanomaterials are attributed to their sharp structures, which function as nanodarts capable of invading bacterial membranes, leading to cell death [76]. On the contrary, alternative viewpoints state that CNTs not only induce mechanical harm, which serves as the principal mechanism of fatality but also induce the development of oxidative stress [79]. AF and antimicrobial activity of CNTs are also dependent on their surface area, diameter, concentration, length, and chemical modifications [79, 80].

Nanomaterials based on graphene are also likely to have both chemical and physical properties. GP sheets also cause membrane damage due to sharp edges [81]. A graphene-based material can also trap bacterial cells owing to its large surface area [82]. A GP is assumed to induce oxidative stress, which causes DNA and protein damage in micro-organisms by forming reactive oxygen species (ROS). In addition to their physicochemical characteristics such as layers, functionalization, and dimensions, a number of factors influence the surface production of GP-based nanomaterials, including graphene loading, aggregation, and nanoparticle dispersion [83–85] (Figure 15.4).

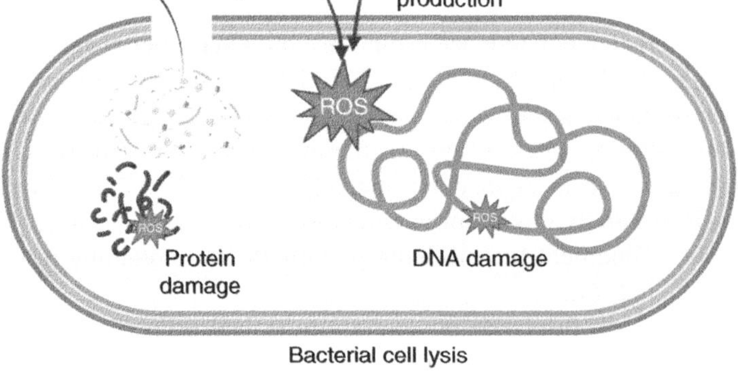

Figure 15.4 A diagrammatic representation of the mechanisms by which carbon nanomaterials, like graphene and CNTs, cause damage to the bacteria. GP causes oxidative stress to micro-organisms by creating ROS, which, in turn, disrupt DNA and protein synthesis in micro-organisms.

15.4.4 Desalination

About 98% of the water on earth comes from brackish water and seawater sources. Hence, desalinating saline water either from seawater or brackish water sources with a number of desalting methods appears to be the most effective way to increase the supply of freshwater. Membranes made from CNTs are utilized as high-flow desalination membranes. The nanotube membrane can be fabricated through catalytic chemical vapor deposition and laser etched to uncap the nanotubes [86]. The water permeability and salt removal properties exhibited by CNT membranes render them highly intriguing for use in desalination processes.

UF membranes remove suspended particles from water, unlike NF and RO technology, which purify water through diffusion. Contrary to that, CNT membranes have proven significant efficacy in the field of gas separation, as well as the removal of both suspended particles and dissolved ions [87]. CNT is an allotropic carbon, unlike RO, UF, and NF, where the basic building blocks are organic polymers. A CNT membrane may have specific physicochemical properties compared to conventional membranes as TiO_2, Ag, or Fe_3O_4 are often added to them [88].

In addition, several wet oxidizing agent-based treatments may be used to modify the CNT surface by incorporating hydrophobic (aromatic) groups, positive ($-NH_3^+$), and negative –COO and sulfonic acids [89]. The modification of CNTs revealed a 400% increase in the capacity for salt adsorption. Furthermore, the efficiency of desalination is twice that of AF [90]. Despite having smaller pores than conventional polycarbonate membranes, nanotube-based membranes possess relatively much higher water permeabilities. In order to properly utilize the inside pores of CNTs, high-density CNT membranes, which are vertically aligned, were synthesized [91]. Furthermore, to remove ions from plasma-treated CNT membranes, charge exclusion was used in order to exclude 98% of the ions added by negatively charged moieties [92].

A novel architectural design aiming to enhance the efficiency of thin-film composite membranes for desalination purposes of seawater and brackish water, RO has been modified by the immobilization of functionalized CNTs within a porous support layer [93]. As the pore characteristics and hydrophilicity of the support layer of the new membranes increase, the water permeability of the membranes increases without affecting the movement of the solvent.

15.5 Carbon Nanomaterials for Sensing and Monitoring

It has become difficult to detect toxic contaminants in water using conventional sensor technologies in a timely manner. Moreover, there is a significant increase in this adversity due to the use of complex wastewater matrices. Due to certain characteristics of CNTs, which include their high adsorbent strength, and their hydrophobicity, the development of highly sensitive and responsible sensors has been spurred by this perspective. In other words, CNTs have high recovery rate, strong electroconductivity, hydrophobicity, and kinetics of the first reaction, which makes them extremely adsorbent to water contaminants and can be used for sensing and trapping them. As a result of the immobilization of enzymes on CNTs, water contaminants are oxidized along with the nuclear electron transfer to produce a high signal level. Additionally, the ratio between the amount of analyte adsorption and the variation in current flux measured during a test may indicate electrical conductance based on the amount of charged or ionic pollutant adsorption into CNTs [10].

A composite electrode (CNTECE-Tyr) consisting of enzyme-integrated CNT epoxy and tyrosinase has been developed [94]. Aiming at the quantitative estimation of catechol and other phenolic water pollutants, CNTECE-Tyr were compared with graphite epoxy composite electrodes (GECE-Tyr) combined with tyrosinase biosensors. As a result, it was observed that the CNTECE-Tyr electrode displayed higher signal levels in comparison to the graphite sheet electrode, indicating that CNTs are more electrocatalytically active. In the case of GECE-Tyr, it was $46\,mA\,mM^{-1}\,cm^2$ and for CNTEC-Tyr, it was $294\,mA\,mM^{-1}\,cm^2$ with an intensity 90% higher than that of GECE-Tyr. There are relatively few risks associated with using CNTs as materials for electrode construction in biosensors because there is very little chance of direct contact between water and CNT electrodes. There may, however, be some risks. Although the use of poly diallyl dimethylammonium chloride (PDDA)-functionalized CNTs for electrochemical biosensors is a widely accepted practice. Yet there is a risk of hemolysis and cell mortality when CNTs-PDDA are employed [43].

15.6 Limitations

Nanotechnology holds considerable promise for continuous clean water supply when it is used in the process of water treatment to reuse water. Although the technology continues to expand into a crucial usage in the treatment of water, there are few barriers which limit its performance to achieve its full potential. In addition to the tremendous growth in the induction of nanotechnology in the purification of water, along with several other limitations, one of the most significant limitations is the lack of efficiency of the technology. Because of the recent boom in industrialization, the production cost of water treatment has also reached a peak. A high performance and affordable carbon nanomaterial are crucial to water treatment. Increasing expenditure on raw materials has come out as hindrances in developing countries from investing in the use of carbon nanomaterials in water treatment. As a result, many poor nations continue to use traditional techniques, such as disinfection, while the developed nations are willing to invest in advanced technologies, such as nanotechnology, to augment the treated water quality. The goal of all welfare states is to produce quality water by eliminating as many pollutants as possible at the lowest possible cost [43].

According to studies, CNTs might cause respiratory issues such as emphysema, bronchitis, asthma, or, lung cancer. Furthermore, there are a few industrial plants where hygiene conditions are lacking, making them potentially dustier [95]. Fine CNT particles in composites or pulverized CNTs may pose an inhalation hazard. Researchers have conducted a number of laboratory experiments on inhaling CNTs in an effort to determine the level of exposure and to better understand how they impact the respiratory tract. In addition, CNTs can cause lung tumors and are carcinogenic. CNTs with shorter fiber lengths will, however, be less carcinogenic [96]. It has been overlooked that nanotoxicity can have detrimental effects on a number of vulnerable populations, including pregnant women, newborns, elder persons, and those with diseases [97]. It is primarily the type of aggregation, stiffness, volume, time of exposure concentration, and others that determine the toxicity of CNTs. The production method, the purity of the formulation, and the preparation of the test samples all contribute to the toxicity of CNTs [96]. Marine ecosystems are at risk of adverse effects from these CNTs due to their interaction with specific biomolecules found in water systems [88].

15.7 Conclusion

Wastewater harbors inorganic contaminants, originating from industrial and residential sites. These contaminants including heavy metals, antibiotics, and chemical pesticides, contribute to many environmental issues. Carbon-based nanomaterials possess the ability to efficiently eliminate environmental contaminants from wastewater through the processes of adsorption and photocatalytic reactions. Nanotechnologies, such as nanomaterials, nanoadsorbents, and nanomembranes, are emerging technologies employed for the remediation of various inorganic contaminants present in wastewater. CNTs, carbon-based quantum dots, graphene and its derivatives, such as GO nanoparticles are prominent carbon-based nanomaterials. Graphene, its derivatives, and carbon-based nanomaterials have been reported by various workers to efficiently degrade pollutants from industrial effluents. The utilization of carbon-based nanomaterials and nanoparticles could assist in reducing environmental contamination with pollutants and improving human health. Wastewater treatment plants can be used to recover nutrients, energy, and other important resources. They can be a source of green energy. Such treatment plants are self-sustainable and have potential to produce 150–160% of green energy. However, the wastewater treatment plants use a high amount of energy by themselves. Therefore, approaches to reduce energy consumption by wastewater treatment plants are required to be investigated in future.

References

1 Al-Tohamy, R., Ali, S.S., Li, F. et al. (2022). A critical review on the treatment of dye-containing wastewater: ecotoxicological and healsth concerns of textile dyes and possible remediation approaches for environmental safety. *Ecotoxicology and Environmental Safety* 231: 113160. https://doi.org/10.1016/j.ecoenv.2021.113160.

2 Saravanan, A., Deivayanai, V.C., Kumar, P.S. et al. (2022). A detailed review on advanced oxidation process in treatment of wastewater: mechanism, challenges and future outlook. *Chemosphere* 308 (Pt3): 136524.

3 Fu, T., Zhang, B., Gao, X. et al. (2023). Recent progresses, challenges, and opportunities of carbon-based materials applied in heavy metal polluted soil remediation. *Science of The Total Environment* 856 (Pt 1): 158810. https://doi.org/10.1016/j.scitotenv.2022.158810.

4 Dutta, V., Verma, R., Gopalkrishnan, C. et al. (2022). Bio-Inspired synthesis of carbon-based nanomaterials and their potential environmental applications: a state-of-the-art review. *Inorganics* 10 (10): 169. https://doi.org/10.3390/inorganics10100169.

5 Singla, S., Sharma, S., Basu, S. et al. (2021). Photocatalytic water splitting hydrogen production via environmental benign carbon based nanomaterials. *International Journal of Hydrogen Energy* 46 (68): 33696–33717.

6 Liu, Z., Ling, Q., Cai, Y. et al. (2022). Synthesis of carbon-based nanomaterials and their application in pollution management. *Nanoscale Advances* 4 (5): 1246–1262.

7 Fei, J., Peng, X., Jiang, L. et al. (2021). Recent advances in graphitic carbon nitride as a catalyst for heterogeneous Fenton-like reactions. *Dalton Transactions* 50 (46): 16887–16908.

8 Ferrari, A.C. and Robertson, J. (2000). Interpretation of Raman spectra of disordered and amorphous carbon. *Physical Review B* 61 (20): 14095. https://doi.org/10.1103/PhysRevB.61.14095.

9 Quan, C., Zhou, Y., Wang, J. et al. (2023). Biomass-based carbon materials for CO2 capture: A review. *Journal of CO_2 Utilization* 68: 102373. https://doi.org/10.1016/j.jcou.2022.102373.

10 Mauter, M.S. and Elimelech, M. (2008). Environmental applications of carbon-based nanomaterials. *Environmental Science & Technology* 42 (16): 5843–5859.

11 Dutta, M., Rajak, P., and Roy, S. (2017). Determination of chronic median lethal concentration of sodium fluoride in *Drosophila melanogaster* and exploring effect of sub-lethal concentrations on differential hemocyte count. *Proceedings of the Zoological Society (Calcutta)* 72: 111–117.

12 Rajak, P., Roy, S., Podder, S. et al. (2022). Synergistic action of organophosphates and COVID-19 on inflammation, oxidative stress, and renin-angiotensin system can amplify the risk of cardiovascular maladies. *Toxicology and Applied Pharmacology* 456: 116267.

13 Sarkar, S., Rajak, P., and Roy, S. (2018). Toxicological evaluation of a new lepidopteran insecticide, flubendiamide, in non-target drosophila melanogaster meigen (diptera: drosophilidae). *Iran Journal of Toxicology* 12: 45–50.

14 Brady-Estévez, A.S., Kang, S., and Elimelech, M. (2008). A single-walled-carbon-nanotube filter for removal of viral and bacterial pathogens. *Small* 4 (4): 481–484. https://doi.org/10.1002/smll.200700863.

15 Balasubramanian, K. and Burghard, M. (2005). Chemically functionalized carbon nanotubes. *Small* 1 (2): 180–192. https://doi.org/10.1002/smll.200400118.

16 Li, Y.H., Zhu, Y., Zhao, Y. et al. (2006). Different morphologies of carbon nanotubes effect on the lead removal from aqueous solution. *Diamond and Related Materials* 15 (1): 90–94. https://doi.org/10.1016/j.diamond.2005.07.004.

17 Gotovac, S., Honda, H., Hattori, Y. et al. (2007). Effect of nanoscale curvature of single-walled carbon nanotubes on adsorption of polycyclic aromatic hydrocarbons. *Nano Letters* 7 (3): 583–587. https://doi.org/10.1021/nl0622597.

18 Yang, K. and Xing, B. (2007). Desorption of polycyclic aromatic hydrocarbons from carbon nanomaterials in water. *Environmental Pollution* 145 (2): 529–537. https://doi.org/10.1016/j.envpol.2006.04.020.

19 Thines, R.K., Mubarak, N.M., Nizamuddin, S. et al. (2017). Application potential of carbon nanomaterials in water and wastewater treatment: a review. *Journal of the Taiwan Institute of Chemical Engineers* 72: 116–133. https://doi.org/10.1016/j.jtice.2017.01.018.

20 Zou, G., Zhang, D., Dong, C. et al. (2006). Carbon nanofibers: Synthesis, characterization, and electrochemical properties. *Carbon* 44 (5): 828–832. https://doi.org/10.1016/j.carbon.2005.10.035.

21 Park, C., Engel, E.S., Crowe, A. et al. (2000). Use of carbon nanofibers in the removal of organic solvents from water. *Langmuir* 16 (21): 8050–8056. https://doi.org/10.1021/la9916068.

22 Zhang, Y., Li, Y., Wang, M. et al. (2022). Adsorption of methylene blue from aqueous solution using gelatin-based carboxylic acid-functionalized carbon nanotubes@metal–organic framework composite beads. *Nanomaterials* 12: 2533. https://doi.org/10.3390/nano12152533.

23 Greiner, A. and Wendorff, J.H. (2007). Electrospinning: a fascinating method for the preparation of ultrathin fibers. *Angewandte Chemie (International Ed. in English)* 46 (30): 5670–5703. https://doi.org/10.1002/anie.200604646.

24 Xie, K., Fang, J., Li, L. et al. (2022). Progress of graphite carbon nitride with different dimensions in the photocatalytic degradation of dyes: a review. *Journal of Alloys and Compounds* 901: 163589. https://doi.org/10.1016/j.jallcom.2021.163589.

25 Cui, Y., An, X., Zhang, S. et al. (2021). Emerging graphitic carbon nitride-based membranes for water purification. *Water Research* 200: 117207. https://doi.org/10.1016/j.watres.2021.117207.

26 Shiraishi, Y., Kanazawa, S., Kofuji, Y. et al. (2014). Sunlight-driven hydrogen peroxide production from water and molecular oxygen by metal-free photocatalysts. *Angewandte Chemie, International Edition* 53: 13454–13459. https://doi.org/10.1002/anie.201407938.

27 Peng, J., Xue, J., Li, J. et al. (2017). Catalytic effect of low concentration carboxylated multi-walled carbon nanotubes on the oxidation of disinfectants with Cl-substituted structure by a Fenton-like system. *Chemical Engineering Journal* 321: 325–334. https://doi.org/10.1016/j.cej.2017.03.133.

28 Lv, D., Wang, R., Tang, G. et al. (2019). Ecofriendly electrospun membranes loaded with visible-light-responding nanoparticles for multifunctional usages: highly efficient air filtration, dye scavenging, and bactericidal activity. *ACS Applied Materials & Interfaces* 11 (13): 12880–12889. https://doi.org/10.1021/acsami.9b01508.

29 Rivera-Utrilla, J., Sánchez-Polo, M., Gómez-Serrano, V. et al. (2011). Activated carbon modifications to enhance its water treatment applications. an overview. *Journal of Hazardous Materials* 187 (1–3): 1–23. https://doi.org/10.1016/j.jhazmat.2011.01.033.

30 Bhatnagar, A., Hogland, W., Marques, M., and Sillanpää, M. (2013). An overview of the modification methods of activated carbon for its water treatment applications. *Journal of Chemical Engineering* 219: 499–511. https://doi.org/10.1016/j.cej.2012.12.038.

31 Yang, R.T. (2003). *Adsorbents: Fundamentals and Applications*. Wiley.

32 Monser, L. and Adhoum, N. (2002). Modified activated carbon for the removal of copper, zinc, chromium and cyanide from wastewater. *Separation and Purification Technology* 26 (2–3): 137–146. https://doi.org/10.1016/s1383-5866(01)00155-1.

33 Pietrowski, P., Ludwiczak, I., and Tyczkowski, J. (2012). Activated carbons modified by Ar and CO2 plasmas–acetone and cyclohexane adsorption. *Materials Science* 18 (2): 158–162. https://doi.org/10.5755/j01.ms.18.2.1919.

34 Babić, B., Milonjić, S., Polovina, M. et al. (2002). Adsorption of zinc, cadmium and mercury ions from aqueous solutions on an activated carbon cloth. *Carbon* 40 (7): 1109–1115. https://doi.org/10.1016/s0008-6223(01)00256-1.

35 Liang, P., Wang, Q., Kang, J. et al. (2018). Dual-metal zeolitic imidazolate frameworks and their derived nanoporous carbons for multiple environmental and electrochemical applications. *Journal of Chemical Engineering* 351: 641–649. https://doi.org/10.1016/j.cej.2018.06.140.

36 Karim, A.V. and Selvaraj, A. (2021). Graphene composites in photocatalytic oxidation of aqueous organic contaminants–a state of art. *Process Safety and Environment Protection* 146: 136–160. https://doi.org/10.1016/j.psep.2020.08.042.

37 Upadhyay, R.K., Soin, N., and Roy, S.S. (2014). Role of graphene/metal oxide composites as photocatalysts, adsorbents and disinfectants in water treatment: a review. *RSC Advances* 4 (8): 3823–3851. https://doi.org/10.1039/c3ra45013a.

38 Quanlong, X., Zhu, B., Cheng, B. et al. (2019). Photocatalytic H2 evolution on graphdiyne/g-C3N4 hybrid nanocomposites. *Applied Catalysis. B, Environmental* 117770. https://doi.org/10.1016/j.apcatb.2019.117770.

39 Zhang, J., Dai, M., Zhang, S. et al. (2022). Recent progress on carbon-nanotube-based materials for photocatalytic applications: a review. *Solar RRL* 6: 2200243. https://doi.org/10.1002/chem.201602199.

40 Crini, G., Lichtfouse, E., Wilson, L.D., and Morin, N. (2019). Conventional and non-conventional adsorbents for wastewater treatment. *Environmental Chemistry Letters* 17: 195–213. https://doi.org/10.1007/s10311-018-0786-8.

41 Varghese, A.G., Annie, S., and Latha, P.M.S. (2019). Remediation of heavy metals and dyes from wastewater using cellulose-based adsorbents. *Environmental Chemistry Letters* 17: 867–877. https://doi.org/10.1007/s10311-018-00843-z.

42 Demirkıran, N., Özdemir, G.D.T., Saraç, M., and Dardağan, M. (2017). Adsorption of methylene blue from aqueous solutions by pyrolusite ore. *Mongolian Journal of Chemistry* 18: 5–11. https://doi.org/10.5564/mjc.v18i44.880.

43 Liu, X., Wang, M., Zhang, S., and Pan, B. (2013). Application potential of carbon nanotubes in water treatment: a review. *Journal of Environmental Sciences* 25: 1263–1280. https://doi.org/10.1016/s1001-0742(12)60161-2.

44 Burakov, A.E., Galunin, E.V., Burakova, I.V. et al. (2018). Adsorption of heavy metals on conventional and nanostructured materials for wastewater treatment purposes: a review. *Ecotoxicology and Environmental Safety* 148: 702–712. https://doi.org/10.1016/j.ecoenv.2017.11.034.

45 Alijani, H. and Shariatinia, Z. (2017). Effective aqueous arsenic removal using zero valent iron doped MWCNT synthesized by in situ CVD method using natural α-Fe2O3 as a precursor. *Chemosphere* 171: 502–511. https://doi.org/10.1016/j.chemosphere.2016.12.106.

46 Sun, W., Jiang, B., Wang, F., and Xu, N. (2015). Effect of carbon nanotubes on Cd (II) adsorption by sediments. *Chemical Engineering Journal* 264: 645–653. https://doi.org/10.1016/j.cej.2014.11.137.

47 Wang, H., Shang, H., Sun, X. et al. (2020). Preparation of thermo-sensitive surface ion-imprinted polymers based on multi-walled carbon nanotube composites for selective adsorption of lead (II) ion. *Colloids and Surfaces A: Physicochemical and Engineering Aspects* 585: 124139. https://doi.org/10.1016/j.colsurfa.2019.124139.

48 Li, Y.H., Ding, J., Luan, Z. et al. (2003). Competitive adsorption of Pb2+, Cu2+ and Cd2+ ions from aqueous solutions by multiwalled carbon nanotubes. *Carbon* 41 (14): 2787–2792. https://doi.org/10.1016/S0008-6223(03)00392-0.

49 Rodríguez, C. and Leiva, E. (2019). Enhanced heavy metal removal from acid mine drainage wastewater using double-oxidized multiwalled carbon nanotubes. *Molecules* 25: 111. https://doi.org/10.3390/molecules25010111.

50 Salam, M.A. (2017). Preparation and characterization of chitin/magnetite/multiwalled carbon nanotubes magnetic nanocomposite for toxic hexavalent chromium removal from solution. *Journal of Molecular Liquids* 233: 197–202. https://doi.org/10.1016/j.molliq.2017.03.023.

51 Ren, G., Han, H., Wang, Y. et al. (2021). Recent advances of photocatalytic application in water treatment: a review. *Nanomaterials (Basel)* 11: 1804. 10.3390/nano11071804.

52 Di Paola, A., García-López, E., Marcì, G., and Palmisano, L. (2012). A survey of photocatalytic materials for environmental remediation. *Journal of Hazardous Materials* 211–212: 3–29. https://doi.org/10.1016/j.jhazmat.2011.11.050.

53 Baby, R., Saifullah, B., and Hussein, M.Z. (2019). Carbon nanomaterials for the treatment of heavy metal-contaminated water and environmental remediation. *Nanoscale Research Letters* 14: 341. https://doi.org/10.1186/s11671-019-3167-8.

54 Yan, R., Luo, D., Fu, C. et al. (2020). Simultaneous removal of Cu(II) and Cr(VI) Ions from wastewater by photoreduction with TiO2–ZrO2. *Journal of Water Process Engineering* 33: https://doi.org/10.1016/j.jwpe.2019.101052.

55 Murruni, L., Conde, F., Leyva, G., and Litter, M.I. (2008). Photocatalytic reduction of Pb(II) over TiO2: New insights on the effect of different electron donors. *Applied Catalysis B: Environmental* 84: 563–569. https://doi.org/10.1016/j.apcatb.2008.05.012.

56 Thomas, B. and Alexander, L.K. (2020). Removal of Pb2+ and Cd2+ toxic heavy metal ions driven by Fermi level modification in NiFe2O4-Pd nano hybrids. *Journal of Solid State Chemistry* 288: 121417. https://doi.org/10.1016/j.jssc.2020.121417.

57 Kadi, M.W., Mohamed, R.M., Ismail, A.A., and Bahnemann, D.W. (2020). Soft and hard templates assisted synthesis mesoporous CuO/g-C3N4 heterostructures for highly enhanced and accelerated Hg(II) photoreduction under visible light. *Journal of Colloid and Interface Science* 580: 223–233. https://doi.org/10.1016/j.jcis.2020.07.001.

58 Garcia, J., Gomes, H.T., Serp, P. et al. (2005). Platinum catalysts supported on MWNT for catalytic wet air oxidation of nitrogen containing compounds. *Catalysis Today* 102: 101–109. https://doi.org/10.1016/j.cattod.2005.02.013.

59 Gomes, H.T., Samant, P.V., Serp, P. et al. (2004). Carbon nanotubes and xerogels as supports of well-dispersed Pt catalysts for environmental applications. *Applied Catalysis B: Environmental* 54: 175–182. doi: 10.1016/j.apcatb.2004.06.009.

60 Taboada, C.D., Batista, J., Pintar, A., and Levec, J. (2009). Preparation, characterization and catalytic properties of carbon nanofiber-supported Pt, Pd, Ru monometallic particles in aqueous-phase reactions. *Applied Catalysis B: Environmental* 89: 375–382. https://doi.org/10.1016/j.apcatb.2008.12.016.

61 Yang, S., Li, X., Zhu, W. et al. (2008). Catalytic activity, stability and structure of multi-walled carbon nanotubes in the wet air oxidation of phenol. *Carbon* 46: 445–452. https://doi.org/10.1016/j.carbon.2006.05.035.

62 Yang, S., Zhu, W., Li, X. et al. (2007). Multi-walled carbon nanotubes (MWNTs) as an efficient catalyst for catalytic wet air oxidation of phenol. *Catalysis Communications* 8: 2059–2063. https://doi.org/10.1016/j.catcom.2007.04.015.

63 Gu, B.E., Huang, C.Y., Shen, T.H., and Lee, Y.L. (2018). Effects of multiwall carbon nanotube addition on the corrosion resistance and underwater acoustic absorption properties of polyurethane coatings. *Progress in Organic Coatings* 121: 226–235. https://doi.org/10.1016/j.porgcoat.2018.04.033.

64 Fu, Y. (2022). Synergism of carbon nanotubes and graphene nanoplates in improving underwater sound absorption stability under high pressure. *ChemistrySelect* 7: e202103222. https://doi.org/10.1002/slct.202103222.

65 Ng, K.-W., Lam, W.-H., and Pichiah, S. (2013). A review on potential applications of carbon nanotubes in marine current turbines. *Renewable and Sustainable Energy Reviews* 28: 331–339. https://doi.org/10.1016/j.rser.2013.08.018.

66 Novoa, A.F., Vrouwenvelder, J.S., and Fortunato, L. (2021). Membrane fouling in algal separation processes: a review of influencing factors and mechanisms. *Frontiers in Chemical Engineering* 3: 687422. https://doi.org/10.3389/fceng.2021.687422.

67 Meng, S., Wang, R., Zhang, K. et al. (2020). Transparent exopolymer particles (TEPs)-associated protobiofilm: a neglected contributor to biofouling during membrane filtration. *Frontiers of Environmental Science & Engineering* 15: 64. https://doi.org/10.1007/s11783-020-1361-7.

68 Wei, S., Du, L., Chen, S. et al. (2020). Electro-assisted CNTs/ceramic flat sheet ultrafiltration membrane for enhanced antifouling and separation performance. *Frontiers of Environmental Science & Engineering* 15: 11. https://doi.org/10.1007/s11783-020-1303-4.

69 Faria, S.I., Teixeira-Santos, R., Gomes, L.C. et al. (2020). Experimental assessment of the performance of two marine coatings to curb biofilm formation of microfoulers. *Coatings* 10: 893. https://doi.org/10.3390/coatings10090893.

70 Ferreira, O., Rijo, P., Gomes, J. et al. (2021). Antimicrobial ceramic filters for water bio-decontamination. *Coatings* 11: 323. https://doi.org/10.3390/coatings11030323.

71 Kumar, S., Ye, F., Dobretsov, S., and Dutta, J. (2021). Nanocoating Is a new way for biofouling prevention. *Frontiers in Nanotechnology* 3: 771098. https://doi.org/10.3389/fnano.2021.771098.

72 Dustebek, J., Kandemir-Cavas, C., Nitodas, S.F., and Cavas, L. (2016). Effects of carbon nanotubes on the mechanical strength of self-polishing antifouling paints. *Progress in Organic Coatings* 98: 18–27. https://doi.org/10.1016/j.porgcoat.2016.04.020.

73 Upadhyayula, V.K.K. and Gadhamshetty, V. (2010). Appreciating the role of carbon nanotube composites in preventing biofouling and promoting biofilms on material surfaces in environmental engineering: a review. *Biotechnology Advances* 28: 802–816. https://doi.org/10.1016/j.biotechadv.2010.06.006.

74 Al-Jumaili, A., Alancherry, S., Bazaka, K., and Jacob, M. (2017). Review on the antimicrobial properties of carbon nanostructures. *Materials* 10: 1066. https://doi.org/10.3390/ma10091066.

75 Lukowiak, A., Kedziora, A., and Strek, W. (2016). Antimicrobial graphene family materials: progress, advances, hopes and fears. *Advances in Colloid and Interface Science* 236: 101–112. https://doi.org/10.1016/j.cis.2016.08.002.

76 Liu, S., Wei, L., Hao, L. et al. (2009). Sharper and faster "nano darts" kill more bacteria: a study of antibacterial activity of individually dispersed pristine single-walled carbon nanotube. *ACS Nano* 3: 3891–3902. https://doi.org/10.1021/nn901252r.

77 Teixeira-Santos, R., Gomes, M., Gomes, L.C., and Mergulhão, F.J. (2021). Antimicrobial and anti-adhesive properties of carbon nanotube-based surfaces for medical applications: a systematic review. *iScience* 24: 102001. https://doi.org/10.1016/j.isci.2020.102001.

78 Szunerits, S. and Boukherrou, R. (2016). Antibacterial activity of graphene-based materials. *Journal of Materials Chemistry B* 4: 6892–6912.

79 Kang, S., Herzberg, M., Rodrigues, D.F., and Elimelech, M. (2008). Antibacterial effects of carbon nanotubes: size does matter! *Langmuir* 24: 6409–6413. https://doi.org/10.1021/la800951v.

80 Gomes, M., Gomes, L.C., Teixeira-Santos, R. et al. (2021). Optimizing CNT loading in antimicrobial composites for urinary tract application. *Applied Sciences* 11: 4038. https://doi.org/10.3390/app11094038.

81 Liu, L., Zhang, M., Zhang, Q., and Jiang, W. (2020). Graphene nanosheets damage the lysosomal and mitochondrial membranes and induce the apoptosis of RBL-2H3 cells. *Science of The Total Environment* 734: 139229. https://doi.org/10.1016/j.scitotenv.2020.139229.

82 Perreault, F., de Faria, A.F., Nejati, S., and Elimelech, M. (2015). Antimicrobial properties of graphene oxide nanosheets: why size matters. *ACS Nano* 9: 7226–7236. https://doi.org/10.1021/acsnano.5b02067.

83 Radhi, A., Mohamad, D., Abdul Rahman, F.S. et al. (2021). Mechanism and factors influence of graphene-based nanomaterials antimicrobial activities and application in dentistry. *Journal of Materials Research and Technology* 11: 1290–1307. https://doi.org/10.1016/j.jmrt.2021.01.093.

84 Fatima, N., Qazi, U.Y., Mansha, A. et al. (2021). Recent developments for antimicrobial applications of graphene-based polymeric composites: a review. *Journal of Industrial and Engineering Chemistry* 100: 40–58. https://doi.org/10.1016/j.jiec.2021.04.050.

85 Azizi-Lalabadi, M., Hashemi, H., Feng, J., and Jafari, S.M. (2020). Carbon nanomaterials against pathogens; the antimicrobial activity of carbon nanotubes, graphene/graphene oxide, fullerenes, and their nanocomposites. *Advances in Colloid and Interface Science* 284: 102250. https://doi.org/10.1016/j.cis.2020.102250.

86 Holt, J.K., Park, H.G., Wang, Y. et al. (2006). Fast mass transport through sub-2-nanometer carbon nanotubes. *Science* 312: 1034–1037. https://doi.org/10.1126/science.1126298.

87 Sanip, S.M., Ismail, A.F., Goh, P.S. et al. (2011). Gas separation properties of functionalized carbon nanotubes mixed matrix membranes. *Separation and Purification Technology* 78: 208–213. https://doi.org/10.1016/j.seppur.2011.02.003.

88 Das, R., Leo, B.F., and Murphy, F. (2018). The toxic truth about carbon nanotubes in water purification: a perspective view. *Nanoscale Research Letters* 13 (1): 183. https://doi.org/10.1186/s11671-018-2589-z.

89 Kar, S., Bindal, R.C., and Tewari, P.K. (2012). Carbon nanotube membranes for desalination and water purification: challenges and opportunities. *Nano Today* 7 (5): 385–389. https://doi.org/10.1016/j.nantod.2012.09.002.

90 Yang, H.Y., Han, Z.J., Yu, S.F. et al. (2013). Carbon nanotube membranes with ultrahigh specific adsorption capacity for water desalination and purification. *Nature Communications* 4: 2220. https://doi.org/10.1038/ncomms3220.

91 Yu, M., Funke, H.H., Falconer, J.L., and Noble, R.D. (2009). High density, vertically-aligned carbon nanotube membranes. *Nano Letters* 9 (1): 225–229. https://doi.org/10.1021/nl802816h.

92 Fornasiero, F., Park, H.G., Holt, J.K. et al. (2008). Ion exclusion by sub-2-nm carbon nanotube pores. *Proceedings of the National Academy of Sciences* 105 (45): 17250–17255. https://doi.org/10.1073/pnas.0710437105.

93 Son, M., Choi, H.G., Liu, L. et al. (2015). Efficacy of carbon nanotube positioning in the polyethersulfone support layer on the performance of thin-film composite membrane for desalination. *Journal of Chemical Engineering* 266: 376–384. https://doi.org/10.1016/j.cej.2014.12.108.

94 Zhang, X., Teng, Y., Fu, Y. et al. (2011). Lectin-based electrochemical biosensor constructed by functionalized carbon nanotubes for the competitive assay of glycan expression on living cancer cells. *Chemical Science* 2 (12): 2353–2360. https://doi.org/10.1039/c1sc00562f.

95 Lam, C.W., James, J.T., McCluskey, R. et al. (2006). A review of carbon nanotube toxicity and assessment of potential occupational and environmental health risks. *Critical Reviews in Toxicology* 36 (3): 189–217. https://doi.org/10.1080/10408440600570233.

96 Francis, A.P. and Devasena, T. (2018). Toxicity of carbon nanotubes: a review. *Toxicology and Industrial Health* 34: 200–210. https://doi.org/10.1177/0748233717747472.

97 Jafar, A., Alshatti, Y., and Ahmad, A. (2016). Carbon nanotube toxicity: the smallest biggest debate in medical care. *Cogent Medicine* 3: 1217970. https://doi.org/10.1080/2331205X.2016.1217970.

16

Role of Carbon-Based Nanomaterials in CO_2 Reduction and Capture Reaction Process

Shyam Raj Yadav and Jai Prakash

Department of Chemistry, S. P. Jain College (Veer Kunwar Singh University, Ara), Sasaram, Bihar, India

16.1 Introduction

Enormous growth of industries after the Industrial Revolution has led to a vast increase in atmospheric carbon dioxide (CO_2) concentration. Fossil fuels are still contributing major share in global energy consumption. Currently nearly 80% energy requirement for industries and other human necessities is fulfilled by fossil fuels. Massive uses of fossil fuels are directly responsible for enhancing environmental concentration of CO_2, a major component of greenhouse gases. According to the 2022 annual report of NOAA's Global Monitoring Lab, worldwide average atmospheric CO_2 reached nearly 417 parts per million (ppm), which is far beyond the safe limit of 350 ppm [1]. If we assess the volume of this rapid CO_2 growth, it is about 5.5% in previous decade. The growth in CO_2 concentration causes an increase in earth's surface temperature, which plays a vital role in several climate crises such as extreme weather conditions, floods, droughts, wildfires, sea level rise, and disappearance of glaciers. Therefore, there is a dire need for immediate actions to mitigate CO_2 emissions. Generation of different useful chemicals and alternative fuels from CO_2 will be beneficial for both energy demand as well as environmental crisis.

Currently, the processes to reduce the amount of CO_2 can be classified into physical and chemical groups. The physical process includes geological sequestrations but the high cost and intensive labor limit its scope [2]. Chemical processes include photocatalytic, biochemical, electrocatalytic, radioreduction,

Carbon-based Nanomaterials for Green Applications, First Edition. Edited by Upendra Kumar, Piyush Kumar Sonkar, and Suman Lata Tripathi.

Published 2025 by John Wiley & Sons, Inc.

and thermochemical [3–7]. Among the mentioned chemical processes, the ECR of CO_2 has been proven to be one of the most efficient methods available. This reduction does not require harsh temperature and pressure conditions, is easy to control, and limits the high amount of chemical intake. ECR of CO_2 also generates carbon-based fuels like methane (CH_4), methanol (CH_3OH), ethanol (C_2H_5OH), etc., which are presently obtained by petrochemicals. Therefore, generation of these alternative fuels will also alleviate world fossil fuel demand. Many valuable industrial chemical feedstocks are also produced via this reduction [1, 2]. Several catalysts have been utilized to perform ECR of CO_2. Majority of catalysts performing the ECR of CO_2 contain metal oxides, their alloys, or metal-organic frameworks [8–10]. But, low abundance, high cost, and low stability limit the use of these catalysts at a bulk scale. Recently, carbon-derived nanomaterials have emerged as a better choice over and above described catalysts. These nanomaterials possess high catalytic capacity and thermal stability, are readily available, have low cost and larger surface area, and possess better electrical conductance, which make them a promising option for the ECR of CO_2 [11].

The present chapter summarizes the different aspects of carbon-based nanomaterials toward electrocatalytic CO_2 reduction and capture. This chapter also deals with fundamentals, possible pathways, mechanism of electrocatalytic CO_2 reduction, and obtained C_1 and C_{2+} products via metal-free carbon-based as well as metal-carbon composite nanomaterials in CO_2 reduction. Finally, contemporary challenges related to the existing methodologies for electrocatalytic CO_2 reduction will also be incorporated.

16.2 Parameters Affecting Electrocatalytic CO_2 Reduction

Cyclic voltammetry (CV) and linear sweep voltammetry (LSV) are most commonly used methods for the evaluation of CO_2 ECR (Figure 16.1a, b). Some basic parameters associated with ECR of CO_2 in the above methods are as follows:

16.2.1 Onset Potential

Onset potential refers to the minimum potential at which a certain product formation is initiated by ECR of CO_2 (Figure 16.1c). This is one of the general matrices associated with CO_2 reduction activity. A low onset potential value denotes higher catalytic efficiency.

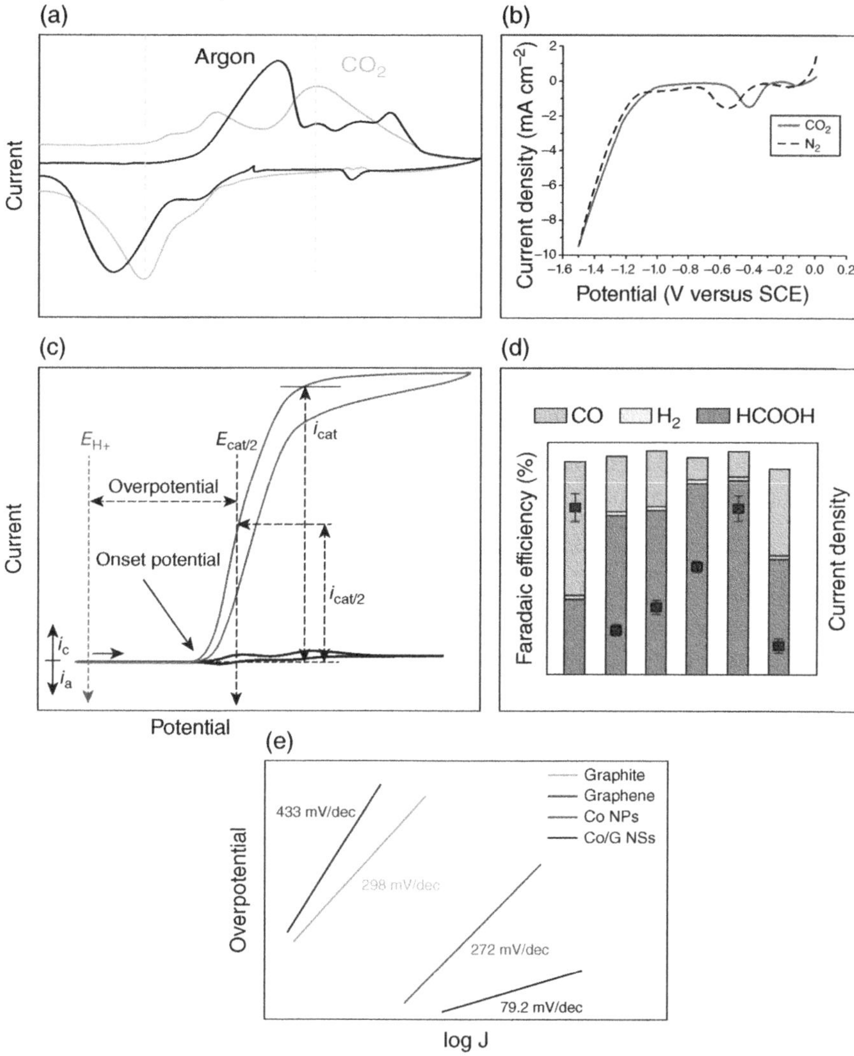

Figure 16.1 Parameters affecting ECR of CO_2. (a) CV pattern. *Source:* [12]/IOP Publishing, (b) LSV pattern. *Source:* [13]/American Chemical Society/CC BY 4.0, (c) Onset potential pattern. *Source:* [14]/American Chemical Society, (d) Faradaic efficiency pattern. *Source:* [15]/The Royal Society of Chemistry, (e) Tafel analysis pattern. *Source:* [16]/ Springer Nature/CC BY 4.0.

16.2.2 Overpotential (η)

The potential required above the theoretical thermodynamic potential to carry out an electrocatalytic reaction at a certain rate is denoted as overpotential (Figure 16.1c). The overpotential value of a reaction is given as:

$$\eta = E - Eeq \tag{16.1}$$

where E is the actual potential required to carry out a certain electrocatalytic reaction and Eeq is the standard theoretical thermodynamic potential to carry out a certain electrocatalytic reaction. A catalyst with a low overpotential value is more efficient for ECR of CO_2.

16.2.3 Current Density (j)

It is the amount of current generated per unit area of cathode for ECR of CO_2. Current density is given by:

$$j = I/A \tag{16.2}$$

where I = current applied and A = surface area of the particular electrode. Current density helps in estimation of the average reaction rate related to ECR of CO_2. It is a key parameter for evaluation of efficiency of any given catalyst. An electrode with a higher value of current density leads to a higher reaction rate.

16.2.4 Faradaic Efficiency (FE)

The Faradaic efficiency (FE) is also an important parameter for ECR of CO_2. It is a ratio of actual utilized charges to generate a certain product and total charges consumed on surface of the catalyst for ECR of CO_2 (Figure 16.1d). It is a measure of selectivity of a catalyst for the production of a certain desired product from ECR of CO_2. FE is calculated by following the equation:

$$\text{FE product} = \frac{\alpha n F}{Q} \tag{16.3}$$

where α = Electrons transferred to generate a desired molecule, n = Moles of a desired molecule obtained, F = Faraday's constant (96485 C mol^{-1}), and Q = Overall charges consumed on surface of the catalyst.

16.2.5 Tafel Analysis

This analysis is performed for the assessment of reaction kinetics of a particular product obtained from ECR of CO_2. The Tafel slope curve is obtained from LSV method by using the linear function:

$$\eta = a + b \log j \tag{16.4}$$

where η = overpotential, j = current density, and a&b are Tafel slope constants (Figure 16.1e). A catalyst with lower value of Tafel slope is more efficient for ECR of CO_2. The interpretation of Tafel slope gives insight into reaction mechanism and it helps determine the rate-limiting step of ECR of CO_2. Usually, Tafel slope value near 118 mV dec^{-1} for formation of $CO_2^{\cdot-}$ from CO_2 via single-electron transfer is the rate-determining step for ECR of CO_2 [17].

16.2.6 Turnover Frequency (TOF) and Turnover Number (TON)

These matrices indicate the catalytic performance of a catalyst. Turnover Frequency (TOF) is defined as the number of molecules consumed to furnish various products in unit time on a single active site of catalyst. The value of TOF of a catalyst is specific for concentration of reactant as well as conditions at which reaction is carried out. TON gives an idea about maximum number of reactant molecules that are consumed to furnish various products in unit time on a single active site of catalyst.

16.3 CO_2 ECR-Derived Products

ECR of CO_2 is mediated through transfer of electrons and protons. Electrons transferred for the generation of a particular product may vary from 2 to 18 depending on the carbon-based electrocatalyst utilized. CO_2 ECR-derived products can be classified on the basis of number of carbons present in it. Products containing one carbon atom are called C_1 products and products having 2 or more than 2 carbon atoms are called C_{2+} products. Details of these products formed are given in Figure 16.2.

Yield of products formed by ECR of CO_2 in aqueous medium is affected by H_2 produced from water reduction at cathode because water acts as the proton source for this reduction. Hydrogen evolution reaction (HER) is prominent due to requirement of low overpotential for production of H_2, leading to the decrease in

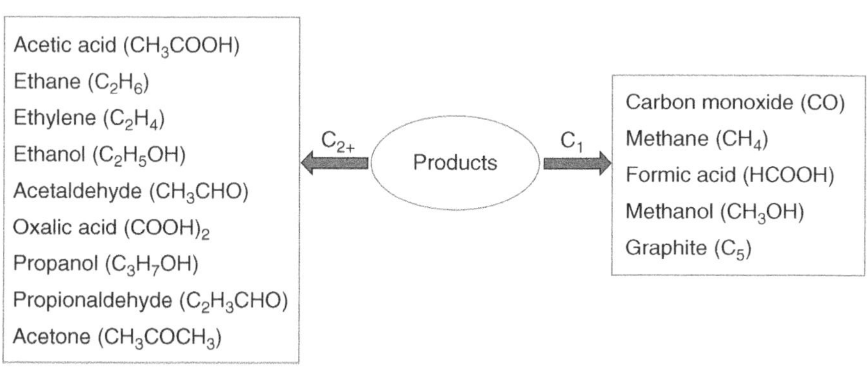

Figure 16.2 CO_2 ECR-derived products.

FE of a desired product [18]. Therefore, efficiency of a catalyst depends upon its low overpotential for CO_2 reduction, which mitigates H_2 evolution at electrode.

Details of the cathodic-half-cell reaction for the reduction of CO_2 as well as formation of some products via reduction of CO_2 are given in Figure 16.3. CO_2 is a non-polar as well as linear molecule having a strong C=O bond, bond energy of the C=O bond is very high as compared to C—O, C—C, and C—H bonds formed in the product. Therefore, reduction of CO_2 requires a high energy barrier and to counter this energy barrier, more negative potential is needed [1]. Initiation of the ECR of CO_2 occurs with the formation of an intermediate $CO_2^{\cdot-}$ radical via single electron transfer at a potential of −1.9 V versus standard hydrogen electrode (SHE). This high negative value of reduction potential denotes high overpotential, which refers to the requirement of high energy for the $CO_2^{\cdot-}$ formation [19]. Further, this highly reactive radical takes part in various proton capture and electron transfer reactions, leading to multiple thermodynamically favorable products.

Figure 16.3 reveals that the close values of the standard reduction potentials for various products obtained by ECR of CO_2 make the selectivity of a particular product too difficult. Therefore, development of different catalysts is needed to achieve better selectivity toward a particular product in ECR of CO_2. Usually, formation of C_{2+} products requires more energy than C_1 products because of involvement of higher numbers of electrons during C_{2+} product formation.

Half-reaction	Potential E^0 (V)
$CO_2 + e^- \rightarrow CO_2^-$	−1.9
$CO_2 + 2H^+ + 2e^- \rightarrow HCOOH$	−0.61
$CO_2 + 2H^+ + 2e^- \rightarrow CO + H_2O$	−0.53
$2CO_2 + 2H^+ + 2e^- \rightarrow (COOH)_2$	−0.88
$CO_2 + 4H^+ + 4e^- \rightarrow HCHO + H_2O$	−0.48
$CO_2 + 6H^+ + 6e^- \rightarrow CH_2OH + H_2O$	−0.38
$CO_2 + 8H^+ + 8e^- \rightarrow CH_4 + 2H_2O$	−0.24
$2CO_2 + 8H^+ + 8e^- \rightarrow CH_3COOH + 2H_2O$	−0.30
$2CO_2 + 10H^+ + 10e^- \rightarrow CH_3CHO + 3H_2O$	−0.35
$2CO_2 + 12H^+ + 12e^- \rightarrow C_2H_4 + 4H_2O$	−0.34
$2CO_2 + 12H^+ + 12e^- \rightarrow C_2H_5OH + 3H_2O$	−0.33
$2CO_2 + 14H^+ + 14e^- \rightarrow C_2H_6 + 4H_2O$	−0.27
$3CO_2 + 16H^+ + 16e^- \rightarrow C_2H_5CHO + 5H_2O$	−0.32
$3CO_2 + 18H^+ + 18e^- \rightarrow C_3H_7OH + 5H_2O$	−0.32
$2H+ + 2e^- \rightarrow H_2$	−0.42

Reference to standard hydrogen electrode at pH 7.

Figure 16.3 Standard potential of CO_2 reduction in aqueous medium (V versus SHE) at 1.0 atm and 25 °C. *Source:* [1]/American Chemical Society.

16.4 Plausible Mechanism for ECR of CO_2

The general reaction of ECR of CO_2 occurring at the cathode of an electrocatalytic cell can be represented as follows (Equation 16.5):

$$xCO_2 + nH^+ + ne^- \rightarrow \text{Product} + yH_2O \quad (16.5)$$

Electrolysis of water produces oxygen molecules at the anode (Eq. 16.6).

$$2H_2O \rightarrow O_2 + 4H^+ + 4e^- \quad (16.6)$$

Therefore, in general, the overall reaction for ECR of CO_2 will be represented as follows: (Eq. 16.7)

$$xCO_2 + yH_2O \rightarrow \text{Product} + yO_2 \quad (16.7)$$

These reactions infer that the protons and electrons utilized during ECR of CO_2 are furnished by water only [20]. Determination of actual mechanism for the production of different molecules from ECR of CO_2 is quite uncertain. However, Density Functional Theory (DFT) studies and characterization of intermediates formed in situ give insight into possible pathways followed during ECR of CO_2 [21, 22]. According to DFT, reduction starts with formation of an intermediate $CO_2^{\cdot-}$ via one-electron transfer. Possible pathways for the production of C_1 and C_{2+} molecules are described in the subsequent sections.

In general, ECR of CO_2 proceeds through three steps as given below:

1) ECR of CO_2 begins with chemisorption of CO_2 at electrocatalyst.
2) Cleavage of C=O bond and generation of C—H bond occur by proton and electron transfer to generate various adsorbed intermediates according to the products.
3) Rearrangement of the adsorbed intermediates followed by desorption from the surface of electrocatalyst results in product.

16.4.1 Pathways for the Formation of C_1 Products

A list of various C_1 products is given in Figure 16.2. Usually, pathways consisting of these product formations require a lesser amount of energy as compared to C_{2+} products. Details of pathways related to some C_1 product formation have been highlighted in subsequent sections.

16.4.1.1 Production of Formic Acid and Formate

Production of formic acid/formate starts with the formation of an intermediate $CO_2^{\cdot-}$ via one-electron transfer (Figure 16.4). This highly reactive intermediate binds to catalyst either through carbon atom (I) or oxygen atoms (bidentate form III).

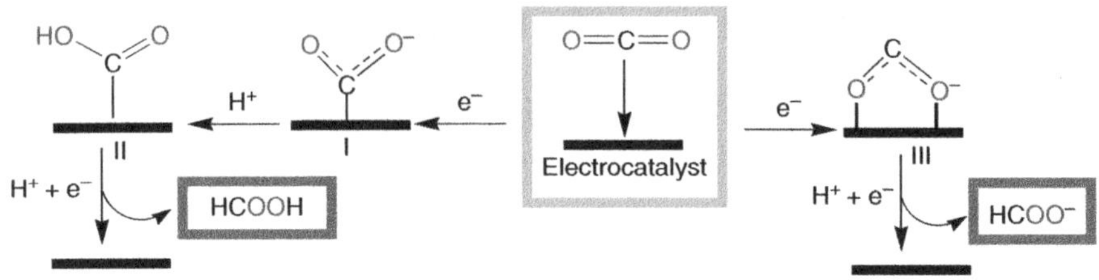

Figure 16.4 Formation of formic acid and formate [21].

This adsorbed bidentate intermediate III furnishes formate via a concerted proton-electron ($H^+ + e^-$) capture followed by desorption. On the other hand, intermediate I takes a proton and forms adsorbed intermediate II, which undergoes a concerted proton-electron ($H^+ + e^-$) capture followed by desorption to produce formic acid.

16.4.1.2 Formation of Carbon Monoxide (CO)

Overall, the formation of carbon monoxide (CO) from CO_2 involves utilization of two electrons (Figure 16.5). There are two possible routes for the formation of adsorbed intermediate II from CO_2. In the first route, there is a decoupled electron-proton transfer to CO_2, resulting in intermediate II via formation of intermediate I. In the second route, there is a concerted electron-proton transfer to CO_2, resulting in intermediate II. This common intermediate from both routes undergoes concerted proton-electron ($H^+ + e^-$) capture followed by removal of

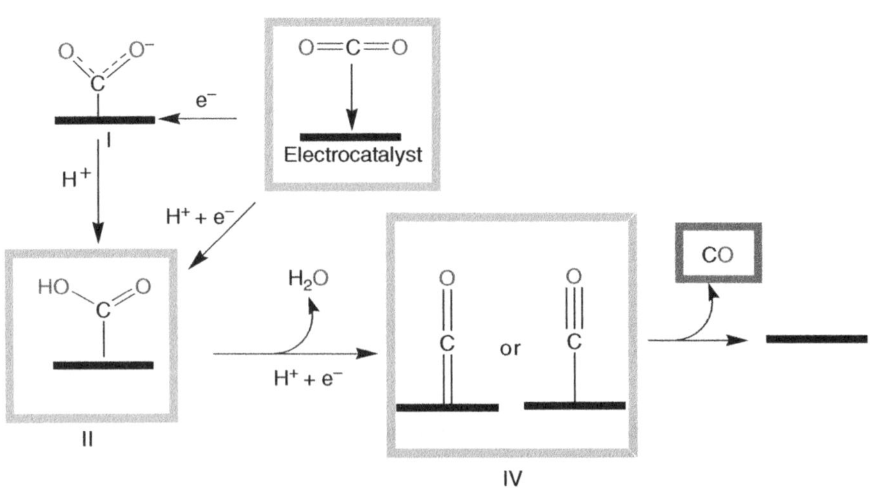

Figure 16.5 Formation of CO.

water molecule, which forms intermediate IV. Desorption from adsorbed intermediate IV leads to formation of CO. Binding energy of CO with electrocatalyst in intermediate IV plays an important role in deciding the extent of formation of CO and other products obtained from ECR of CO_2. Electrocatalysts having weak binding energy with CO, form desorbed CO and other CO_2 reduction products easily by utilizing CO. However, electrocatalysts with strong binding energy with adsorbed CO have poor selectivity toward the formation of CO and other products obtained from ECR of CO_2 because of low availability of catalyst surface area due to more occupancy of CO. This hindered CO_2 reduction promotes the formation of H_2. Therefore, a promising catalyst should have appropriate binding energy with adsorbed CO to form CO and other products obtained from ECR of CO_2.

16.4.1.3 Formation of Methane (CH_4), Formaldehyde (HCHO), and Methanol (CH_3OH)

Formations of all these valuable molecules occur via the formation of adsorbed CO intermediate IV (Figure 16.6). This intermediate undergoes a simultaneous proton-electron ($H^+ + e^-$) transfer and forms intermediate X, which again undergoes a simultaneous proton-electron ($H^+ + e^-$) to produce either HCHO or a methoxy intermediate XI. This intermediate takes a proton and an electron to produce CH_3OH or releases CH_4 resulting in intermediate XII.

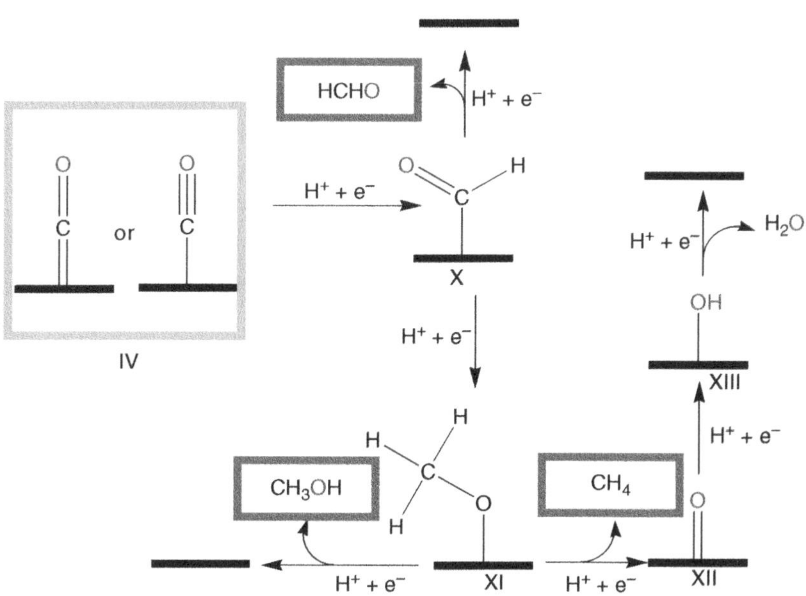

Figure 16.6 Formation of CH_4, HCHO, and CH_3OH [23].

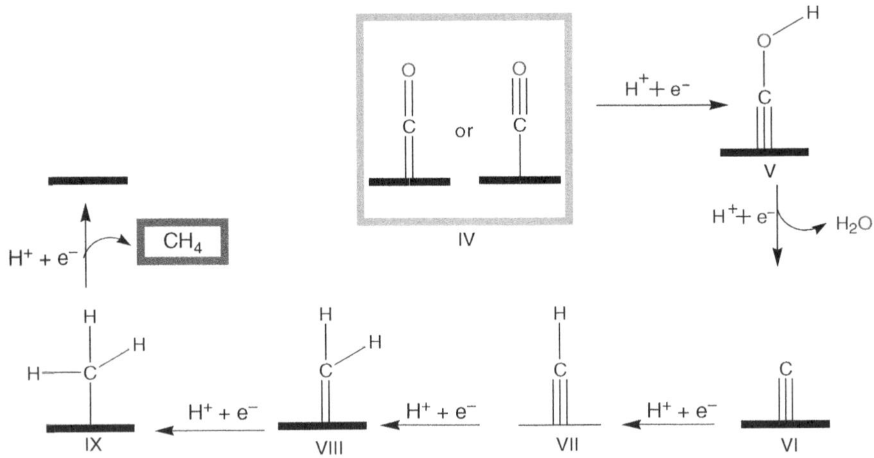

Figure 16.7 Formation of CH_4 [23].

An alternative approach for the formation of CH_4 is described in Figure 16.7. Here intermediate IV undergoes a sequential proton-electron ($6H^+ + 6e^-$) transfer to produce CH_4 via various adsorbed intermediates.

16.4.2 Pathways for the Production of C_{2+} Products

A list of various C_{2+} products is given in Figure 16.2. Usually, pathways consisting of these product formations require more amount of energy as compared to C_1 products. Details of pathways related to some valuable C_{2+} product (hydrocarbons and oxygenates) formations have been highlighted in subsequent sections.

16.4.2.1 Formation of Acetaldehyde (CH_3CHO), Ethanol (C_2H_5OH), and Ethene (C_2H_4)

Description of the formation of CH_3CHO, C_2H_5OH, and C_2H_4 is depicted in Figure 16.8. Intermediate IV is converted to adsorbed CH_2 (carbene-like) intermediate VIII via multiple steps by utilizing four protons and four electrons. This intermediate may adopt two routes. Either these carbene-like intermediates dimerize to form C_2H_4 or a concerted CO insertion and one electron capturing by intermediate VIII leads to formation of the intermediate XV. Further, this intermediate may adopt three pathways:

a) It can have four protons and three electrons transfer followed by removal of water to produce C_2H_4.

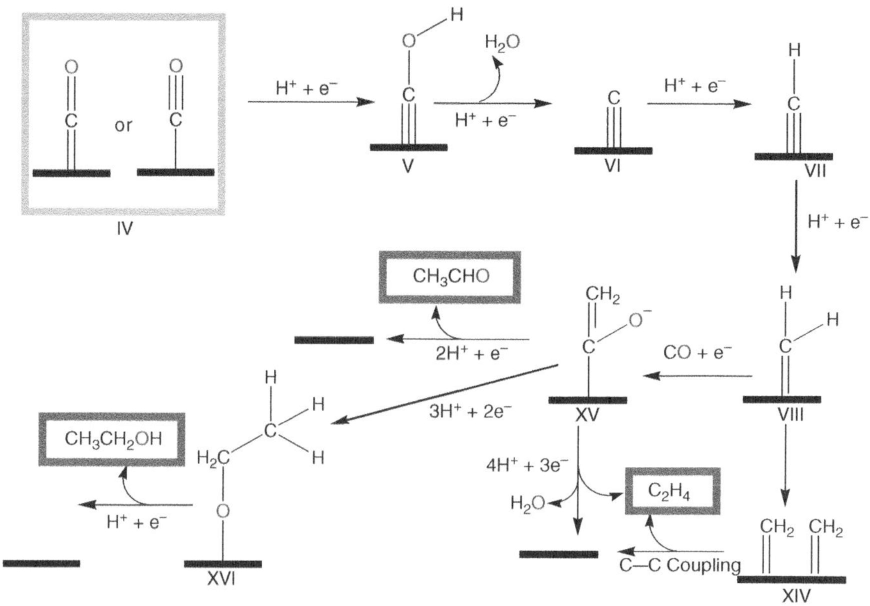

Figure 16.8 Formation of CH_3CHO, C_2H_5OH, and C_2H_4 [23, 24].

b) It can have two protons and one electron transfer to produce desorbed CH_3CHO.
c) It can undergo four protons and three electrons transfer to produce C_2H_5OH via ethoxy intermediate XVI.

16.4.2.2 Formation of Acetic Acid (CH_3COOH)

CH_3COOH is also an important C_{2+} product obtained by ECR of CO_2. A detail of the route for the formation of CH_3COOH is given in Figure 16.9.

Intermediate IV proceeds through capturing of five protons and five electrons followed by removal of water to produce adsorbed methyl intermediate IX. Further, this reacts with $CO_2^{\cdot-}$ to produce acetate ion which subsequently takes a proton to furnish CH_3COOH.

16.4.2.3 Formation of Acetone (CH_3COCH_3) and *n*-propanol ($CH_3CH_2CH_2OH$)

CH_3COCH_3 and *n*-$CH_3CH_2CH_2OH$ are important C_{2+} products having three carbon atoms. C_3 oxygenates are less prominent products formed via ECR of CO_2. CH_3COCH_3 and *n*-$CH_3CH_2CH_2OH$ are only two products that have been well explored C_3 oxygenates having FE to reasonable extent.

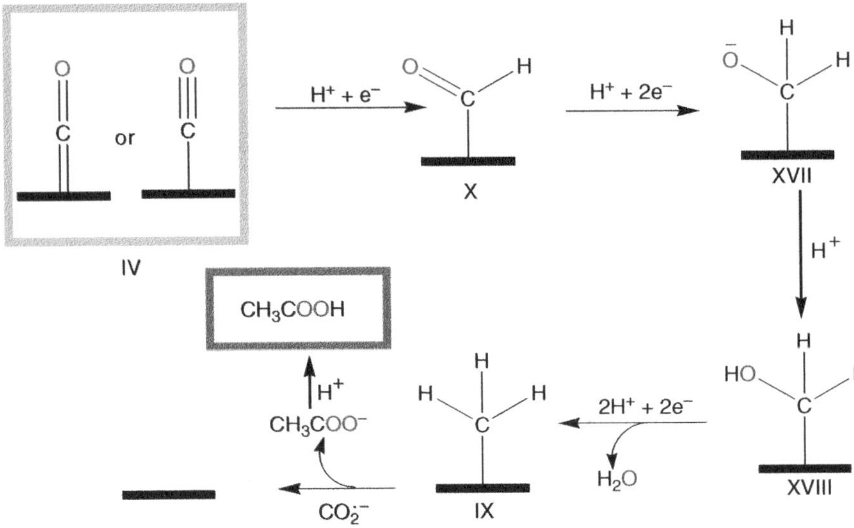

Figure 16.9 Formation of CH_3COOH [25].

General route for the formation of these molecules is sketched in Figure 16.10, which shows that the formation occurs via the common intermediate XV. This intermediate takes a proton and forms adsorbed enol intermediate XIX. This intermediate may take two paths:

a) In the first path, this intermediate may take two protons and two electrons followed by removal of water to furnish intermediate XXIV. This intermediate undergoes concerted CO insertion and hydrogenation to form carbonyl intermediate XXV, which further takes two protons and two electrons to generate *n*-$CH_3CH_2CH_2OH$.
b) In the second path, enol intermediate XIX changes to keto intermediate XX. This intermediate undergoes CO insertion followed by proton-electron ($H^+ + 2e^-$) capture to generate intermediate XXI. Further, this intermediate is converted to carbonyl intermediate XXIII by sequential proton-electron ($3H^+ + 4e^-$) transfer and removal of water molecules. This adsorbed intermediate undergoes a coupled proton-electron transfer to produce desorbed CH_3COCH_3.

16.5 Carbon-Based Nanomaterials in CO_2 Reduction

Worldwide ongoing research related to ECR of CO_2 reveals that catalysts based on carbon nanomaterials have emerged as a better choice over traditional catalysts made of either transition metals or compounds derived from transition metals.

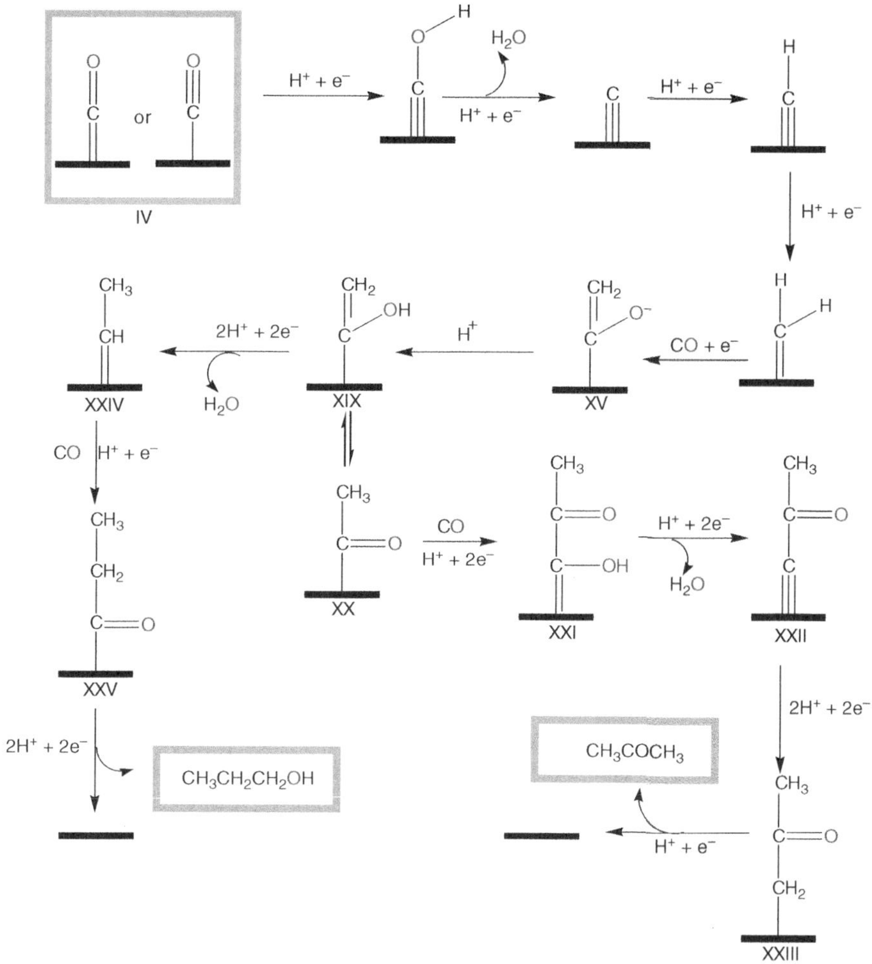

Figure 16.10 Formation of CH_3COCH_3 and n-$CH_3CH_2CH_2OH$ [1, 24].

Carbon-based electrocatalysts possess high onset potential for HER on cathode as compared to metal-based catalysts. Therefore, competing HER might be suppressed on the carbon-based electrocatalysts, which may enhance activity and selectivity in ECR of CO_2 [26]. The assembling property of carbon atoms is responsible for producing carbon-based electrocatalyst materials of different shapes, dimensions, and structures, leading to generation of catalysts with more surface area, better chemical and thermal stability, high conductivity, mechanical durability, better activity, and selectivity. Moreover, these materials are low-cost, non-hazardous, and readily available in bulk amounts [17, 27]. Literature survey reveals those catalysts containing various forms of carbon-derived materials like graphite, graphene, porous carbon, and carbon nanotubes (CNTs) have shown better selectivity with good FE toward obtained products from ECR of CO_2 [28–30]. Earlier discussed perspective reveals that this class of materials can be utilized to develop an efficient and commercially viable electrocatalyst for obtaining CO_2 ECR products with better FE.

16.5.1 Applications of Various Metal-Free Carbon-Based Nanomaterials in CO_2 Reduction

ECR of CO_2 may occur on pure nanocarbon materials but it requires very high negative potentials to generate a reactive intermediate $CO_2^{\cdot-}$ which initiates CO_2 reduction. These materials also have no specific selectivity toward a particular CO_2 reduction product. These two limitations associated with pure nanocarbon materials restrict their practical viability [31]. These associated limitations can be mitigated by doping heteroatoms (B, N, P, S, and F) in pure carbon-based nanomaterials [17, 32]. The difference in electronegativity value of these dopants from carbon induces polarization and disrupts neutrality of carbon atoms. This disruption of neutrality of carbon atoms enhances activity due to generation of profound active sites for CO_2 adsorption [33]. Besides heteroatom doping, shape and structure of allotropic forms also affect reactivity and selectivity toward ECR. Amorphous forms of carbon show very poor ECR activity as compared to crystalline forms because of disordered structure as well as low electrical conductivity. Therefore, an efficient carbon-based electrocatalyst for ECR of CO_2 can be developed by heteroatom doping and different carbon forms. Carbon materials exist in many forms like fullerenes, CNTs, graphite, nanodiamond, carbon quantum dots (CQDs), graphene, porous carbon, and carbon nanofibers (CNFs) [34]. Applications of some catalysts based on these metal-free carbon materials related to formation of various products by ECR of CO_2 are given in Table 16.1. In the subsequent sections, details of some of these metal-free carbon-based nanomaterials for production of various C_1 and C_{2+} products via CO_2 reduction have been highlighted.

16.5.1.1 Carbon Nanofibers

Pure CNFs have a smaller number of catalytic active sites for ECR of CO_2. Therefore, heteroatomic doping is carried out to enhance its catalytic properties. First, Kumar et al. reported the production of CO from CO_2 N-doped CNFs [35]. They synthesized CNFs from polyacrylonitrile (PAN) by electrospinning and subsequent carbonization. This prepared CNF catalyst was able to generate near about 13 times more reduction current density as compare to bulk Ag catalyst for

Table 16.1 Metal-free carbon catalysts for ECR of CO_2.

Catalyst	Electrolyte	Product	Potential (V)	FE	Reference
N-doped CNFs	EMIM-BF_4	CO	−0.573 versus SHE	98%	[35]
NS-CNF	$KHCO_3$	CO	−0.7 versus RHE	94%	[36]
N-CNTs	$KHCO_3$	CO	−0.82 versus RHE	80%	[37]
N-CNTs	$KHCO_3$	CO	−0.6 versus RHE	100%	[38]
N-NPC/CNT membranes	$KHCO_3$	HCOOH	−0.80 versus RHE	81%	[39]
N-CNTs	$KHCO_3$	CO	−0.90 versus SHE	90%	[40]
PEI-NCNT	$KHCO_3$	HCOOH	−1.8 versus SHE	87%	[41]
O-CNTs	$KHCO_3$	CH_3COOH	—	—	[42]
g-C_3N_4/CNT	$KHCO_3$	CO	−0.75 versus RHE	60%	[43]
SN-NPC	$KHCO_3$	CO	−0.99 V versus RHE	11.3%	[44]
SN-NPC	$KHCO_3$	CH_4	−0.99 V versus RHE	0.18%	[44]
N- carbon micropores	$KHCO_3$	C_2H_5OH	−0.56 V versus RHE	78%	[45]
SeBN-NPC	$KHCO_3$	CO	−0.6 V versus RHE	95.2%	[46]
OSP-NPCs	$NaHCO_3$	CO	−0.90 V versus SHE	81%	[47]
Mesoporous N-NPC	$KHCO_3$	CO	−0.58 V versus RHE	92%	[48]
NB-NPC	$KHCO_3$	CO	−0.4 V versus RHE	83%	[49]
N-NPC from coal	$KHCO_3$	CO	−0.6 V versus RHE	92%	[50]
cN-NPC	$KHCO_3$	C_2H_5OH	−0.56 V versus SHE	77%	[51]
N-Containing graphene	$KHCO_3$	CO	−1.4 V versus RHE	8.6%	[52]
B-doped graphene	$KHCO_3$	HCOOH	−1.4 V versus SCE	66%	[53]
N-graphene nanoribbon	$KHCO_3$	CO	−0.49 V versus. SHE	87.6%	[54]
N-graphene	$KHCO_3$	HCOOH	−0.84 V versus SHE	73%	[55]
B-ND	CH_3OH	HCHO	−1.7 V versus Ag/Ag^+	74%	[56]
N-ND/Si RA	$NaHCO_3$	CH_3COOH	−1.0 V versus RHE	91.8%	[57]
BN-ND	$NaHCO_3$	C_2H_5OH	−1.0 V versus RHE)	93.2%	[58]
N-CQD	$KHCO_3$	C_2H_4	−0.75 V versus SHE	31%	[59]

formation of CO from ECR of CO_2. They have found that active site for the reduction consists of carbon atom, not the nitrogen atom and the ionic liquid, 1-ethyl-3-methylimidazolium tetrafluoroborate (EMIM-BF4) as electrolyte promotes reduction via production of EMIM–CO_2 complex. N is the most frequently used atom for carbon doping because of its comparable size with carbon and higher electronegativity. But other elements are also doped with nitrogen in order to tune catalytic properties of CNFs. In this regard, Yang et al. have developed nitrogen and sulfur co-doped CNFs (NS-CNF) by electrospinning method utilizing ZIF-8 nanoparticles, trithiocyanuric acid (TA) and PAN followed by carbonization at 900 °C in an inert atmosphere. They have reported that CO formation occurs with 94% FE by utilizing NS-CNF catalyst from ECR of CO_2 [36]. Literature survey reveals that fewer reports are available about carbon-based non-metallic CNF catalysts. Therefore, an increased focus of research is required to develop a commercially viable carbon-based non-metallic CNF catalyst.

16.5.1.2 Carbon Nanotubes

CNTs are hollow tubular structures consisting of graphitic layers. It can be divided into two groups on the basis of the number of graphitic layers: single-walled nanotubes (SWCNTs) and multi-walled nanotubes (MWCNTs). Unique structure of CNTs offers profound sites for the incorporation of different functional groups. Moreover, their unusual geometry enhances surface area, electron conductivity, sturdiness, and chemical stability. These parameters may enhance the activity as well as selectivity of electrocatalysts toward ECR of CO_2 for a certain product [60]. In comparison to pristine CNTs, N-doped CNTs (N-CNTs) produce local defect that influences catalytic activities. Wu et al. have produced N-CNTs by liquid chemical vapor deposition (CVD) method (Figure 16.11a). They have reported a large difference in current density between pure CNTs and NCNTs, inferring vital role of N as dopant in the ECR of CO_2 (Figure 16.11b). They have also shown that the maximum FE of CO reaches near to 80% at overpotential of −0.26 V in case of N-CNTs (Figure 16.11c).

On the other hand, pure CNTs are showing less activity and selectivity toward ECR of CO_2 to CO. On the basis of CHE model-based free energy calculations for ECR of CO_2 to CO, they have found that N-CNTs lower the activation energy for the production of an intermediate *COOH from CO_2 in comparison to pure CNTs (Figure 16.11d). Moreover, Figure 16.11d reveals that this activation energy varies according to graphitic N, pyrrolic N, and pyridinic N sites on N-CNTs. The alteration in arc of the CNTs can also affect the electronic environment around the active sites of the electrocatalyst for ECR of CO_2. Related to this, recently Zhao et al. have reported the curvature effect on different pyridinic N-modified CNTs [38]. They have synthesized different pyridinic N-modified CNTs via oxidizing-reducing method. First, surface-oxidized CNTs (O-CNTs) were obtained from oxidation of

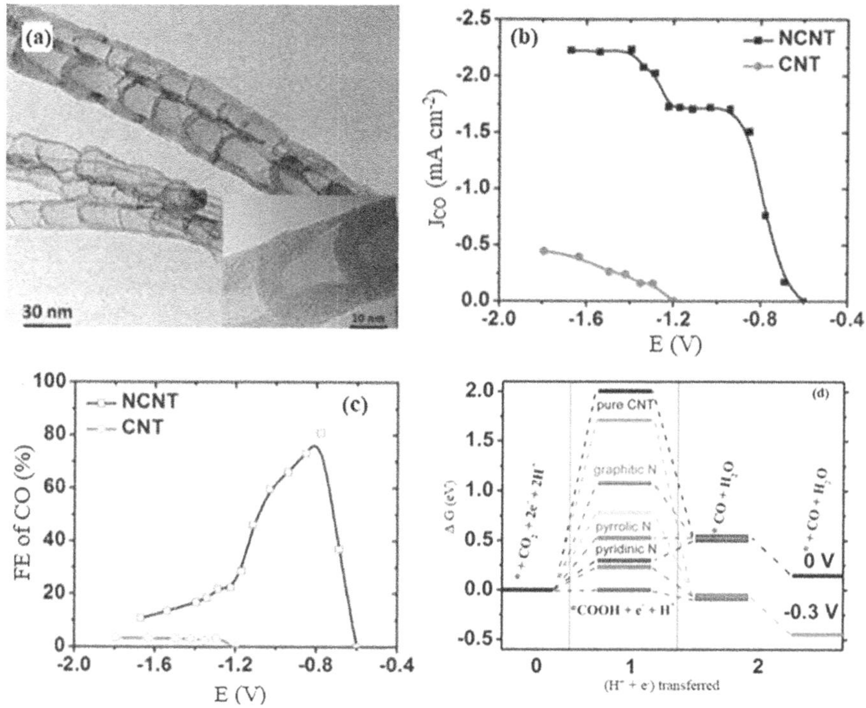

Figure 16.11 (a) TEM image of N-CNTs, (b) FE comparison between pure CNT and N-CNT, (c) current density comparison between pure CNT and N-CNT, (d) free energy diagram for ECR of CO_2 to CO on pure CNT and N-CNT. *Source:* [37]/American Chemical Society.

CNTs. These O-CNTs and dicyandiamide were mixed in sonicator and reaction mixture was freeze-dried and further exposed to annealing process at 600 °C for two hours. They have obtained three types of N-CNTs with average curvature values (K) 0.08 (N-CNTs-1), 0.04 (N-CNTs-2), and <0.02 (N-CNTs-3) based on diameters of utilized CNT precursors 8–15 nm, 20–30 nm, and >50 nm, respectively (Figures 16.12a–c, 16.14a). Among different pyridinic N-modified CNTs, N-CNTs-2 shows almost 100% FE at −0.6 V versus RHE (Figure 16.13a). N-CNTs-2 also has the highest current density in comparison to other pyridinic N-modified CNTs, depicting larger active sites associated with it (Figure 16.13b).

They have also found that free energy for the production of *COOH, a key intermediate during ECR of CO_2 utilizing different pyridinic N-modified CNTs is lowest for N-CNTs-2 and their order is as follows: N-CNTs-2 < N-CNTs-1 < N-CNTs-3 < N-CNTs-4 (Figure 16.14b).

Calculated free energy for pyridinic N-modified CNTs by DFT method associated with HER shows that N-CNTs-2 has the lowest *H adsorption, leading to less

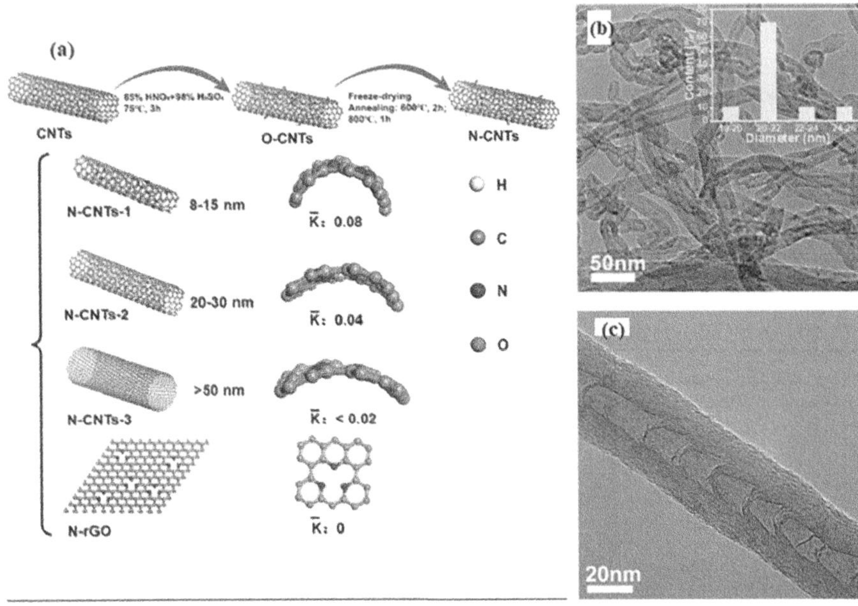

Figure 16.12 (a) Synthesis route of N-CNTs, (b) HR-TEM image of N-CNTs-2, (c) magnified HR-TEM image of N-CNTs-2. *Source:* [38]/American Chemical Society.

affinity toward HER and high affinity for *COOH and *CO adsorption (Figure 16.14c). Wang et al. have developed N-doped nanoporous carbon/carbon nanotube (N-NPC/CNT) membranes containing large surface area, high conductivity, profound active sites, ordered porous architecture as well as membrane stiffness. These nanoporous membrane structures make it to act as diffusion electrodes with high activity and selectivity toward the formation of formate via ECR of CO_2 [39]. Xu et al. have synthesized various N-CNTs. They have shown that according to the DFT calculations, N-CNTs having quaternary nitrogen are the most suitable among all the N-CNT catalysts due to their ability to stabilize CO_2^{-}, a key intermediate for the formation of CO via CO_2 ECR [40]. Zhou et al. have synthesized peanut-shaped N-CNTs. They have reported that pyridinic N, pyrrolic N, and octatomic N sites in these nanotubes show even lower overpotential than commonly used Cu catalysts for the production of CH_3OH by ECR of CO_2 [61]. Zhang et al. have found that the use of polyethylenimine (PEI) as a co-catalyst with N-CNTs significantly reduces the overpotential and increases current density and efficiency, which leads to the stabilization of $CO_2^{\cdot-}$ as well as high concentration of CO_2 in the vicinity of PEI over layer for the formation of formate via CO_2 ECR [41]. Lu et al. have reported the production of multi-wall CNTs by incorporating a graphitic carbon nitride. This catalyst (g-C_3N_4/MWCNTs) selectively reduces CO_2 to CO because of enhanced surface area, better conductivity, and creation of active carbon–nitrogen bonds [43].

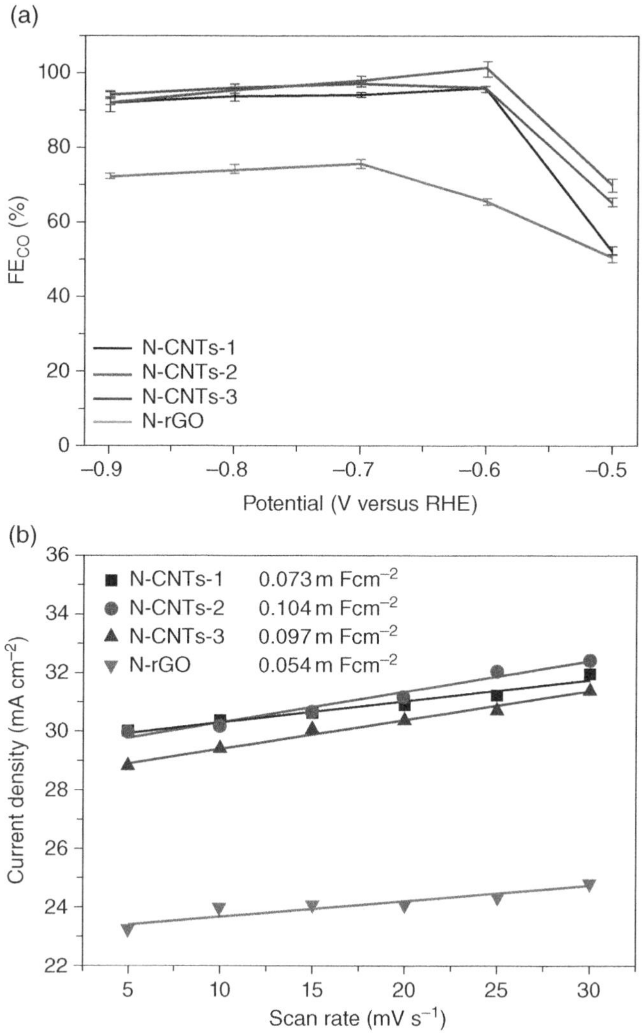

Figure 16.13 (a) FE comparison of different N-CNTs, (b) current density comparison of different N-CNTs. *Source:* [38]/American Chemical Society.

16.5.1.3 Nanoporous Carbon

Nanoporous Carbon (NPC) materials are three-dimensional (3D) arrangements of carbons having porous surfaces. Because of the porous surfaces, these materials possess high surface area leading to the availability of a larger number of active sites for ECR and capture of CO_2. Other than these properties, NPC materials also possess high wettability that eases the diffusion of electrolytes inside the pores [46, Figure 16.15]. NPC materials are classified on the basis of diameters (d) of constituting pores: microporous ($d < 2$ nm), mesoporous ($d = 2$–50 nm), and macroporous

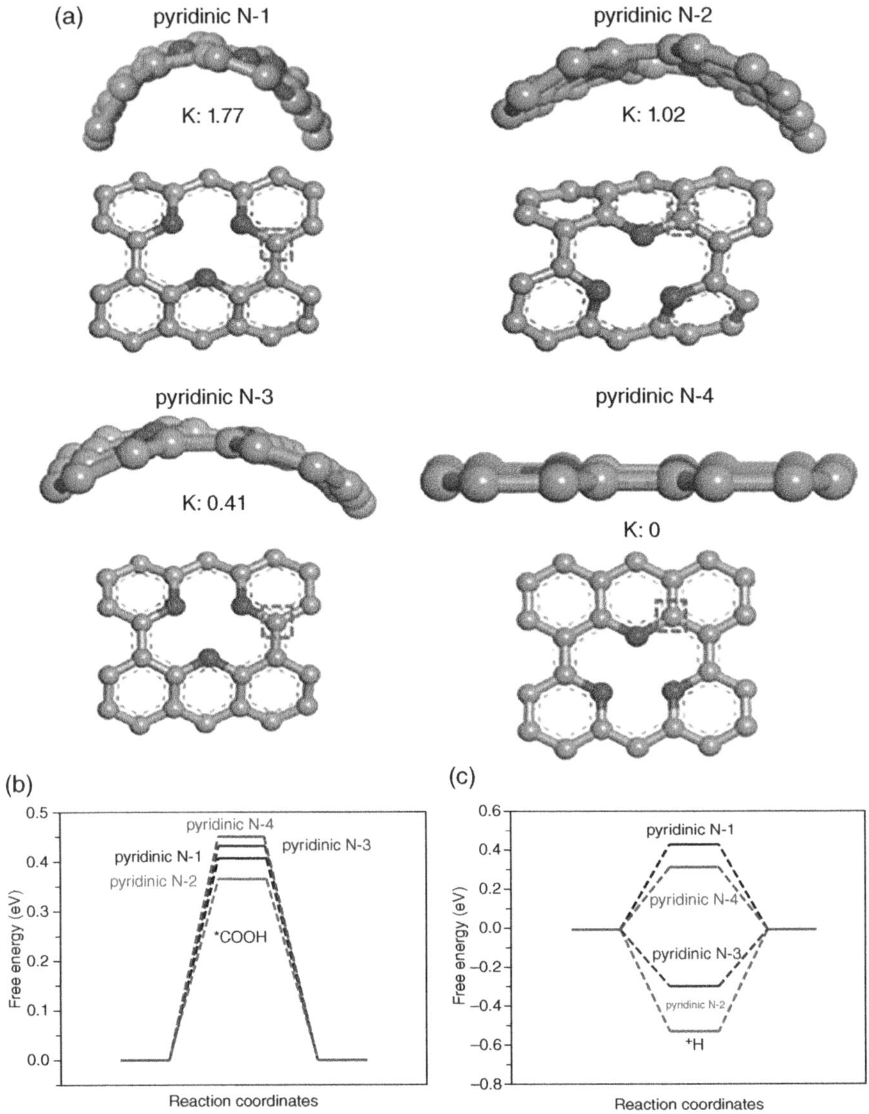

Figure 16.14 (a) Models of different curvatures associated with pyridinic N-modified CNTs, (b) calculated free energy for pyridinic N-modified CNTs by DFT method associated with CO_2 ECR to CO, (c) calculated free energy for pyridinic N-modified CNTs by DFT method associated with HER. *Source:* [38]/American Chemical Society.

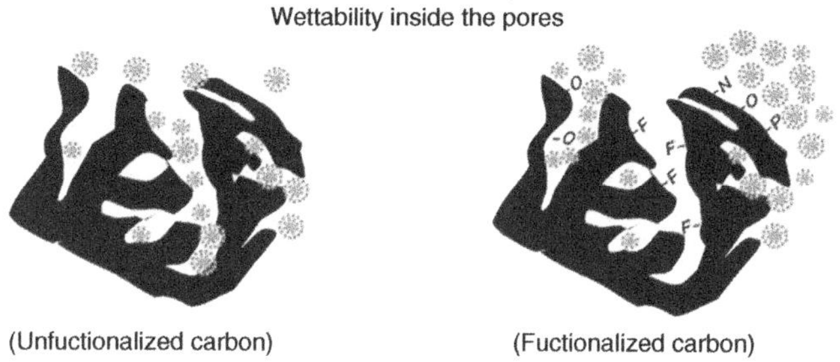

Figure 16.15 Wettability in nanoporous carbon. *Source:* [62]/with permission of Elsevier.

(d > 50 nm). CO_2 adsorption on micropores occurs at atmospheric pressure, whereas mesopores require high pressure [62]. These NPC materials have both sp^2 and sp^3 carbons. Due to this, the quantitative modulation of these carbon atoms may alter the electronic environment as well as pore distribution in NPC materials, which leads to obtaining an improved catalyst with better activity and selectivity toward ECR of CO_2. Literature survey also reveals that NPCs having a size between 0.7 and 2 nm show selectivity toward the formation of CH_4 and C_2H_5OH [44, 45].

Recently, Wang et al. have reported a non-metallic Se, B, and N-doped NPCs (SeBN-NPCs) for ECR of CO_2 [46]. These SeBN-NPCs have been synthesized by mixing chitosan, Se powder, $(C_6H_5)_4BNa$, NH_4Cl (pore maker), and homogenizing the reaction mixture by ball-milling treatment (Figure 16.16). Initially, reaction mixture was heated at 300 °C for one hour under inert atmosphere. Further, it was heated at different temperatures to obtain various SeBN-NPCs. TEM image of SeBN-NPC-1100 (obtained at 1100 °C) is shown in Figure 16.17a.

They have found that SeBN-NPC synthesized at 1100 °C is the most efficient electrocatalyst for ECR of CO_2 to CO with FE of 95.2% at −0.6 V among the various SeBN-NPCs synthesized at different temperatures (Figure 16.17b). A minute variation in FE and current density was visualized while continuous operational monitoring for up to 10 hours showed reasonable stability of the SeBN-NPC catalyst (Figure 16.17c, d).

Liu et al. reported onion-shaped phosphorus doped-NPCs (OSP-NPCs) for the formation of CO via ECR of CO_2 with better or comparable efficiency than earlier reported N-doped carbon catalysts [47]. Kuang et al. have described the production of mesoporous N-NPCs by secondary doping method utilizing N,N-dimethylformamide (DMF) solvent. These DMF-mediated N-NPCs possess extra pyridinic N sites and local defects that surprisingly increase FE as well as partial current density for ECR of CO_2 to CO [48]. Jia et al. have synthesized a nitrogen and boron (B) co-doped 3D hierarchical NPCs (NB-NPC) by salt–sugar method. They have found that these dual-doped NPCs are more efficient as compared to

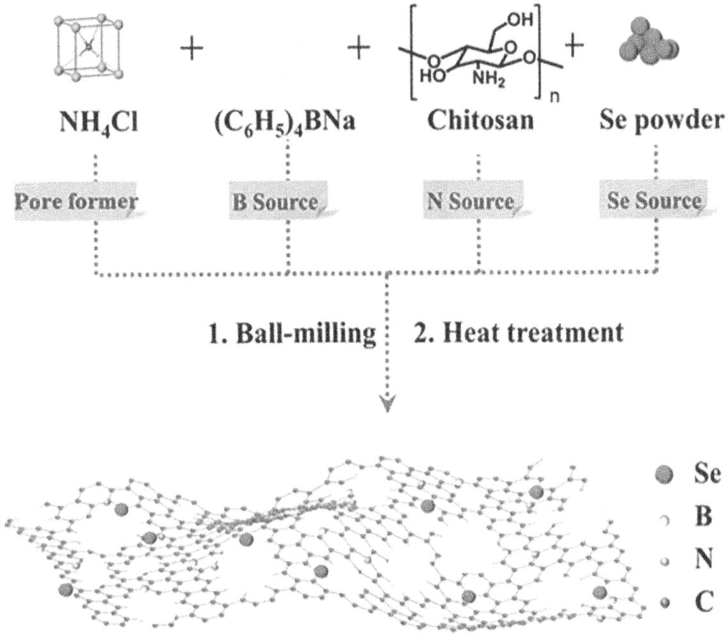

Figure 16.16 Synthesis of SeBN-NPCs. *Source:* [46]/American Chemical Society/CC BY 4.0.

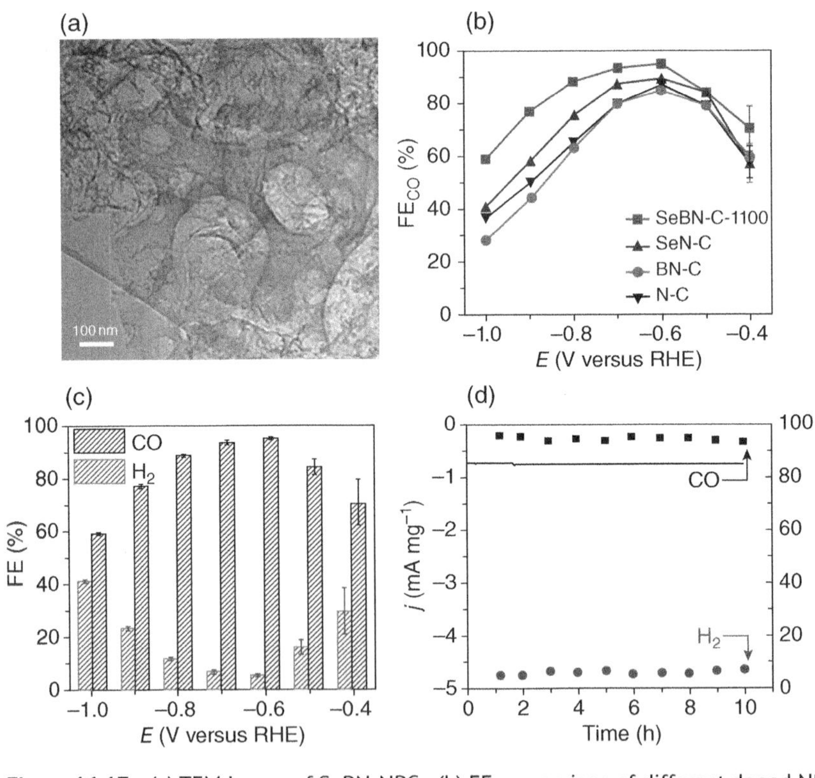

Figure 16.17 (a) TEM image of SeBN-NPCs, (b) FE comparison of different doped NPCs, (c) FE of products versus the applied potential for SeBN-NPCs (obtained at 1100 °C), (d) stability test of SeBN-NPCs for CO_2 ECR. *Source:* [46]/American Chemical Society/CC BY 4.0.

mono-doped (N or B) NPCs [49]. Song et al. have synthesized a cylindrical N-doped mesoporous carbon (cN-NPC) framework (Figure 16.18a–c). They have demonstrated by DFT calculations that cN-NPC contains a network of high electron density with pyridinic/pyrrolic N sites, highly favoring *CO dimerization and leading to conversion of C_2H_5OH with almost 100% selectivity, inferring complete hindrance of the formation of CO [54, Figure 16.18d].

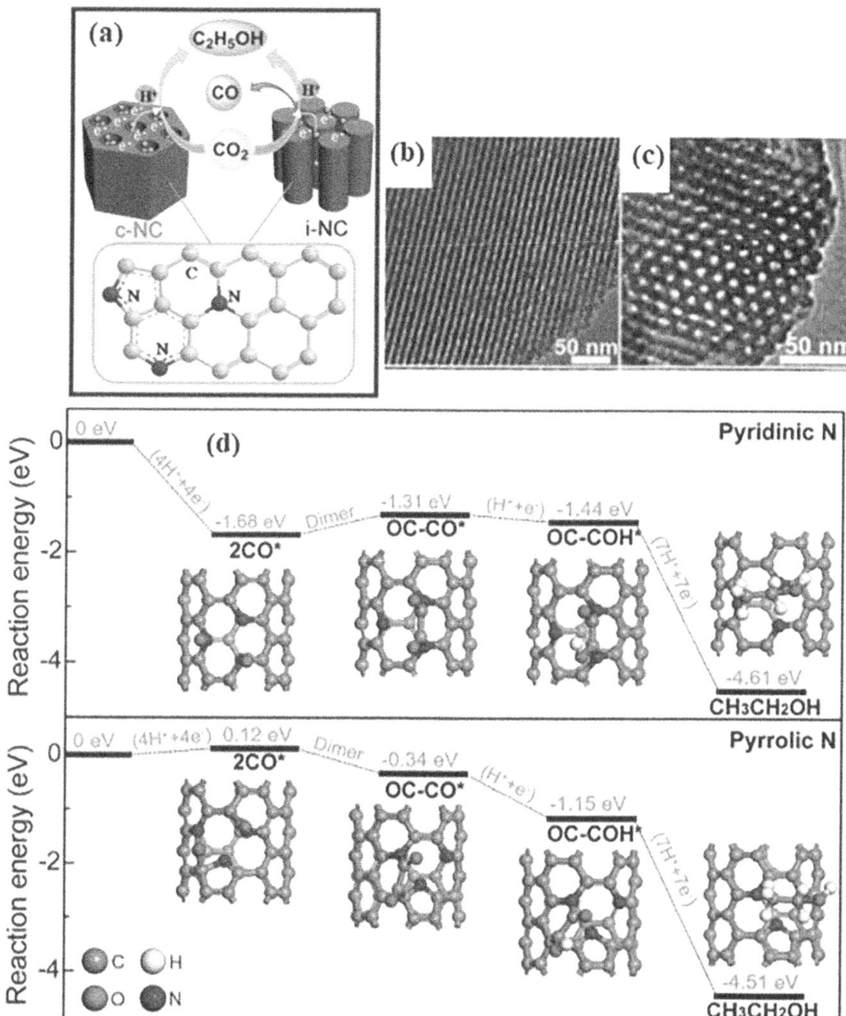

Figure 16.18 (a) Schematic illustration of cN-NPC (b, c) TEM image of cN-NPCs (d) Free energy diagram for the ECR of CO_2 to C_2H_5OH. *Source:* [54] / John Wiley & Sons.

16.5.1.4 Graphene

It is an allotropic form of carbon atoms having a 2D structure. It contains sp^2-hybridized carbon atoms in planar hexagonal arrangement. It is utilized as efficient electrocatalyst for ECR of CO_2 because of its robust chemical stability, high surface area, and better electrical conductivity. Pristine graphene has poor catalytic activity for ECR of CO_2 due to its inability to bind CO_2 molecules and large free energy barrier to create and stabilize the *COOH intermediate. Moreover, introduction of different heteroatom dopants to pristine graphene mitigates the above limitations related to ECR of CO_2 because introduction of these dopants changes not only its structure but also properties like wettability, electron densities, polarity, and edge strain [63]. In this regard, Mattarozzi et al. have synthesized N-containing or O-containing groups on the graphene surface. They have found that surface functionalization plays a vital role in selectivity of ECR of CO_2 because N-containing groups on graphene surfaces had 20 times greater selectivity toward CO as compared to O-containing groups [52]. Narayanaru et al. have synthesized B-doped graphene and they found that incorporation of B into graphene introduces asymmetric spin density, leading to efficient conversion of CO_2 to formate as compared to pristine graphene [53].

16.5.1.5 Nanodiamond

This is also an allotropic form of carbon having tetrahedral geometry with sp^3-hybridized carbon atoms. Pristine nanodiamond displays very poor electrical conductivity. Therefore, it has lower catalytic performance toward ECR of CO_2 but incorporation of heteroatoms substantially increases electrical conductivity, operational stability, and highly unfavorable potential window for HER [1]. Literature survey reveals that B and N are the most utilized dopants among the heteroatoms. In this regard, Nakata et al. have synthesized B-doped nanodiamond (B-ND). This B-ND exhibits high FE for the formation of HCHO [56]. Liu et al. have reported N-doped nanodiamond/Si rod array (N-ND/Si RA)-mediated synthesis of CH_3COOH. N-ND/Si RA overcomes usual low selectivity toward C_2 product formation and, therefore, it forms CH_3COOH over HCOOH with better FE [57]. Moreover, Liu et al. reported co-doped nanodiamonds having both B and N being used as a potential catalyst for the formation of C_2H_5OH via ECR of CO_2. Collaborative outcome of co-doping of B and N, high overpotential and more N contents lead to better catalytic performance of these catalysts [58].

16.5.2 Applications of Various Metal-Carbon Composite Nanomaterials in CO_2 Reduction

Metal-carbon composite materials are derived from combining metal with the different allotropes of carbon. In these materials, metal atoms are dispersed homogeneously in the matrices of carbon. These materials have been extensively employed to obtain various products by ECR of CO_2. Although non-metallic carbon-based electrocatalysts have been extensively used for ECR of CO_2 due to their unique properties but available literature reports reveal that mostly the metal is required to further reduce CO to obtain different products other than CO from ECR of CO_2 [64]. These metal-carbon composite nanomaterials not only reduce the amount of metals utilization but also generate exclusive active sites with synergistic effects of metal and carbon materials that enable to increase in the catalytic performance as well as selectivity toward product formation by ECR of CO_2 [65]. Metals in metal-carbon composite nanomaterials promote adsorption as well as activation of CO_2 molecules that lead to lowering the barrier of activation energy and the overpotential for ECR of CO_2. On the other hand, carbon materials offer a high surface area with tunable structure that may enhance diffusion of reactants and mass transfer, leading to expediting the ECR of CO_2. Metal-carbon composite nanomaterials also offer electrons to generate *H, which speeds up hydrogenation of *CO and forms a triple phase boundary at interface of electrode and solution, significantly enhancing the proximity CO_2 pressure on the surface of the catalyst, which is helpful in making a C—C bond [1]. In recent years, several metals like Sn, Au, Co, Ag, Fe, Ni, Cu, Pt, and Pd have been utilized for the ECR of CO_2. Out of these, Cu distinctively forms various hydrocarbons and oxygenated products. Cu offers moderate *CO-binding energy and a positive *H-binding energy [66]. This energy for *CO facilitates CO activation over CO poisoning, favoring multi-electron transfer that leads to formation of products requiring greater than two electrons. The positive binding energy for *H favors ECR of CO_2 over HER. However, selectivity toward a particular product with Cu is poor. This limitation can be mitigated by incorporation of different atoms as well as allotropes of carbon. Lim et al. have reported on the basis of DFT studies of the defective graphene-supported Cu nanoparticles that geometrical flexibility for adsorbates and anchoring property of defect sites associated with this material promotes ECR of CO_2 to obtain hydrocarbon fuels [67]. Recently, Wu et al. have developed copper–barium-decorated carbon-nanotube composite (CuBa-CNT) for the formation of C_2 products (C_2H_4 and C_2H_5OH) [68]. Partial current density obtained for C_2 products of CuBa-CNT was higher than that of $CuCO_3$ and $CuBaCO_3$ and also greater than the standard sputtered Cu. CuBa-CNT shows exceptional activity with high FE even at a lower potential of −0.7V (versus RHE) than the standard sputtered Cu (Figure 16.19a, b).

It has been found that in case of Fe, HER predominates over ECR of CO_2 because of lower overpotential for HER and high adsorption affinity toward CO. These limitations can be mitigated by incorporating Fe with different carbon allotropes. Selectivity toward formation of different products via ECR of CO_2 can also be tuned by changing both Fe forms and functional groups on the surface of carbon allotropes [69]. Genovese et al. have developed an electrocatalyst for ECR of CO_2 having

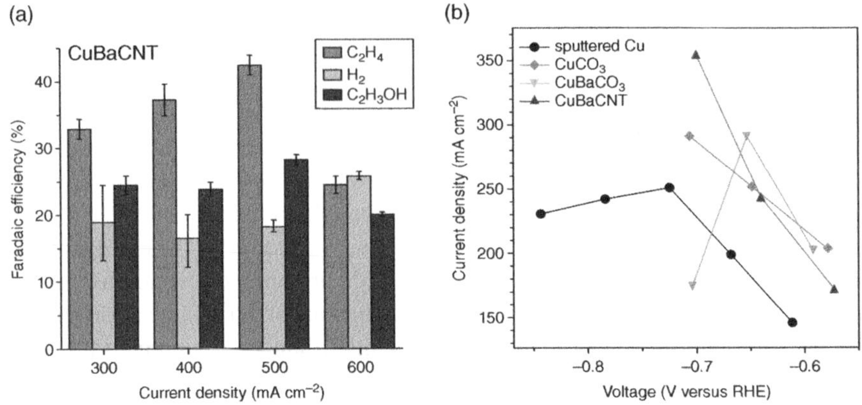

Figure 16.19 (a) FE of CuBa-CNT at various current densities, (b) current densities for C_2 molecules at different potentials. *Source:* [68]/The Royal Society of Chemistry.

Fe and Pt incorporated on oxidized CNTs. They have found that in case of Fe, CH_3OH and C_2H_5OH are formed as major products while Pt favors C_2H_5OH and isopropanol. Moreover, Fe-incorporated oxygen-functionalized CNTs favor C_2 products while Fe-incorporated nitrogen-functionalized CNTs favor C_3 products [69, 70]. On the other hand, oxidation state of Fe also affects the activity of electrocatalyst toward ECR of CO_2. Li et al. have developed phthalocyanine-modified graphene nanocomposites containing Fe(II) and Fe(III) and DFT analysis reveals that Fe(II)-containing electrocatalysts show greater activity over Fe(III) and Fe(II)/Fe(III) dimer [71]. Besides, Cu, Pt, and Fe, other metals like Ag, Au, Zn, Cd, Sn, Pd, etc., have been also utilized for the ECR of CO_2 with carbon-based materials. Among these metals, Zn, Cd, Sn, and Pd require relatively high overpotential and favor HER over ECR of CO_2. Some metals like Au possess low overpotential for ECR of CO_2 but its commercial application is hindered due to high cost and low availability.

Applications of some catalysts based on these metals associated with various carbon materials related to formation of various products by ECR of CO_2 are given in Table 16.2.

16.6 Imminent Challenges

Many metal-based as well as metal-free carbon nanomaterials have been established to obtain various C_1 and C_{2+} products via the ECR of CO_2. Various applications of these catalysts have been presented in Tables 16.1 and 16.2. Although various approaches utilizing these catalysts have been explored to synthesize

Table 16.2 Metal carbon composites-based catalysts for ECR of CO_2.

Catalyst	Electrolyte	Product	Potential (V)	FE	Reference
CuBa-CNT	KOH	C_2H_5OH and C_2H_4	−0.7 versus RHE	71%	[68]
Copper Nanoparticle/ N-Doped Graphene	$KHCO_3$	C_2H_5OH	−1.2 versus RHE	63%	[72]
Cu-N-doped carbon material	$CsHCO_3$	C_2H_5OH	−1.2V versus RHE	55%	[73]
CuN-NPC	$KHCO_3$	CH_3COCH_3	−0.36V versus RHE	36.7%	[74]
Fe on oxidized CNTs	$KHCO_3$	C_2H_5OH	−1.4V versus SHE	70.49%	[69]
Pt on oxidized CNTs	$KHCO_3$	Isopropanol	−1.4V versus SHE	34.82%	[69]
NFe-CNTs	$KHCO_3$	CO	−0.48V versus RHE	69%	[75]
phthalocyanine-modified graphene nanocomposites containing Fe(II)	$KHCO_3$	CO	−0.5V versus RHE	>90%	[71]
Graphene-confined Sn nanosheets	$NaHCO_3$	HCOOH	−0.96V versus RHE	85%	[76]
Bi_2O_3N-doped graphene	$KHCO_3$	HCOOH	−0.90V versus RHE	95%	[77]
Ag-CNTs	KOH	CO	−0.80V versus RHE	100%	[78]
AuNPs/graphene	$KHCO_3$	CO	−0.70V versus RHE	75%	[79]
Ni single-atom/ CNT	$KHCO_3$	CO	−0.70V versus RHE	~90	[80]
CoO-anchored N-doped mesoporous carbon and CNTs	$KHCO_3$	C_2H_5OH	−0.32V versus RHE	60.1%	[81]
Graphene/ZnO/ Cu_2O	$KHCO_3$	n-propanol	−0.9V versus Ag/AgCl	30%	[82]
Pd-MWNTs	$KHCO_3$	HCOOH	—	34.5%	[83]
Pd-MWNTs	$KHCO_3$	CH_3COOH	—	52.3%	[83]

C_1 and C_{2+} products, but still selectivity issues of desired products with high FE at low overpotential are a major obstacle toward commercialization of these catalysts. Some challenges associated with developing efficient catalysts are addressed below.

A literature survey reveals that currently available strategies for ECR of CO_2 are mostly efficient in furnishing C_1 products like CO and HCOOH but the synthesis of C_{2+} products is still a difficult task to accomplish. Formation of C_1 products undergoes via a simple two-electron pathway, whereas C_{2+} product formation requires transfer of more than two electrons, making the whole reaction pathway too difficult. Therefore, the quest for the development of an efficient electrocatalyst for the formation of various C_{2+} products still remains an objective of ongoing research in this area. Some metals like Cu possess low operational stability in aqueous electrolyte solution because of greater mobility and aggregation of atoms. Therefore, development of electrocatalysts having high operational stability is still required [1].

One of the limitations associated with the catalysts concerned with ECR of CO_2 is that these catalysts cannot be used at bulk level to produce commercially viable products. To overcome this problem, exploration of such catalysts is required that may be utilized for formation of ECR products at bulk level.

Another important aspect that needs to be paid attention to is that most of the products obtained via ECR of CO_2 are either mixed with an electrolytic solution or in case of gaseous products with residual CO_2 and H_2. Therefore, isolation and purification of desired products become a cumbersome exercise, making practical viability of these methods too difficult. Therefore, methods need to be evolved that furnish desired products in higher yields, fetching commercial viability of products obtained by ECR of CO_2 [84].

One of the major obstacles is to understand the actual mechanistic pathways involved during the formation of various products via ECR of CO_2. Formation of different products by ECR of CO_2 undergoes through involvement of various intermediates. The most common method employed for assessment of mechanistic pathways is via DFT analysis. However, the DFT calculations are based on simplistic model that considers only ideal situations. It does not consider structural reconstruction of electrocatalyst and change in pH value in proximity to catalyst [84]. Therefore, a major insight is required to understand the actual mechanistic pathway related to ECR of CO_2.

16.7 Conclusion

In summary, carbon-based electrocatalysts are preferred over metal-based traditional catalysts employed for ECR of CO_2 due to high onset potential for HER, profound surface area, tunable structure, high operational stability, high conductivity, better activity, and selectivity. These materials are cost-effective, environment-friendly, and easily available in bulk amounts. Therefore, these materials can be utilized for developing an efficient electrocatalyst to produce clean fuels and other industrially important feedstocks via ECR of CO_2. However, most of the available approaches for ECR of CO_2 satisfactorily produce only C_1 products. Thorough analysis of available literature reveals that among various C_1 products, the most common are CO and HCOOH. It has been found that FE related to formation of CO by utilizing various metal-free carbon and metal-carbon composite catalysts is up to 100% and in case of HCOOH, it is up to 95%. However, the generation of C_{2+} products with better discernment, low overpotential, and high current density is still a challenge. Among various C_{2+} products, the most explored one is C_2H_5OH with FE up to 93%. In addition to C_2H_5OH, the other C_{2+} products are CH_3COOH, *n*-propanol, CH_3COCH_3, and C_2H_4 with FE up to 92%, 30%, 37%, and 31%, respectively. Additionally, this chapter also depicts several important perspectives related to development and application of various carbon-based electrocatalysts to generate several important products by ECR of CO_2.

References

1 Zhao, K. and Quan, X. (2021). Carbon-based materials for electrochemical reduction of CO_2 to C_{2+} oxygenates: recent progress and remaining challenges. *ACS Catalysis* 11 (4): 2076–2097. https://doi.org/10.1021/acscatal.0c04714?ref=pdf.

2 Jin, S., Hao, Z., Zhang, K. et al. (2021). Advances and challenges for the electrochemical reduction of CO_2 to CO: from fundamentals to industrialization. *Angewandte Chemie (International Ed. in English)* 60 (38): 20627–20648. https://doi.org/10.1002/anie.202101818.

3 Habisreutinger, S.N., Schmidt-Mende, L., and Stolarczyk, J.K. (2013). Photocatalytic reduction of CO_2 on TiO_2 and other semiconductors. *Angewandte Chemie International Edition* 52 (29): 7372–7408. https://doi.org/10.1002/anie.201207199.

4 Sabiha, S., Sahoo, P.C., Martha, S., and Parida, K. (2016). A review of harvesting clean fuels from enzymatic CO_2 reduction. *RSC Advances* 6 (50): 44170–44194. https://doi.org/10.1039/C6RA05472B.

5 Lim, R.J., Xie, M., Sk, M.A. et al. (2014). Review on the electrochemical reduction of CO_2 in fuel cells, metal electrodes and molecular catalysts. *Catalysis Today* 233: 169–180. https://doi.org/10.1016/j.cattod.2013.11.037.

6 Grodkowski, J. and Neta, P. (2001). Copper-catalyzed radiolytic reduction of CO_2 to CO in aqueous solutions. *The Journal of Physical Chemistry B* 105 (21): 4967–4972. https://doi.org/10.1021/jp004567d.

7 Wang, W., Wang, S., Ma, X., and Gong, J. (2011). Recent advances in catalytic hydrogenation of carbon dioxide. *Chemical Society Reviews* 40 (7): 3703–3727. https://doi.org/10.1039/C1CS15008A.

8 Janáky, C., Hursán, D., Endrődi, B. et al. (2016). Electro- and photoreduction of carbon dioxide: the twain shall meet at copper oxide/copper interfaces. *ACS Energy Letters* 1 (2): 332–338. https://doi.org/10.1021/acsenergylett.6b00078.

9 Bai, S., Shao, Qi., Wang, P., Dai, Q., Wang, X.,Huang, X., "Highly active and selective hydrogenation of CO_2 to ethanol by ordered Pd–Cu nanoparticles" *Journal of the American Chemical Society*, 2017, 139, 20, 6827–6830, https://doi.org/https://doi.org/10.1021/jacs.7b03101.

10 Narváez-Celada, D. and Varela, A.S. (2022). CO_2 electrochemical reduction on metal–organic framework catalysts: current status and future directions. *Journal of Materials Chemistry A* 10 (11): 5899–5917. https://doi.org/10.1039/D1TA10440C.

11 Wu, J., Sharifi, T., Gao, Y. et al. (2019). Emerging carbon-based heterogeneous catalysts for electrochemical reduction of carbon dioxide into value-added chemicals. *Advanced Materials* 31: 1804257. https://doi.org/10.1002/adma.201804257.

12 Grimaud, A., Yin, W., Lepoivre, F., and Tarascon, J.M. (2018). Controlling the specific CO_2 adsorption on electrochemically formed metallic copper surfaces. *Journal of the Electrochemical Society* 165 (3): H163. https://doi.org/10.1149/2.0031805jes.

13 Serafini, M., Mariani, F., Fasolini, A. et al. (2021). Nanostructured copper-based electrodes electrochemically synthesized on a carbonaceous gas diffusion membrane with catalytic activity for the Electroreduction of CO_2. *ACS Applied Materials & Interfaces* 13 (48): 57451–57461. https://doi.org/10.1021/acsami.1c18844.

14 Appel, A.M. and Helm, M.L. (2014). Determining the overpotential for a molecular electrocatalyst. *ACS Catalysis* 4: 630–633. https://doi.org/10.1021/cs401013v.

15 Zhao, J., Sun, L., Canepa, S. et al. (2017). Phosphate tuned copper electrodeposition and promoted formic acid selectivity for carbon dioxide reduction. *Journal of Materials Chemistry A* 5 (23): 11905–11916. https://doi.org/10.1039/C7TA01871A.

16 Mehmood, R., Tariq, N., Zaheer, M. et al. (2018). One-pot synthesis of graphene-cobalt hydroxide composite nanosheets (Co/G NSs) for electrocatalytic water oxidation. *Scientific Reports* 8: 13772. https://doi.org/10.1038/s41598-018-32177-9.

17 Zhu, D.D., Liu, J.L., and Qiao, S.Z. (2016). Recent advances in inorganic heterogeneous electrocatalysts for reduction of carbon dioxide. *Advanced Materials* 28: 3423–3452. https://doi.org/10.1002/adma.201504766.

18 Cheng, Y., Yang, S., Jiang, S.P., and Wang, S. (2019). Supported single atoms as new class of catalysts for electrochemical reduction of carbon dioxide. *Small Methods* 3: 1800440. https://doi.org/10.1002/smtd.201800440.

19 Chen, J., Kamran, D., Wenhao, R. et al. (2009). Carbon-based catalysts for electrochemical CO2 reduction. *The Royal Society of Chemistry* 3 (11): 2890–2906. https://doi.org/10.1039/C9SE00527G.

20 Nitopi, S., Bertheussen, E., Scott, S.B. et al. (2019). Progress and perspectives of electrochemical CO_2 reduction on copper in aqueous electrolyte. *Chemical Reviews* 119 (12): 7610–7672. https://doi.org/10.1021/acs.chemrev.8b00705.

21 Kortlever, R., Shen, J., Schouten, K.J.P. et al. (2015). Catalysts and reaction pathways for the electrochemical reduction of carbon dioxide. *Journal of Physical Chemistry Letters* 6: 4073–4082. https://doi.org/10.1021/acs.jpclett.5b01559.

22 Nie, X., Esopi, M.R., Janik, M.J., and Asthagiri, A. (2013). Selectivity of CO_2 reduction on copper electrodes: the role of the kinetics of elementary steps. *Angewandte Chemie, International Edition* 52: 2459–2462. https://doi.org/10.1002/anie.201208320.

23 Sun, Z., Tao, M., Tao, H. et al. (2017). Fundamentals and challenges of electrochemical CO_2 reduction using two-dimensional materials. *Chem* 3: 560–587. https://doi.org/10.1016/j.chempr.2017.09.009.

24 Hori, Y., Takahashi, R., Yoshinami, A.Y., and Murata, A. (1997). Electrochemical reduction of CO at a copper electrode. *The Journal of Physical Chemistry. B* 101 (36): 7075–7081. https://doi.org/10.1021/jp970284i.

25 Genovese, G., Ampelli, C., Perathoner, S., and Centi, G. (2017). Mechanism of C–C bond formation in the electrocatalytic reduction of CO_2 to acetic acid. A challenging reaction to use renewable energy with chemistry. *Green Chemistry* 19: 2406–2415. https://doi.org/10.1039/C6GC03422E.

26 Liu, S., Yang, H.B., Su, X. et al. (2019). Rational design of carbon-based metal-free catalysts for electrochemical carbon dioxide reduction: a review. *Journal of Energy Chemistry* 36: 95–105. https://doi.org/10.1016/j.jechem.2019.06.013.

27 Liu, X. and Dai, L. (2016). Carbon-based metal-free catalysts. *Nature Reviews Materials* 1: 16064. https://doi.org/10.1038/natrevmats.2016.64.

28 Su, J., Liu, Y., Song, Y. et al. (2022). Recent development of nanomaterials for carbon dioxide electroreduction. *SmartMat* 3: 35–53. https://doi.org/10.1002/smm2.1106.

29 Yang, M., Wang, M., Zhang, M. et al. (2022). Nanostructured carbon electrocatalysts for clean energy conversion and storage: a mini review on the structural impact. *Frontiers in Materials* 9: 1090412. https://doi.org/10.3389/fmats.2022.1090412.

30 Jia, C., Dastafkan, K., Ren, W.H. et al. (2019). Carbon-based catalysts for electrochemical CO2 reduction. *Sustainable Energy and Fuels* 3: 2890–2906. https://doi.org/10.1039/C9SE00527G.

31 Yang, N., Waldvogel, S.R., and Jiang, X. (2016). Electrochemistry of carbon dioxide on carbon electrodes. *ACS Applied Materials and Interfaces* 8 (42): 28357–28371. https://doi.org/10.1021/acsami.5b09825.

32 Masood ul Hasan, I., Peng, L., Mao, J. et al. (2021). Carbon-based metal-free catalysts for electrochemical CO_2 reduction: activity, selectivity, and stability. *Carbon Energy* 3: 24–49. https://doi.org/10.1002/cey2.87.

33 Liu, J., Song, P., Ning, Z. et al. (2015). Recent advances in heteroatom-doped metal-free electrocatalysts for highly efficient oxygen reduction reaction. *Electrocatalysis* 6: 132–147. https://doi.org/10.1007/s12678-014-0243-9.

34 Duan, X., Xu, J., Wei, Z. et al. (2017). Metal-free carbon materials for CO_2 electrochemical reduction. *Advanced Materials* 29: 1701784. https://doi.org/10.1002/adma.201701784.

35 Kumar, B., Asadi, M., Pisasale, D. et al. (2013). Renewable and metal-free carbon nanofibre catalysts for carbon dioxide reduction. *Nature Communications* 4: 2819. https://doi.org/10.1038/ncomms3819.

36 Yang, H., Wu, Y., Lin, Q. et al. (2018). Composition tailoring via N and S Co-doping and structure tuning by constructing hierarchical pores: metal-free catalysts for high-performance electrochemical reduction of CO_2. *Angewandte Chemie International Edition* 57: 15476. https://doi.org/10.1002/anie.201809255.

37 Wu, J., Yadav, R.M., Liu, M. et al. (2015). Achieving highly efficient, selective and stable CO_2 reduction on nitrogen-doped carbon nanotubes. *ACS Nano* 9 (5): 5364–5371. https://doi.org/10.1021/acsnano.5b01079.

38 Zhao, Y., Yuan, Q., Sun, K. et al. (2023). Curvature effect of pyridinic N-modified carbon atom sites for electrocatalyzing CO_2 conversion to CO. *ACS Applied Materials and Interfaces* 15 (31): 37593–37601. https://doi.org/10.1021/acsami.3c08853.

39 Wang, H., Jia, J., Song, P. et al. (2017). Efficient electrocatalytic reduction of CO_2 by nitrogen-doped nanoporous carbon/carbon nanotube membranes: a step towards the electrochemical CO_2 refinery. *Angewandte Chemie (International Ed. in English)* 56 (27): 7847–7852. https://doi.org/10.1002/anie.201703720.

40 Xu, J., Kan, Y., Huang, R. et al. (2016). Revealing the origin of activity in nitrogen-doped nanocarbons towards electrocatalytic reduction of carbon dioxide. *ChemSusChem* 9: 1085–1089. https://doi.org/10.1002/cssc.201600202.

41 Zhang, S., Kang, P., Ubnoske, S.M. et al. (2014). Polyethylenimine-enhanced electrocatalytic reduction of CO_2 to formate at nitrogen-doped carbon nanomaterials. *Journal of the American Chemical Society* 136: 7845–7848. https://doi.org/10.1021/ja5031529.

42 Marepally, B.C., Ampelli, C., Genovese, C. et al. (2017). Role of small Cu nanoparticles in the behaviour of nanocarbon-based electrodes for the electrocatalytic reduction of CO_2. *Journal of CO_2 Utilization* 21: 534–542. https://doi.org/10.1016/j.jcou.2017.08.008.

43 Lu, X., Tan, T.H., Ng, Y.H., and Amal, R. (2016). Highly selective and stable reduction of CO_2 to CO by a graphitic carbon nitride/carbon nanotube composite electrocatalyst. *Chemistry* 22 (34): 11991–11996. https://doi.org/10.1002/chem.201601674.

44 Li, W., Seredych, M., Rodríguez-Castellón, E., and Bandosz, T.J. (2016). Metal-free nanoporous carbon as a catalyst for electrochemical reduction of CO_2 to CO and CH_4. *ChemSusChem* 9: 606–616. https://doi.org/10.1002/cssc.201501575.

45 Song, Y., Wang, S., Chen, W. et al. (2020). Enhanced ethanol production from CO_2 electroreduction at micropores in nitrogen-doped mesoporous carbon. *ChemSusChem* 13: 293–297. https://doi.org/10.1002/cssc.201902833.

46 Wang, W., Han, J., Sun, Y. et al. (2022). Metal-free SeBN ternary-doped porous carbon as efficient electrocatalysts for CO_2 reduction reaction. *ACS Applied Energy Materials* 5 (9): 10518–10525. https://doi.org/10.1021/acsaem.2c01201.

47 Liu, T., Ali, S., Lian, Z. et al. (2018). Phosphorus-doped onion-like carbon for CO_2 electrochemical reduction: the decisive role of the bonding configuration of phosphorus. *Journal of Materials Chemistry A* 6 (41): 19998–12000. https://doi.org/10.1039/C8TA06649C.

48 Kuang, M., Guan, A., Gu, Z. et al. (2019). Enhanced N-doping in mesoporous carbon for efficient electrocatalytic CO_2 conversion. *Nano Research* 12: 2324–2329. https://doi.org/10.1007/s12274-019-2396-6.

49 Jia, C., Ren, W., Chen, X. et al. (2020). (N, B) dual heteroatom-doped hierarchical porous carbon framework for efficient electroreduction of carbon dioxide. *ACS Sustainable Chemistry and Engineering* 8 (15): 6003–6010. https://doi.org/10.1021/acssuschemeng.0c00739.

50 Li, C., Wang, Y., Xiao, N. et al. (2019). Nitrogen-doped porous carbon from coal for high efficiency CO_2 electrocatalytic reduction. *Carbon* 151: 46–52. https://doi.org/10.1016/j.carbon.2019.05.042.

51 Song, Y., Chen, W., Zhao, C. et al. (2017). Metal-free nitrogen-doped mesoporous carbon for electroreduction of CO_2 to ethanol. *Angewandte Chemie, International Edition* 56: 10840. https://doi.org/10.1002/anie.201706777.

52 Mattarozzi, F., Tapia, R.M., van de Poll, R.C.J. et al. (2023). Surface-modified carbon materials for CO_2 electroreduction. *European Journal of Inorganic Chemistry* 26: e202300152. https://doi.org/10.1002/ejic.202300152.

53 Narayanaru, S., Azeezulla, N.M., Veettil, V.T. et al. (2015). Metal-free boron-doped graphene for selective electroreduction of carbon dioxide to formic acid/formate. *Chemical Communications* 51 (89): 16061–16064. https://doi.org/10.1039/C5CC06051F.

54 Liu, S., Yang, H., Huang, X. et al. (2018). Identifying active sites of nitrogen-doped carbon materials for the CO_2 reduction reaction. *Advanced Functional Materials* 28: 1800499. https://doi.org/10.1002/adfm.201800499.

55 Wang, H.X., Chen, Y.B., Hou, X.L. et al. (2016). Nitrogen-doped graphenes as efficient electrocatalysts for the selective reduction of carbon dioxide to formate in aqueous solution. *Green Chemistry* 18: 3250–3256. https://doi.org/10.1039/C6GC00410E.

56 Nakata, K., Ozaki, T., Terashima, C. et al. (2014). High-yield electrochemical production of formaldehyde from CO_2 and seawater. *Angewandte Chemie (International Ed. in English)* 53 (3): 871–874. https://doi.org/10.1002/anie.201308657.

57 Liu, Y.M., Chen, S., Quan, X., and Yu, H.T. (2015). Efficient electrochemical reduction of carbon dioxide to acetate on nitrogen-doped nanodiamond. *Journal of the American Chemical Society* 137: 11631–11636. https://doi.org/10.1021/jacs.5b02975.

58 Liu, Y., Zhang, Y., Cheng, K. et al. (2017). Selective electrochemical reduction of carbon dioxide to ethanol on a boron- and nitrogen-co-doped nanodiamond. *Angewandte Chemie, International Edition* 56: 15607–15611. https://doi.org/10.1002/anie.201706311.

59 Wu, J., Ma, S., Sun, J. et al. (2016). A metal-free electrocatalyst for carbon dioxide reduction to multi-carbon hydrocarbons and oxygenates. *Nature Communications* 7: 13869. https://doi.org/10.1038/ncomms13869.

60 Genovese, C., Ampelli, C., Perathoner, S., and Centi, G. (2013). Electrocatalytic conversion of CO_2 to liquid fuels using nanocarbon-based electrodes. *Journal of Energy Chemistry* 22: 202–213. https://doi.org/10.1016/S2095-4956(13)60026-1.

61 Zhou, W., Shen, H., Wang, Q. et al. (2019). N-doped peanut-shaped carbon nanotubes for efficient CO_2 electrocatalytic reduction. *Carbon* 152: 241–246. https://doi.org/10.1016/j.carbon.2019.05.078.

62 Gomis-Berenguer, A., Amaterz, E., Torres, S. et al. (2023). Nanoporous carbons for the electrochemical reduction of CO_2: challenges to discriminate the roles of nanopore confinement and functionalization. *Current Opinion in Electrochemistry* 40: 101323. https://doi.org/10.1016/j.coelec.2023.101323.

63 Wang, L., Lira, P., Hu, G. et al. (2023). Graphene-based electrodes and catalysts for electroreduction of CO_2 to low-carbon alcohols. *Materials Reports Energy* 3 (2): 100192. https://doi.org/10.1016/j.matre.2023.100192.

64 Siahrostami, S. (2019). Designing carbon-based materials for efficient electrochemical reduction of CO_2. *Industrial and Engineering Chemistry Research* 58 (2): 879–885. https://doi.org/10.1021/acs.iecr.8b05253.

65 Daiyan, R., Saputera, W.H., Masood, H. et al. (2020). A disquisition on the active sites of heterogeneous catalysts for electrochemical reduction of CO_2 to value-added chemicals and fuel. *Advanced Energy Materials* 10: 1902106. https://doi.org/10.1002/aenm.201902106.

66 Peterson, A.A., Abild-Pedersen, F., Studt, F. et al. (2010). How copper catalyzes the electroreduction of carbon dioxide into hydrocarbon fuels. *Energy and Environmental Science* 3 (9): 1311–1315. https://doi.org/10.1039/C0EE00071J.

67 Lim, D.H., Jo, J.H., Shin, D.Y. et al. (2014). Carbon dioxide conversion into hydrocarbon fuels on defective graphene-supported Cu nanoparticles from first principles. *Nanoscale* 6 (10): 5087–5092. https://doi.org/10.1039/C3NR06539A.

68 Wu, F.Y., Tsai, H.J., Lee, T.J. et al. (2023). Copper–barium-decorated carbon-nanotube composite for electrocatalytic CO_2 reduction to C_2 products. *Journal of Materials Chemistry A* 11 (25): 13217–13222. https://doi.org/10.1039/D3TA01912H.

69 Genovese, C., Ampelli, C., Perathoner, S., and Centi, G. (2013). Electrocatalytic conversion of CO_2 on carbon nanotube-based electrodes for producing solar fuels. *Journal of Catalysis* 308: 237–249. https://doi.org/10.1016/j.jcat.2013.08.026.

70 Ampelli, C., Genovese, C., Passalacqua, R. et al. (2012). The use of a solar photoelectrochemical reactor for sustainable production of energy. *Theoretical Foundations of Chemical Engineering* 46: 651–657. https://doi.org/10.1134/S0040579512060012.

71 Li, X., Chai, G., Xu, X. et al. (2020). Electrocatalytic reduction of CO_2 to CO over iron phthalocyanine-modified graphene nanocomposites. *Carbon* 167: 658–667. https://doi.org/10.1016/j.carbon.2020.06.036.

72 Song, Y., Peng, R., Hensley, D.K. et al. (2016). High-selectivity electrochemical conversion of CO_2 to ethanol using a copper nanoparticle/N-doped graphene electrod. *ChemistrySelect* 1: 6055–6061. https://doi.org/10.1002/slct.201601169.

73 Fontecave, M., Karapinar, D., Tran, N.H. et al. (2019). Electroreduction of CO_2 on single-site copper-nitrogen-doped carbon material: selective formation of ethanol and reversible restructuration of the metal sites. *Angewandte Chemie, International Edition* 58: 15098–15103. https://doi.org/10.1002/anie.201907994.

74 Zhao, K., Nie, X., Wang, H. et al. (2020). Selective electroreduction of CO_2 to acetone by single copper atoms anchored on N-doped porous carbon. *Nature Communications* 11: 2455. https://doi.org/10.1038/s41467-020-16381-8.

75 Pan, F., Zhao, H., Deng, W. et al. (2018). A novel N, Fe-decorated carbon nanotube/carbon nanosheet architecture for efficient CO_2 reduction. *Electrochimica Acta* 273: 154–161. https://doi.org/10.1016/j.electacta.2018.04.047.

76 Lei, F., Liu, W., Sun, Y., Xu, J., Liu, K., Liang, L., et al., "Metallic tin quantum sheets confined in grapheme toward high-efficiency carbon dioxide electroreduction", *Nature Communications*, 2016, 2, 7, 12697, https://doi.org/10.1038/ncomms12697.

77 Chen, Z., Mou, K., Wang, X., and Liu, L. (2018). Nitrogen-doped graphene quantum dots enhance the activity of Bi_2O_3 nanosheets for electrochemical reduction of CO_2 in a wide negative potential region. *Angewandte Chemie (International Ed. in English)* 57 (39): 12790–12794. https://doi.org/10.1002/anie.201807643.

78 Ma, S., Luo, R., Gold, J.I. et al. (2016). Carbon nanotube containing Ag catalyst layers for efficient and selective reduction of carbon dioxide. *Journal of Materials Chemistry A* 4: 8573–8578. https://doi.org/10.1039/C6TA00427J.

79 Zhao, Y., Wang, C., Liu, Y. et al. (2018). Engineering surface amine modifiers of ultrasmall gold nanoparticles supported on reduced graphene oxide for improved electrochemical CO_2 reduction. *Advanced Energy Materials* 8: 1801400. https://doi.org/10.1002/aenm.201801400.

80 Cheng, Y., Zhao, S., Johannessen, B. et al. (2018). Atomically dispersed transition metals on carbon nanotubes with ultrahigh loading for selective electrochemical carbon dioxide reduction. *Advanced Materials* 30 (13): 1706287. https://doi.org/10.1002/adma.201706287.

81 Du, J., Li, S., Liu, S. et al. (2020). Selective electrochemical reduction of carbon dioxide to ethanol by relay catalytic platform. *Chemical Science* 11: 5098–5104. https://doi.org/10.1039/D0SC01133A.

82 Geioushy, R.A., Khaled, M.M., Alhooshani, K. et al. (2017). Graphene/ZnO/Cu_2O electrocatalyst for selective conversion of CO_2 into n-propanol. *Electrochimica Acta* 245: 456–462. https://doi.org/10.1016/j.electacta.2017.05.185.

83 Lu, G., Wang, H., Bian, Z.Y., and Liu, X. (2013). Electrochemical reduction of CO_2 to organic acids by a Pd-MWNTs gas-diffusion electrode in aqueous medium. *Scientific World Journal* 2013: 424617. https://doi.org/10.1155/2013/424617.

84 Fan, L., Xia, C., Yang, F.Q. et al. (2020). Strategies in catalysts and electrolyzer design for electrochemical CO_2 reduction toward C_{2+} products. *Science Advances* 6: eaay3111. https://doi.org/10.1126/sciadv.aay3111.

17

Application of Carbon Nanomaterials in CO_2 Capture and Reduction

R. Sanjeevi, J. Anuradh, and Sandeep Tripathi

Nims Institute of Allied Medical Science and Technology, Nims University Rajasthan, Jaipur, Rajasthan, India

17.1 Introduction

As a consequence of human activities including industrial operations and the burning of fossil fuels, carbon dioxide (CO_2) levels in the atmosphere are rising. This has sparked a great deal of research into novel technologies that may be used to moderate and fight climate change. The use of carbon nanoparticles in CO_2 collection and reduction has become one of these technologies' most promising applications. Because of their special structural and chemical characteristics, carbon nanostructures provide innovative ways to absorb and transform CO_2, supporting the worldwide effort to lower greenhouse gas emissions [1].

Because of their unique qualities, carbon nanostructures are ideally adapted to tackle the problems related to CO_2 collection and reduction. These characteristics include large surface area, variable porosity, and excellent catalytic activity. This study examines the most recent advances and important discoveries in the application of carbon nanostructures for the dual purposes of absorbing CO_2 from emissions from industries and promoting its transformation into useful goods. We hope to shed light on the many functions that carbon nanomaterials play in the control of CO_2 and offer insights into their potential as effective and long-lasting strategies for addressing climate change [2, 3].

The extraordinary increase in global CO_2 levels, which is mostly attributable to human activity, has spurred scientists to look for novel approaches to combat climate change. The use of carbon nanoparticles in CO_2 capture and reduction

Carbon-based Nanomaterials for Green Applications, First Edition. Edited by Upendra Kumar, Piyush Kumar Sonkar, and Suman Lata Tripathi.

Published 2025 by John Wiley & Sons, Inc.

has become apparent as a transformational and promising strategy among many others. With their remarkable structural, chemical, and electrical characteristics, carbon nanomaterials provide a special combination of resources to address the dual problems of absorbing CO_2 emissions and turning them into useful goods. This study examines the many uses and current developments of carbon nanomaterials in the vital fields of CO_2 collection and reduction, illuminating their potential to fundamentally alter how we approach climate change mitigation [4, 5].

Rising sea levels, ecological disturbance, and global warming are just a few of the severe environmental effects brought about by the Earth's atmosphere's increasing CO_2 concentrations. There has never been a greater pressing need to reduce CO_2 emissions, which has sparked a group effort to create environmentally friendly technology that can slow down climate change [6]. The usage of carbon nanoparticles is one strategy that has received a lot of momentum because of its special qualities, which can be used for both CO_2 reduction and capture.

17.2 Different Types of Carbon Nanomaterials

The broad class of materials known as carbon nanoparticles has special qualities that make these promising for a variety of uses. Their dimensionality, or the number of dimensions that they occupy, is the basis for their classification [7].

17.2.1 Zero-Dimensional Carbon Nanomaterials

- **Fullerenes**: Fullerenes, also referred to as "buckyballs," are spherical or cage-like molecules composed of hexagonal and pentagonal rings of carbon atoms. With sizes generally ranging from 0.6 to 1 nm, these are among the smallest known nanomaterials made of carbon. Due to their remarkable thermal and electrical qualities, fullerenes are being investigated for usage in a number of applications, including electronic devices, medicine delivery, and energy storage.

17.2.2 One-Dimensional Carbon Nanomaterials

- **Carbon Nanotubes (CNTs)**: CNTs are cylindrical nanostructures composed of graphene sheets that have been rolled up. Depending on how many graphene layers they contain, they are categorized as single-walled (SWCNTs), double-walled (DWCNTs), or multi-walled (MWCNTs) nanotubes. The mechanical strength, electrical conductivity, and thermal conductivity of CNTs are remarkable. Numerous applications, such as composite materials, sensors, and electrical devices, are being investigated for their potential usage.

17.2.3 Two-Dimensional Carbon Nanomaterials

- **Graphene**: A single sheet of carbon atoms organized in a hexagonal lattice is called graphene. In addition to being one of the strongest and most conductive materials known, it is also the thinnest. Numerous possible uses for graphene exist, such as in sensors, batteries, and electrical devices.

17.2.4 Three-Dimensional Carbon Nanomaterials

- **Nanodiamonds**: Diamond nanoparticles, or nanodiamonds, usually have a diameter of two to ten nanometers. Their optical qualities, mechanical strength, and biocompatibility are outstanding. Numerous uses for nanodiamonds are being investigated, such as delivering drugs, biosensing, and nanomedicine.
- **Carbon Nanohorns**: Sp2 hybridized carbon atoms form the horn-shaped nanoparticles known as carbon nanohorns. They are exceptionally strong mechanical, electrical, and thermal conductors. Researchers are looking at using carbon nanohorns in composite materials, sensors, and electrical devices, among other uses.

17.2.5 Other Carbon Nanomaterials

Numerous more carbon nanomaterials are being researched and created in addition to the well-known ones mentioned earlier. Carbon nanofibers, carbon nanodots, and carbon nano-onions are a few of them [8, 9].

Research and development on carbon nanomaterials is expanding quickly, and it has the potential to completely transform a variety of sectors. We anticipate seeing even more cutting-edge uses for these materials in the years to come as our understanding of them deepens.

Carbon nanostructures are a dynamic and adaptable class of materials that have attracted a great deal of interest from scientists in many different fields. These materials, which differ in their structural and chemical makeup, have shown remarkable qualities that make them indispensable in a wide range of applications. We explore prominent instances of carbon nanomaterials in this thorough review, such as fullerenes, graphene, CNTs, nanodiamonds, and carbon nanohorns. These materials' unique qualities, which are frequently used at the nanoscale, make them invaluable for tackling a variety of issues, especially those pertaining to CO_2 control [10–12].

17.2.5.1 Fullerenes: Spherical Marvels

Fullerenes, also known as buckyballs or buckytubes, are an intriguing family of carbon nanomaterials distinguished by their unique cage-like or spherical shapes in Figure 17.1. Buckminsterfullerene (C60), the most prevalent fullerene, is made

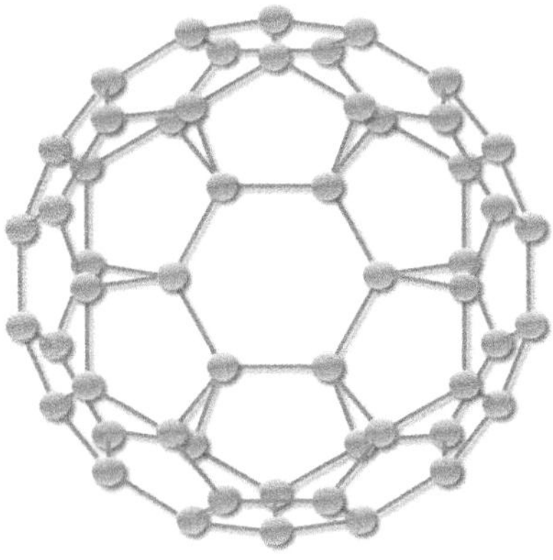

Figure 17.1 Structure of fullerenes.

up of 60 carbon atoms grouped in pentagonal and hexagonal rings, much like a football. The discovery of fullerenes in 1985 brought the researchers a Chemistry Nobel Prize. Fullerenes are distinguished by their distinct three-dimensional structure, which provides a broad surface area and a range of chemical reactivity [13, 14]. Fullerenes are being investigated for a variety of uses due to their excellent thermal and electrical conductivity and capacity to include other molecules inside their structure.

Fullerenes have demonstrated potential for gas capture and storage within their hollow structures in the context of CO_2 control. The possibility of functionalized fullerenes for improved CO_2 adsorption is being explored by researchers, opening the door for novel strategies in carbon capture systems.

17.2.5.2 Carbon Nanotubes (CNTs): Cylindrical Wonders

Like a continuous cylinder, CNTs are cylindrical nanostructures made of rolled-up graphene sheets in Figure 17.2. The number of concentric layers separates them into three main types: single-walled carbon nanotubes (SWCNTs), double-walled carbon nanotubes (DWCNTs), and multi-walled carbon nanotubes (MWCNTs). Because of the sp2-hybridized carbon atoms that make up its hexagonal lattice, CNTs have exceptional mechanical strength, thermal conductivity, and electrical conductivity [15, 16]. Because of their cylindrical shape, CNTs have a high aspect ratio, which increases their flexibility and tensile strength.

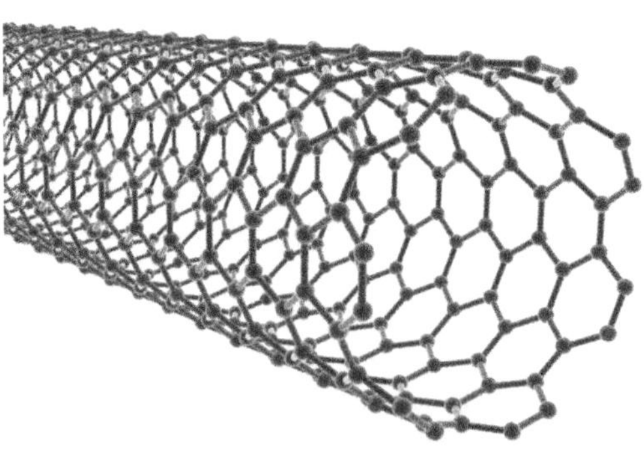

Figure 17.2 Structure of carbon nanotubes (CNTs).

CNTs are essential for gas separation and capture in the field of CO_2 management. Because of their porous nature, which permits selective CO_2 molecule adsorption, they are attractive candidates for gas storage and separation membranes. Furthermore, functionalized CNTs are being

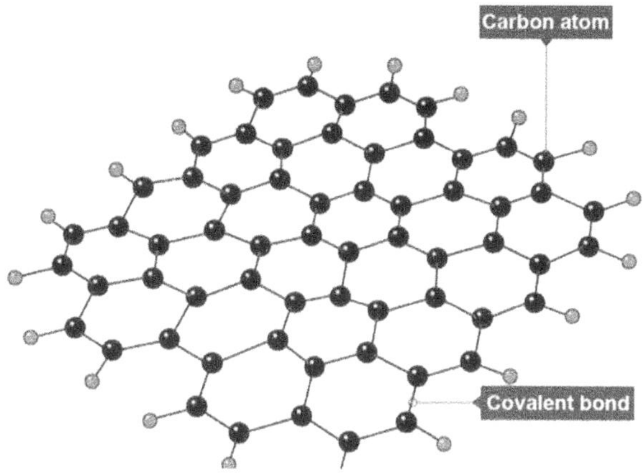

Figure 17.3 Structure of graphene.

investigated for their potential to catalyze CO_2 reduction reactions, which could advance the sustainable energy field.

17.2.5.3 Graphene: The Thinnest Marvel

A single sheet of carbon atoms organized in a hexagonal lattice to produce a two-dimensional structure is called graphene, and it is frequently heralded as the "wonder material." It is the basic component of all other carbon allotropes, such as fullerenes, graphite, and CNTs [17, 18] in Figure 17.3. The remarkable mechanical strength, thermal conductivity, and electrical conductivity of graphene are only a few of its remarkable qualities. Its large surface area and two-dimensional structure make it a desirable material for a wide range of uses, including energy storage and electronics.

Graphene shows potential in several areas when it comes to managing CO_2. Because of its large surface area, it can efficiently adsorb gases, which helps with carbon capture methods. Moreover, graphene-based catalysts have proven effective in CO_2 reduction processes, providing a long-term method of turning CO_2 into useful goods like fuels.

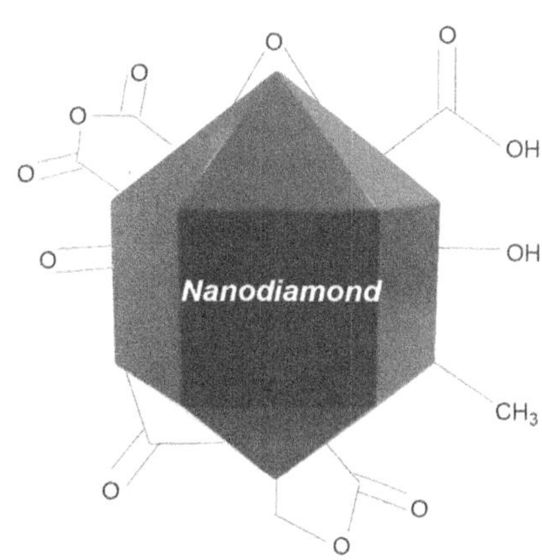

Figure 17.4 Structure of nanodiamonds.

17.2.5.4 Nanodiamonds: The Small, Shining Gems

In Figure 17.4, diamond nanoparticles, or nanodiamonds, usually have sizes

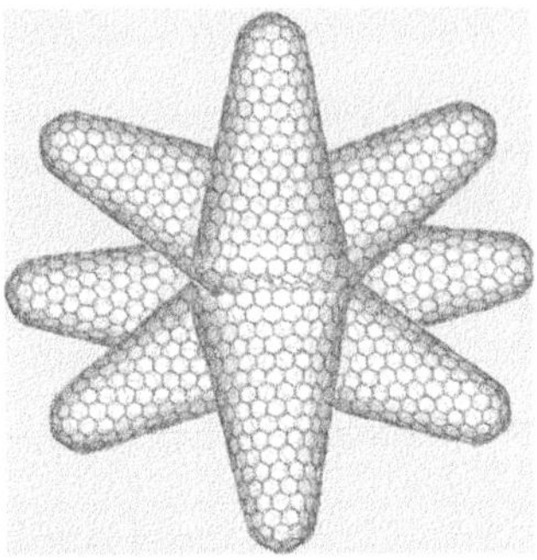

Figure 17.5 Structure of carbon nanohorns.

between two and ten nanometers. Because they are smaller than their larger counterparts, nanodiamonds have special qualities. Their remarkable mechanical strength, optical qualities, and biocompatibility render them advantageous for a wide range of uses [19–21]. Nanodiamonds are being investigated for their potential in CO_2 conversion and capture within the field of CO_2 management. Gas adsorption sites are provided by their high surface area and functional groups. Furthermore, the fact that they are compatible with biological systems creates opportunities for medicine administration and carbon capture-related biosensing applications.

17.2.5.5 Carbon Nanohorns: Horn-Shaped Marvels

Carbon nanohorns are horn-shaped nanoparticles made of carbon atoms that have undergone sp2 hybridization. These structures shown in Figure 17.5, which have drawn attention due to their remarkable mechanical strength, electrical conductivity, and thermal conductivity, resemble minuscule horns or cups [22–24]. The potential applications of carbon nanohorns in CO_2 management are fascinating. They support carbon capture methods because of their high surface area and porous nature, which make them ideal for gas adsorption. Furthermore, because of their adjustable electrical characteristics, nanohorns are a prospective option for catalyzing CO_2 reduction reactions, which will support environmentally friendly methods of using carbon.

17.3 Applications in CO_2 Management: Leveraging Unique Properties

By utilizing these carbon nanoparticles' special qualities, CO_2 management can effectively tackle the problems related to carbon capture and reduction. The high surface areas of carbon nanohorns, fullerenes, CNTs, graphene, and nanodiamonds offer plenty of places for CO_2 molecules to adsorb, making it easier for them to be captured from a variety of sources.

These nanoparticles' remarkable mechanical and electrical qualities help explain why they work so well as catalysts for CO_2 reduction processes [25]. These substances can act as catalysts to accelerate the transformation of captured CO_2 into useful products, providing a long-term solution for carbon utilization and aiding in the fight against climate change.

17.4 CO_2 Capture

Carbon capture technologies are becoming essential tools in the fight against climate change as the world community steps up its efforts. These technologies help reduce greenhouse gas emissions. Using carbon nanostructures' special qualities to trap CO_2 has drawn a lot of interest in these technologies. The function of carbon nanoparticles in absorbing CO_2 emissions from factories, power plants, and other point sources is examined in this article. Examples of these materials include activated carbon, graphene oxide, and metal-organic frameworks (MOFs). We explore the mechanisms underlying the effectiveness of these materials in CO_2 collection, highlighting the dimensions at the nanoscale that contribute to this efficiency [26, 27]. Additionally, current studies focused on modifying the exterior chemistry and functional structures of carbon nanomaterials for enhanced CO_2 adsorption capabilities are presented, emphasizing the possibility of selective CO_2 capture and its subsequent conversion into useful chemicals.

17.4.1 The Imperative for CO_2 Capture Technologies

Effective carbon capture solutions are urgently needed because of the increasing amounts of CO_2 in the atmosphere, which are mostly caused by human activity such as industrial processes and the burning of fossil fuels. By preventing CO_2 from being released into the atmosphere, CO_2 capture technologies seek to lessen the effect of CO_2 on climate change. Using the characteristics of carbon nanoparticles has become a viable strategy among the several methods for absorbing CO_2 [15, 16, 28].

17.4.2 Carbon Nanomaterials: Building Blocks for Capture

Because of their nanoscale dimensions, carbon nanomaterials come in a wide range of structures, each with its own set of unique features. MOFs, graphene oxide, and activated carbon are a few notable examples. These materials offer strong options for CO_2 emission capture due to their large surface area, adjustable porosity, and distinct chemical makeup [29].

17.4.3 High Surface Area and Porosity: Key Features for CO_2 Adsorption

The high surface area and porosity of carbon nanomaterials are among the main factors that make them suitable for CO_2 capture. These materials have a large number of active sites that are available for gas adsorption due to their nanoscale size. This property is best shown by activated carbon, which is made from different carbonaceous sources and treated to have a porous structure [30, 31]. Graphene oxide is a graphene derivative that has several functional sites on a large surface area due to the presence of oxygen-containing functional groups in the nanoscale layers. Conversely, MOFs are crystalline materials that have an ordered configuration of metal ions or clusters connected by organic ligands, resulting in a precisely designed porosity structure.

Carbon nanoparticles are more effective at trapping CO_2 molecules because of their large surface area and porosity. This makes them especially useful for absorbing emissions from sources where CO_2 concentrations vary, including power plants or industrial flue gases.

17.4.4 Nanoscale Efficiency: Enhanced CO_2 Capture

The effectiveness of CO_2 capture technologies is significantly improved by the nanoscale size of carbon compounds. These materials' porous properties allow gases to diffuse through, greatly increasing the likelihood of contact with active areas. Selectivity and quick adsorption are critical when extracting CO_2 from complicated gas mixtures, and this diffusion efficiency is critical to the process [29]. Activated carbon is a great illustration of how nanoscale efficiency leads to improved CO_2 extraction because of its complex network of pores. Activated carbon's mesopores and micropores combine to form a labyrinthine structure that maximizes CO_2 molecule exposure to adsorption sites, promoting a quick and effective capture process.

Similarly, graphene oxide provides a broad reactive surface for gas adsorption because of its two-dimensional structure and functional groups. The efficient collection of CO_2 is facilitated by the nanoscale architecture created by the planar arrangement of carbon atoms in graphene sheets.

Within the field of MOFs, the carefully designed porous structure at the nanoscale enables adjustable characteristics, rendering them adaptable options for CO_2 selective entrapment. Under certain circumstances, the framework's composition and pore size can be altered to improve CO_2 capture efficiency.

17.4.5 Tailoring Surface Chemistry for Enhanced CO_2 Adsorption

In order to improve CO_2 adsorption capabilities, carbon nanomaterials for CO_2 capture have recently made strides in this area by customizing their surface chemistry and functional groups. By optimizing the chemical interactions between the nanomaterials and CO_2 molecules, this strategic alteration seeks to increase the selectivity and efficiency of capture.

The functionalization of graphene surfaces with amine groups is one prominent example. Adding functional groups to amines produces a chemically active surface that increases graphene's affinity for CO_2. The formation of chemical bonds between amine groups and CO_2 molecules results in a more selective capture mechanism [32, 33]. In addition to exhibiting enhanced CO_2 adsorption capabilities, the amine-functionalized graphene also makes later procedures easier, such as turning absorbed CO_2 into useful chemicals. This dual purpose demonstrates how carbon nanoparticles with customized surface chemistry can be used for all-encompassing and sustainable CO_2 management.

17.4.6 Dual Functionality

17.4.6.1 From Capture to Conversion

Applications for carbon nanomaterials are made more versatile by their dual functionality, which includes effective CO_2 capture and the possibility of further conversion. While trapping CO_2 is the main goal to stop it from escaping into the atmosphere, there are other ways to convert the trapped gas into useful goods. Rather than only being sequestered, captured CO_2 can be used as a feedstock to produce materials, chemicals, or fuels, supporting the idea of carbon utilization. Carbon materials' specific surface chemistry and nanoscale dimensions are key factors in catalyzing these conversion processes [34, 35]. In electrochemical or photochemical reduction reactions, for example, amine-functionalized graphene can function as an effective catalyst in addition to its role in selective CO_2 collection. These processes provide a sustainable method of using CO_2 as a resource by converting it into useful molecules like ethylene (C_2H_4) or methane (CH_4).

17.5 Catalytic Conversion of CO_2: Nanomaterials as Agents of Change

Within the global energy and environmental responsibility movement, CO_2 is no longer only seen as pollution but also as a possible resource. Because of their special electrical qualities and catalytic capacities, carbon nanostructures are becoming increasingly important in the catalytic conversion of collected CO_2 into useful chemicals. After CO_2 is first captured, nanomaterials help with effective catalysis in chemical reactions, which converts CO_2 into chemicals, fuels, and other products with added value. The important role that carbon nanomaterials—metal nanoparticles supported on materials such as graphene—play in catalyzing CO_2 reduction processes through electrochemical and photocatalytic means is examined in this article. Because of their special qualities, nanoparticles are effective catalysts that help develop a sustainable strategy for using CO_2 as a resource for useful uses [36].

17.5.1 The Paradigm Shift: From Pollutant to Resource

The need to cut greenhouse gas emissions has caused a paradigm change in our understanding of CO_2 as the effects of human activity on the climate become more apparent. Rather than being exclusively thought of as a pollutant that needs to be sequestered, CO_2 is beginning to be seen as a resource that may be used for productive purposes. Catalytic CO_2 conversion appears to be a transformational strategy in this scenario. Researchers are looking into ways to use catalytic processes to transform CO_2 into useful substances, such as chemicals or fuels, as opposed to just absorbing and storing it. With their special nanoscale characteristics, carbon nanomaterials are leading this effort and serving as change agents in the search for sustainable CO_2 management.

17.5.2 Carbon Nanomaterials as Catalysts: Unlocking Potential

Carbon nanoparticles are effective in catalytic processes because of their high surface area, active sites, and special electrical characteristics. Graphene and metal nanoparticles supported on different carbon substrates have shown exceptional catalytic activity in the conversion of CO_2 among the multitude of carbon nanomaterials [37, 38]. A single sheet of carbon atoms organized in a hexagonal lattice, graphene has a high surface area and remarkable electrical conductivity. These characteristics, along with graphene oxide's adjustable nature, make it the perfect medium for encapsulating metal nanoparticles and boosting their catalytic activity.

When carbon nanomaterials are used to support metal nanoparticles, these materials act as catalysts, enabling the electrochemical and photocatalytic reduction of CO_2. The carbon substrate and the metal nanoparticles work in concert to generate catalytic sites, which allow CO_2 to be converted into useful compounds.

17.5.3 Electrochemical CO_2 Reduction: Harnessing Electrical Energy

Electrochemical CO_2 reduction is a process that converts CO_2 into a variety of products by means of electrical energy. In this regard, carbon nanomaterials have shown great potential, especially metal nanoparticles supported by graphene [2, 39]. The procedure usually entails submerging an electrode in a solution containing dissolved CO_2, with the electrode frequently being made of metal nanoparticles supported by graphene. Several products are formed when an electrical potential is given to the electrode, causing a sequence of reduction processes.

Effective electron transfer during the electrochemical reduction process is ensured by graphene's high electrical conductivity. As catalysts, the metal nanoparticles help reduce CO_2 to produce products including C_2H_4, CH_4, and carbon monoxide (CO) [40].

One major benefit of utilizing graphene-supported metal nanoparticles is the ability to control the product distribution selectively. The electrochemical cell's unique circumstances and the metal nanoparticles' makeup and morphology affect the product selectivity.

17.5.4 Photocatalytic CO_2 Reduction: Harvesting Solar Energy

Another way that carbon nanomaterials, and especially graphene-supported catalysts, might show off their catalytic abilities is through photocatalytic CO_2 reduction. Through this method, CO_2 is transformed into useful chemicals by harnessing solar energy. Because of its remarkable electrical characteristics, graphene is a great support for metal nanoparticles and an effective way to transfer charge during photocatalysis. Under the effect of solar radiation, the metal nanoparticles operate as active sites to catalyze the reduction of CO_2. Semiconductor materials are frequently used in the photocatalytic reduction of CO_2, and the addition of graphene improves the catalyst's overall effectiveness [32]. Graphene's effective ability to separate and transfer charges leads to increased photocatalytic activity, which in turn enhances the conversion of CO_2 into desirable chemicals.

17.5.5 Metal Nanoparticles on Graphene: Catalysts for Sustainable CO_2 Conversion

An intriguing feature of research on sustainable energy is the synergy of metal nanoparticles and graphene in catalyzing CO_2 conversion processes. Because of its two-dimensional structure, graphene offers a lot of surface area on which metal nanoparticles can be supported, forming catalytic sites that facilitate effective CO_2 reduction.

17.5.5.1 Tunable Catalytic Activity

Because metal nanoparticles supported on graphene have variable catalytic activity, scientists can customize the catalyst for particular CO_2 conversion pathways. The selection of the metal, its size, and the shape of the nanoparticles can affect the selectivity and kinetics of the reaction, giving the producer composition control.

For instance, the electrochemical conversion of CO_2 to C_2H_4 has demonstrated great potential when using copper nanoparticles supported on graphene. The catalytic activity is increased by the contact between copper and graphene, and by modifying the experimental setup, the selectivity toward C_2H_4 can be maximized.

17.5.5.2 Enhanced Electron Transfer

The remarkable electrical conductivity of graphene improves the effectiveness of electron transport during catalytic reactions. This is especially important for electrochemical and photocatalytic CO_2 reduction, as the redox reactions in both processes depend on quick and effective electron transfer.

The catalyst supported on graphene serves as a link to enable electron transmission from the metal nanoparticles to the CO_2 molecules' reaction sites. This improved electron transport increases the catalytic process's overall efficiency.

17.5.5.3 Stability and Durability

Catalysts' stability and endurance are crucial for their usefulness in industrial processes. When juxtaposed with unsupported metal nanoparticles or alternative catalytic systems, metal nanoparticles supported on graphene demonstrate superior stability and endurance.

The overall stability of the catalyst is aided by the mechanical strength and resistance to the chemical degradation of graphene. Because of the support that graphene provides, metal nanoparticles do not aggregate and can continue to be catalytically active for longer periods of time.

17.5.6 Selective CO_2 Reduction: Tailoring Products for Specific Applications

Selective distribution of the product is one of the major benefits of using carbon nanomaterials as catalysts in CO_2 conversion. The selection of metal nanoparticles, their dimensions, and the experimental setup enable scientists to regulate the synthesis of particular compounds, customizing the result for intended uses.

17.5.6.1 Methane Production

Catalysts based on graphene, especially those that include metals like copper or nickel, have been effective in the electrochemical reduction of CO_2 to create CH_4. In the search for sustainable energy sources, selective CH_4 generation from CO_2 is of great interest. CH_4 is a valuable fuel and chemical feedstock.

Researchers can increase the selectivity toward CH_4 by fine-tuning the catalyst composition and reaction conditions, providing a path toward the production of a flexible and clean energy carrier from CO_2.

17.5.6.2 Ethylene Synthesis

Graphene-supported catalysts can be used to selectively synthesize C_2H_4 from CO_2, a vital building component in the chemical industry. For example, copper nanoparticles supported on graphene have demonstrated potential in facilitating the electrochemical conversion of CO_2 to C_2H_4.

The capacity to generate C_2H_4 with specificity is important for using it in the synthesis of polymers and other chemicals, which helps create a more sustainable and circular economy.

17.5.6.3 Carbon Monoxide Generation

CO can also be produced by the electrochemical reduction of CO_2 with graphene-supported catalysts. Among its uses in industry are the synthesis of different compounds and the metallurgical processes where it acts as a reducing agent.

The capacity to produce CO with specificity broadens the possible uses for CO_2 conversion technologies and offers flexibility in applying CO_2 that has been captured.

17.5.7 Challenges and Future Directions

Although there has been progress in using carbon nanomaterials for the catalytic conversion of CO_2, there are still a number of obstacles to overcome and areas that might need further investigation.

17.5.7.1 Catalyst Efficiency

Increasing the catalytic processes' efficiency is still a research priority. To maximize the catalytic activity of metal nanoparticles supported on graphene, one must comprehend the complex interactions that occur between the metal nanoparticles, the graphene support, and the reaction environment.

To improve catalytic efficiency, researchers are investigating new approaches to catalyst design, such as adding more functional groups to graphene and using alloyed nanoparticles.

17.5.7.2 Reaction Selectivity

It is difficult to tailor the selectivity of CO_2 conversion reactions because different metal nanoparticles have varied catalytic preferences. A greater comprehension of the underlying response mechanisms and the variables controlling selectivity is necessary to achieve exact control over the product distribution.

The creation of sophisticated catalyst design principles and computational modelling to forecast and direct the selectivity of CO_2 conversion reactions may be future research avenues.

17.5.7.3 Scalability and Practical Applications

Scalability and pragmatic issues need to be taken into account in order for carbon nanomaterial-based catalysts to go from the lab to industrial uses. The widespread application requires scalable synthesis techniques for graphene-supported catalysts that guarantee repeatability and economy.

Furthermore, interdisciplinary cooperation between materials scientists, engineers, and industry participants is necessary to incorporate these catalysts into workable devices and systems for large-scale CO_2 conversion.

17.5.7.4 Environmental Impact

Ensuring the overall sustainability of CO_2 conversion technologies requires evaluating the environmental impact of catalysts based on carbon nanomaterials. Life cycle evaluations and in-depth investigations of these catalysts' environmental impact will support their appropriate use in industrial processes.

17.5.7.5 Cross-Disciplinary Collaboration

Working across disciplines is essential to advancing the field of carbon nanomaterial-based catalysts for CO_2 conversion. Collaboration is necessary among researchers in materials science, the field of chemistry engineering, and ecological science to meet the many potential problems in this rapidly developing subject.

17.6 Challenges and Future Directions

The use of carbon nanomaterials in CO_2 collection and reduction has great potential to help solve the urgent problem of growing CO_2 levels in the atmosphere. Like any emerging subject, there are obstacles that need to be overcome before these technologies can be widely used. This section delves into the primary obstacles and delineates the forthcoming paths for investigation and advancement.

17.6.1 Scalability of Production Processes

The scalability of production processes is a major obstacle in the use of carbon nanomaterials for CO_2 collection and reduction. The effectiveness of nanomaterials is demonstrated by several laboratory-scale studies. However, transferring these procedures to industrial sizes necessitates overcoming formidable obstacles. High-quality carbon nanomaterials, including graphene and CNTs, need to be produced using synthesis techniques and production procedures that are optimized for large-scale manufacturing without sacrificing their structural and functional qualities.

To improve scalability, researchers are investigating a range of production processes, such as liquid-phase exfoliation, chemical vapor deposition, and template-assisted procedures. Developing economical and sustainable production methods that may satisfy the requirements of broad implementation in sectors looking for carbon capture solutions is essential.

17.6.2 Long-Term Stability of Nanomaterials

A crucial factor in the practical application of carbon nanomaterials is their long-term stability under abrasive operation environments. High pressures, temperatures, and corrosive conditions are common in industrial operations, which over time may put nanomaterials' structural integrity and functionality in jeopardy. It is crucial to comprehend the processes of degradation and create plans to improve these materials' stability in harsh environments.

Degradation problems can be reduced by functionalizing or coating carbon nanostructures with protective coatings. Furthermore, the durability of composite materials or strong support matrices can be improved by adding nanoparticles. The dependability and durability of carbon nanomaterial-based CO_2 collection and reduction systems will be dependent on long-term stability experiments conducted under practical operating settings.

17.6.3 Economic Viability

Any technology needs to be profitable in order to be widely used. A major obstacle is the cost of producing carbon nanomaterials, which is further compounded by the expenses of setting up and maintaining capture and reduction systems. These technologies' viability from an economic standpoint is contingent upon various aspects, including the cost of raw materials, the energy needed for synthesis, and the overall effectiveness of the collection and conversion processes.

Optimizing synthesis routes, investigating substitute precursor materials, and creating effective nanomaterial recycling techniques are some of the steps taken to lower production costs. Enhancing economic viability will also be greatly aided by advancements in the design of capture and reduction systems to reduce energy consumption and operational expenses.

17.7 Future Directions

The future of carbon nanomaterials in CO_2 collection and reduction depends on purposeful research projects targeted at improving performance, comprehending underlying interactions, and developing the field as we negotiate the obstacles. Future research can be directed in several important ways:

17.7.1 Optimization of Synthesis and Engineering

In order to attain improved performance and durability, future research efforts should concentrate on optimizing the synthesis and engineering of carbon nanomaterials. Getting exact control over the size, shape, and functionalization of nanomaterials entails improving on already-used synthesis methods and creating brand-new ones. Tailored nanomaterials with enhanced capabilities for CO_2 collection and reduction can be produced using rational design approaches, supported by sophisticated characterization methods and computational modelling.

For example, in the case of graphene, continuous attempts are being made to address production issues and regulate the number of layers in a scalable way. To increase a nanomaterial's catalytic activity in CO_2 reduction processes, methods such as chemical functionalization and doping can be investigated to add certain active sites to the surface of the material.

17.7.2 Exploration of Novel Catalytic Mechanisms

Investigating novel catalytic processes for CO_2 reduction is necessary to advance the area. Although conventional catalysts have demonstrated potential, novel catalytic pathways can be unlocked by comprehending the precise interactions that exist between carbon nanomaterials and CO_2 at the atomic and molecular levels. Research on modifying the electronic structure of nanomaterials to improve their stability, selectivity, and catalytic activity is very promising.

Advanced spectroscopic methods and computational modelling can shed light on the minute features of catalytic processes. Through the intricate interaction between CO_2 molecules and nanomaterials during catalysis, scientists can create more effective and focused catalysts, advancing the creation of environmentally friendly CO_2 conversion technology.

17.7.3 Fundamental Interactions between Nanomaterials and CO_2

To advance the research, a better comprehension of the underlying interactions between carbon nanostructures and CO_2 is necessary. This entails investigating the chemical routes, adsorption processes, and diffusion kinetics that take place at the nanomaterial interface during CO_2 capture and reduction. Under operational conditions, real-time insights into these interactions can be obtained through experimental techniques like microscopy and in situ spectroscopy.

Clarifying how external elements like temperature and pressure affect the functionality and stability of nanomaterials should be another goal of research. With this thorough knowledge, scientists will be able to create customized, resilient nanomaterials that function well in a variety of environments, increasing the dependability of CO_2 capture and reduction systems.

17.7.4 Integration of Nanomaterials into Multi-Functional Systems

Future research on CO_2 collection and reduction strategies based on carbon nanoparticles may need the integration of nanomaterials into multi-purpose systems. The synergistic effects that result from combining nanoparticles with other materials, including polymers or MOFs, can improve overall performance. Customized locations for CO_2 capture, effective reactant transportation, and enhanced catalytic activity can all be obtained using multi-functional systems.

Beyond only the materials themselves, this integrated approach is used in the design of entire systems, such as reactors and separation units. In order to develop comprehensive and effective CO_2 capture and conversion processes, future research should examine the synergies that may be obtained by integrating various materials and engineering solutions.

17.7.5 Techno-Economic and Life Cycle Assessments

In order to examine the economic feasibility of CO_2 capture and reduction methods based on carbon nanomaterials, thorough techno-economic and life cycle analyses should be incorporated into future research. These evaluations can offer a comprehensive picture of these technologies' long-term viability from an economic standpoint, as well as their environmental effect and sustainability. Researchers need to account for the costs of system installation, operation, and maintenance in addition to the production costs of nanomaterials.

Researchers can find cost-effective solutions for large-scale implementation with the use of techno-economic modelling. Life cycle assessments will assist in assessing the technologies' environmental impact and ensuring that the advantages of CO_2 capture and reduction are not outweighed by unfavorable environmental effects.

17.7.6 Collaboration and Interdisciplinary Research

Interdisciplinary cooperation is required due to the complexity of the opportunities and problems associated with CO_2 capture and reduction using carbon nanomaterials. To address the complex issues in this sector, scientists from materials science, chemistry, engineering, environmental science, and economics must work together. Collaboration across disciplines can result in creative fixes, all-encompassing strategies, and a deeper comprehension of the intricate details.

Academic institutions, business associates, and government organizations should support cooperative research projects that unite specialists from many disciplines. Research initiatives and funding sources for multi-disciplinary projects can hasten development and aid in the creation of comprehensive solutions.

17.8 Conclusion

Our strategy for reducing CO_2 emissions and combating climate change has undergone a paradigm leap with the use of carbon nanomaterials. In their pursuit of sustainable technology, researchers are breaking new ground by utilizing the special qualities of nanomaterials. Unprecedented changes to turn CO_2, a once-pervasive environmental concern, into a resource for a greener and more sustainable future arise as we explore further into the nanodimensions of carbon compounds. The revolutionary potential of carbon nanomaterials in CO_2 collection and reduction is set to redefine the environmental and energy technology landscape as this field develops, providing a glimmer of hope in the battle against climate change.

References

1 Ali, E.Y. (2023). Chapter 16—environmental impact of the use of nanomaterials for CO2 capture and conversion technologies. In: *Nanomaterials for Carbon Dioxide Capture and Conversion Technologies* (ed. S.A. Mazari, N.M. Mubarak, and M. Tripathi). Micro and Nano Technologies, 401–422. Elsevier. https://doi.org/10.1016/B978-0-323-89851-5.00012-3.

2 Shingdilwar, S., Kumar, D., Sahu, B., and Banerjee, S. (2022). Straightforward synthesis of multifunctional porous polymer nanomaterials for CO2 capture and removal of contaminants. *Polymer Chemistry* 13 (15): 2165–2172. https://doi.org/10.1039/d2py00067a.

3 Sanjeevi, R., Sathvara, P., Tripathi, S., and Anuradha, J. (2024). Methodology and approaches for biodiversity conservation. In: *Impact of Societal Development and Infrastructure on Biodiversity Decline*, 271–299. IGI Global. https://doi.org/10.4018/979-8-3693-6950-0.ch015.

4 Castro-Muñoz, R., Zamidi Ahmad, M., Malankowska, M., and Coronas, J. (2022). A new relevant membrane application: CO2 direct air capture (DAC). *Chemical Engineering Journal* 446: 137047. https://doi.org/10.1016/j.cej.2022.137047.

5 Cedeño Morales, E.M., Kharisov, B.I., and Méndez-Rojas, M.A. (2021). CO2 photoreduction by MOF-derived carbon nanomaterials: a review. *Materials Today Proceedings* 46: 2982–2997. https://doi.org/10.1016/j.matpr.2020.12.702.

6 Chakrabortty, S., Kumar, R., Nayak, J. et al. (2023). Green synthesis of MeOH derivatives through in situ catalytic transformations of captured CO2 in a membrane integrated photo-microreactor system: a state-of-art review for carbon capture and utilization. *Renewable and Sustainable Energy Reviews* 182: 113417. https://doi.org/10.1016/j.rser.2023.113417.

7 Mazri, N.A.F., Arifutzzaman, A., Aroua, M.K. et al. (2023). Graphene and its tailoring as emerging 2D nanomaterials in efficient CO2 absorption: a state-of-the-art interpretative review. *Alexandria Engineering Journal* 77: 479–502. https://doi.org/10.1016/j.aej.2023.06.070.

8 Firdaus, R.M., Desforges, A., Rahman Mohamed, A., and Vigolo, B. (2021). Progress in adsorption capacity of nanomaterials for carbon dioxide capture: a comparative study. *Journal of Cleaner Production* 328: 129553. https://doi.org/10.1016/j.jclepro.2021.129553.

9 Kim, C., Talapaneni, S.N., and Dai, L. (2023). Porous carbon materials for CO2 capture, storage and electrochemical conversion. *Materials Reports: Energy* 3 (2): 100199. https://doi.org/10.1016/j.matre.2023.100199.

10 Li, H. (2023). CO2 capture by various nanoparticles: Recent development and prospective. *Journal of Cleaner Production* 414: 137679. https://doi.org/10.1016/j.jclepro.2023.137679.

11 Li, S., Li, S., Wu, Z. et al. (2024). Rational design of highly efficient carbon-based materials for electrochemical CO2 reduction reaction. *Fuel* 357: 129760. https://doi.org/10.1016/j.fuel.2023.129760.

12 Sathvara, P., Anuradha, J., Tripathi, S., and Sanjeevi, R. (2024). Current scenario of biodiversity loss due to developmental activities. In: *Impact of Societal Development and Infrastructure on Biodiversity Decline*, 50–69. IGI Global. https://doi.org/10.4018/979-8-3693-6950-0.ch005.

13 Chin, B.L.F., Loy, A.C.M., Cheah, K.W. et al. (2023). Chapter 8—graphene-based nanomaterials for CO2 capture and conversion. In: *Nanomaterials for Carbon Dioxide Capture and Conversion Technologies* (ed. S.A. Mazari, N.M. Mubarak, and M. Tripathi). Micro and Nano Technologies, 211–243. Elsevier. https://doi.org/10.1016/B978-0-323-89851-5.00011-1.

14 Ramakrishnan, S., Sathvara, P., Tripathi, S., and Jayaraman, A. (2024). Water pollutants, sensor types, and their advantages and challenges. In: *Sensors for Environmental Monitoring, Identification, and Assessment*, 78–101. IGI Global. https://doi.org/10.4018/979-8-3693-1930-7.ch005.

15 Wang, J., Zhou, Y., Hu, X., and Liu, J. (2022). Adsorption of CO2 by amine-modified novel nanomaterials. *Physics and Chemistry of the Earth, Parts A/B/C* 126: 103154. https://doi.org/10.1016/j.pce.2022.103154.

16 Wang, R., Ashkanani, H.E., Li, B., and Morsi, B.I. (2022). Development of an innovative process for post-combustion CO2 capture to produce high-value NaHCO3 nanomaterials. *International Journal of Greenhouse Gas Control* 120: 103761. https://doi.org/10.1016/j.ijggc.2022.103761.

17 Lu, C., Shi, X., Liu, Y. et al. (2021). Nanomaterials for adsorption and conversion of CO2 under gentle conditions. *Materials Today* 50: 385–399. https://doi.org/10.1016/j.mattod.2021.03.016.

18 Ma, G., Qin, L., Liu, Y. et al. (2023). A review of CO2 reduction reaction catalyzed by atomical-level Ag nanomaterials: atom-precise nanoclusters and atomically dispersed catalysts. *Surfaces and Interfaces* 36: 102555. https://doi.org/10.1016/j.surfin.2022.102555.

19 Pajdak, A., Skoczylas, N., Dębski, A. et al. (2019). CO2 and CH4 sorption on carbon nanomaterials and coals—comparative characteristics. *Journal of Natural Gas Science and Engineering* 72: 103003. https://doi.org/10.1016/j.jngse.2019.103003.

20 Quan, C., Zhou, Y., Wang, J. et al. (2023). Biomass-based carbon materials for CO2 capture: a review. *Journal of CO_2 Utilization* 68: 102373. https://doi.org/10.1016/j.jcou.2022.102373.

21 Abbasi, T., Sanjeevi, R., Makhija, M., and Abbasi, S.A. (2012). Role of vitamins B-3 and C in the fashioning of granules in UASB reactor sludge. *Applied Biochemistry and Biotechnology* 167 (2): 348–357. https://doi.org/10.1007/s12010-012-9691-y.

22 Ruhaimi, A.H., Hitam, C.N.C., Aziz, M.A.A. et al. (2022). The role of surface and structural functionalisation on graphene adsorbent nanomaterial for CO2 adsorption application: recent progress and future prospects. *Renewable and Sustainable Energy Reviews* 167: 112840. https://doi.org/10.1016/j.rser.2022.112840.

23 Shen, Y. (2022). Preparation of renewable porous carbons for CO2 capture—a review. *Fuel Processing Technology* 236: 107437. https://doi.org/10.1016/j.fuproc.2022.107437.

24 Sathvara, P.B., Jayaraman, A., Tripathi, S., and Ramakrishnan, S. (2024). Air sensors and their capabilities. In: *Sensors for Environmental Monitoring, Identification, and Assessment*, 10–24. IGI Global. https://doi.org/10.4018/979-8-3693-1930-7.ch002.

25 Sathvara, P.B., Anuradha, J., Tripathi, S., and Sanjeevi, R. (2024). Biodiversity and its importance: an introduction. In: *Biodiversity Loss Assessment for Ecosystem Protection*, 1–17. IGI Global. https://doi.org/10.4018/979-8-3693-3330-3.ch001.

26 Rathod, A.B., Anuradha, J., Sathvara, P.B. et al. (2023). Vegetational change detection using machine learning in GIS technique: a case study from Jamnagar (Gujarat). *Journal of Data Acquisition and Processing* 38 (1): 1046–1061. https://doi.org/10.5281/zenodo.7700655.

27 Rathod, A.B., Sathvara, P.B., Tripathi, A. et al. (2022). Vegetational cartography analysis utilizing multi-temporal ndvi data series: a case study from Rajkot district (Gujarat), India. *Journal of Tianjin University Science and Technology* 55 (4): 490–497. https://doi.org/10.17605/OSF.IO/UGJYM.

28 Sathvara, P.B., Anuradha, J., Tripathi, S., and Sanjeevi, R. (2023). Impact of climate change and its importance on human performance. In: *Insights on Impact of Climate Change and Adaptation of Biodiversity*, 1e, 1–9. KD Publication.

29 Zhang, F., Li, Y.H., Qi, M.Y. et al. (2021). Photothermal catalytic CO2 reduction over nanomaterials. *Chem Catalysis* 1 (2): 272–297. https://doi.org/10.1016/j.checat.2021.01.003.

30 Wang, X., Bao, Z., Akhmedov, N.G. et al. (2023). Unique biological amino acids turn CO2 emission into novel nanomaterials with three switchable product pathways. *Environmental Technology & Innovation* 32: 103279. https://doi.org/10.1016/j.eti.2023.103279.

31 Wu, J., Wang, S., Qi, J. et al. (2022). Research advances in the light-driven conversion of CO2 to valuable chemicals by two-dimensional nanomaterials. *Materials Today Energy* 28: 101065. https://doi.org/10.1016/j.mtener.2022.101065.

32 Zhang, W., Yang, Y., Li, Y. et al. (2023). Recent progress on integrated CO2 capture and electrochemical upgrading. *Materials Today Catalysis* 2: 100006. https://doi.org/10.1016/j.mtcata.2023.100006.

33 Zhu, X., Du, C., Gao, B., and He, B. (2023). Strategies to improve the mass transfer in the CO2 capture process using immobilized carbonic anhydrase. *Journal of Environmental Management* 332: 117370. https://doi.org/10.1016/j.jenvman.2023.117370.

34 Abbasi, T., Sanjeevi, R., Anuradha, J., and Abbasi, S.A. (2013). Impact of Al 3+ on sludge granulation in UASB reactor. *Indian Journal of Biotechnology* 12 (2): 254–259.

35 Jain, R., Chauhan, N.S., Anuradha, J. et al. (2016). Evaluation of water quality of a reservoir in Sanganer (Pink City) Rajasthan: a multivariate study. *Journal of Chemistry and Chemical Sciences* 6 (11).

36 Sanjeevi, R., Haruna, M., Tripathi, S. et al. (2017). Impacts of global carbon foot print on the marine environment. *International Journal of Engineering Research and Technology* 5: 51–54.

37 Ramakrishnan, S. and Jayaraman, A. (2019). Global warming and pesticides in water bodies. In: *Handbook of Research on the Adverse Effects of Pesticide Pollution in Aquatic Ecosystems*, 421–436. IGI Global. https://www.igi-global.com/chapter/global-warming-and-pesticides-in-water-bodies/213519.

38 Ramakrishnan, S. and Jayaraman, A. (2019). Pesticide contaminated drinking water and health effects on pregnant women and children. In: *Handbook of Research on the Adverse Effects of Pesticide Pollution in Aquatic Ecosystems*, 123–136. IGI Global. https://www.igi-global.com/chapter/pesticide-contaminated-drinking-water-and-health-effects-on-pregnant-women-and-children/213500.

39 Luo, W., Wang, C., Jin, M. et al. (2023). Research progress on nanoconfined ILs in two-dimensional composite membranes for CO2 capture. *ScienceDirect* 16: 125406. https://www.sciencedirect.com/science/article/abs/pii/S1383586623023146.

40 Zunita, M. (2023). Integration of graphene oxide/ionic liquid for CO2 capture: recent progress. *Case Studies in Chemical and Environmental Engineering* 8: 100486. https://doi.org/10.1016/j.cscee.2023.100486.

18

Industrial Applications of Carbon Nanomaterials

Y. Pathania[1], P. K. Ahluwalia[2], and Pooja Kapoor[3]

[1] *Department of Physics, DAV Post graduate College, Sector-10, Chandigarh, India*
[2] *Department of Physics, Himachal Pradesh University, Shimla, Himachal Pradesh, India*
[3] *School of Basic and Applied Sciences, Maharaja Agrasen University, Baddi, Himachal Pradesh, India*

18.1 Introduction

Carbon nanomaterials (CNs) are nanomaterials primarily containing carbon atoms. This group includes graphene, graphene nanoribbon (GNR), CNTs, fullerene, CQDs, CNFs, and many others [1, 2]. These materials possess distinctive properties which make them highly valuable for various applications [2, 3]. Their nanoscopic size and larger surface area contribute to their notable characteristics. Additionally, their exclusive carbon composition ensures high stability, low toxicity, and environmentally friendly attributes. By modifying the surfaces of these materials, a whole new range of reagents can be developed, offering significant potential for novel applications [4, 5]. The behavior of carbon as an insulator, semiconductor, or conductor depends on its structure and carbon-to-carbon linkage. Furthermore, carbon exhibits a remarkable ability to absorb various wavelengths. These nanomaterials also demonstrate excellent biocompatibility. Figure 18.1 provides an overview of the numerous features of carbon-based nanostructures that have the potential to revolutionize various industries.

The properties of nanomaterials are tuned by varying their synthesis method. The process of synthesizing nanomaterials, such as the chemical vapor deposition technique, and sol-gel technique, involves the generation of toxins that may harm living beings, such as humans, animals, and the environment. Furthermore, the

Carbon-based Nanomaterials for Green Applications, First Edition. Edited by Upendra Kumar, Piyush Kumar Sonkar, and Suman Lata Tripathi.

Published 2025 by John Wiley & Sons, Inc.

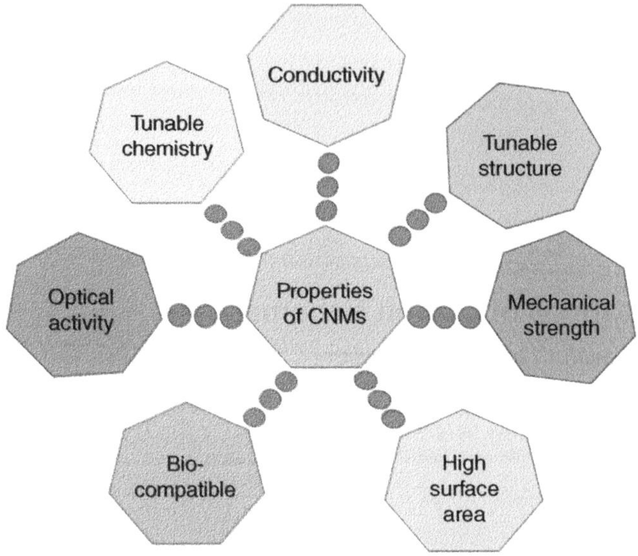

Figure 18.1 Various features of carbon nanomaterials for industrial applications.

initial products required to manufacture these nanostructures are also harmful [6]. Hence, novel eco-friendly pathways for the production of these nanomaterials are being explored. These methods ensure safety and affordability and offer significant advantages in the synthesis of nanomaterials, including enhanced productivity and economic viability [7].

To fully understand the unique properties and potential applications of CNs, theyare classified as (1) zero-dimensional (fullerenes, CNPs, and CQDs), (2) one-dimensional (CNTs and CNFs), (3) two-dimensional (graphene and carbon nanosheets), and (iii) three-dimensional materials (nanodiamonds [NDs] and carbon aerogels) [1, 2]. On the basis of their morphological characteristics, they are divided into two types. The first contains nanostructures with hollow space such as fullerenes and CNTs. The second are robust nanostructures without internal spacings such as NDs, graphene, and CQDs.

The diverse carbon nanostructures exhibit varied hybridization states. The distinct characteristics exhibited by different carbon nanostructures are due to their different hybridized states. So, based on hybridization, there are two classes. One consists of graphene nanostructures, primarily made up of sp^2 hybridized carbon atoms, although sp^3 carbon atoms may be present at the edges. The second class mainly contains sp^3 hybridized carbon atoms, but both sp^2 and sp^3 carbon atoms may be present in various ratios. NDs belong to the second class.

This chapter is organized into five sections. Section 18.1 introduces the CNs, their properties, synthesis, and classifications. Section 18.2 delves into different forms of CNs, discussing their distinctive properties that play a significant role in enabling diverse applications. In Section 18.3, a comprehensive description of several industrial applications utilizing these nanostructures is provided. The challenges associated with the utilization of CNs are depicted in the next section. Finally, last section of the chapter presents the conclusions and future insights.

18.2 Different Forms of Carbon-Based Nanomaterials

CNs are an ever-growing family of materials. Different CNs used for various industrial applications are described here and are shown in Figure 18.2.

Fullerenes, also called Buckyballs (C60), possess a structural resemblance to a football. These molecules exhibit weak intermolecular bonding, spherical shape, and strong intramolecular bonding. Consequently, their lubricating capabilities have garnered significant interest [8, 9]. Fullerenes possess an appealing characteristic of serving as a durable enclosure for various substances [10, 11].

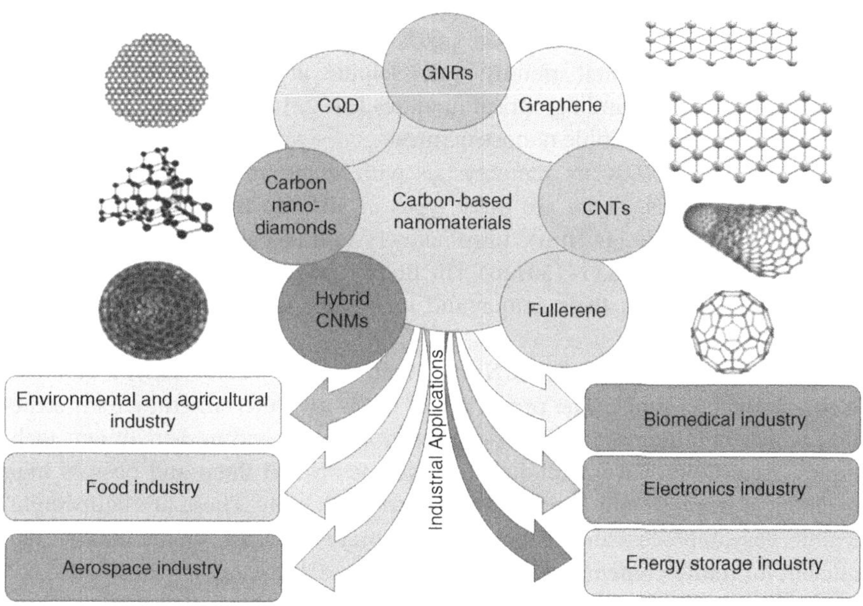

Figure 18.2 Different types of carbon nanomaterials and their industrial applications.

Graphene, a hexagonal 2D atomic layer of carbon atoms, exhibits distinct properties [12, 13] and has two types of edge arrangements. These edges can either be in a zigzag or armchair direction. By cutting the graphene sheet, armchair and zigzag *nanoribbons* can be formed. *Zigzag nanoribbons* display a metal-like behavior, whereas *armchair nanoribbons* can exhibit either metallic or semiconducting behavior. Larger GNRs show smaller energy gaps [14–16].

Graphene sheets with size in the nanometer range are called *nanographene* [17, 18]. Graphene containing oxygen groups is *graphene oxide (GO)* and when the number of oxygen atoms is reduced,it is called *reduced graphene oxide (rGO)*. Both *GO* and *rGO* are derivatives of graphene, which are easier to prepare and are highly useful.

CNTs are long and thin cylindrical tubes of carbon atoms and can be formed by folding carbon nanosheets along their axes. These nanotubes are one-dimensional and possess a high aspect ratio, which gives them unique and exceptional properties [19, 20]. Armchair, zigzag, and chiral nanotubes are three types of nanotubes formed by rolling a graphene sheet in different directions. The terms zigzag and armchair describe the different arrangements of hexagons along the circumference of CNT, depending on the rolling axis. Armchair CNTs show metallic behavior, whereas chiral and zigzag nanotubes are either semi-metallic or semi-conducting. Additionally, CNTs are categorized as single-walled or multi-walled, depending on the layers they contain either single or multiple layers.

Carbon quantum dots (*CQDs*) are most recently discovered CN, which can be synthesized and functionalized easily. CQDs are zero-dimensional, spherical, biocompatible, environmental friendly, water-soluble, and slightly toxic with high conductivity and outstanding optical qualities [21, 22]. CQDs have been studied as fluorescent biocompatible nanostructures.

Nanodiamonds (ND)s are nanocrystals with shape and electrical properties similar to diamonds. There are three types of NDs on the basis of their size, namely diamondoids (1–2 nm), ultra-nanocrystalline particles (2–10 nm), and nanocrystalline particles (1–150 nm). The fluorescence of NDs is attributed to the vacant space found at their center and makes it useful for the biomedicinal field [20, 23].

Carbon nanofibers (CNFs) are CNs having thread-like structures with diameter in the nanometer range. The properties of CNFs are determined by their structure, which is influenced by the production methods and post-treatment techniques used. CNFs have a conductive envelope around them and possess high mechanical strength and thermal and electrical stability. These are lightweight, flexible, easy to process and shape, and resistant to corrosion, which makes them efficient for reinforcement purposes [24, 25].

Nanocomposites are complex materials consisting of multiple phases, with atleast one phase having dimensions in the nanometer range. Carbon nanocomposites contain carbon nanostructures like graphite, nanotubes, and C_{60} as a matrix, combined with reinforced materials such as metal oxides, sulfides, and nitrides. CNT-based nanocomposites are used in supercapacitors, gas and biosensors, absorbers, and photovoltaic cells [26, 27].

Carbon nano-onions (CNOs) are zero-dimensional onion-like shells having multiple layers overlapping each other. These are useful for carrying drugs and in cancer treatment [28, 29].

18.3 Applications of Carbon Nanomaterials

18.3.1 Biomedical Industry

Applications of CNs in biomedical field are divided into four sectors, namely biomedicine, biosensors, drug delivery, and bioimaging and are explained below and depicted in Figure 18.3.

18.3.1.1 Biosensors

Sensors can sense a small change and present each change with a particular response by variation in its property. The desired features of a sensor are higher selectivity, faster response, cost-effectiveness, accuracy, and sustainability. Rigorous research worldwide has led to the development of novel materials and technologies for sensing.

To diagnose various ailments and to make timed decisions about the procedure, it is important to detect the biomolecule very accurately. CNT-based biosensors are useful for biomolecule detection [30, 31]. CNT-based nanosensors show enhanced performance in health monitoring and various biomedicinal applications. Inculcated nanosensors can monitor the heart's functioning and regulate its proper working. To explore the DNA sequence of humans, CNT-based nanobiosensors are employed. Pressure sensors are employed in surgeries of the eye, beds

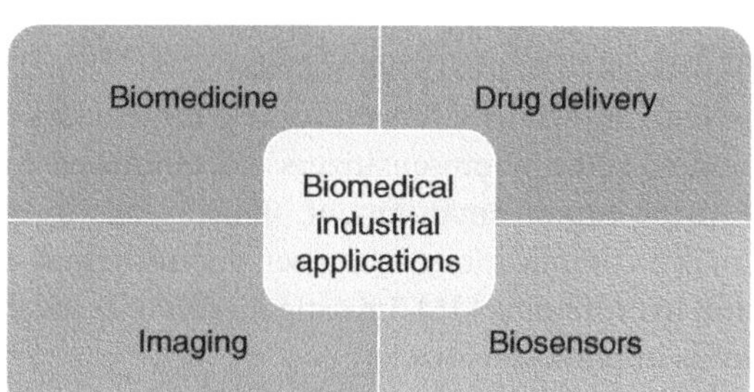

Figure 18.3 Biomedical industrial applications.

in hospitals, machines used for respiration, monitoring the patients, and devices for dialysis of the kidney [32, 33].

Graphene-based materials have been employed for sensing purposes from some time because of their excellent properties. The stretched graphene sheet has high sensibility [34]. Being immune to stress indicates its ability to be unaffected by external disturbances. The sensitivity of biosensors improves by adding nanoparticles, CNTs, or graphene-based materials [35].

Fullerenes are biofriendly, sensible, and show quick responses. So,they operate in biosensing techniques [36]. A highly sensitive and fast-responding Fullerene-based tyrosinase enzyme inhibition biosensor is formed by combining fullerenes and conjugated polymers [37]. A fullerene-based biosensor is employed to sense genetic alterations. A biosensor to detect CYP2C19*2 gene was proposed by Zhang and his colleagues in 2018 [4]. A glucose oxidase-based catalyst was designed to measure glucose concentration accurately by Christwardana and his team in 2017 [31].

A graphene quantum dot-based electrochemiluminescence (ECL) biosensor was fabricated for miRNA analysis by Yaqin Chai and his team. [38]. A graphene sensor was employed to differentiate the growing tumor cells [39]. Various drugs have been distinguished by using graphene sensors [40]. Modified graphene electrodes showed enhanced sensitivity for drug molecule detection. Highly sensitive and selective nanographene-based biosensors are designed due to their high conductance and high absorbance for enzymes and proteins. A novel electrochemical aptasensor using graphene-modified carbon electrode was reported by F. Li and his team [41]. This sensor detects biomolecules accurately in serum without any immobilization.

18.3.1.2 Drug Delivery

Giannopoulos and his team fabricated a nanocarrier using fullerene and loaded it with molnupiravir to deliver it to the human body because it is effective in reduction of hospitalization period especially during COVID-19 [42]. Raza and his co-workers investigated the use of fullerene for the delivery of docetaxel to infectious cells [43]. The fullerene enhanced the ability of a drug to be absorbed by the body by 4.2 times and reduced clearance of the drug to half. It was useful for the controlled release of drugs and to treat erythrocytes.

Photoluminescence shown by CNs is favorable for cancer cell drug therapy. In cancer phototherapy, some photosensitizers are employed to generate reactive oxygen species in the tumor environment, which leads to the death of tumor cells [44]. The major complications in cancer photo therapy are that it must be target-specific and hydrophobic [45]. Investigations are being done to employ carbon nanostructures in photosensitizers.

Nitrogen-doped CQDs forms a bond with Qucbl (7-(3-bromopropoxy)-2-quinol ylmethyl chlorambucil) and release drug in a regularized manner to illuminate and demolish cancer cells [46]. In 2015, Beack et al. [47] showed that a CQD-Cl-e6-HA system was also effective in the phototherapy of mouse melanoma and used for skin cancer treatment.

CNOs, on functionalization, become water-soluble and are employed to deliver drugs at specific sites. CNOs are desirable for drug distribution because they are less toxic and show minimal inflammation. These are efficient drug carriers due to their larger surface area and free pi-electrons. CNOs loaded with acetylcholinesterase inhibitors were prepared [48] to increase the efficiency of an anti-Alzheimer drug and release it continuously. CNTs are used as carriers for delivery of drugs to brain. Polymers and lipophilic moieties-functionalized NDs enhance the delivery of drugs to tumors [49]. Diazoxide-containing fluorescent NDs prepared by Tian et al. [50] are useful for both imaging and drug delivery.

Nanographene derivatives like GO are preferred and studied more for drug delivery than nanographene due to its cytotoxicity. Polymer-functionalized nanographene is more effective for drug delivery. PEG (Polyethylene glycol)-modified nGO is more hydrophilic and bio-friendly. The drug SN38 on being bonded to PEG-nGO becomes water-soluble and very toxic for cancer cells [51]. Polyethylenimine (PEI), even being toxic to cells, is generally employed to deliver genes. nGO-PEI is formed when PEI gets bonded to nanographene oxide and can carry plasmid and show better performance [52–54].

18.3.1.3 Biomedicine

CNTs are explored rigorously as they have shown calibre as theranostic and carrier to deliver drugs as they easily penetrate cell membranes [55]. In a study by Golubewa et al. [56], glioblastoma cells effectively accumulate within SWCNTs and lead to the light-initiated demolition of cancer cells by laser irradiation. Various amorphous carbon nanocomposites like amorphous carbon dots, amorphous carbon nanofoam, amorphous CNFs, and amorphous carbon films have been synthesized recently [57–59]. These nanocomposites offer huge possibilities in the biomedicinal field. Gao and his team studied fluorescent CQD as a theranostic nanoprobe [60]. Nanographene has shown huge potential for employment in the field of biomedicine [61].

18.3.1.4 Imaging

The transport of drugs within the body can be visualized using CQDs. A CQD-based nanocarrier loaded with cisplatin(IV) was developed by Feng and his team [44], which could deliver drugs and send images during transport. CQDs functionalized with P and loaded with HA-Dox are efficient for image formation and visualization, detecting hyaluronidase, and delivering the drug. CNT ring decorated with gold nanoparticles was used as a contrast media for visual-based cancer detection and therapy [60].

Due to their distinct optical and chemical properties and their bio-friendliness, NDs are desired candidates for developing imaging techniques. Mohan and his collaborators used fluorescent NDs to visualize the C. elegans cells [62]. Si-containing NDs were made to do imaging of stem cells by Merson and team in 2013 [63].

18.3.2 Energy Storage Industry

The utilization of CNs is observed in various energy storage systems, namely supercapacitors, fuel cells, and batteries as given in Figure 18.4.

18.3.2.1 Batteries

CNTs can store greater amount of lithium than conventional forms of carbon. CNTs are employed for making anodes and add-ons for cathodes of alkali metal ion batteries and have enhanced performance due to their beneficial properties (electrical, mechanical, and structural). CNTs have been modified to tune their properties related to ion intercalation, adsorption, and diffusion and to use them as potent electrode materials [64, 65].

Different researchers studied the ion storage procedure in CNTs [66]. The ion storage capacity depends on the diameter of CNT and defects present in it [66]. These defects may be introduced by ball milling or nitric acid treatment or are present inherently. It is noted that holes present on the CNT wall increase its ability by diffusing the ions into the nanotube and reducing the diffusion path [67, 68]. CNTs have two different types of behavior, metallic or semiconducting. Metallic CNTs can store more lithium ions approximately five times more than semiconducting CNTs [69]. The MWCNT film retained its capacity for more number of cycles than other CNT films [70]. CNTs are functionalized and doped with (N, B) to get enhanced ion storage [71]. N-doped CNTs showed double reversible capacity than conventional CNTs [72].

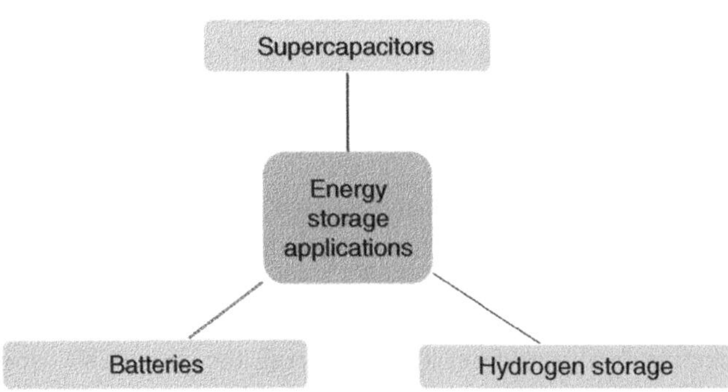

Figure 18.4 Energy storage systems incorporating carbon nanomaterials.

CNT is utilized in silicon-based anode [73] either as a conducting additive or an active component of the cell. The defect-containing and disordered CNTs are synthesized and used in sodium-ion batteries as anode to intensify the sodium storage and show better performance. Extremely defective and disordered porous CNTs prepared by the ethanol flame method are used as electrodes and show excellent performance [74]. Different amorphous carbons are used as additives in the anode to suppress the sulfation process. Aurbach and his group used SWCNT to suppress the non-manageable sulfation process in lead-acid battery electrodes [75]. CNTs are also used as conducting supplements for catalysts in metal-air batteries [76].

Functionalized CNTs are employed as air electrodes in metal-air batteries. CNTs doped with nitrogen (N-CNTs) raise redox reactions in acidic and alkaline solutions [77]. Also CNTs can also be employed as a current collector in batteries. CNT-based carbon papers are manufactured and utilized as anode and current collector for batteries. The electrodes in which CNT is employed as an adhesive have more thermal stability than conventional adhesives [78].

18.3.2.2 Supercapacitors

CNTs are used as favorable material supercapacitor electrodes due to their outstanding electrical properties and structure. CNs employed in supercapacitors show higher capacitance, enhanced densities, and longer life cycles. It is observed that defect formation and activated open ends increase the surface area of CNTs [79] and lead to a higher value of capacitance. SWCNT electrodes have more capacitance than MWCNT electrodes because of their larger surface area. N-doped CNTs boost the performance of supercapacitors. The electrical conductivity and quantum capacitance of CNTs increase as the doped nitrogen atoms lead to modification in electronic band structure as the conduction band is altered by the presence of nitrogen bands [80]. Recently, researchers have fabricated a wire-type supercapacitor using CNT, which gives better performance and increased densities [81].

18.3.2.3 Hydrogen Storage

In a fuel cell, a catalyst is used for oxidation of fuel, which generates electrons and ions at the anode. Ion movement across the electrolyte leads to the generation of electricity. CNT-supported fuel cells exhibit better electrochemical performances [82]. Heteroatom-doped CNTs or transition metal-filled CNTs improve the storage capacity of fuel cells. Researchers have reported that when Ag, Fe, Co, CuSe, and Ni-based compounds are encapsulated into pure CNTs, it leads to the better performance of CNTs in fuel cells [83]. The use of CNT in a fuel cell enhances its stability and makes it resistant to corrosion during electrochemical reactions [83, 84]. CNTs enhance the catalytic ability and increase the corrosion resistance in a fuel cell. CNTs also enhance the mass transmission capacity of electrodes in a fuel cell. Various carbon nanostructures like nanofibers, nanotubes, graphene, and fullerenes and their hybrids combined with metals are used in hydrogen fuel cells for improved efficiency [85].

18.3.3 Electronic Industry

CNs have the potential to replace conventional semiconducting materials in the realm of the electronic industry due to their exceptional sp^2 hybridized electronic structure and the ability to fine-tune the band gap. Figure 18.5 illustrates the key sectors of the electronic industry where CNs find applications, namely FETS, displays, wearable technology, and digital electronics.

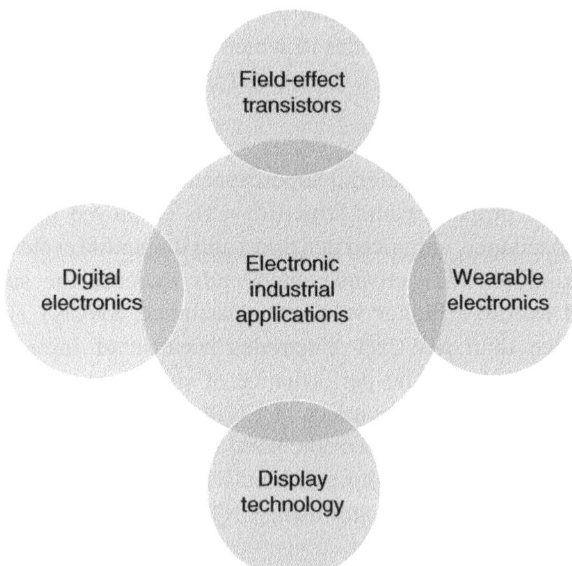

Figure 18.5 Main areas of the electronic industry employing carbon nanomaterials.

18.3.3.1 Field-Effect Transistors and Digital Electronics

CNs-based field-effect transistors (FETs) offer a promising platform for optoelectronic devices and sensors because of their exceptional conductivity, chemical stability, and accessibility [86–88]. Moreover, their environmental ultrasensitivity further enhances their potential in these applications. CNT-based FETs show promising characteristics for future digital logic circuits and memory chips. Internet of Things (IoT), thin-film transistors (TFTs), and high-performance computing chips are among the most notable potential uses of CNs-based FETs [89, 90].

FETs employ semiconducting CNTs as channel materials, while transparent conductors hold promise in the form of metallic CNT thin films [91].

Nevertheless, the primary obstacle lies in the efficient manufacturing of CNT-based FET chips with channel lengths below 10 nm, while maintaining the desired precision and purity levels of the channel array. IBM's persistent endeavors have yielded a chipset comprising 10,000 nanotube transistors. However, to render it commercially feasible, this count needs to escalate to a billion transistors.

18.3.3.2 Wearable Electronics

The combination of high electrical conductivity, good flexibility, supreme electrochemical properties, and superior mechanical strength make CNT an excellent choice for a fiber-shaped electrode [92] in wearable electronics [93]. Much attention has been paid to multi-functional wearable supercapacitors that offer self-healing [94], shape-memory [95], and electrochromic capabilities [96]. Chen et al. [97] conducted a study where they utilized a CNT film to encase pre-stretched elastic wires. These wires were subsequently twisted to form a wire-shaped supercapacitor. This device showcased a significant capacitance of 30.7 Fg^{-1} and possessed the capability to stretch up to 350% strain. Meanwhile, a fiber-shaped supercapacitor was developed by Zhang et al., utilizing two aligned CNT/polyaniline composite sheets as electrodes [98]. Notably, this groundbreaking device demonstrated a specific capacitance of approximately 79.4 Fg^{-1}, even after enduring a strain of 300% throughout 5000 cycles. Furthermore, nanographene and its derivatives are also considered the optimal materials for creating flexible electronics and wearables [99]. The utilization of heteroatom doping in nanographene films, as reported by Wu et al. [100], can result in the development of microscale supercapacitors that display remarkable pseudocapacitive behavior. A wearable strain sensor composed of elastomer-filled graphene-woven fabric was proposed by Liu et al. [101] to monitor human physiological signals. Romero et al. [102] devised a versatile electrode to facilitate universal electrocardiogram monitoring.

18.3.3.3 Display Technology

The versatile applications of CNTs span across various fields. They act as polarizers in liquid crystal displays (LCDs), as field emitters in field emission devices (FEDs), and as transparent electrodes in organic light-emitting devices (OLEDs). Additionally, CNTs have also found utility in emerging display technologies like flexible fiber displays, thermochromic displays, and incandescence displays. By utilizing field emitters composed of MWCNTs, Saito et al. [103] fabricated cathode ray tubes (CRTs) with adequate luminance, long life, and stable electron emission. Additionally, transparent CNT film and CNT wire were employed as the anode and cathode, respectively, in the construction of luminescent tubes. Field electron emitters were developed by Wei et al. [104] by fragmenting a continuous MWCNT yarn into segments, which acted as field emitters. The resulting emission current was highly stable and could reach several milliamperes. CNT displays characteristics similar to those of a liquid crystal (LC) because of its structural similarity to a large rod-like molecule [105]. Schindler et al. [106] established that CNT networks possess the necessary flexibility for use in displays and that SWCNT-based FETs offer a practical solution for integrating into LCDs. Freitag et al. [107] fabricated top-emitting white OLEDs by employing CNT. These devices exhibited a well-balanced, almost Lambertian white emission with color rendering indices of 70. Furthermore, they exhibited negligible spectral variation when observed from various perspectives. To fabricate touch screens, large-scale synthesis of graphene-based transparent conductive electrodes can be achieved by utilizing the roll-to-roll technique [108, 109].

18.3.4 Food Industry

For enhancing food quality, safety, and sustainability, the use of CNs is in the early stages of research and development and is playing a crucial role in food processing, food safety, and food packaging as illustrated in Figure 18.6.

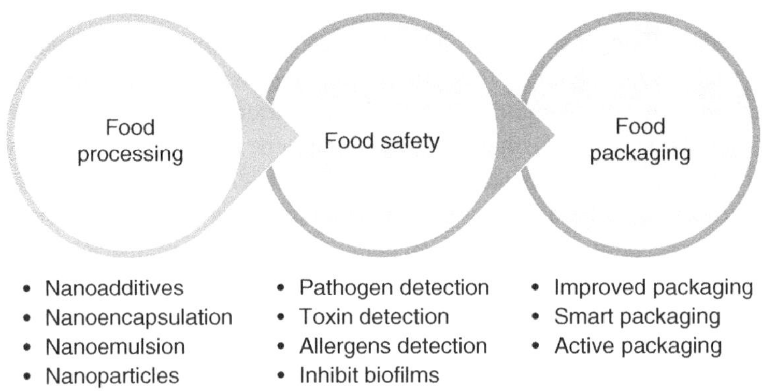

Figure 18.6 Application of carbon nanomaterials in food processing, food safety, and food packaging.

18.3.4.1 Food Processing

The application of carbon-based nanoparticles in food processing has proven beneficial in improving the nutritional quality, flow properties, flavor, color, and stability of food products, as well as extending their shelf life. For delivering bioactive compounds or nutrients, graphene has found application as a nano-carrier. CNFs are also used as carriers for encapsulating and delivering flavors, nutrients, or bioactive compounds. Activated carbon is also used in food processing for the removal of contaminants, such as heavy metals, pesticides, or mycotoxins, from food ingredients or water. Ngafwanet al. [110] investigated CQD-based fluorescence sensors as a valuable means of detecting a wide range of substances in food samples. The identification of food additives, heavy metals, bacteria, insecticide residues, antibiotics, and nutritional components is effectively carried out by these sensors.

18.3.4.2 Food Safety

The utilization of sensors and biosensors has become prevalent in the food industry, serving as a means to ensure the safety and quality control of food products, while also enabling the detection of any potential food contamination. CNT-based biosensors can be used in meat freshness evaluation. For detecting contaminants or spoilage in food, graphene nanosensors have been used. Shah [111] investigated a glassy carbon electrode (GCE) modified with calixarene and gold nanoparticles for detecting two toxic food dyes (metanil yellow and fast green). Dalal et al. [112] utilized carboxylated multi-walled CNTs (c-MWCNTs)-modified screen-printed electrode-based bionanosensor for detecting the time of ripening of tomato concerning its malic acid concentration.

18.3.4.3 Food Packaging

Due to excellent mechanical and thermal properties, graphene is used as a food packaging material, where it can provide barrier properties against gases and moisture, thereby extending the shelf life of food products. Ag-ZnO-rGO-PEG nanocomposites were fabricated by Naskar et al. [113] to investigate their antibiofilm activities and potential use in food packaging. Wang et al. [114] developed an antibacterial MTAC-rGO/EVOH film for active food packaging. Arfat et al. [115] integrated GO into polylactic acid to fabricate antimicrobial films with effective antibacterial activity against *E. coli* and *S. aureus*. The primary purpose of these films was to serve as materials for packaging food products. Utilizing Carbon Dots, Salimi et al. [116] fabricated an antimicrobial bacterial nanocellulose (BNC) sheet that was environmentally friendly. The resulting nanopaper exhibited strong UV-blocking capabilities and demonstrated effectiveness against *L. Monocytogenes*. Yu et al. [117] fabricated bionanocomposites by combining poly (3-hydroxybutyrate-co-3-hydroxyvalerate) with functionalized CNTs. Their study revealed that the introduction of CNTs into the polymeric matrix significantly improved the film's thermal and mechanical properties. Consequently, these prepared materials exhibit promising prospects for utilization in food preservation and packaging.

18.3.5 Aerospace Industry

The next-generation aircraft, rotorcraft, unmanned aerial vehicles (UAVs), and missiles will face more stringent requirements regarding their visual and thermal signatures, lightweight design, increased speed, and improved maneuverability.

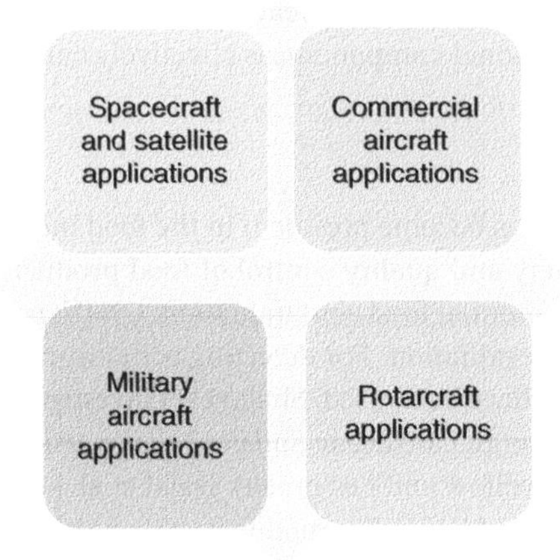

Figure 18.7 Carbon nanomaterials in aviation industry.

Owing to high mechanical strength, lightweight nature, and excellent thermal and electrical conductivity, CNs offer numerous benefits in the aerospace industry, such as weight reduction, increased energy efficiency, enhanced durability, and improved performance. CNs find significant applications in three key sectors of the Aviation Industry: Airframe Structure, Aero-Engine Components, and Aircraft Electro-Communication Systems. Figure 18.7 illustrates the aerospace industrial benefits in different areas.

18.3.5.1 Spacecraft and Satellite Applications

CNTs have the potential to revolutionize the satellite industry by reducing the weight of a satellite to one-third of its current weight [118]. This is a significant advantage for large satellite applications. Additionally, patch antennas made from CNTs have a planar structure and low profile and offer a gain of 7 dB and a bandwidth of 5–20% [119]. Furthermore, CNTs can be used for heat flow distribution in satellites and thermal protection systems [120]. Graphene is also utilized to improve performance and efficiency, electromagnetic shielding, and thermal management in spacecraft and satellites.

18.3.5.2 Commercial Aircraft Applications

The potential cost savings that can be achieved through the adoption of CNT-reinforced polymer (CNRP) technology are illuminated by the research carried out by O'Donnell et al. [121]. This research involved simulation analysis on four conceptual aircraft structures, namely the Airbus A320, Boeing 747–400, Boeing 757–200, and Embraer E145. The substitution of current copper wiring with CNTs offersa promising opportunity for weight reduction and decreased fuel consumption [122–124]. Furthermore, this replacement would effectively prevent premature failures and overheating caused by vibration fatigue, oxidization, and corrosion of the copper wires [125].

18.3.5.3 Military Aircraft Applications

In numerous military scenarios, stealth technology is vital in ensuring the system's survival by reducing its radar detectability. Nanotechnologies can be employed as components in radar-absorbing structures to attain the desired stealth characteristics. Kim and Lee [126] carried out research on a radar-absorbing structure incorporating CNTs. They employed a nanocomposite consisting of E-glass fabric, epoxy resin, and CNTs, which were affixed to the outer surface of a sandwich construction for the purpose of absorbing electromagnetic waves. The composite sandwich construction utilized in the study achieved a remarkable absorption rate of up to 90% for incident electromagnetic waves in the X-band frequency, while also exhibiting peel strength approximately three times greater than that of the equivalent aluminum foil.

18.3.5.4 Rotorcraft Applications

Peck et al. studied the issue of icing in rotor aircraft and found that ice adhesion can cause vibrations in rotor blades, leading to forced landings, impaired visibility for pilots, and unreliable instrument readings [127]. To tackle this challenge, Chu et al. developed self-heating CNT-reinforced composites that exhibit excellent electrical heating capabilities at ambient temperature [128]. Yoon et al. also contributed to the field by developing thin and transparent films based on MWCNTs for deicing applications [129]. These coatings offer protection against icing and enhance the barrier resistance on the aircraft surfaces without compromising the thrust-to-weight ratio [130].

18.3.5.5 Unmanned Aerial Vehicle (UAV) Applications

UAVs exhibit a wide range of designs, varying from sizable aircraft to compact drones referred to as nanoair vehicles (NAVs). NAV has become a reality in recent times due to advancements in nanofabrication methods [131]. NAVs encompass a multitude of components, such as embedded sensors, power storage devices, and propulsion and movement systems, which collectively contribute to their operational capabilities [132].

To incorporate all the essential components for NAVs, smaller chips are necessary, and the advancements made in bottom-up fabrication methods could potentially offer a solution. One of the primary challenges lies in fuel storage. Even if the components can be miniaturized, resulting in smaller NAVs, the availability of space for fuel storage remains limited. This is precisely where nanoscale solid-state energy storage systems could potentially provide a solution.

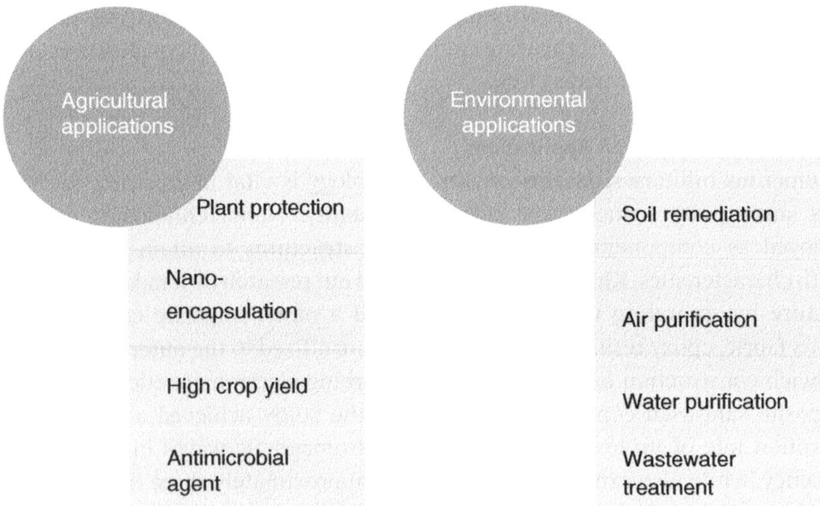

Figure 18.8 Applications of carbon nanomaterials in environmental and agricultural sectors.

18.3.6 Environmental and Agricultural Sectors

The benefits of CNs in environmental and agricultural sectors are illustrated in Figure 18.8.

18.3.6.1 Environmental Applications

The environment is in peril due to the rapid increase in urbanization, industrialization, and population growth, which is releasing pollutants at an alarming rate. Discharge of contaminated water with excessive heavy metals (Cd, As, Pb, Cr, Hg, and Ni), metalloids, and harmful gases into the environment has degraded the water bodies, agricultural land, and the air. The presence of these substances in the environment poses a substantial risk to the health of living organisms and the overall ecosystem. In recent years, CNs have demonstrated remarkable potential for the treatment of pollutants (air, water, and soil) as well as in improving crop yields in the agricultural sector, offering sustainable and effective solutions.

18.3.6.1.1 Pollutant Treatment CNs are used to remove (1) water pollutants such as heavy metals, and organic dyes from wastewater released by industrial waste and domestic sources; (2) soil pollutants such as pesticides, heavy metals, chemicals, and volatile organic compounds (VOCs); and (3) air pollutants such as toxic gasses, aromatic hydrocarbons, VOCs released from the industries and vehicles.

To address the presence of organic dyes and heavy metals in water, researchers have turned to CNTs, graphene, activated carbon, CNFs, and hybrid nanomaterials as effective solutions for their removal. Functionalized CNTs with the introduction of specific functional groups offer enhanced adsorption properties that can interact with the organic dyes [133]. Graphene, GOs, and reduced GOs due to a large π conjugated system, are used as adsorbents for dye removal [134, 135]. Through electrostatic interaction, the carboxyl and hydroxyl groups present on the GO surface can establish interactions with heavy metal ions [136]. Surface modifications of activated carbon, such as chemical functionalization or impregnation with specific chemicals, can enhance its heavy metal removal efficiency [137]. With interconnected porous structures and high aspect ratios, CNFs have also shown excellent adsorption capacity for heavy metals [138]. Carbon-based hybrid nanomaterials can combine the advantages of different CNs and provide synergistic effects for pollutant removal. Various techniques, including batch adsorption, column adsorption, and membrane filtration, can be utilized to exploit the adsorption properties of CNs for pollutant removal application from water [139].

For the removal of pesticides from soil, CNs have also shown promising potential. Activated carbon, Biochar, CNTs, GO, and their hybrid nanomaterials use various techniques, including in-situ treatment, soil amendment, and filtration to utilize the adsorption of pesticides [140]. The adsorption of VOCs onto the AC surface occurs through physical interaction such as Vander Waals forces and π–π stacking. CNTs can adsorb these molecules through physisorption, as well as chemical interactions facilitated by their functionalized surfaces. Various porous CNs, such as carbon foams, carbon aerogel, and carbon nitride materials have been explored for VOC removal [141].

The detection and treatment of toxic gases can be efficiently carried out using CNs, specifically CNTs and carbon nanodots, which offer high sensitivity, flexibility, and cost-effectiveness. The disposable nitrogen dioxide gas sensors using carbon adhesive tape (CAT) as a gas sensing element have also been reported [142]. Zhang et al. created a reduced GO/gold nanocomposite-based NO_2 sensor. Ag/CNT hybrids have been tested for antimicrobial air filtration [143]. Miyawaki et al. have confirmed the validity of a hybrid catalyst for the clean removal of formaldehyde gas [144]. Optical sensors for toxic metal ion detection using graphene quantum dots-based materials have also been investigated [145].

18.3.6.2 Agricultural Applications

Various forms of CNs, including fullerenes, fullerols, carbon nanoparticles (CNPs), and CNTs, have recently gained attention due to their potential for controlling plant development. Numerous studies have investigated the effects of CNs on plants, considering factors such as CN types, concentration, exposure duration, plant species, and growth conditions.

Graphene has a favorable impact on seed sprouting, suggesting its potential to facilitate water absorption by puncturing and breaking the seedling husks. Consequently, this leads to enhanced seedling development and a notable increase in seed germination rates [146]. Higher quantities of graphene have resulted in adverse outcomes, while the duration of plants' interaction with graphene influences their phytotoxicity [147].

By incorporating CNPs, the growth of plants was greatly enhanced, and the uptake and deposition of both micro- and macro-elements were stimulated [148]. This led to an increased effectiveness of fertilizers and improved viability of crops. As a result, water-soluble CNPs may offer a more advantageous solution compared to manure or fertilizers, as they allow for a gradual and regulated absorption of nutrients by crops, thereby enhancing digestion.

Carbon nanodots also promote seed germination rates, increase root and shoot length, and increase plant biomass in various plant species [149]. They serve as nanocarriers for the delivery of nutrients, pesticides, and agricultural chemicals to plants and improve stress tolerance [150]. At high concentrations or under certain experimental conditions, some adverse effects on plants have also been reported.

18.3.7 Automotive Industry

Graphene-based nanosensors are employed to gain the details about particulars of the vehicle like pressure, heat, speed, exhaust gases, and engine. A pressure sensor is used to determine if an air filter is dirty [151]. The main fields where CNT-based sensors can be applied are vehicle dynamic control (VDC), rollover detection, and anti-theft systems [152]. CNT-based carbon monoxide (CO) sensors are utilized in car parks for controlling ventilation systems [153]. CNT-based gas sensors show outstanding performance and are used to monitor flammable gases, detect gas leakage, monitor explosives such as TNT or RDX, monitor environmental pollution, and cooking controls, etc.

18.3.8 Green Energy Applications

As industries strive to reduce their environmental impact, dependence on fossil fuels and improve energy efficiency, the adoption of green energy harvesting technologies has become increasingly crucial. Solar energy harvesting does not create any carbon emissions or heat-trapping "greenhouse" gases. There is always a need for fabrication of devices with better photoconversion efficiency and cheaper manufacturing costs. Carbon-based nanomaterials such as graphene show an immense potential for next-generation high-performance electronic and optoelectronic devices. The distinct band structure and intrinsic driving potential to transfer carriers have enabled different architectures of graphene to be interfaced with various semiconductor structures [154–156]. Graphene derivatives are employed in solar harvesting devices as transparent conducting film [157], active light-harvesting material [158], cathode electrode for dye-sensitized PVs [159], and hole-transporting material for hybrid perovskite PVs [160]. The main purpose of using graphene derivatives in these devices is either bandgap engineering of the adjacent semiconductors or conductive illumination platforms. The origin of the photoresponse in graphene junctions has been studied extensively and is attributed to either thermoelectric or photovoltaic effects. Graphene-based heterojunction solar cells exhibit a good efficiency (18.5%) when combined with GaAs structure [161]. In graphene-assisted heterojunctions, e.g. MoF_6/graphene [162] and MoO_3/graphene [163], doping effects have been reported. Doping with organic compounds also improves the properties of graphene, e.g. doping with trifluoromethanesulfonic amide (TFSA) led to an increased conversion efficiency from 1.9% to 8.6%, increased current density from 14.2 to 25.3 $mA\,cm^{-2}$, and increased open circuit voltage from 0.43 to 0.54 V, in Graphene/Silicon device, using doped and undoped graphene, respectively [164].

18.4 Challenges

Carbon nanostructures are hydrophobic, so assemble in body tissues. The process of conversion of nanostructure is dependent on its surroundings. There are various kinds of research on dissimilar conversion processes of CNTs by drugs in distinct tissues and in the presence of unlike enzymes [165]. The uptake of nanostructures is limited by body cells as blood proteins bind to the surface of carbon nanostructures. Carbon nanostructures aggregate in a particular organ for a large interval of time and release reactive oxygen, leading to the death of cells [166].

CNs are not completely unreactive. They may intervene in the cell-signaling pathways and create toxicity. Nanographene reacts within the human body and may cause stress injuries to cells [167]. The nanographene-based wearables are not durable because they absorb moisture. The residues generated during the production of CNs are toxic to the environment. Another challenge in using carbon nanostructures for drug carriage is its poor solubility.

Another hurdle for using CN for drug delivery is it must be target-specific. Specific nanocarriers must be designed for the target-oriented delivery of drugs to different body organs. Carbon nanostructures cause toxicity and fibrosis in the lungs [168]. The combination of nanostructures with various polymers which are sensitive to pH change, magnetism, and redox reactions may be investigated to predict their response in biological media and utilize them if they show improved performance. Suitable functionalization of CNs must be done to avoid damage to healthy cells and to ensure safety [169]. Proper safety protocols must be followed for patients, health professionals, and workers. CNs like NDs are not biodegradable in the human body.

Most graphene-based sensors are still in the lab stage and are not yet commercialized because large-scale production of good-quality graphene film cannot be achieved. The fabrication method for graphene QDs is not fully developed due to size limitation because, with size enlargement [170], the attraction increases and leads to a decrease in solution solubility.

The production of pure nanotubes is not cost-effective. It is difficult to control the growth of defects in nanotubes [171]. To obtain extensive nanowire structures, control over nanotube growth on surfaces is desired [172]. At large strains, CNTs become buckled, kinks are formed, and structure becomes unstable [173].

The performance of graphene nanocomposite-based devices is noted to be low. Undesired side reactions are going on within the system that degrade carbon electrodes and hinder the flow of ions across the systems [174]. The use of hybrid nanocomposites along with conducting polymer and metals may reduce resistances, improve contacts, and provide an easy path for the flow of ions. There are concerns about harmful effects on the environment caused by CNs.

18.5 Conclusions and Future Scope

The immense potential of CNs in various industries has been highlighted through their industrial applications. In the realm of electronics, their distinct electrical properties have brought about a revolution in the development of smaller, faster, and more efficient devices. Their large thermal conductance makes them ideal for managing heat in electronics, thereby enhancing device performance. Carbon nanostructures are increasingly employed for delivery of drugs to specific body organs, bioimaging, and sensing of tumors and ailments in the biomedicinal field. The utilization of CNs in energy storage devices has greatly improved their performance, leading to advancements in batteries and supercapacitors. The multi-walled CNT (MWCNT)-based anode has a high lithium storage capacity of 980 $mAh\,g^{-1}$ in its first cycle. Cerium-MOF/MWCNT composites coating on Li–S batteries has a capacity of 993.5 $mAh\,g^{-1}$. The cell with GO/MWCNTs has a capacity of 1370 $mAh\,g^{-1}$ [175]. A supercapacitor with electrodes made from multi-walled nanotubes, helical CNTs, vertically aligned CNTs, and nanotubes coated with nickel hydroxide has a capacitance of 102, 125, 365, and 1065 $F\,g^{-1}$, respectively. The electrodes made from graphene 2D nanosheets show a specific capacitance of 316.8 $F\,g^{-1}$. Under green synthesis approaches, biomass charcoal, waste polypropylene, and flaxseed residues-based carbon electrodes are used and show capacitance values of 175.6, 349, and 369 $F\,g^{-1}$, respectively [176, 177]. A fuel cell with 3D carbon fiber composite material formed from CNT, and graphite and CNTs, gives maximum power efficiency of 362 ± 20, and 495 $mW\,cm^{-2}$, respectively. Nitrogen-doped carbon aerogels and Pt nanoparticle-coated carbon aerogels, when used as catalysts in fuel cells, give a power density of 1048 ± 47 and $536\,mW\,cm^{-2}$, respectively. NAF-functionalized fullerene membrane-based fuel cells show a power density of $146\,mW\,cm^{-2}$ [178]. CNs have shown their potential in the food industry, environment, and agricultural sectors. Additionally, CNs possess exceptional tensile strength, which makes them functional for aerospace and automotive applications [179].

Despite the progress made in the utilization of CNs, there are some hurdles which need to be addressed for their wider adoption in industrial applications, as discussed in Section 18.4. To fully exploit the potential of CNs, it is imperative to conduct further research and make technological advancements to overcome these existing challenges. Areas that require exploration include enhancing the scalability and cost-effectiveness of production methods, developing efficient purification techniques, and improving the capability and durability of CN-based devices. Additionally, synergetic research by material scientists, physicists, chemists, and engineers will be vital in integrating CNs into various industrial sectors and realizing their complete potential. The combination of diverse expertise and perspectives will facilitate the generation of innovative solutions and drive the development of novel applications. Proper safety protocols must be developed for patients, health professionals, and workers. Environmental safety needs to be assured.

Acknowledgment

Dr. Pooja Kapoor wishes to acknowledge the cooperation and support received from University Management at Maharaja Agrasen University, Baddi, (H.P.).

Declarations

The authors declare that they have no competing interests and conflicts to declare.

Funding

This work did not receive any specific grant from funding agencies in the public, commercial, or not-for-profit sectors.

References

1 Kaushik, N., Borkar, S.B., Nandanwar, S.K. et al. (2022). Nanocarrier cancer therapeutics with functional stimuli-responsive mechanisms. *Journal of Nanbiotechnology* 20: 152. https://doi.org/10.1186/s12951-022-01364-2.

2 Singh, J., Nayak, P., Singh, G. et al. (2023). Carbon nanostructures as therapeutic cargoes: recent developments and challenges. *Journal of Carbon Research* 9: 3. https://doi.org/10.3390/c9010003.

3 Jha, R., Singh, A., Sharma, P.K., and Fuloria, N.K. (2001). Smart carbon nanotubes for drug delivery system: a comprehensive study. *Journal of Drug Delivery Science and Technology* 2020 (58): 101811. https://doi.org/10.1016/j.jddst.2020.101811.

4 Liu, H., Chen, J., Qiao, S., and Zhang, W. (2021). Carbon-based nanomaterials for bone and cartilage regeneration: a review. *ACS Biomaterials Science & Engineering* 7 (10): 4718–4735. https://doi.org/10.1021/acsbiomaterials.1c00759.

5 Rauti, R., Musto, M., Bosi, S. et al. (2019). Properties and behavior of carbon nanomaterials when interfacing neuronal cells: how far have we come? *Carbon* 143: 430–446. https://doi.org/10.1016/j.carbon.2018.11.026.

6 Noah, N.M. and Ndangili, P.M. (2021). Green synthesis of nanomaterials from sustainable materials for biosensors and drug delivery. *Sensors International* 3: 100166. https://doi.org/10.1016/j.sintl.2022.100166.

7 Adorinni, S., Cringoli, M.C., Perathoner, S. et al. (2021). Green approaches to carbon nanostructure-based biomaterials. *Applied Sciences* 11 (6): 2490. https://doi.org/10.3390/app11062490.

8 Kuznetsov, V. (2020). Stereochemistry of simple molecules inside nanotubes and fullerenes: unusual behavior of usual systems. *Molecules* 25 (10): 2437. https://doi.org/10.3390/molecules25102437.

9 Kazemzadeh, H. and Mozafari, M. (2019). Fullerene-based delivery systems. *Drug Discovery Today* 24 (3): 898–905. https://doi.org/10.1016/j.drudis.2019.01.013.

10 Lao, F., Li, W., Han, D. et al. (2009). Fullerene derivatives protect endothelial cells against NO induced damage. *Nanotechnology* 20 (22): 225103. https://doi.org/10.1088/0957-4484/20/22/225103.

11 Shu, C., Corwin, F.D., Zhang, J. et al. (2009). Facile preparation of a new gadofullerene-based magnetic resonance imaging contrast agent with high 1H relaxivity. *Bioconjugate Chemistry* 20 (6): 1186–1193. https://doi.org/10.1021/bc900051d.

12 Novoselov, K.S., Geim, A.K., Morozov, S.V. et al. (2004). Electric field effect in atomically thin carbon films. *Science* 306 (5696): 666–669. https://doi.org/10.1126/science.1102896.

13 Lee, C., Wei, X., Kysar, J.W., and Hone, J. (2008). Measurement of the elastic properties and intrinsic strength of monolayer graphene. *Science* 321 (5887): 385–388. https://doi.org/10.1126/science.1157996.

14 Martin, I. and Blanter, Y.M. (2009). Transport in disordered graphene nanoribbons. *Physical Review B: Condensed Matter and Materials Physics* 79: 235132(1–5), https://doi.org/10.1103/PhysRevB.79.235132.

15 Son, Y.W., Cohen, M.L., and Louie, S.G. (2006). Energy gaps in graphene nanoribbons. *Physical Review Letters* 97 (21): 1–4. https://doi.org/10.1103/PhysRevLett.97.216803.
16 Zhang, X., Xin, J., and Ding, F. (2013). The edges of graphene. *Nanoscale* 5 (7): 2556–2569. https://doi.org/10.1039/C3NR34009K.
17 Tan, Y.-Z., Yang, B., Parvez, K. et al. (2013). Atomically precise edge chlorination of nanographenes and its application in graphene nanoribbons. *Nature Communications* 4: 2646. https://doi.org/10.1038/ncomms3646.
18 Fujii, S. and Enoki, T. (2013). Nanographene and graphene edges:electronic structure and nanofabrication. *Accounts of Chemical Research* 46 (10): 2202–2210. https://doi.org/10.1021/ar300120y.
19 Geckeler, K.E. and Premkumar, T. (2011). Carbon nanotubes: are they dispersed or dissolved in liquids? *Nanoscale Research Letters* 6 (1): 136. https://doi.org/10.1186/1556-276X-6-136.
20 Hosnedlova, B., Kepinska, M., Fernandez, C. et al. (2019). Carbon nanomaterials for targeted cancer therapy drugs: a critical review. *Chemical Record* 19: 502–522. https://doi.org/10.1002/tcr.201800038.
21 Devi, P., Saini, S., and Kim, K.H. (2019). The advanced role of carbon quantum dots in nanomedical applications. *Biosensors & Bioelectronics* 141: 111158. https://doi.org/10.1016/j.bios.2019.02.059.
22 Li, M., Chen, T., Gooding, J.J., and Liu, J. (2019). Review of carbon and graphene quantum dots for sensing. *ACS Sensors* 4 (7): 1732–1748. https://doi.org/10.1021/acssensors.9b00514.
23 Debnath, S.K. and Srivastava, R. (2021). Drug delivery with carbon-based nanomaterials as versatile nanocarriers: progress and prospects. *Frontiers in Nanotechnology* 3: 15. https://doi.org/10.3389/fnano.2021.644564.
24 Tibbetts, G., Lake, M., Strong, K., and Rice, B. (2007). A review of the fabrication and properties of vapor-grown carbon nanofiber/polymer composites. *Composites Science and Technology* 67 (7–8): 1709–1718. https://doi.org/10.1016/j.compscitech.2006.06.015.
25 Jong, D., Krijn, P., and Geus John, W. (2007). Carbon nanofibers: catalytic synthesis and applications. *Catalysis Reviews* 42 (4): 481–510. https://doi.org/10.1081/CR-100101954. hdl:1874/2326. S2CID 97230458.
26 Mara, N., Bhattacharyya, D., Dickerson, P. et al. (2008). Deformability of ultrahigh strength 5 nm Cu/ Nbnanolayered composites. *Applied Physics Letters* 92 (23): 231901. https://doi.org/10.1063/1.2938921.
27 Mohan, A.N. and Panicker, S. (2019). Facile synthesis of graphene-tin oxide nanocomposite derived from agricultural waste for enhanced antibacterial activity against Pseudomonas aeruginosa. *Scientific Reports* 9 (1): 4170. https://doi.org/10.1038/s41598-019-40916-9.

28 Bartkowski, M. and Giordani, S. (2020). Supramolecular chemistry of carbon nano-onions. *Nanoscale* 12: 9352–9358. https://doi.org/10.1039/D0NR01713B.

29 Mamidi, N., Delgadillo, R.M.V., Ortiz, A.G., and Barrera, E.V. (2020). Carbon nano-onions reinforced multilayered thin film system for stimuli-responsive drug release. *Pharmaceutics* 12 (12): 1208. https://doi.org/10.3390/pharmaceutics12121208.

30 Gergeroglu, H., Yildirim, S., and Ebeoglugil, M.F. (2020). Nano-carbons in biosensor applications: an overview of carbon nanotubes (CNTs) and fullerenes (C60). *SN Applied Sciences* 2: 603. https://doi.org/10.1007/s42452-020-2404-1.

31 Christwardana, M., Ji, J., Chung, Y., and Kwon, Y. (2017). Highly sensitive glucose biosensor using new glucose oxidase based biocatalyst. *Korean Journal of Chemical Engineering* 34 (11): 2916–2921. https://doi.org/10.1007/s11814-017-0224-9.

32 Romero, M., Figueroa, R., and Madden, C. (2000). Pressure sensing systems for medical devices. *Medical Device and Diagnostic Industry* 22: 124. http://www.devicelink.com/mddi/archive/00/10/004.html.

33 Joseph, H., Swafford, B., and Terry, S. (1997). MEMS in the medical world. *Sensing Magazine* 14: 47–51.

34 Xiao, X., Fan, S.-C., Li, C., and Xing, W.-W. (2019). Stress-insensitive resonant graphene mass sensing via frequency ratio. *Sensors* 19 (13): 3027. https://doi.org/10.3390/s19133027.

35 Kveton, F., Blsakova, A., Lorencova, L. et al. (2019). A graphene-based glycan biosensor for electrochemical label-free detection of a tumor-associated antibody. *Sensors* 19 (24): 5409. https://doi.org/10.3390/s19245409.

36 Patil, A., Mishra, V., Thakur, S. et al. (2019). Nanotechnology derived nanotools in biomedical perspectives: an update. *Current Nanoscience* 15 (2): 137–146. https://doi.org/10.2174/1573413714666180426112851.

37 Kurbanoglu, S., Cevher, S.C., Toppare, L. et al. (2022). Electrochemical biosensor based on three components random conjugated polymer with fullerene (C60). *Bioelectrochemistry* 147: 108219. https://doi.org/10.1016/j.bioelechem.2022.108219.

38 Zhang, P., Zhuo, Y., Chang, Y. et al. (2015). Electrochemiluminescent graphene quantum dots as a sensing platform: a dual amplification for MicroRNA assay. *Analytical Chemistry* 87 (20): 10385–10391. https://doi.org/10.1021/acs.analchem.5b02495.

39 Han, S.I. and Han, K.H. (2015). Electrical detection method for circulating tumor cells using graphene nanoplates. *Analytical Chemistry* 87: 10585–10592. https://doi.org/10.1021/acs.anal chem.5b03147.

40 Afkhami, A., Khoshsafar, H., Bagheri, H., and Madrakian, T. (2014). Preparation of $NiFe_2O_4$/graphene nanocomposite and its application as a modifier for the fabrication of an electrochemical sensor for the simultaneous determination of tramadol and acetaminophen. *AnalyticaChimica Acta* 831: 50–59. https://doi.org/10.1016/j.aca.2014.04.061.

41 Ge, L., Wang, W., Sun, X. et al. (2016). Affinity-mediated homogeneous electrochemical aptasensor on a graphene platform for ultrasensitive biomolecule detection via exonuclease-assisted target-analog recycling amplification. *Analytical Chemistry* 88 (4): 2212–2219. https://doi.org/10.1021/acs.analchem.5b03844.

42 Giannopoulos, G.I. (2022). Fullerene derivatives for drug delivery against COVID-19: a molecular dynamics investigation of Dendro [60] fullerene as nanocarrier of molnupiravir. *Nanomaterials* 12 (15): 2711. https://doi.org/10.3390/nano12152711.

43 Raza, K., Thotakura, N., Kumar, P. et al. (2015). C60-fullerenes for delivery of docetaxel to breast cancer cells: a promising approach for enhanced efficacy and better pharmacokinetic profile. *International Journal of Pharmaceutics* 495 (1): 551–559. https://doi.org/10.1016/j.ijpharm.2015.09.016.

44 Feng, T., Ai, X., An, G. et al. (2016). Charge-convertible carbon dots for imaging-guided drug delivery with enhanced in vivo cancer therapeutic efficiency. *ACS Nano* 10 (5): 5587. PMID: 2699743110.1021/acsnano.6b00043.

45 Engelberth, S.A., Hempel, N., and Bergkvist, M. (2014). Development of nanoscale approaches for ovarian cancer therapeutics and diagnostics. *Critical Reviews in Oncogenesis* 19 (3–4): 281–315. https://doi.org/10.1615/critrevoncog.2014011455.

46 Karthik, S., Saha, B., Ghosh, S.K., and Singh, P. (2013). N.D. Photoresponsivequinoline tethered fluorescent carbon dots for regulated anticancer drug delivery. *Chemical Communications* 49: 10471–10473. https://doi.org/10.1039/C3CC46078A.

47 Beack, S., Kong, W.H., Jung, H.S. et al. (2015). Photodynamic therapy of melanoma skin cancer using carbon dot-chlorin E6-hyaluronate conjugate. *Acta Biomaterialia* 26: 295–305. https://doi.org/10.1016/j. actbio. 2015.08.027.

48 Ahlawat, J., Masoudi, A.S., Guillama, B.G. et al. (2021). Application of carbon nano onions in the biomedical field: recent advances and challenges. *Biomaterials Science* 9: 626–644. https://doi.org/10.1039/D0BM01476A.

49 You, Y., Wang, N., He, L. et al. (2019). Designing dual-functionalized carbon nanotubes with high blood–brain-barrier permeability for precise orthotopic glioma therapy. *Dalton Transactions* 48: 1569–1573. https://doi.org/10.1039/C8DT03948H.

50 Zhong, B., Mateu-Roldán, A., Fanarraga, M.L. et al. (2022). Graphene-encapsulated magnetic nanoparticles for safe and steady delivery of ferulic acid in diabetic mice. *Chemical Engineering Journal* 435: 134466. https://doi.org/10.1016/j.cej.2021.134466.

51 Tian, Y., Nusantara, A.C., Hamoh, T. et al. (2022). Functionalized fluorescent nanodiamonds for simultaneous drug delivery and quantum sensing in HeLa cells.

ACS Applied Materials & Interfaces 14 (34): 39265–39273. https://doi.org/10.1021/acsami.2c11688.

52 Zhang, L., Lu, Z., Zhao, Q. et al. (2011). Enhanced chemotherapy efficacy by sequential delivery of siRNA and anticancer drugs using PEI grafted graphene oxide. *Small* 7 (4): 460–464. https://doi.org/10.1002/smll.201001522.

53 Yang, H.-W., Huang, C.-Y., Lin, C.-W. et al. (2014). Gadolinium functionalized nanographene oxide for combined drug and microRNA delivery and magnetic resonance imaging. *Biomaterials* 35 (24): 6534–6542. https://doi.org/10.1016/j.biomaterials.2014.04.057.

54 Zhao, H., Ding, R., Zhao, X. et al. (2017). Graphene-based nanomaterials for drug and/or gene delivery, bioimaging, and tissue engineering. *Drug Discovery Today* 22 (9): 1302–1317. https://doi.org/10.1016/j. drudis. 2017.04.002.

55 Liu, Z., Davis, C., Cai, W. et al. (2008). Circulation and long-term fate of functionalized, biocompatible single-walled carbon nanotubes in mice probed by raman spectroscopy. *Proceedings. National Academy of Sciences. United States of America* 105 (5): 1410–1415. https://doi.org/10.1073/pnas. 0707654105.

56 Golubewa, L., Timoshchenko, I., Romanov, O. et al. (2020). Single-walled carbon nanotubes as a photo-thermo-acoustic cancer theranostic agent: theory and proof of the concept experiment. *Scientific Reports* 10 (1): 22174. https://doi.org/10.1038/s41598-020-79238-6.

57 Li, X., Zhao, L., Li, P. et al. (2017). In-situ electron microscopy observation of electrochemical sodium plating and stripping dynamics on carbon nanofiber current collectors. *Nano Energy* 42: 122–128. https://doi.org/10.1016/j.nanoen.2017.10.050.

58 Li, X., Zhang, D., Xu, X., and Lee, K.R. (2020). Tailoring the nanostructure of graphene as an oil-based additive: toward synergistic lubrication with an amorphous carbon film. *ACS Applied Materials & Interfaces* 12 (38): 43320–43330. https://doi.org/10.1021/acsami.0c12890.

59 Zhao, T., Ji, X., Jin, W. et al. (2017). Electromagnetic wave absorbing properties of aligned amorphous carbon nanotube/$BaFe_{12}O_{19}$nanorod composite. *Journal of Alloys and Compounds* 703: 424–430. https://doi.org/10.1016/j.jallcom.2017.02.014.

60 Gao, N., Yang, W., Nie, H. et al. (2017). Turn-on theranostic fluorescent nanoprobe by electrostatic self-assembly of carbon dots with doxorubicin for targeted cancer cell imaging, in vivo hyaluronidase analysis, and targeted drug delivery. *Biosensors & Bioelectronics* 96: 300–307. https://doi.org/10.1016/j.bios.2017.05.019.

61 Sanchez, V.C., Jachak, A., Hurt, R.H. et al. (2012). Biological interactions of graphene-family nanomaterials: an interdisciplinary review. *Chemical Research in Toxicology* 25 (1): 15–34. https://doi.org/10.1021/tx200339h.

62 Mohan, N., Chen, C.S., Hsieh, H.H. et al. (2010). In vivo imaging and toxicity assessments of fluorescent nanodiamonds in caenorhabditis elegans. *Nano Letters* 10 (9): 3692–3699. https://doi.org/10.1021/nl1021909.

63 Merson, T.D., Turnley, A.M., Turbic, A. et al. (2013). Nanodiamonds with silicon vacancy defects for nontoxic photostable fluorescent labeling of neural precursor cells. *Optics Letters* 38 (20): 4170–4173. https://doi.org/10.1364/OL.38.004170.

64 Xiong, Z., Yun, Y.S., and Jin, H.-J. (2013). Applications of carbon nanotubes for lithium ion battery anodes. *Materials* 6 (3): 1138–1158. https://doi.org/10.3390/ma6031138.

65 de Las Casas, C. and Li, W. (2012). A review of application of carbon nanotubes for lithium ion battery anode material. *Journal of Power Sources* 208: 74–85. https://doi.org/10.1016/j.jpowsour.2012.02.013.

66 Shimoda, H., Gao, B., Tang, X.P. et al. (2001). Lithium intercalation into opened single-wall carbon nanotubes: storage capacity and electronic properties. *Physical Review Letters* 88 (1): 015502. https://doi.org/10.1103/PhysRevLett.88.015502.

67 Mi, C.H., Cao, G.S., and Zhao, X.B. (2004). A non-GIC mechanism of lithium storage in chemical etched MWNTs. *Journal of Electroanalytical Chemistry* 562 (2): 217–221. https://doi.org/10.1016/j.jelechem.2003.09.004.

68 Yang, S., Huo, J., Song, H., and Chen, X. (2008). A comparative study of electrochemical properties of two kinds of carbon nanotubes as anode materials for lithium ion batteries. *Electrochimica Acta* 53 (5): 2238–2244. https://doi.org/10.1016/j.electacta.2007.09.040.

69 Hosam, M.S. and Martin, K. (2019). Introductory chapter: carbon nanotubes. *IntechOpen* https://doi.org/10.5772/intechopen.85387.

70 Chew, S.Y., Ng, S.H., Wang, J. et al. (2009). Flexible free-standing carbon nanotube films for model lithium-ion batteries. *Carbon* 47 (13): 2976–2983. https://doi.org/10.1016/j.carbon.2009.06.045.

71 Wang, L., Guo, W., Lu, P. et al. (2019). A flexible and boron-doped carbon nanotube film for high-performance Li storage. *Frontiers in Chemistry* 7 (832): https://doi.org/10.3389/fchem.2019.00832.

72 Li, X., Liu, J., Zhang, Y. et al. (2012). High concentration nitrogen doped carbon nanotube anodes with superior Li^+ storage performance for lithium rechargeable battery application. *Journal of Power Sources* 197: 238–245. https://doi.org/10.1016/j.jpowsour.2011.09.024.

73 Wang, B., Li, X., Luo, B. et al. (2013). Intertwined network of Si/C nanocables and carbon nanotubes as lithium-ion battery anodes. *ACS Applied Materials & Interfaces* 5 (14): 6467–6472. https://doi.org/10.1021/am402022n.

74 Han, W., Zhou, Y., Zhu, T., and Chu, H. (2020). Combustion synthesis of defect-rich carbon nanotubes as anodes for sodium-ion batteries. *Applied Surface Science* 520: 146317. https://doi.org/10.1016/j.apsusc.2020.146317.

75 Banerjee, A., Ziv, B., Shilina, Y. et al. (2017). Single-wall carbon nanotube doping in lead-acid batteries: a new horizon. *ACS Applied Materials & Interfaces* 9 (4): 3634–3643. https://doi.org/10.1021/acsami.6b13377.

76 Zhu, Q.-C., Du, F.-H., Xu, S.-M. et al. (2016). Hydroquinone resin induced carbon nanotubes on Ni foam As binder-free cathode for Li–O2 batteries. *ACS Applied Materials & Interfaces* 8 (6): 3868–3873. https://doi.org/10.1021/acsami.5b10669.

77 Zhu, S., Chen, Z., Li, B. et al. (2011). Nitrogen-doped carbon nanotubes as air cathode catalysts in zinc-air battery. *Electrochimica Acta* 56 (14): 5080–5084. https://doi.org/10.1016/j.electacta.2011.03.082.

78 Ruhul, A., Petla, R.K., and Ilias, B. (2020). *Carbon Nanotubes: Applications to Energy Storage Devices*, vol. 73737. IntechOpen. https://doi.org/10.5772/intechopen.94155.

79 Yun, Y.S., Yoon, G., Kang, K., and Jin, H.-J. (2014). High-performance supercapacitors based on defect-engineered carbon nanotubes. *Carbon* 80 (1): 246–254. https://doi.org/10.1016/j.carbon. 2014.08.063.

80 Lim, S.H., Elim, H.I., Gao, X.Y. et al. (2006). Electronic and optical properties of nitrogen-doped multiwalled carbon nanotubes. *Physical Review B* 73 (4): 045402. https://doi.org/10.1103/PhysRevB.73.045402.

81 Song, C., Yun, J., Keum, K. et al. (2018). High performance wire-type supercapacitor with Ppy/CNT-ionic liquid/AuNP/carbon fiber electrode and ionic liquid based electrolyte. *Carbon* 144: 639–648. https://doi.org/10.1016/j.carbon.2018.12.100.

82 Girishkumar, G., Rettker, M., Underhile, R. et al. (2005). Single-wall carbon nanotube-based proton exchange membrane assembly for hydrogen fuel cells. *Langmuir* 21 (18): 8487–8494. https://doi.org/10.1021/la051499j.

83 Tan, L., Liu, Z.-Q., Li, N. et al. (2016). CuSe decorated carbon nanotubes as a high performance cathode catalyst for microbial fuel cells. *Electrochimica Acta* 213: 283–290. https://doi.org/10.1016/j.electacta.2016.07.099.

84 Deng, D., Yu, L., Chen, X. et al. (2013). Iron encapsulated within pod-like carbon nanotubes for oxygen reduction reaction. *Angewandte Chemie, International Edition* 52 (1): 371–375. https://doi.org/10.1002/anie.201204958.

85 Liu, H. and Li, Y. (2020). Modified carbon nanotubes for hydrogen storage at moderate pressure and room temperature. *Fullerenes, Nanotubes, and Carbon Nanostructures* 28 (8): 663–670. https://doi.org/10.1080/1536383x.2020.1738396.

86 Novoselov, K.S., Colombo, L., Gellert, P.R. et al. (2012). A roadmap for graphene. *Nature* 490 (7419): 192–200. https://doi.org/10.1038/nature11458.

87 Sun, D., Timmermans, M.Y., Tian, Y. et al. (2011). Flexible high-performance carbon nanotube integrated circuits. *Nature Nanotechnology* 6 (3): 156–161. https://doi.org/10.1038/nnano.2011.1.

88 Huang, X., Zeng, Z., Fan, Z. et al. (2012). Graphene-based electrodes. *Advanced Materials* 24 (45): 5979–6004. https://doi.org/10.1002/adma.201201587.

89 Franklin, A.D., Hersam, M.C., and Wong, S.P. (2022). Carbon nanotube transistors: making electronics from molecules. *Science* 378 (6621): 726–732. https://doi.org/10.1126/science.abp8278.

90 Franklin, A.D. (2015). Nanomaterials in transistors: from high-performance to thin-film applications. *Science* 349: aab2750. https://doi.org/10.1126/science.aab2750.

91 Franklin, A.D., Luisier, M., Han, S. et al. (2012). Sub-10 nm carbon nanotube transistor. *Nano Letters* 12 (2): 758–762. https://doi.org/10.1021/nl203701g.

92 Lu, W., Zu, M., Byun, H. et al. (2012). State of the art of carbon nanotube fibers: opportunities and challenges. *Advanced Materials* 24 (14): 1805–1833. https://doi.org/10.1002/adma.201104672.

93 Chen, T., Qiu, L., Yang, Z., and Peng, H. (2013). Novel solar cells in a wire format. *Chemical Society Reviews* 42: 5031–5041. https://doi.org/10.1039/c3cs35465b.

94 Sun, H., You, X., Jiang, Y. et al. (2014). Self-healable electrically conducting wires for wearable microelectronics. *Angewandte Chemie, International Edition* 53 (36): 9526–9531. https://doi.org/10.1002/anie.201405145.

95 Deng, J., Zhang, Y., Zhao, Y. et al. (2015). A shape-memory supercapacitor fiber. *Angewandte Chemie, International Edition* 54 (51): 15419–15423. https://doi.org/10.1002/anie.201508293.

96 Chen, X., Lin, H., Deng, J. et al. (2014). Electrochromic fiber-shaped supercapacitors. *Advanced Materials* 26 (48): 8126–8132. https://doi.org/10.1002/adma.201403243.

97 Chen, X., Qiu, L., Ren, J. et al. (2013). Novel electric double-layer capacitor with a coaxial fiber structure. *Advanced Materials* 25 (44): 6436–6441. https://doi.org/10.1002/adma.201301519.

98 Wang, K., Meng, Q., Zhang, Y. et al. (2013). High-performance two-ply yarn supercapacitors based on carbon nanotubes and polyaniline nanowire arrays. *Advanced Materials* 25 (10): 1494–1498. https://doi.org/10.1002/adma.201204598.

99 Koh, W.S., Lee, K.M., Toh, P.Y., and Yeap, S.P. (2019). Nano-graphene and its derivatives for fabrication of flexible electronic devices: a quick review. *Advanced Materials Letters* 10 (10): 676–681. https://doi.org/10.5185/amlett.2019.0050.

100 Wu, Z.S., Tan, Y.Z., Zheng, S. et al. (2017). Bottom-Up fabrication of sulfur-doped graphene films derived from sulfur-annulated nanographene for ultrahigh volumetric capacitance micro-supercapacitors. *Journal of the American Chemical Society* 139 (12): 4506–4512. https:www.doi.org/10.1021/jacs.7b00805.

101 Liu, X., Liú, D., Lee, J. et al. (2019). Spider-web-inspired stretchable graphene woven fabric for highly sensitive, transparent, wearable strain sensors.

ACS Applied Materials & Interfaces 11: 2282–2294. https://doi.org/10.1021/acsami.8b18312.

102 Romero, F.J., Castillo, E., Rivadeneyra, A. et al. (2019). Inexpensive and flexible nanographene-based electrodes for ubiquitous electrocardiogram monitoring. *npj Flexible Electronics* 3 (1): 1–6. https://doi.org/10.1038/s41528-019-0056-2.

103 Saito, Y., Uemura, S., and Hamaguchi, K. (1998). Cathode ray tube lighting elements with carbon nanotube field emitters. *Japanese Journal of Applied Physics* 37 (3): L346–L348. https://doi.org/10.1143/JJAP.37.L346.

104 Wei, Y., Weng, D., Yang, Y. et al. (2006). Efficient fabrication of field electron emitters from the multiwalled carbon nanotube yarns. *Physics Letters* 89 (6): 063101. https://doi.org/10.1063/1.2236465.

105 Song, W. (2003). Nematic liquid crystallinity of multiwall carbon nanotubes. *Science* 302 (5649): 1363. https://doi.org/10.1126/science.1089764.

106 Schindler, A., Brill, J., Fruehauf, N. et al. (2007). Solution-deposited carbon nanotube layers for flexible display applications. *Physica E: Low-Dimensional Systems and Nanostructures* 37 (1–2): 119–123. https://doi.org/10.1016/j.physe.2006.07.016.

107 Freitag, P., Zakhidov, A.A., Luessem, B. et al. (2012). Lambertian white top-emitting organic light emitting device with carbon nanotube cathode. *Journal of Applied Physics* 112 (11): 114505. https://doi.org/10.1063/1.4767439.

108 Kim, K.S., Zhao, Y., Jang, H. et al. (2009). Large-scale pattern growth of graphene films for stretchable transparent electrodes. *Nature* 457 (7230): 706–710. https://doi.org/10.1038/nature07719.

109 Ryu, J., Kim, Y., Won, D. et al. (2014). Fast synthesis of high-performance graphene by rapid thermal chemical vapor deposition. *ACS Nano* 8 (1): 950–956. https://doi.org/10.1021/nn405754d.

110 Ngafwan, N., Rasyid, H., Abood, E.S. et al. (2021). Study on novel fluorescent carbon nanomaterials in food analysis. *Food Science and Technology* 42: e37821. https://doi.org/10.1590/fst.37821.

111 Shah, A. (2020). A novel electrochemical nanosensor for the simultaneous sensing of two toxic food dyes. *ACS Omega* 5: 6187–6193. https://doi.org/10.1021/acsomega.0c00354.

112 Dalal, A., Rana, J.S., and Kumar, A. (2017). Ultrasensitive nanosensor for detection of malic acid in tomato as fruit ripening indicator. *Food Analytical Methods* 10: 3680–3686. https://doi.org/10.1007/s12161-017-0919-x.

113 Naskar, A., Khan, H., Sarkar, R. et al. (2018). Anti-biofilm activity and food packaging application of room temperature solution process based polyethylene glycol capped Ag-ZnO-graphene nanocomposite. *Materials Science and Engineering: C* 91: 743–753. https://doi.org/10.1016/j.msec.2018.06.009.

114 Wang, H., Chen, M., Jin, C. et al. (2018). Antibacterial [2-(methacryloyloxy) ethyl] trimethylammonium chloride functionalized reduced graphene oxide/poly (ethylene-co-vinyl alcohol) multilayer barrier film for food packaging. *Journal of Agricultural and Food Chemistry* 66: 732–739. https://doi.org/10.1021/acs.jafc.7b04784.

115 Arfat, Y.A., Ahmed, J., Ejaz, M., and Mullah, M. (2018). Polylactide/graphene oxide nanosheets/clove essential oil composite films for potential food packaging applications. *International Journal of Biological Macromolecules* 107: 194–203. https://doi.org/10.1016/j.ijbiomac.2017.08.156.

116 Salimi, F., Moradi, M., Tajik, H., and Molaei, R. (2021). Optimization and characterization of eco-friendly antimicrobial nanocellulose sheet prepared using carbon dots of white mulberry (Morus Alba L.). *Journal of the Science of Food and Agriculture* 101: 3439–3447. https://doi.org/10.1002/jsfa.10974.

117 Yu, H.Y., Qin, Z.Y., Sun, B. et al. (2014). Reinforcement of transparent poly (3-hydroxybutyrate-co-3-hydroxyvalerate) by incorporation of functionalized carbon nanotubes as a novel bionanocomposite for food packaging. *Composites Science and Technology* 94: 96–104. https://doi.org/10.1016/j.compscitech.2014.01.018.

118 Meador, M. (2012). Preface: forum on polymers for aerospace applications. *ACS Applied Materials and Interfaces* 4: 491. https://doi.org/10.1021/am300102x.

119 Keller, S. and Zaghloul, A. (2014). Carbon nanotube yarn and sheet antennas. In: *Nanotube Superber Materials—Changing Engineering Design* (ed. W. Andrew), 749–787,. http://dx.doi.org/focus.lib.kth.se/10.1016/B978-1-4557-7863-8.00026-8. Boston: Elsevier [chapter 26].

120 Baur, J. and Silverman, E. (2007). Challenges and opportunities in multifunctional nanocomposite structures for aerospace applications. *MRS Bulletin* 32 (4): 328–334. https://doi.org/10.1557/mrs2007.231.

121 O'Donnell, S., Sprong, K., and Haltli, B.(2004). Potential impact of carbon nanotube reinforced polymer composite on commercial aircraft performance and economics.*AIAA Fourth Aviation Technology, Integration and Operations (ATIO) Forum*, Chicago, IL (20–22 September 2004). Project no. 02MSR055-F3.

122 Roco, M., Mirkin, C., and Hersam, M. (2011). Nanotechnology research directions for societal needs in 2020: summary of international study. *Journal of Nanoparticle Research* 13: 897–919. https://doi.org/10.1007/s11051-011-0275-5.

123 Van Noorden, R. (2011). Chemistry: the trails of new carbon. *Nature* 469: 14–16. https://doi.org/10.1038/469014a.

124 Kukowski, T.(2012). Lightweight CNT cables for aerospace.*NESC NDE Technology Assessment NASA NDI Workshop*, Johnson SpaceCenter Houston, TX (29 February–1 March 2012).

125 McKenna, E. (2012). Product focus: wire and cable. Avionics Magazine, 2012 ⟨http://www.aviationtoday.com/av/military/Product-Focus-Wire-and-Cable-79536.html>.

126 Kim, P.C. and Lee, D.G. (2009). Composite sandwich constructions for absorbing the electromagnetic waves. *Composite Structures* 87 (2): 161–167. https://doi.org/10.1016/j.compstruct.2008.05.015.

127 Peak, L., Ryerson, C., and Martel, C. (2002).Army aircraft icing.UShttps://api.semanticscholar.org/CorpusID:108988009.

128 Chu, H., Zhang, Z., Liu, Y., and Leng, J. (2014). Self-heating fiber reinforced polymer composite using meso/macropore carbon nanotube paper and its application in deicing. *Carbon* 66: 154–163. https://doi.org/10.1016/j.carbon.2013.08.053.

129 Yoon, H., Song, W., Kim, D. et al. (2007). Transparent film heater using single-walled carbon nanotubes. *Advanced Materials* 19 (23): 4284–4287. https://doi.org/10.1002/adma.200701173.

130 Yao, X., Falzon, B., Hawkins, S., and Tsantzalis, S. (2018). Aligned carbon nanotube webs embedded in a composite laminate: a route towards a highly tunable electro-thermal system. *Carbon* 129: 486–494. https://doi.org/10.1016/j.carbon.2017.12.045.

131 ElFaham, M.M., Mostafa, A.M., and Nasr, G. (2020). Unmanned aerial vehicle (UAV) manufacturing materials: synthesis, spectroscopic characterization and dynamic mechanical analysis (DMA). *Journal of Molecular Structure* 1201: 127211. https://doi.org/10.1016/j.molstruc.2019.127211.

132 Zhang, Z. and Zhu, L. (2023). A review on unmanned aerial vehicle remote sensing: platforms, sensors, data processing methods, and applications. *Drones* 7 (6): 398. https://doi.org/10.3390/drones7060398.

133 Schütt, F., Postica, V., Adelung, R., and Lupan, O. (2017). Single and networked ZnO-CNT hybrid tetrapods for selective room-temperature high-performance ammonia sensors. *ACS Applied Materials & Interfaces* 9 (27): 23107–23118. https://doi.org/10.1021/acsami.7b03702.

134 Ali, I., Basheer, A.A., Mbianda, X. et al. (2019). Graphene based adsorbents for remediation of noxious pollutants from wastewater. *Environment International* 127: 160–180. https://doi.org/10.1016/j.envint.2019.03.029.

135 Rout, D.R. and Jena, H.M. (2020). Removal of malachite green dye from aqueous solution using reduced graphene oxide as an adsorbent. *Materials Today Proceedings* 47: 1173–1182. https://doi.org/10.1016/j.matpr.2021.03.406.

136 Ahmad, S.Z.N., Salleh, W.N.W., Ismail, A.F. et al. (2020). Adsorptive removal of heavy metal ions using graphene-based nanomaterials: toxicity, roles of functional groups and mechanisms. *Chemosphere* 248: 126008. https://doi.org/10.1016/j.chemosphere.2020.126008.

137 Wang, B., Lan, J., Bo, C. et al. (2023). Adsorption of heavy metal onto biomass-derived activated carbon: review. *RSC Advances* 13 (7): 4275–4302. https://doi.org/10.1039/d2ra07911a.

138 Shukla, J. (2021). Carbon nanofiber to remove heavy metals from aqueous medium. In: *Carbon Nanofibers: Fundamentals and Applications*, chapter 10 (ed. M. Sharon and M. Sharon), 251–289. Willey International Publication. https://doi.org/10.1002/9781119769149.ch10.

139 Mansouri, F., Chouchene, K., Roche, N., and Ksibi, M. (2021). Removal of pharmaceuticals from water by adsorption and advanced oxidation processes: state of the art and trends. *Applied Sciences* 11 (14): 6659. https://doi.org/10.3390/app11146659.

140 Fallah, Z., Zare, E.N., Ghomi, M. et al. (2021). Toxicity and remediation of pharmaceuticals and pesticides using metal oxides and carbon nanomaterials. *Chemosphere* 275: 130055. https://doi.org/10.1016/j.chemosphere.2021.130055.

141 Shah, K.W. and Li, W. (2019). A review on catalytic nanomaterials for volatile organic compounds VOC removal and their applications for healthy buildings. *Nanomaterials* 9 (6): 910. https://doi.org/10.3390/nano9060910.

142 Lee, S.W., Lee, W., Lee, D. et al. (2018). A simple and disposable carbon adhesive tape-based NO2 gas sensor. *Sensors and Actuators B: Chemical* 266: 485–492. https://doi.org/10.1016/j.snb.2018.03.161.

143 Jung, J.H., Hwang, G.B., Lee, J.E., and Bae, G.N. (2011). Preparation of airborne AG/CNT hybrid nanoparticles using an aerosol process and their application to antimicrobial air filtration. *Langmuir* 72 (16): 10256–10264. https://doi.org/10.1021/la201851r.

144 Miyawaki, J., Lee, G., Yeh, J. et al. (2012). Development of carbon-supported hybrid catalyst for clean removal of formaldehyde indoors. *Catalysis Today* 185 (1): 278–283. https://doi.org/10.1016/j.cattod.2011.09.036.

145 Anas, N.A.A., Fen, Y.W., Omar, N.A.S. et al. (2019). Development of graphene quantum dots-based optical sensor for toxic metal ion detection. *Sensors* 19: 3850. https://doi.org/10.3390/s19183850.

146 Park, S., Choi, K.S., Kim, S. et al. (2020). Graphene oxide-assisted promotion of plant growth and stability. *Nanomaterials* 10: 1–11. https://doi.org/10.3390/nano10040758.

147 Verma, S.K., Das, A.K., Gantait, S. et al. (2019). Applications of carbon nanomaterials in the plant system: a perspective view on the pros and cons. *Science of the Total Environment.* 667: 485–499. https://doi.org/10.1016/j.scitotenv.2019.02.409.

148 Zhao, F., Xin, X., Cao, Y. et al. (2021). Use of carbon nanoparticles to improve soil fertility, crop growth, and nutrient uptake by corn (Zea mays l.). *Nanomaterials* 11 (10): 678–690. https://doi.org/10.3390/nano11102717.

149 Li, Y., Xu, X., Wu, Y. et al. (2020). A review on the effects of carbon dots in plant systems. *Materials Chemistry Frontiers* 4 (2): 437–448. https://doi.org/10.1039/C9QM00614A.

150 Mahmoud, N.E. and Abdelhameed, R.M. (2021). Superiority of modified graphene oxide for enhancing the growth, yield, and antioxidant potential of pearl millet (Pennisetumglaucum L.) under salt stress. *Plant Stress* 2: 100025. https://doi.org/10.1016/j.stress.2021.100025.

151 Hsu, T.R. (2002). *MEMS and Microsystems: Design and Manufacture*. Boston: McGraw-Hill.

152 Weinberg, H. (2002). Using the ADXL202 in pedometer and personal navigation applications. http://www.sensorsmag.com/articles/0202/36/.

153 Chopra, S., McGuire, K., Gothard, N. et al. (2003). Selective gas detection using a carbon nanotube sensor. *Applied Physics Letters* 83 (11): 2280–2282. https://doi.org/10.1063/1.1610251.

154 Tongay, S., Schumann, T., Miao, X. et al. (2011). Tuning Schottky diodes at the many-layer-graphene/semiconductor interface by doping. *Carbon* 4962033–4962038. https://doi.org/10.1016/j.carbon.2011.01.029.

155 Chen, C.-C., Aykol, M., Chang, C.-C. et al. (2011). Graphene-silicon Schottky diodes. *Nano Letters* 1151863–1151867. https://doi.org/10.1021/nl104364c.

156 Zhong, H., Liu, Z., Xu, G. et al. (2012). Self-adaptive electronic contact between graphene and semiconductors. *Applied Physics Letters* 100: 122108. https://doi.org/10.1063/1.3696671.

157 Vaianella, F., Rosolen, G., and Maes, B. (2015). Graphene as a transparent electrode for amorphous silicon-based solar cells. *Journal of Applied Physics* 117. https://doi.org/10.1063/1.4923232.

158 Shim, J.P., Choe, M., Jeon, S.R. et al. (2011). InGaN-based p–i–n solar cells with graphene electrodes. *Applied Physics Express* 4: 4. https://doi.org/10.1143/APEX.4.052302.

159 Kavan, L., Yum, J.H., and Grätzel, M. (2011). Optically transparent cathode for dye-sensitized solar cells based on graphene nanoplatelets. *ACS Nano* 5 (1): 165–172. https://doi.org/10.1021/nn102353h.

160 Yang, Q.D., Li, J., Cheng, Y. et al. (2017). Graphene oxide as an efficient hole-transporting material for high-performance perovskite solar cells with enhanced stability. *Journal of Materials Chemistry A* 5: 9852. https://doi.org/10.1039/C7TA01752A.

161 Li, X., Chen, W., Zhang, S. et al. (2015). 18.5% efficient graphene/GaAs van der Waals heterostructure solar cell. *Nano Energy* 16: 310–319. https://doi.org/10.1016/j.nanoen.2015.07.003.

162 Walter, A.L., Jeon, K.J., Bostwick, A. et al. (2011). Highly -doped epitaxial graphene obtained by fluorine intercalation. *Applied Physics Letters* 98 (18): 184102. https://doi.org/10.1063/1.3586256.

163 Chandramohan, S., Janardhanam, V., Seo, T.H. et al. (2019). Improved photovoltaic effect in graphene/silicon solar cell using MoO_3/Ag/MoO_3

multilayer coating. *Materials Letters* 246: 103–106. https://doi.org/10.1016/j.matlet.2019.03.024.

164 Miao, X., Tongay, S., Petterson, M.K. et al. (2012). High efficiency graphene solar cells by chemical doping. *Nano Letters* 12 (6): 2745–2750. https://doi.org/10.1021/nl204414u.

165 Yang, S.T., Luo, J., Zhou, Q., and Wang, H. (2012). Pharmacokinetics, metabolism and toxicity of carbon nanotubes for biomedical purposes. *Theranostics* 2 (3): 271–282. https://doi.org/10.7150/thno.3618.

166 Sandova, J., Ventura-Sobrevilla, J., Boone-Villa, D. et al. (2019). Carbon nanomaterials as pharmaceutic forms for sustained and controlled delivery systems. In: *Nanomaterials for Drug Delivery Therepy* (ed. A.M. Grumezescu), 403–434. Amsterdam, The Netherlands: Elsevier.

167 Li, Y., Feng, L., Shi, X. et al. (2014). Surface coating-dependent cytotoxicity and degradation of graphene derivatives: towards the design of non-toxic, degradable nano graphene. *Small* 10 (8): 1544–1554. https://doi.org/10.1002/smll.201303234.

168 R. F. Service (2004). Nanotoxicology. Nanotechnology grows up. *Science* 304 (5678): 1732–1734. https://doi.org/10.1126/science.304.5678.1732.

169 Malina, T., Poláková, K., Hirsch, C. et al. (2021). Toxicity of carbon nanomaterials-towards reliable viability assessment via new approach in flow cytometry. *International Journal of Molecular Sciences* 22: 7750. https://doi.org/10.3390/ijms22147750.

170 Baker, S.N. and Baker, G.A. (2010). Luminescent carbon nanodots: emergent nanolights. *Angewandte Chemie, International Edition* 49: 6726–6744. https://doi.org/10.1002/anie.200906623.

171 Dai, H. (2001). *Carbon Nanotubes in Topics in Applied Physics*, vol. 80 (ed. M.S. Dresselhaus, G. Dresselhaus, and P. Avouris), 29–52. New York: Springer.

172 Dai, H. (2002). Carbon nanotubes: opportunities and challenges. *Surface Science* 500: 218–241. https://doi.org/10.1016/S0039-6028(01)01558-8.

173 Ruoff, R.S., Qian, D., and Liu, W.K. (2003). Mechanical properties of carbon nanotubes: theoretical predictions and experimental measurements. *ComptesRendus Physique* 4: 9933–1008. https://doi.org/10.1016/j.crhy.2003.08.001.

174 Pender, J.P., Jha, G., Youn, D.H. et al. (2020). Electrode degradation in lithium-ion batteries. *ACS Nano* 14 (2): 1243–1295. https://doi.org/10.1021/acsnano.9b04365.

175 Wei, H., Liu, Y., Zhai, X. et al. (2020). Application of carbon nanotube-based materials as interlayers in high-performance lithium-sulfur batteries: a review. *Frontiers in Energy Research* 8: 585795. https://doi.org/10.3389/fenrg.2020.585795.

176 Chen, T. and Dai, L. (2013). Carbon nanomaterials for high-performance supercapacitors. *Materials Today* 16 (7–8): 272–280. https://doi.org/10.1016/j.mattod.2013.07.002.

177 Zhong, M., Zhang, M., and Li, X. (2022). Carbon nanomaterials and their composites for supercapacitors. *Carbon Energy* 4 (5): 665–1001. https://doi.org/10.1002/cey2.219.

178 Chen, T.-W., Kalimuthu, P., Veerakumar, P. et al. (2022). Recent developments in carbon-based nanocomposites for fuel cell applications: a review. *Molecules* 27 (3): 761. https://doi.org/10.3390/molecules27030761.

179 Tripathi, S.L. and Padmanabhan, S. (2020). *Green Energy: Fundamentals, Concepts, and Applications*. USA: Scrivener Publishing, Wiley. https://doi.org/10.1002/9781119760801.

19

Carbon-Based Nanomaterials and Their Green Energy Applications: Carbon Nanotubes

Smita Singh, Varsha Singh, Vikram Rathour, and Vellaichamy Ganesan

Department of Chemistry, Institute of Science, Banaras Hindu University, Varanasi, Uttar Pradesh, India

19.1 Introduction

The need for energy is increasing exponentially and it is anticipated that by 2050, the energy requirement will increase almost 50% of the present energy needs [1–4]. To combat the energy crisis, a significant amount of research and devices based on renewable sources of energy like wind, solar [5], and hydro power have been done [5]. However, their geographical confinement, cost of production, and interrupted supply pose major challenges, which affect their extensive adoption [2]. Aligning with the goals of environment sustainability and development, electrochemical energy production and storage devices have been boosted in recent years owing to their promising use in various electronic devices [6]. In the last few years, advancements in materials science, particularly in the domain of carbon nanomaterials, have significantly influenced the performance of energy-related devices [7, 8]. The unique properties of carbon nanomaterials, like fullerenes, carbon nanotubes (CNTs), graphene, *etc.*, have been exploited to enhance the efficiency of energy conversion and storage [6, 9]. These materials exhibit size- and surface-dependent characteristics, including structural, electrical, optical, and mechanical features, which are essential for improving the overall efficiency of the devices [10]. The extensive research and development efforts focus on the exploitation of the distinct attributes of carbon nanomaterials for highly efficient energy-conversion and storage devices. Notably, the growth is remarkable over the past two decades in utilizing fullerenes, CNTs, and graphene in advanced energy technologies [6, 9]. This book chapter deals with a detailed review of the

Carbon-based Nanomaterials for Green Applications, First Edition. Edited by Upendra Kumar, Piyush Kumar Sonkar, and Suman Lata Tripathi.

Published 2025 by John Wiley & Sons, Inc.

progress made based on carbon nanomaterials, shedding light on their potential applications in advanced energy conversion technologies such as fuel cells, in addition to energy storage devices like supercapacitors and batteries. The discussion comprehends the achievements emphasizing the positive impact of carbon nanomaterials on the efficiency and functionality of energy devices. Acknowledging that the progress has been significant, this chapter also explores the challenges encountered in this field and the hurdles that necessitate innovative solutions.

19.1.1 Carbon Nanomaterials

Carbon is recognized to occur in several forms like amorphous carbon, graphite, fullerenes, graphene, CNTs, and diamond [11]. The distinctive properties of these forms stem from the arrangement of carbon atoms (C-atoms). Graphite is characterized by a soft and black composition, representing the stable and prevalent type of carbon [6]. This stability arises from the robust covalent bonding within the carbon plane supplemented by comparatively feeble van der Waals interactions between the layers [9]. On the other hand, diamond exhibits hardness and transparency. In a regular lattice structure, individual C-atoms are intricately attached to four other C-atoms [9]. This results in the remarkable physical properties of diamond, including its hardness and clarity. The stark contrast in properties between graphite and diamond highlights the significance of carbon's atomic arrangement in influencing its characteristics, making it a fascinating element with diverse applications across a range of fields. The characteristics of fullerenes, graphene, and CNTs are explained in detail in the subsequent segments as these forms of carbon are being explored extensively in the green energy field. Figure 19.1 shows the schematic illustration of various carbon nanomaterials (fullerenes, graphene, and CNTs) with their key applications.

19.1.2 Fullerenes (C_{60})

In 1985, Kroto et al. made a groundbreaking discovery, revealing that buckminsterfullerene. C_{60} possesses a distinctive soccer ball-like structure [12]. This molecular structure features a radius of 35.5 nm and nearly 100 nm center-to-center distance among adjacent molecules in a single crystal [9, 10, 12]. Notably, this structure comprises 12 pentagons and 20 hexagons symmetrically arranged, marking the inception of a wholly new branch of chemistry [12]. However, it wasn't until 1990 when Kratschmer and Huffman conveyed a straightforward method for producing macroscopic quantities of fullerenes that scientists understood the physicochemical properties of C_{60} [6]. This breakthrough enabled the application of spectroscopic methods, allowing researchers to utilize fullerenes as

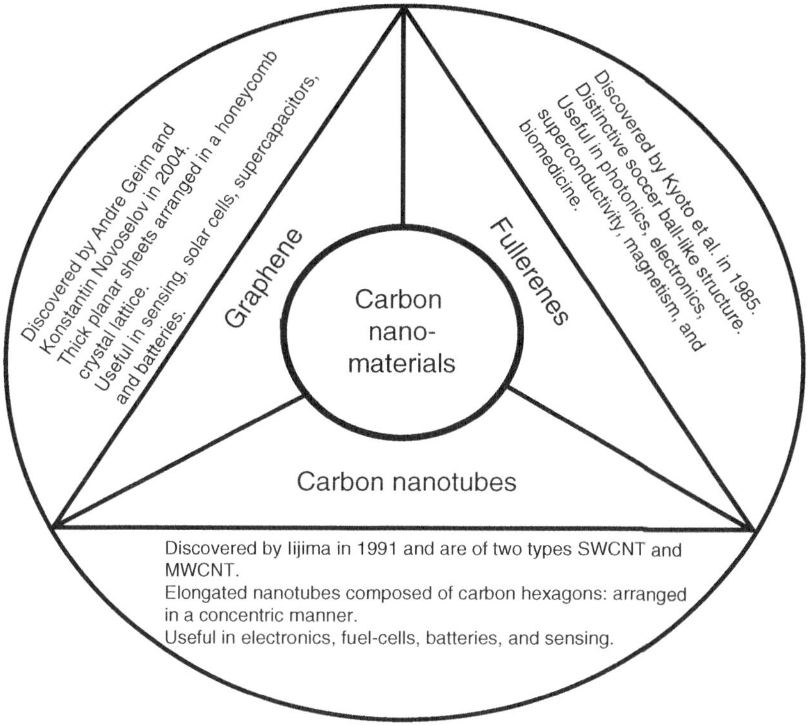

Figure 19.1 Illustration of carbon nanomaterials and their key applications.

valuable reagents in synthetic chemistry. After this key development, scientists successfully synthesized various spherical fullerenes, known as buckyballs, featuring varying numbers of hexagonal faces. The unique molecular structures of fullerenes and their derivatives swiftly revealed captivating properties in the areas of photonics [13], electronics [14], superconductivity [15], magnetism [16], and biomedicine [17]. Building upon these findings, diverse arrays of uses have been suggested for fullerenes and their derivatives. The exceptional properties exhibited by these carbon clusters have fuelled exploration of the potential uses across multiple disciplines showcasing the versatility and promise of fullerenes in various scientific and technological realms.

19.1.3 Carbon Nanotubes

The finding of CNTs by Iijima is marked as the commencement of a new period in materials science and nanotechnology [18]. Worldwide, the commercial fascination with CNTs is evident through their production capability. It is projected to surpass several thousand tons annually and will continue to grow further [18]. These elongated nanotubes exhibit a unique structure composed of carbon hexagons organized in a co-axial way, and frequently, both ends are confined by structures resembling fullerenes, comprising pentagonal elements [6]. This distinctive arrangement imparts exceptional properties to CNTs, leading to their widespread exploration and application in diverse fields, extending from advanced materials to cutting-edge nanotechnology. CNTs are conceptualized as a graphene sheet folded into a tube, giving rise to a single-walled CNT (SWCNT) [18]. When examined under a transmission electron microscope (TEM), they appear as two distinct planes. Furthermore, there is the possibility of additional graphene coaxial tubes surrounding the core of the SWCNT, resulting in the formation of a multi-walled CNT (MWCNT) [18, 19]. This hierarchical structure, characterized by concentric layers, contributes to the unique properties and versatile applications of CNTs in various scientific and technological domains [20].

19.1.4 Graphene

In 2004, Andre Geim and Konstantin Novoselov ascertained graphene by repeatedly cleaving graphite crystals with adhesive tape until they attained single atomic layers [6, 21]. This groundbreaking innovation earned them the Physics Nobel Prize in 2010 and sparked a surge of interest in graphene. It consists of one-atom-thick planar sheets formed by densely packed sp^2-bonded carbon atoms, arranged into a honeycomb structure [21, 22]. Graphene, recognized as the primary unit of the graphitic structure, serves as the precursor for carbon materials across various dimensionalities. This comprises spherical buckyballs with zero dimensions, tubular nanotubes with one dimension, and voluminous graphite with three dimensions [9]. Having properties similar to CNTs, such as a high aspect ratio, high surface area, diverse electronic states, and robust mechanical features [6], graphene arises as an interesting candidate for potent applications in numerous fields such as in sensing [23], solar cells [24], supercapacitors [25], and batteries [26]. Figure 19.2 shows the structures of fullerene, CNT, and graphene.

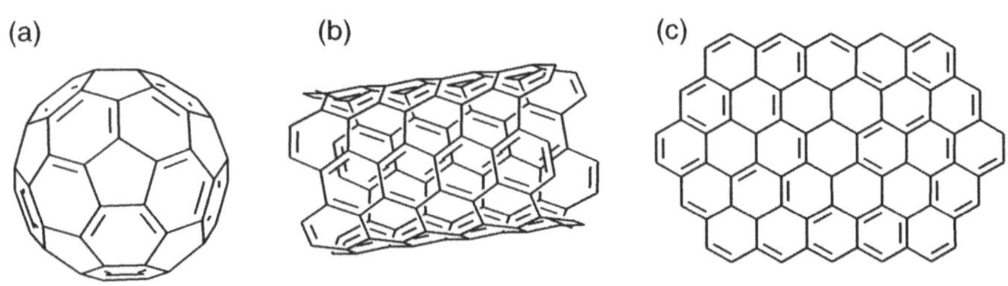

Figure 19.2 Structures of fullerene (a), CNT (b), and graphene (c).

19.2 Synthesis of CNTs

For the first time, CNTs were produced accidentally by Iijima while he was trying to synthesize fullerene using arc discharge method. There are two widely used methods for the synthesis of CNTs: (1) plasma-based synthesis (comprises arc discharge and laser ablation) and (2) thermal-based synthesis (includes chemical vapor deposition (CVD) and plasma-enhanced chemical vapor deposition [PECVD]).

19.2.1 Plasma-Based Synthesis

Arc discharge method is the popular technique to produce CNTs. This process includes the growth of CNTs on graphite electrodes by vaporizing carbon through direct current arc discharge, utilizing an inert gas such as helium or argon. To initiate the process, anode and cathode are gradually brought closer until they are separated by a distance of fewer than 1 mm. Around 100 A current elapses across these electrodes, creating plasma within the space. This plasma can reach incredibly high temperatures, reaching up to approximately 4000 K. Under such extreme heat, the carbon material on the anode vaporizes and then accumulates on the cathode. Typically, the anode has a smaller diameter compared to the cathode. Both electrodes are equipped with cooling systems to manage the heat generated during the process, usually employing water-cooling mechanisms. Additionally, continuous movement of the anode is necessary to maintain a consistent gap between the electrodes. The yield is around 75% in this method and temperature requirement is high (above 1700 °C). The arc-discharge method can be used to achieve two different synthesis processes: one with a catalyst for producing MWCNTs and the other without a catalyst for producing SWCNTs. Typically, SWCNTs are produced using catalyst precursors and a sophisticated anode composed of a combination of graphite and a metal such as Gd, Co, Ni, Fe, Ag, Pt, and Pd. [27]. On the other hand, MWCNTs can be synthesized without the need for catalyst precursors. The incorporation of a metallic catalyst during synthesis process can introduce impurities into the CNTs emphasizing the necessity for purifying the final product [15, 24].

Laser ablation method reduces the yield of CNTs by 70% and uses limited quantity of Ni and Co at a temperature of 1200 °C. The primary objective of laser ablation is to vaporize the graphite material enclosed within a quartz container. The energy from the laser pulses causes the nanotube diameters to decrease, resulting in thinner tubes. Laser ablation maintains a stable pressure by utilizing helium or argon gas. SWCNTs synthesized using this method are more pure and have better quality [15, 25]. Both the arc-discharge and laser ablation methods share certain similarities. However, in the laser ablation technique, the necessary power is supplied by a laser that heats a pure graphite pellet containing a catalyst.

19.2.2 Thermal-Based Synthesis

In 1996, the CVD method was adopted for the synthesis of CNTs. This technique allocates a specific control over the composition and enables the synthesis of large quantities of CNTs. By changing the volatile chemicals, it is possible to induce reactions that transform them into the desired solid films [28]. Figure 19.3 depicts the TEM images of CNTs synthesized using Ar and acetylene (C_2H_2) carrier gases. The vapor phase growth, an advanced adaptation of CVD, facilitates the synthesis of CNTs within a chamber containing a catalyst, eliminating the need for a substrate. This method employs ferrocene as a catalyst within the reaction chamber with two low-temperature furnaces to support the process [19].

Flame synthesis method is another alternative approach to synthesizing CNTs, which involves the utilization of a flame. This method employs the flame as a synthesis medium, supplying together the necessary energy and chemical species in several carbon forms at the nanoscale. The flame is generated by combining fuels (CH_4, C_2H_4, C_2H_2), among others, with an oxidizer to produce a diverse gaseous mixture comprising gases (CO_2, H_2O, CO, H_2), hydrocarbons (C_2H_2, C_2H_4, C_2H_6), and radicals. Catalytic agents are essential to impart active sites for the accumulation of solid carbon. Notably, the flame synthesis method operates as an autothermal process, self-sustaining in generating the requisite temperatures to achieve the desired synthesis conditions, distinguishing it from other methods [19].

The PECVD method offers a relatively low-temperature synthesis enabling precise control over the growth mechanism of CNTs. Through PECVD, gas molecules undergo efficient dissociation due to the plasma energy, initiating CNT formation at lower temperatures. This method showcases growth rates substantially higher, up to two orders of magnitude compared to conventional approaches [15, 26].

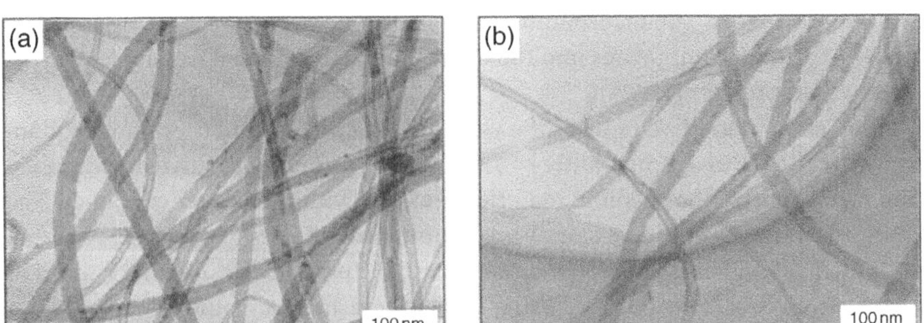

Figure 19.3 TEM image of CNTs produced by Ar (a) and C_2H_2 (b) carrier gas. *Source:* Reproduced with permission from [29]/Elsevier.

19.3 Properties of Carbon Nanotubes

The properties of CNTs are profoundly affected by issues like the degree of graphitization, chirality, and diameter. These characteristics significantly impact the mechanical strength, electrical conductivity, thermal properties, and other key attributes of CNTs. CNTs exhibit structural flexibility, low mass density, chemical stability, high aspect ratio (length-to-diameter ratio), high conductivity, and large surface area [5, 27]. Characteristics of CNTs like mechanical, electrical, thermal, optical, and chemical are discussed briefly in the following section.

19.3.1 Mechanical Properties of CNTs

Researchers have extensively examined the mechanical properties of CNTs through both theoretical analysis and experimental investigations over numerous years. Due to the remarkable strength of C—C bonds within the honeycomb lattice structure, it is necessary to investigate the mechanical characteristics of CNTs. Moreover, if a graphene sheet is turned into a SWCNT, the axial bonding between carbon atoms notably intensifies. This heightened bonding is the primary factor contributing to the exceptional stiffness of CNTs, surpassing that of diamond, and resulting in their remarkably high Young's modulus and tensile strength [25, 26]. CNTs possess extraordinary tensile strength, being among the strongest materials known. They can withstand high tensile forces without deformation, owing to the strong covalent bonds within their structure. CNTs exhibit remarkable stiffness or Young's modulus, which measures their resistance to deformation under stress. They are extremely rigid and can maintain their shape and structural integrity even under substantial mechanical loads. When combined with their lower density, CNTs have been discovered to possess Young's modulus of up to 1000 GPa, making them approximately five times stiffer than steel. Additionally, their tensile strength can reach up to 100 GPa, which is nearly 50 times greater than that of steel. These two qualities make the nanotube ideal for structural applications [30]. Despite their incredible stiffness, CNTs also display a degree of flexibility and elasticity. They can undergo reversible deformations, allowing them to bend or stretch without permanent damage under certain conditions. CNTs are highly resistant to buckling, meaning that they can maintain their structural stability even when subjected to compressive forces, making them valuable for reinforcing materials in various applications [31].

19.3.2 Electrical Properties of CNTs

CNTs exhibit exceptional electrical properties owing to their unique structure. Their one-dimensional structure made from rolled-up graphene sheets imparts remarkable electrical conductivity and other relevant properties. SWCNTs show exceptional conductance, even surpassing that of copper, making them promising candidates for various applications involving electrochemical sensors and energy conversion and storage. This high conductivity arises from the strong σ-bonds and π-electrons along the length of the nanotube. The electronic properties of CNTs can be categorized into metallic, semiconducting, or a combination of both (known as semiconducting with a small bandgap). This behavior is determined by the chirality (the arrangement of carbon atoms in the graphene sheet turned up to develop the nanotube) and it significantly affects their electrical properties. CNTs' electrical conductivity is influenced by temperature. Typically, metallic CNTs exhibit a decrease in resistance with decreasing temperature, following a metallic behavior. In contrast, the conductivity of semiconducting CNTs may change with temperature due to variations in the bandgap and carrier mobility. The conductance of CNTs is quantized due to their nanoscale diameter. This quantization leads to discrete energy levels for electrons, resulting in unique conductance properties. The conductance of metallic CNTs is quantized in integer multiples of $2e^2 h^{-1}$ (e and h represent the electron charge and Planck's constant, respectively), whereas semiconducting CNTs display quantization in terms of $4e^2 h^{-1}$ for each band. Achieving high electrical contacts between CNTs and other materials is crucial for exploiting their electrical properties effectively [25, 26].

19.3.3 Thermal Properties of CNTs

The thermal conductivity of CNTs is another fascinating aspect of their properties. CNTs possess remarkable thermal transport characteristics due to their unique structure, which impacts how heat is conducted through these materials. CNTs exhibit anisotropic thermal conductivity, meaning their thermal conductance varies with direction. Thermal conductivity along the length of the nanotube (axial direction) is exceptionally high, even surpassing that of most conventional materials like copper or diamond. Thermal conductivity across the nanotube's diameter (radial direction) is significantly lower. The exceptional thermal conductivity along the axial direction of CNTs is attributed to the robust sp^2 carbon–carbon bonds and the highly ordered crystalline structure. These characteristics enable efficient phonon (vibrational energy carrier) transport along the length of the nanotube, leading to high intrinsic thermal conductivity values. Longer nanotubes generally exhibit higher thermal conductivity due to fewer boundary scattering events. Moreover, thinner-diameter nanotubes may possess higher thermal conductivity along their axial direction compared to wider-diameter nanotubes. Optimizing CNT synthesis methods and understanding their thermal properties are ongoing areas of research to further exploit them in thermal applications [5, 32].

19.3.4 Optical Properties of CNTs

CNTs possess unique optical characteristics owing to their one-dimensional structure and high aspect ratio. These properties render them promising candidates for diverse applications in optics, electronics, and materials science. CNTs exhibit strong absorption and transmission properties above 200 nm. The absorption spectrum of CNTs depends on their diameter, chirality, and structural defects. They can absorb light efficiently across a wide range of wavelengths. They can emit light when excited with photons, a phenomenon known as photoluminescence. The emitted light's wavelength (or energy) is determined by the nanotube's structure. They can emit light in the visible or near-infrared spectrum, making them useful in optoelectronic devices like light-emitting diodes (LEDs) and sensors [19].

19.3.5 Chemical Properties of CNTs

CNTs can be chemically functionalized by introducing different functional groups or chemical moieties onto their surface. This process alters their properties, such as solubility, reactivity, and interaction with other materials. The reactivity increases as the diameter of the CNTs decreases and reaction occurs on the sidewall and end-top of the CNTs. Functionalization enhances CNTs' compatibility with certain polymers, biomolecules, or other nanomaterials for various applications. CNTs can undergo oxidation when exposed to strong oxidizing agents (e.g. nitric acid and ozone). This process initiates oxygen-containing functional groups on the CNT surface, altering their properties and enabling further chemical modifications. They can selectively sorb certain gases and molecules onto their surface because of their high surface area and exceptional structure. This property is of interest in gas sensing, purification, and storage applications. CNTs' surface properties can be modified to interact with biomolecules such as DNA, proteins, and drugs. Functionalized CNTs have been employed for drug delivery systems, biosensors, and biomedical uses due to their ability to transport biomolecules and penetrate cell membranes. CNTs naturally aggregate due to van der Waals forces. Chemical modifications can improve their dispersibility in solvents or matrices, enhancing their compatibility and utility in various applications, including composite materials and coatings. CNTs exhibit redox behavior due to their ability to accept or donate electrons. This property is exploited in applications involving energy storage devices like supercapacitors and batteries [19].

19.4 Green Energy Applications of CNTs

CNTs have shown immense potential in various green energy applications because of their prominent electrical conductivity, excellent mechanical strength, and remarkable thermal characteristics. Some of the notable green energy applications of CNTs include hydrogen evolution reaction (HER), oxygen evolution reaction (OER), oxygen reduction reaction (ORR), electrochemical-sensing applications, and materials for supercapacitors.

19.4.1 Supercapacitors

Supercapacitors store electrochemical charge through electrostatic forces generated by applying a potential across two parallel electrodes. Despite their ability to store more energy than conventional capacitors, their energy capacity remains lower than that of batteries. Importantly, supercapacitors demonstrate significantly higher charge/discharge stability than batteries. Supercapacitors find applications in various domains, such as hybrid electric vehicles, to provide auxiliary power, short-term backup energy sources for UPS, and to power portable electronic devices [33]. Electrochemical supercapacitors comprise two similar electrodes divided by a dielectric membrane, facilitating the transport of electrolyte ions between the electrodes [34]. The supercapacitive behavior can be categorized into two categories: electrochemical double-layer capacitance (EDLC) and pseudocapacitance. EDLC operates based on the Helmholtz double-layer theory, which relies on the buildup of electrostatic charges at the electrode-electrolyte interface for energy storage. In contrast, pseudocapacitors function through the Faradaic processes, i.e. charge transfer among the electrode and the electrolyte. The exceptional features of CNTs, such as their high surface area, distinctive porosities, chemical inertness, and superior electrical conductivity, render them well-suited as electrode materials for supercapacitors [29, 35]. Notably, CNTs exhibit superior catalytic activity in comparison to fullerenes and activated carbons. The heightened catalytic efficiency of CNTs is commonly credited to the spacious/accessible regions that facilitate the unrestricted movement of electrolytes between the electrodes and subsequently increase the capacitance. However, in the activated carbon, since it has non-uniform pores, the movement of electrolyte is hindered between the electrodes resulting in a relatively lower capacitance [36]. Niu et al. stated that CNTs can be utilized in supercapacitors. They functionalized the surface of MWCNTs using nitric acid. The resulting MWCNTs possessed a specific surface area of 430 $m^2\,g^{-1}$ and showed an energy density of 0.5 $Wh\,kg^{-1}$ when H_2SO_4 was used as the electrolyte. The functional groups and residual catalysts could induce pseudocapacitance, resulting in the contribution of both Faradaic and non-Faradaic processes in the CNT-based supercapacitor [36]. The redox behavior detected in the cyclic voltammetry (CV) of the SWCNT-based electrodes suggests that pseudocapacitance is influenced by the presence of functional groups and impurities [37]. Functionalization represents an effective strategy for boosting the supercapacitor performance of CNTs by enhancing pseudocapacitance. Zhang et al. functionalized the CNTs with varying amounts of oxygen functionalities, which exhibited a specific capacitance within the range of 24–58/Fg in the organic electrolytes. Although no notable alteration occurred in the surface area due to the functionalization, the interaction of electrolyte and electrodes varied based on the degrees of functionalities. A higher amount of functionalities may augment pseudocapacitive properties while potentially diminishing overall capacitance. Hence, achieving an optimal content of oxygen is crucial to prevent excessive charge transfer amid the electrode and electrolyte, thereby boosting double-layer capacitance [38]. Enhancing supercapacitor performance can also be achieved through the strategic formation of composites with various materials. The distinctive blend of pseudocapacitive elements, such as oxides, and metal oxyhydroxides of transition metals, carbon nitrides, metal nitrides, and conducting polymers, represent an alternative approach to optimizing supercapacitor functionality [39–43]. Zuttel and co-workers compared activated carbon with CNTs for ECDL and demonstrated a remarkable areal capacitance ranging from 0.0008 to 0.28 $F\,cm^{-2}$. The performance of CNTs is higher than that of activated carbon, which typically exhibits areal capacitances in the range of 0.0004–0.0031 $F\,cm^{-2}$ [44]. Kalra and co-workers studied capacitance for polyaniline (PANI) fiber and PANI/CNTs, which presented a large specific capacitance of 386 $F\,g^{-1}$ relative to PANI which is 308 $F\,g^{-1}$. The interconnected nanofiber network between PANI and CNTs nanofiber enhances electron conduction and the inter- and intra-fiber porosity facilitates exceptional electrolyte diffusion within the polymer matrix. This enables rapid ion movement to and from the active sites [45]. Zhao and co-workers synthesized PANI/CNT-COOH composite to show the effect of functional groups present on CNTs toward the supercapacitor applications. The PANI/CNT-COOH displays greater capacitance (381 $F\,g^{-1}$) than PANI/CNTs (without —COOH functionalization, 238 $F\,g^{-1}$) and pure PANI (186 $F\,g^{-1}$) [46].

19.4.2 HER

Intensive research efforts are currently focused on exploring other energy resources to decrease our reliance on fossil fuels and diminish the environmental effects related to their combustion [1, 3]. Hydrogen is recognized as a potential clean energy source and transport medium owing to its multitude of advantages, such as its high energy density by weight, sustainability, recyclability, CO_2 neutrality, and emission-free usage. However, the main hurdle revolves around the scaling up of hydrogen production for storage and practical use [42, 43]. The extraction of hydrogen directly from water is feasible through methods like electrolysis, photoelectrolysis, or solar thermochemical processes. Among these methods, electrolysis has emerged as the most commercially successful process. During water electrolysis, hydrogen and oxygen are produced through a sequence of electrode-based reactions. Despite the prolonged period of research in water electrolysis, its efficiency remains constrained, typically ranging from 56% to 73%, attributed to a deficit from overpotential and less turnover frequency, among other factors [6]. Electrochemical water splitting via the HER presents a viable solution toward this goal. Efficient electrocatalysts for HER play a pivotal role in achieving optimal performance. Currently, platinum-based compounds stand out as the most efficient and common electrocatalysts for HER, exhibiting remarkable catalytic activity with minimal overpotential.

Electrolysis of water in different electrolytes leads to varying pathways. In acidic electrolytes, a higher concentration of protons facilitates HER catalysis, making it more achievable at lower overpotentials. Conversely, achieving HER in a basic medium needs additional energy to produce H^+ by breaking down H_2O, thereby impacting the overall reaction rates [6]. HER mechanism in basic medium involves three steps: (1) the Volmer step, (2) water dissociation, and (3) the formation of reactive hydrogen intermediates. This mechanism is distinct from the one observed in acidic mediums, which typically follow either the Heyrovsky pathway or the Tafel recombination step [45, 46]. Likewise, in an alkaline environment, the HER pathway can follow either the Volmer–Heyrovsky step or the Volmer–Tafel step, as depicted in Eqs. (19.1–19.3). By referencing the Tafel slope values provided in Eqs. (19.1–19.3), it is possible to evaluate the reaction pathway occurring on a particular catalyst within alkaline electrolytes:

$$\left(\text{Volmer: Tafel slope} - 120\,\text{mV dec}^{-1}\right) \rightarrow H_2O + e^- \rightarrow H^* + HO^- \quad (19.1)$$

$$\left(\text{Heyrovsky: Tafel slope} - 40\,\text{mV dec}^{-1}\right) \rightarrow H_2O + e^- + H^* \rightarrow H_2 + HO^- \quad (19.2)$$

$$\left(\text{Tafel: Tafel slope} - 30\,\text{mV dec}^{-1}\right) \rightarrow 2H^* \rightarrow H_2 \quad (19.3)$$

There are two main approaches extensively explored to address the difficulties linked with noble metal-based catalysts and to enhance the productivity of water electrolysis. These strategies entail enhancing the use of catalysis centered around non-noble metals and utilizing catalysts devoid of metal components. Carbon-based materials have been a focal point in HER research owing to their characteristics such as large surface area, rapid charge transfer kinetics, excellent conductivity, and robust chemical stability. These attributes enable carbon-based materials to facilitate HER processes effectively, making them suitable materials for serving as support materials and/or active HER catalysts in both acidic and alkaline electrolysis [47–50].

CNTs were activated to introduce oxygen functionalities which significantly enhance their performance in electrocatalysis. These oxygen functionalities act as catalytic sites, facilitating multiple electrocatalytic reactions. In a study by Cui and co-workers [51], it was showed that an activated MWCNT serves as a superior metal-devoid catalyst for the HER compared to unactivated MWCNTs and glassy carbon electrodes. This activation process involved treating the MWCNTs in nitric acid at 120 °C for varying durations extending from 0.5 to 8 hours. This treatment led to oxidation and the introduction of terminal acidic groups within the MWCNTs. Ramasamy and co-workers [52] developed a highly efficient electrocatalyst for HER by synthesizing platinum nano-islands on functionalized MWCNTs through an electrodeposition technique. This catalyst showcased an overpotential of $-84\,\text{mV}$ at the $-10\,\text{mA}\,\text{cm}^{-2}$ current density and a Tafel slope value of $21.2\,\text{mV}\,\text{dec}^{-1}$, indicating that the reaction undergoes through the Volmer-Tafel mechanism. To enhance wettability and introduce oxygen functional groups, pristine MWCNTs were treated with acid as part of the functionalization process. Doping of CNTs results in improving the HER activity rather than undoped CNTs. In a notable investigation, nitrogen-doped MWCNTs (N-MWCNTs) were synthesized through ex situ doping approach. This involved adsorbing emeraldine salt onto MWCNTs with ultrasonication for more than 20 hours, followed by pyrolysis at 800 °C in Ar environment to yield the N-MWCNTs. These N-MWCNTs were assessed as bifunctional metal-free catalysts for both the ORR and HER. The pyrolysis process resulted in an augmented presence of pyridinic nitrogen within the hybrid structure, consequently enhancing the HER activity demonstrated by the N-MWCNTs [53].

CNTs have been expansively studied as substrates for anchoring several materials like metal oxides, sulfides, nitrides, phthalocyanines, porphyrins, and corroles, effectively enhancing conductivity, charge transfer, electron transfer, and surface area. The porous nature of CNTs contributes significantly to boosting the catalytic efficiency of composites involving transition metals. In a study by Premnath et al. [54], a composite of nickel sulfide with tungsten (NiWS) was synthesized through a one-step coprecipitation-reduction process. The catalytic activity of this composite was notably enhanced by incorporating CNT supports, showcasing elevated current densities at lower overpotentials with increased electrochemical surface area, and long-term stability across various pH electrolytes. Surface modification of CNTs leads to increased electrocatalytic performance when compared to unmodified CNTs. This process alters the CNTs by introducing various functional groups like hydroxyl, carboxyl, and carbonyl, among others, thereby enabling the CNTs to gain various oxidation states and expand their surface chemistry [55, 56]. Sannegowda and co-workers [55] designed uranium phthalocyanine

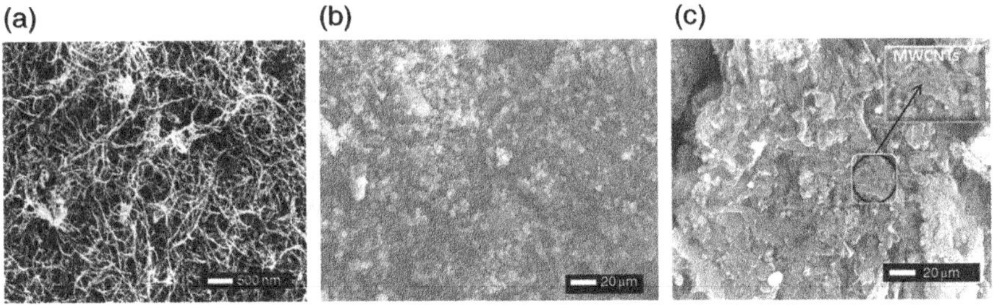

Figure 19.4 SEM images of the acid-functionalized MWCNTs (a), UTBrImPc (b), and composites of MWCNTs-UTBrImPc (c). *Source:* Reproduced with permission from [55]/ American Chemical Society.

(UTBrImPc) immobilized onto acid-functionalized MWCNTs, which functioned as an extremely effective electrocatalyst for HER and OER. The improved outcomes seen in these processes within the combined organic hybrid catalyst were attributed to the collaborative effect of acid-functionalized MWCNTs and UTBrImPc. This combination is likely responsible for the enhancement in the electronic conductivity and surface area, leading to superior performance. Figure 19.4 displays the SEM images of acid-functionalized MWCNTs, UTBrImPc, and the composite, MWCNTs-UTBrIm.

19.4.3 OER

Electrochemical water splitting is a clean and cost-effective method to produce H_2 [57]. Overall water splitting comprises anodic OER and cathodic HER [57]. OER serves as the obstruction of whole water splitting owing to its slow kinetics and requires a thermodynamic potential of 1.23 V versus RHE [58]. OER is a four-electron-proton-coupled reaction and follows either adsorbate evolution mechanism (AEM) or lattice oxygen mechanism (LOM) [59]. In AEM mechanism, the process initiates with the adsorption of OH^- onto the surface of the vacancy site (Eq. 19.4). The adsorbed HO* species then undergoes sequential deprotonation, resulting in the formation of O* (Eq. 19.5). The subsequent step involves O—O bond development, enabling O* to react with another OH^- and create the HOO* intermediate (Eq. 19.6). In the ultimate stage, O_2 is generated as HOO* and undergoes deprotonation, leading to the regeneration of the active site (Eq. 19.7) [58].

$$OH^- + {}^* \rightarrow HO^* + e^- \tag{19.4}$$

$$HO^* \rightarrow O^* + e^- + H^+ \tag{19.5}$$

$$O^* + HO^- \rightarrow HOO^* + e^- \tag{19.6}$$

$$HOO^* \rightarrow * + O_2(g) + e^- + H^+ \tag{19.7}$$

* indicates the active site

In the LOM, the active sites involved in the OER are considered to be oxygen atoms within the crystal lattice of the electrocatalytic material [59]. The different stages of the LOM can be shown in Eqs. (19.8–19.11). In the beginning, the surface lattice oxygen undergoes nucleophilic reaction with H_2O to form —OOH intermediate (Eq. 19.8). In the subsequent step (Eq. 19.9) —OOH intermediate undergoes deprotonation and subsequently O_2 is released from the lattice (Eq. 19.10) creating oxygen vacancy on the surface. H_2O then replenishes the resulting oxygen vacancy sites, restoring the surface of the material (Eq. 19.11) [59].

$$\text{'O} - \text{M} - \text{O} + \text{H}_2\text{O} \rightarrow \text{'O} - \text{M} - \text{OOH} + \text{H}^+ + \text{e}^- \tag{19.8}$$

$$\text{'O} - \text{M} - \text{OOH} \rightarrow \text{'O} - \text{M} - \text{OO} + \text{H}^+ + \text{e}^- \tag{19.9}$$

$$\text{'O-M-OO} \longrightarrow \text{'O-M-}\bigcirc + \text{O}_2 \tag{19.10}$$

$$\text{'O-M-}\bigcirc + \text{H}_2\text{O} \longrightarrow \text{'O-M-OH} + \text{H}^+ + \text{e}^- \tag{19.11}$$

where 'O is the oxygen of the bulk material; M is the metal center and $\bigcirc$ depicts the oxygen vacancy.

The type of active sites governs the OER pathway during the electrochemical water splitting. The metal center, vacant site, positively charged species, or all combined together are generally the active center of the OER [59]. The standard catalysts for OER are RuO_2 and IrO_2. However, due to their expensive nature and scarcity, alternative catalysts are continuously being synthesized and tested by researchers all over the world [60]. Transition metal oxides [54, 55], MOFs [3, 4], and metal chalcogenides [2] have shown promising activity in catalyzing the OER effectively. Further, to improve their activity toward OER catalysis, carbon supports like C_2H_2 black, Vulcan carbon, graphene (and doped graphenes), and CNTs (and doped CNTs) are employed [61].

Pristine CNTs have been utilized in OER catalysis. However, they have exhibited less actively toward the OER [61]. Nevertheless, Cheng et al. demonstrated the effective use of stable pristine MWCNTs for OER in KOH electrolytes [56]. The MWCNT's inner tube functioned as a pathway for electronic conduction, enabling the charge transfer process by electron tunneling among the inner and the outer walls, while the outer wall worked as the location for the reaction between OH— and oxygen intermediates (OOH*). Notably, the catalytic performance of these MWCNTs surpassed that of the traditional Ru/C and Pt/C electrocatalysts. The excellent efficiency of the catalyst was ascribed to the proficient electron transfer occurring within the inner tubes of the MWCNTs, possibly facilitated by electron tunneling. This mechanism significantly enhanced the charge transfer at the outer wall's surface, contributing to the overall catalytic efficacy of the pristine MWCNTs [56]. Similarly, Keskin and co-workers developed free-standing MWCNTs by isolating vertically grown MWCNTs over the glass surface [62]. The obtained MWCNTs without any surface modification and doping were utilized in the OER catalysis. The OER onset potential was found to be 1.60V versus RHE, which was better than the Pt-based electrode (1.72 V versus RHE), thus showing the proficiency of MWCNTs in catalyzing the OER [62]. Doping of the CNTs also improves the activity toward the OER, for example, Yan and co-workers synthesized boron-doped MWCNTs, which showed an onset potential of 0.78 V versus SCE in 0.1 M KOH, which was better than the undoped MWCNTs [63]. The introduction of nitrogen doping in MWCNTs also increases OER activity mainly due to the electron-withdrawing effects of the formed pyridinic nitrogen [63]. This was successfully demonstrated by Davodi et al. who synthesized nitrogen-doped MWCNTs, which exhibited a superior OER activity (onset potential 1.54 V versus RHE) than the pristine MWCNTs [53]. Ali et al. synthesized fluorine-doped CNTs (F-CNTs) using XeF_2 as a fluorine source by solid-state fusion technique. The F-CNTs showed an overpotential of 280 mV at10 mA cm^{-2} in comparison to the pristine CNTs (340 mV at 10 mA cm^{-2}) and Ir/C (305 at 10 mA cm^{-2}) in basic medium [64].

Further, transition metal-based catalysts are incorporated into CNTs to improve the catalytic activity of composite toward the OER [61]. While transition metal-based catalysts are the active centers for the OER, the CNTs serve as electrical conductors, thus synergistically facilitating the OER catalysis [9]. Huang et al. synthesized nickel iron tannic acid polyphenol (Ni_3Fe-TA) and incorporated it on MWCNTs to form the composite (Ni_3Fe-TA/MWCNTs), which resulted in 287 mV at 10 mA cm^{-2} overpotential in 1.0 M KOH [65]. They attributed the OER activity to the electron transfer occurring from CNTs to Ni_3Fe/TA on the hetero-interface, which brings about a modification in the local electronic environment of catalysts at the atomic level. This modification leads to a reduction in the binding energies of adsorbed species, consequently expediting the OER [65]. Das and co-workers synthesized NiO and MWCNT composite (NiO-MWCNTs) via one-step wet chemical process and performed the alkaline OER activity [66]. An overpotential of 301 mV at 10 mA cm^{-2} was observed. They ascribed this OER activity to the synergistic effect between NiO and MWCNTs facilitating better charge transfer and prevention of volume shrinkage of oxides [66]. NiS incorporated on the surface of MWCNTs (NiS@MWCNTs) shows efficient OER catalysis. The NiS@MWCNT composite needed an overpotential of 320 mV to attain 10 mA cm^{-2} in 1.0 M KOH [67]. Similarly, Prabakaran et al. synthesized flower-like CoS on MWCNTs by using a hydrothermal technique. The composite, CoS on MWCNTs reached the current density of 15 mA cm^{-2} at 330 mV overpotential [68].

MOFs have been used as a precatalyst for the OER. However, MOFs suffer from poor conductivity and slow electron transfer [4]. Integrating CNTs with MOFs results in CNT@MOF hybrids that establish conductive paths for electron movement facilitated by the interconnected pores within the hybrid allowing electrolyte access [69]. The inclusion of CNTs improves the long-range conductivity of the CNT@MOF hybrid. Additionally, CNTs play a crucial role in generating catalytic sites within the hybrids, enhancing the percentage of active sites and surface area. This augmentation significantly boosts the electrocatalytic activity of MOFs [69]. Talha and co-workers synthesized iron nickel and 2-aminoterepthallic acid MOF on CNTs ($FeNiNH_2BDC$-CNTs) [70]. This composite showed an overpotential of 220 V at 10 mA cm^{-2}. The high activity toward the OER was mainly due to the synergistic effect between MOF and CNTs [70]. Further, MOFs can be carbonized to overcome their drawbacks [69]. For example, Li et al. synthesized cobalt and benzimidazole MOF and then immobilized the MOF on MWCNTs [71]. This MOF@MWCNTs was carbonized in the argon atmosphere to form Co_3O_4 nanocrystals rooted in N-doped mesoporous graphitic carbon layer/MWCNTs (Co_3O_4@C-MWCNTs) for which the alkaline OER overpotential was found to be 320 mV at 10 mA cm^{-2} and stability up to 25 hours [71]. In another work, by Ma and co-workers, $ZnCo_2O_4$@C-MWCNTs hybrid was synthesized from dual ZIF-8/ZIF-67 (zeolitic imidazole framework of Zn (ZIF-8) and Co (ZIF-67)) MOF immobilized on MWCNT by carbonizing the composite at 900 °C in N_2 atmosphere for two hours. The as obtained $ZnCo_2O_4$@C-MWCNTs hybrid displayed an overpotential of 327 mV at 10 mA cm^{-2} with long-term stability of 25 hours in 1.0 M KOH [72].

19.4.4 ORR

Fuel cell is a device which can convert chemical energy into electrical energy. Remarkably, fuel cells exhibit exceptional energy conversion efficiency and maintain low emissions. The depletion of fossil fuel reserves has prompted researchers to explore fuel cells as promising alternatives. The cathodic ORR assumes a significant role in regulating the overall performance of the fuel cells. The ORR can occur through either a two-electron process, resulting in the formation of H_2O_2, or a four-electron process, heading to the production of H_2O. Selective four-electron reduction process is necessary for the fuel cells and the two-electron process is deleterious to fuel cells. Carbon-supported platinum (Pt-C) catalysts currently aid as the most extensively employed electrocatalyst for ORR in fuel cells, despite the high cost and scarcity. The practical applicability of fuel cells critically depends on the presence of cost-effective and efficient ORR electrocatalysts [67, 73–78]. The enhancement of the catalyst performance can be achieved through the utilization of catalyst supports. Catalyst supports not only provide a high surface area and conductivity but also decrease the loading amount of the active catalyst, thus decreasing the overall cost of the catalytic material. During the ORR, agglomeration is a common phenomenon which decreases the amount of active sites and, in turn, the electrocatalytic activity. However, catalyst supports reduce aggregation and enhance the stability of the catalyst for longer duration. Typical catalyst support materials include mesoporous carbon, graphene, CNTs, and various other forms of carbon. CNTs stand out as one of the extensively explored carbon materials for electrocatalysts. They possess notable characteristics, including high strength, high specific surface area, and durability in both basic and acidic environments. Despite these inherent properties, significant emphasis is placed on manipulating their electronic characteristics to boost chemical durability and catalytic activity. The electronic properties can be modified by surface tailoring of CNTs, involving the oxygen-containing functional groups or by the integration of heteroatoms into the CNT structure. Treatment of CNTs with potent acids or bases results in the generation of oxygen functionalities, like —COOH, —OH, and epoxides on the surface. This modification generally enhances the electrocatalytic properties of CNTs, making them more promising for catalytic activities. Chen and co-workers synthesized nitrogen-doped CNTs (NCNTs) via a one-step CVD process for the ORR. The nitrogen precursors utilized in the synthesis included ethylenediamine (EDA-NCNT) and pyridine (Py-NCNT). EDA-NCNT demonstrated comparable ORR performance to that of Pt-C, exhibiting a half-wave potential ($E_{1/2}$) of −0.15 V (versus Ag/AgCl) while Pt-C shows an $E_{1/2}$ of −0.16 V (versus Ag/AgCl). Furthermore, EDA-NCNT exhibited excellent ORR efficacy in terms of limiting current density and H_2O selectivity. Py-NCNT showed an $E_{1/2}$ of −0.33 V(versus Ag/AgCl) inferior to EDA-NCNT due to the low amount of nitrogen doping [79]. Recently, our group, immobilized iron phthalocyanine (FePc) on $-SO_3H$-functionalized MWCNTs (MWCNTs-SO_3H) forming in the composite of FePc@MWCNTs-SO_3H, which shows an $E_{1/2}$ of 0.90 V (versus RHE). The $-SO_3H$ functional groups present on the MWCNT interact with the oxygen reduction intermediate via a second sphere mechanism and increase the selectivity toward the four-electron process [76]. In earlier work from our group, certain cobalt macrocycles, meso-tetraphenylporphyrinatocobalt(II), meso-tetrakis(4-hydroxyphenyl) porphyrinato cobalt(II), and meso-tetrakis(4-carboxy-phenyl) porphyrinatocobalt(II) were immobilized on MWCNTs and the ORR activity were accessed. MWCNTs improve the conductivity and offer a high surface area for the cobalt macrocycles to enhance the catalytic activity of the ORR [75]. Jeon and coworkers synthesized a carbon nanostructure through the covalent grafting of MWCNTs with sulfur-functionalized platinum (S–Pt). This nanostructure contains a minimal amount of Pt, yet demonstrates a significantly improved catalytic activity for the oxygen reduction compared to the catalysts with a higher content of unbound Pt on MWCNTs [80]. Sun and co-workers synthesized N-doped CNTs supported with cobalt selenide ($CoSe_2$). This alteration culminates in an improved overall catalytic efficacy, yielding a noteworthy conversion efficiency from O_2 to H_2O_2. In a 0.1 M perchloric acid solution, the $CoSe_2$@NCNTs hybrid demonstrates a remarkable H_2O_2 selectivity of 93.2% and a significant H_2O_2 yield rate of 172 ppm h^{-1}, exhibiting outstanding stability over a span of 24 hours [81].

19.4.5 Electrochemical Sensors

Nanomaterials represent a distinctive class of substances in the realm of chemicals and biosensors, owing to their reduced dimensions and expansive surface-to-volume ratio. CNTs, in particular, have emerged as highly promising entities for advancing and diversifying the landscape of chemical, molecular, and biological sensors [67, 78]. Extensive knowledge accumulated over centuries regarding carbon structures and CNTs have been greatly encouraged for their potential in this field [82]. The primary operational principle of CNT-based sensors involves the utilization of pristine, high-quality nanotubes. SWCNTs on GCE were immobilized by drop casting method [83] and the fabricated electrochemical sensor was effectively utilized for the sensitive determination of a biologically relevant molecule, acetaminophen. CNTs can be either chemically treated by employing appropriate coatings or by ascribing specific functional groups. Such modifications are instrumental in enhancing the selectivity and sensitivity of CNT-based sensors, ensuring their effectiveness in detecting and responding to targeted chemical, molecular, or biological entities. CNTs demonstrate wide applicability in cancer treatment and serve as nano-carriers for drug delivery. Leng and co-workers [84] studied chitosan-functionalized MWCNTs for delivery of antitumor drug, doxorubicin hydrochloride to evaluate its use in cancer treatments. Integration of CNT-based chemical sensors and an interface chip can take place at the PCB (printed circuit board) level. The system's functionality is validated by subjecting the CNT-based sensors to NO_2 exposure. These sensors are positioned within a gas chamber and the data acquisition from the fabricated chip occurs via a logic analyzer [85].

Further, CNTs are studied as potential materials for transistors, interconnects, and other electronic devices. CNTs show changes in electrical conductivity when exposed to specific gases, making them suitable for gas-sensing applications. They are also used in the development of biosensors for detecting biomolecules, such as DNA and proteins in which CNTs are functionalized to carry and deliver drugs to specific targets in the body. In addition, CNTs have been explored for their potential to remove pollutants and contaminants from water due to their adsorption properties.

19.5 Challenges Associated with CNTs

While CNTs have exclusive characteristics that make them fascinating for various uses, including energy storage and conversion devices, there are several challenges associated with their use. The challenges encompass synthesis of impurity-free material, high cost associated with large-scale production, aggregation during functionalization, and high resistance during the electrode fabrication [28].

19.5.1 Synthesis and Purity

Contamination: The synthesis of CNTs can lead to the presence of impurities or other types of carbon structures, affecting the performance of energy storage/conversion devices.

Chirality Control: Achieving precise control over the chirality (structural characteristics) of CNTs is still a significant challenge. Different chiralities can have different electronic properties, and achieving a specific chirality is crucial for optimizing device performance.

19.5.2 Scaling Up Production

Cost: Large-scale production of high-quality CNTs can be expensive. Reducing the production cost is crucial for the widespread commercial adoption of green energy applications.

19.5.3 Functionalization and Dispersion

Aggregation: CNTs tend to aggregate which can reduce their effective surface area and negatively impact the performance of energy storage/conversion devices. Ensuring proper functionalization and dispersion of CNTs is essential.

19.5.4 Electrode Fabrication

Contact Resistance: Integrating CNTs into electrodes can be challenging due to the formation of high contact resistance between CNTs and other materials. This can limit the efficiency of devices based on CNTs.

19.5.5 Chemical and Environmental Stability

Oxidation: CNTs can be susceptible to oxidation, especially at high temperatures, which may limit their long-term stability in certain energy conversion environments.

Corrosion: In some applications, CNTs may be exposed to harsh chemical environments, leading to corrosion or degradation over time.

19.6 Conclusion

This chapter deals with the synthesis, properties, and green energy applications of CNTs. CNTs have extraordinary mechanical, electrical, thermal, and chemical properties making them as most promising carbon materials for various green energy applications, be it supercapacitor, HER, OER, or ORR. In HER and OER, CNTs can be utilized to decrease the overpotential for overall water splitting, leading to the development of more efficient water electrolyzers. Additionally, ORR is crucial in fuel cells, and CNTs can be exploited to increase the efficiency and durability of ORR electrocatalysts. In supercapacitors, CNTs can significantly increase power density and lifespan owing to their high electrical conductivity and large surface area. While CNTs themselves or as carbon support are excellent catalysts in catalyzing the above-mentioned reactions, they also suffer from some drawbacks like high cost of production and susceptibility to oxidation and corrosion. Research is ongoing to address these tasks and unlock the full ability of CNTs in energy storage/conversion technologies. Advances in synthesis techniques, functionalization strategies, and material integration are expected to contribute to overcoming these hurdles.

Acknowledgments

Financial supports from the IoE incentive grant, BHU (scheme number 6031) and SERB (CRG/2022/003370) are gratefully acknowledged. SS acknowledges SERB for financial assistance, and VS and VR acknowledge CSIR, New Delhi, India for CSIR Junior and Senior Research fellowships.

References

1 Yadav, D.K., Ganesan, V., Sonkar, P.K., and Gupta, R. (2017). Templated synthesis of nickel–iron layered double hydroxide for enhanced electrocatalytic water oxidation: towards the development of non-precious-metal catalysts. *ChemElectroChem* 4: 3134–3139. https://doi.org/10.1002/celc.201700867.

2 Singh, D.K., Ganesan, V., Yadav, D.K., and Yadav, M. (2020). Facile synthesis of sulfur-doped mesoporous carbon nitride supported defect-rich cobalt sulfide for electrocatalytic water oxidation. *Crystal Growth & Design* 20: 6321–6328. https://doi.org/10.1021/acs.cgd.9b01703.

3 Singh, S., Yadav, M., Singh, D.K. et al. (2022). One step synthesis of a bimetallic (Ni and Co) metal–organic framework for the efficient electrocatalytic oxidation of water and hydrazine. *New Journal of Chemistry* 46: 13422–13430. https://doi.org/10.1039/D2NJ00773H.

4 Yadav, D.K., Gupta, R., Ganesan, V. et al. (2018). Gold nanoparticles incorporated in a zinc-based metal-organic framework as multifunctional catalyst for the oxygen reduction and hydrogen evolution reactions. *ChemElectroChem* 5: 2612–2619. https://doi.org/10.1002/celc.201800519.

5 Kumar, U. and Upadhyay, S. (2020). $Sr_2 SnO_4$ ruddlesden popper oxide: future material for renewable energy applications. In: *Green Energy* (ed. S.L. Tripathi, S. Ganguli, T. Magradze, and A. Kumar), 221–250. Wiley. https://doi.org/10.1002/9781119760801.ch8.

6 Kothandam, G., Singh, G., Guan, X. et al. (2023). Recent advances in carbon-based electrodes for energy storage and conversion. *Advancement of Science* 10: 1–50. https://doi.org/10.1002/advs.202301045.

7 Sonkar, P.K. and Ganesan, V. (2023). *Nanomaterials for Sustainable Energy Applications*. Boca Raton: CRC Press. https://doi.org/10.1201/9781003208709.

8 Tripathi, S.L., Ganguli, S., Magradze, T., and Kumar, A. (2022). *Intelligent Green Technologies for Sustainable Smart Cities*. Wiley. https://doi.org/10.1002/9781119816096.

9 Dai, L., Chang, D.W., Baek, J.B., and Lu, W. (2012). Carbon nanomaterials for advanced energy conversion and storage. *Small* 8: 1130–1166. https://doi.org/10.1002/smll.201101594.

10 Maduraiveeran, G. and Jin, W. (2021). Carbon nanomaterials: synthesis, properties and applications in electrochemical sensors and energy conversion systems. *Materials Science and Engineering B* 272: 115341. https://doi.org/10.1016/j.mseb.2021.115341.

11 Science, P., Roy, J.N., Bose, D.N. et al. (2019). Carbon materials: science and applications by Deborah D.L. Chung. *MRS Bulletin* 44: 661–662. https://doi.org/10.1557/mrs.2019.196.

12 Astefanei, A., Núñez, O., and Galceran, M.T. (2015). Characterisation and determination of fullerenes: a critical review. *Analytica Chimica Acta* 882: 1–21. https://doi.org/10.1016/j.aca.2015.03.025.

13 Zeng, Z., Li, X., Wang, H., and Xie, C. (2023). Optical patterning fullerene nanostructures with high purity and high surface quality. *APL Photonics* 8: https://doi.org/10.1063/5.0152366.

14 Maeyoshi, Y., Saeki, A., Suwa, S. et al. (2012). Fullerene nanowires as a versatile platform for organic electronics. *Scientific Reports* 2: 1–6. https://doi.org/10.1038/srep00600.

15 Schön, J.H., Kloc, C., Siegrist, T. et al. (2001). Superconductivity in single crystals of the fullerene C_{70}. *Nature* 413: 831–833. https://doi.org/10.1038/35101577.

16 Makarova, T.L. (2004). Magnetism in polymerized fullerenes. *Journal of Magnetism and Magnetic Materials* 272–276: E1263–E1268. https://doi.org/10.1016/j.jmmm.2003.12.305.

17 Moussa, F. (2018). *[60]Fullerene and Derivatives for Biomedical Applications*. Elsevier Ltd.10.1016/B978-0-08-100716-7.00005-2.

18 Kumar, S., Nehra, M., Kedia, D. et al. (2018). Carbon nanotubes: a potential material for energy conversion and storage. *Progress in Energy and Combustion Science* 64: 219–253. https://doi.org/10.1016/j.pecs.2017.10.005.

19 Rathinavel, S., Priyadharshini, K., and Panda, D. (2021). A review on carbon nanotube: an overview of synthesis, properties, functionalization, characterization, and the application. *Materials Science and Engineering B* 268: 115095. https://doi.org/10.1016/j.mseb.2021.115095.

20 Shoukat, R. and Khan, M.I. (2021). Carbon nanotubes: a review on properties, synthesis methods and applications in micro and nanotechnology. *Microsystem Technologies* 27: 4183–4192. https://doi.org/10.1007/s00542-021-05211-6.

21 Mbayachi, V.B., Ndayiragije, E., Sammani, T. et al. (2021). Graphene synthesis, characterization and its applications: a review. *Results in Chemistry* 3: 100163. https://doi.org/10.1016/j.rechem.2021.100163.

22 Zhen, Z. and Zhu, H. (2017). *Structure and Properties of Graphene*. Elsevier Inc.10.1016/B978-0-12-812651-6.00001-X.

23 Yadav, M., Ganesan, V., Gupta, R. et al. (2019). Cobalt oxide nanocrystals anchored on graphene sheets for electrochemical determination of chloramphenicol. *Microchemical Journal* 146: 881–887. https://doi.org/10.1016/j.microc.2019.02.025.

24 Gong, K., Hu, J., Cui, N. et al. (2021). The roles of graphene and its derivatives in perovskite solar cells: a review. *Materials and Design* 211: 110170. https://doi.org/10.1016/j.matdes.2021.110170.

25 Yang, W., Ni, M., Ren, X. et al. (2015). Graphene in supercapacitor applications. *Current Opinion in Colloid & Interface Science* 20: 416–428. https://doi.org/10.1016/j.cocis.2015.10.009.

26 Brownson, D.A.C., Kampouris, D.K., and Banks, C.E. (2011). An overview of graphene in energy production and storage applications. *Journal of Power Sources* 196: 4873–4885. https://doi.org/10.1016/j.jpowsour.2011.02.022.

27 Onyancha, R.B., Ukhurebor, K.E., Aigbe, U.O. et al. (2022). A methodical review on carbon-based nanomaterials in energy-related applications. *Adsorption Science and Technology* 2022: https://doi.org/10.1155/2022/4438286.

28 Gupta, N., Gupta, S.M., and Sharma, S.K. (2019). Carbon nanotubes: synthesis, properties and engineering applications. *Carbon Letters* 29: 419–447. https://doi.org/10.1007/s42823-019-00068-2.

29 Lee, C.J., Lyu, S.C., Kim, H.W. et al. (2002). Large-scale production of aligned carbon nanotubes by the vapor phase growth method. *Chemical Physics Letters* 359: 109–114. https://doi.org/10.1016/S0009-2614(02)00648-6.

30 Nasir, S., Hussein, M.Z., Zainal, Z., and Yusof, N.A. (2018). Carbon-based nanomaterials/allotropes: a glimpse of their synthesis, properties and some applications. *Materials (Basel)* 11: 1–24. https://doi.org/10.3390/ma11020295.

31 Dervishi, E., Li, Z., Xu, Y. et al. (2009). Carbon nanotubes: synthesis, properties, and applications. *Particulate Science and Technology* 27: 107–125. https://doi.org/10.1080/02726350902775962.

32 Tabassum, H., Mahmood, A., Zhu, B. et al. (2019). Recent advances in confining metal-based nanoparticles into carbon nanotubes for electrochemical energy conversion and storage devices. *Energy & Environmental Science* 12: 2924–2956. https://doi.org/10.1039/c9ee00315k.

33 Singu, B.S., Goda, E.S., and Yoon, K.R. (2021). Carbon nanotube–manganese oxide nanorods hybrid composites for high-performance supercapacitor materials. *Journal of Industrial and Engineering Chemistry* 97: 239–249. https://doi.org/10.1016/j.jiec.2021.02.002.

34 Benzigar, M.R., Dasireddy, V.D.B.C., Guan, X. et al. (2020). Advances on emerging materials for flexible supercapacitors: current trends and beyond. *Advanced Functional Materials* 30: https://doi.org/10.1002/adfm.202002993.

35 Li, Z., Wang, L., Li, Y. et al. (2019). Carbon-based functional nanomaterials: preparation, properties and applications. *Composites Science and Technology* 179: 10–40. https://doi.org/10.1016/j.compscitech.2019.04.028.

36 Niu, C., Sichel, E.K., Hoch, R. et al. (1997). High power electrochemical capacitors based on carbon nanotube electrodes. *Applied Physics Letters* 70: 1480–1482. https://doi.org/10.1063/1.118568.

37 Barisci, J.N., Wallace, G.G., and Baughman, R.H. (2000). Electrochemical characterization of single-walled carbon nanotube electrodes. *Journal of the Electrochemical Society* 147: 4580. https://doi.org/10.1149/1.1394104.

38 Zhang, Q., Cai, C., Qin, J., and Wei, B. (2014). Tunable self-discharge process of carbon nanotube based supercapacitors. *Nano Energy* 4: 14–22. https://doi.org/10.1016/j.nanoen.2013.12.005.

39 Shi, J., Jiang, B., Li, C. et al. (2020). Review of transition metal nitrides and transition metal nitrides/carbon nanocomposites for supercapacitor electrodes. *Materials Chemistry and Physics* 245: 122533. https://doi.org/10.1016/j.matchemphys.2019.122533.

40 Reddy, A.L.M. and Ramaprabhu, S. (2007). Nanocrystalline metal oxides dispersed multiwalled carbon nanotubes as supercapacitor electrodes. *Journal of Physical Chemistry C* 111: 7727–7734. https://doi.org/10.1021/jp069006m.

41 Chen, P., Shen, G., Shi, Y. et al. (2010). Preparation and characterization of flexible asymmetric supercapacitors. *ACS Nano* 4: 4403–4411.

42 Xavier, J.R., Vinodhini, S.P., and Chandraraj, S.S. (2023). Synthesis and electrochemical characterization of CNTs-based multi metal sulphide nanocomposite for supercapacitor applications. *Journal of Cluster Science* 34: 1805–1817. https://doi.org/10.1007/s10876-022-02352-0.

43 Xu, H., Cui, L., Pan, X. et al. (2022). Carboxymethylcellulose-polyaniline/carbon nanotube (CMC-PANI/CNT) film as flexible and highly electrochemical active electrode for supercapacitors. *International Journal of Biological Macromolecules* 219: 1135–1145. https://doi.org/10.1016/j.ijbiomac.2022.08.141.

44 Emmenegger, C., Mauron, P., Sudan, P. et al. (2003). Investigation of electrochemical double-layer (ECDL) capacitors electrodes based on carbon nanotubes and activated carbon materials. *Journal of Power Sources* 124: 321–329. https://doi.org/10.1016/S0378-7753(03)00590-1.

45 Simotwo, S.K., Delre, C., and Kalra, V. (2016). Supercapacitor electrodes based on high-purity electrospun polyaniline and polyaniline-carbon nanotube nanofibers. *ACS Applied Materials & Interfaces* 8: 21261–21269. https://doi.org/10.1021/acsami.6b03463.

46 He, X., Liu, G., Yan, B. et al. (2016). Significant enhancement of electrochemical behaviour by incorporation of carboxyl group functionalized carbon nanotubes into polyaniline based supercapacitor. *European Polymer Journal* 83: 53–59. https://doi.org/10.1016/j.eurpolymj.2016.08.001.

47 Bhatt, M.D. and Lee, J.Y. (2020). Advancement of platinum (Pt)-free (non-Pt precious metals) and/or metal-free (non-precious-metals) electrocatalysts in energy applications: a review and perspectives. *Energy and Fuels* 34: 6634–6695. https://doi.org/10.1021/acs.energyfuels.0c00953.

48 Jorge, A.B., Jervis, R., Periasamy, A.P. et al. (2020). 3D carbon materials for efficient oxygen and hydrogen electrocatalysis. *Advanced Energy Materials* 10: 1–35. https://doi.org/10.1002/aenm.201902494.

49 Wang, X., Vasileff, A., Jiao, Y. et al. (2019). Electronic and structural engineering of carbon-based metal-free electrocatalysts for water splitting. *Advanced Materials* 31: 1–8. https://doi.org/10.1002/adma.201803625.

50 Huang, H., Yan, M., Yang, C. et al. (2019). Graphene nanoarchitectonics: recent advances in graphene-based electrocatalysts for hydrogen evolution reaction. *Advanced Materials* 31: 1–34. https://doi.org/10.1002/adma.201903415.

51 Cui, W., Liu, Q., Cheng, N. et al. (2014). Activated carbon nanotubes: a highly-active metal-free electrocatalyst for hydrogen evolution reaction. *Chemical Communications* 50: 9340–9342. https://doi.org/10.1039/c4cc02713b.

52 Kalasapurayil Kunhiraman, A., Ramasamy, M., and Ramanathan, S. (2017). Efficient hydrogen evolution catalysis triggered by electrochemically anchored platinum nano-islands on functionalized-MWCNT. *International Journal of Hydrogen Energy* 42: 9881–9891. https://doi.org/10.1016/j.ijhydene.2017.02.169.

53 Davodi, F., Tavakkoli, M., Lahtinen, J., and Kallio, T. (2017). Straightforward synthesis of nitrogen-doped carbon nanotubes as highly active bifunctional electrocatalysts for full water splitting. *Journal of Catalysis* 353: 19–27. https://doi.org/10.1016/j.jcat.2017.07.001.

54 Premnath, K., Madhavan, J., Prasad, S. et al. (2022). Ultra-efficient, low-cost and carbon-supported transition metal sulphide as a platinum free electrocatalyst towards hydrogen evolution reaction at alkaline medium. *International Journal of Hydrogen Energy* 47: 41974–41983. https://doi.org/10.1016/j.ijhydene.2021.09.120.

55 Giddaerappa, K.P.C.H., Puttaningaiah, S.A., Shantharaja, N.K. et al. (2023). Uranium phthalocyanine-anchored acid-functionalized multiwalled carbon nanotubes for water electrolysis. *ACS Applied Nano Materials* 6: 8880–8893. https://doi.org/10.1021/acsanm.3c01328.

56 Cheng, Y., Xu, C., Jia, L. et al. (2015). Pristine carbon nanotubes as non-metal electrocatalysts for oxygen evolution reaction of water splitting. *Applied Catalysis B: Environmental* 163: 96–104. https://doi.org/10.1016/j.apcatb.2014.07.049.

57 Zhang, L., Xiao, J., Wang, H., and Shao, M. (2017). Carbon-based electrocatalysts for hydrogen and oxygen evolution reactions. *ACS Catalysis* 7: 7855–7865. https://doi.org/10.1021/acscatal.7b02718.

58 Song, J., Wei, C., Huang, Z.F. et al. (2020). A review on fundamentals for designing oxygen evolution electrocatalysts. *Chemical Society Reviews* 49: 2196–2214. https://doi.org/10.1039/c9cs00607a.

59 Xie, X., Du, L., Yan, L. et al. (2022). Oxygen evolution reaction in alkaline environment: material challenges and solutions. *Advanced Functional Materials* 32: https://doi.org/10.1002/adfm.202110036.

60 Singh, D.K., Yadav, M., Ganesan, V., and Bhobe, P.A. (2022). $Co_x(VO)_yO_z$ nanocrystal-integrated covalent organic polymers as a highly active and durable catalyst for electrochemical water oxidation: an untold role of the VO^{2+}/ VO_2^+redox couple. *ACS Applied Energy Materials* 5: 2805–2816. https://doi.org/10.1021/acsaem.1c03426.

61 Kadja, G.T.M., Mualliful Ilmi, M., Mardiana, S. et al. (2023). Recent advances of carbon nanotubes as electrocatalyst for in-situ hydrogen production and CO_2 conversion to fuels. *Results in Chemistry* 6: 101037. https://doi.org/10.1016/j.rechem.2023.101037.

62 Ali, A., Akyüz, D., Asghar, M.A. et al. (2018). Free-standing carbon nanotubes as non-metal electrocatalyst for oxygen evolution reaction in water splitting. *International Journal of Hydrogen Energy* 43: 1123–1128. https://doi.org/10.1016/j.ijhydene.2017.11.060.

63 Cheng, Y., Tian, Y., Fan, X. et al. (2014). Boron doped multi-walled carbon nanotubes as catalysts for oxygen reduction reaction and oxygen evolution reactionin in alkaline media. *Electrochimica Acta* 143: 291–296. https://doi.org/10.1016/j.electacta.2014.08.001.

64 Ali, Z., Mehmood, M., Ahmad, J. et al. (2021). Fluorine doped CNTs for efficient OER activity outperforming iridium supported carbon electrocatalyst. *Journal of Applied Electrochemistry* 51: 1573–1581. https://doi.org/10.1007/s10800-021-01600-x.

65 Huang, H., Zhao, J., and Liu, R. (2021). Carbon nanotube boosting electrocatalytic oxygen evolution of NiFe-polyphenol coordination catalyst through donor-acceptor modulation. *Journal of Colloid and Interface Science* 582: 396–404. https://doi.org/10.1016/j.jcis.2020.08.051.

66 Roy, A., Ray, A., Saha, S. et al. (2018). NiO-CNT composite for high performance supercapacitor electrode and oxygen evolution reaction. *Electrochimica Acta* 283: 327–337. https://doi.org/10.1016/j.electacta.2018.06.154.

67 Yadav, M., Singh, D.K., and Ganesan, V. (2023). Template-free hydrothermal synthesis of nickel sulfide nanocrystals on MWCNTs: efficient and stable bifunctional electroactive material for oxygen electrocatalysis. *Sustainable Energy & Fuels* 7: 3077–3087. https://doi.org/10.1039/D3SE00239J.

68 Prabakaran, K., Lokanathan, M., and Kakade, B. (2019). Three dimensional flower like cobalt sulfide (CoS)/functionalized MWCNT composite catalyst for efficient oxygen evolution reactions. *Applied Surface Science* 466: 830–836. https://doi.org/10.1016/j.apsusc.2018.10.015.

69 Chronopoulos, D.D., Saini, H., Tantis, I. et al. (2022). Carbon nanotube based metal–organic framework hybrids from fundamentals toward applications. *Small* 18: https://doi.org/10.1002/smll.202104628.

70 Yaqoob, L., Noor, T., Iqbal, N. et al. (2021). Electrochemical synergies of Fe–Ni bimetallic MOF CNTs catalyst for OER in water splitting. *Journal of Alloys and Compounds* 850: 156583. https://doi.org/10.1016/j.jallcom.2020.156583.

71 Li, X., Fang, Y., Lin, X. et al. (2015). MOF derived Co_3O_4 nanoparticles embedded in N-doped mesoporous carbon layer/MWCNT hybrids: extraordinary bi-functional electrocatalysts for OER and ORR. *Journal of Materials Chemistry A* 3: 17392–17402. https://doi.org/10.1039/c5ta03900b.

72 Liu, J., Xie, Y., Nan, Y. et al. (2017). $ZnCo_2O_4$ nanoparticles derived from dual-metal-organic-frameworks embedded in Multiwalled Carbon Nanotubes: a favorable electrocatalyst for the water splitting. *Electrochimica Acta* 257: 233–242. https://doi.org/10.1016/j.electacta.2017.10.080.

73 Singh, D.K., Ganesan, V., Yadav, D.K., and Yadav, M. (2020). Metal (Mn, Fe, Co, Ni, Cu, and Zn) phthalocyanine-immobilized mesoporous carbon nitride materials as durable electrode modifiers for the oxygen reduction reaction. *Langmuir* 36: 12202–12212. https://doi.org/10.1021/acs.langmuir.0c01822.

74 Singh, V., Thakur, P.S., Ganesan, V., and Sankar, M. (2022). Zn(II) porphyrin-based polymer facilitated electrochemical synthesis of green hydrogen peroxide. *Journal of Electroanalytical Chemistry* 919: 116536. https://doi.org/10.1016/j.jelechem.2022.116536.

75 Sonkar, P.K., Prakash, K., Yadav, M. et al. (2017). Co(II)-porphyrin-decorated carbon nanotubes as catalysts for oxygen reduction reactions: an approach for fuel cell improvement. *Journal of Materials Chemistry A* 5: 6263–6276. https://doi.org/10.1039/c6ta10482g.

76 Singh, D.K., Natchimuthu Karuppusamy, M., Shrivastava, A. et al. (2023). Sulfonic acid functionalization-boosted ultrafast, durable, and selective four-electron oxygen reduction reaction: evidenced by EC-SHINERS and DFT studies#. *ACS Catalysis* 13: 4155–4167. https://doi.org/10.1021/acscatal.3c00719.

77 Yadav, M., Kumar, P., and Prakash, K. (2020). Sciencedirect insight into efficient bifunctional catalysis: oxygen reduction and oxygen evolution reactions using

MWCNTs based composites with. *International Journal of Hydrogen Energy* 45: 9710–9722. https://doi.org/10.1016/j.ijhydene.2020.01.224.

78 Singh, D.K., Ganesan, V., Yadav, D.K. et al. (2021). Mesoporous carbon nitride supported 5,10,15,20-tetrakis(4-methoxyphenyl)-21H,23H-porphine cobalt(ii) as a selective and durable electrocatalyst for the production of hydrogen peroxideviatwo-electron oxygen reduction, Catal. *Science and Technology* 11: 1014–1026. https://doi.org/10.1039/d0cy01801e.

79 Chen, Z., Higgins, D., Tao, H. et al. (2009). Highly active nitrogen-doped carbon nanotubes for oxygen reduction reaction in fuel cell applications. *Journal of Physical Chemistry C* 113: 21008–21013. https://doi.org/10.1021/jp908067v.

80 Ahmed, M.S., Kim, D., and Jeon, S. (2013). Covalently grafted platinum nanoparticles to multi walled carbon nanotubes for enhanced electrocatalytic oxygen reduction. *Electrochimica Acta* 92: 168–175. https://doi.org/10.1016/j.electacta.2012.10.160.

81 Zhang, L., Liang, J., Yue, L. et al. (2022). N-doped carbon nanotubes supported $CoSe_2$ nanoparticles: a highly efficient and stable catalyst for H_2O_2 electrosynthesis in acidic media. *Nano Research* 15: 304–309. https://doi.org/10.1007/s12274-021-3474-0.

82 Tiwari, A., Yadav, M., Kumar Singh, D., and Ganesan, V. (2022). Nafion-multi-walled carbon nanotubes supported tris(bipyridyl)iron(II) for nicotine detection. *Electroanalysis* 1–11. 10.1002/elan.202200106.

83 Yadav, M., Ganesan, V., Maiti, B. et al. (2019). Sensitive determination of acetaminophen in the presence of dopamine and pyridoxine facilitated by their extent of interaction with single-walled carbon nanotubes. *Electroanalysis* 31: 2472–2479. https://doi.org/10.1002/elan.201900178.

84 Dong, X., Sun, Z., Wang, X. et al. (2017). Simultaneous monitoring of the drug release and antitumor effect of a novel drug delivery system-MWCNTs/DOX/TC. *Drug Delivery* 24: 143–151. https://doi.org/10.1080/10717544.2016.1233592.

85 Cho, T.S., Lee, K.J., Kong, J., and Chandrakasan, A.P. (2007). A low power carbon nanotube chemical sensor system. *Proceeding of IEEE 2007 Custom Integrated Circuits Conference CICC 2007*, San Jose, California, USA (16–19 September), pp. 181–184. https://doi.org/10.1109/CICC.2007.4405708.

20

Carbon-Based Nanoparticles as Visible-Light Photocatalysts

Sonam Soni

Department of Chemistry, Sunbeam Women's College Varuna, Varanasi, Uttar Pradesh, India

20.1 Introduction

Pollution, climate change, overpopulation, and the impact of events like COVID-19 are global challenges affecting both developed and underdeveloped nations. Another pressing issue is the global energy crisis, exacerbated by the rising population and escalating demand for energy consumption, while natural (non-renewable) energy resources are finite. The World Energy Council report suggests that by the year 2030, the global energy demand is expected to upsurge to 45–60% (optimistic variant) by 35% (pessimistic variant) as compared to 2010.

The International Energy Agency (IEA) highlights a significant surge in oil consumption. In 2000, the world consumed 205 million barrels of oil, but by 2030, this is projected to increase 1.5 times compared to 2010, reaching approximately 334 million barrels [1]. This alarming trend underscores the urgent need for a shift toward renewable, environmentally friendly, and sustainable energy sources.

In essence, the world faces two major intertwined challenges: pollution in its various forms, alongside the imminent energy crisis. Among these, water and air pollution stand out as particularly detrimental, causing widespread damage to the environment, climate, and all living organisms. The consequences, including high mortality rates, emphasize the critical importance of addressing these environmental issues on a global scale. It is imperative to focus on solutions that promote sustainable and long-term energy resources to mitigate the impact of these interconnected challenges.

Carbon-based Nanomaterials for Green Applications, First Edition. Edited by Upendra Kumar, Piyush Kumar Sonkar, and Suman Lata Tripathi.

Published 2025 by John Wiley & Sons, Inc.

20.1.1 Application of Green Energy

Green energy, also known as renewable energy, plays a crucial role in addressing environmental challenges. There are several forms of green energy such as solar energy, wind energy, geothermal energy, and biomass energy. Among these, solar energy is one of the most common green energy resources. It found its application in solar photovoltaic panels, which capture sunlight and convert it directly into electricity [2]. They power individual buildings, solar farms, and community projects. Alongside, visible-light photocatalysis is a fascinating tool with immense potential for green energy and environmental remediation. By harnessing freely available solar energy, visible-light photocatalysts can drive various processes:

a) Water Splitting generates renewable energy carriers (such as hydrogen and oxygen), contributing to sustainable fuel production.
b) **Pollution Reduction:** Photocatalysis helps decrease pollution by degrading organic pollutants in water and air.
c) **CO_2 Reduction:** These materials can convert carbon dioxide (CO_2)into useful organic substances, offering a pathway for green synthesis.

20.1.2 Importance of Photocatalytic Technique

The global emphasis on addressing critical challenges is essential for the survival of the planet's inhabitants. The escalating problems of energy shortages and pollution pose significant hurdles, with the current energy resources struggling to meet future demands. Pollution, a major concern, claims the lives of around 9 million people annually, surpassing casualties from violence and war, as reported by the World Health Organization (WHO). Necessities like fresh water and clean air underscore the pressing need for effective solutions. Confronting these challenges requires solutions that are environmentally friendly, cost-effective, and sustainable. A promising avenue in this regard is the application of the "photocatalytic technique," which has shown remarkable results in addressing these critical issues. This approach aligns with the global need for solutions that are both environmentally friendly and economically viable.

In the energy sector, photocatalytic techniques offer an eco-friendly and enduring solution for hydrogen (H_2) production, addressing the increasing demand for clean energy. Additionally, in the domain of pollution control, photocatalysis presents an environmentally friendly method for water and air treatment without adverse side effects. Implementing this technique holds the potential to create a pollution-free environment and contribute to the broader objective of sustainability.

20.1.3 Importance of Nanotechnology in Visible Light Photocatalysis

Nanotechnology plays a significant role in visible light photocatalysis (VLPC), which involves the use of light in the visible spectrum to initiate chemical reactions. Photocatalysis, in general, hinges on the properties of certain materials, called photocatalysts, that absorb light and promote reactions at the material's surface. VLPC has become a rapidly advancing domain in organic synthesis. The process of VLPC involves absorption of low-energy visible light by organic substrates and utilizing it in energy transfer (ET) or single-electron transfer (SET) processes. It stands out as a dynamic field, facilitating highly selective approaches for promoting organic reactions involving radical intermediate via distinctive reaction pathways. The exceptionally mild conditions characterize these conversions, along with their adaptability in aqueous environments. Thus, photocatalysis emerges as a crucial empowering technology in the realm of drug discovery. Here are some key aspects of the role of nanotechnology in VLPC:

i) **Enhanced Surface Area:** Nanomaterials often have high surface-to-volume ratios compared to bulk materials, which offer more active sites for catalytic reactions, facilitating a higher efficiency of visible light absorption and utilization.
ii) **Band Gap Engineering:** Nanotechnology allows for the manipulation of the electronic structure of materials at the nanoscale. This is crucial in designing photocatalysts with appropriate band gaps to absorb visible light efficiently. Narrowing the band gap enables materials to absorb visible light, which is crucial for utilizing solar energy.
iii) **Tailored Properties:** The morphology (size and shape) and composition of nanoparticles can be precisely controlled through nanotechnology. This tunability allows researchers to design photocatalysts with specific properties, such as improved charge separation, extended light absorption range, and enhanced stability.
iv) **Heterojunctions and Composite Materials:** Nanotechnology facilitates the creation of heterojunctions (interfaces between different materials) and composite materials. Combining different nanomaterials with complementary properties can lead to synergistic effects, enhancing the overall photocatalytic performance. For example, combining a semiconductor with a metal nanoparticle can improve charge separation and transfer.
v) **Plasmonic Effects:** Metallic nanoparticles can exhibit surface plasmon resonances that enhance the absorption of light. When incorporated into photocatalysts, these plasmonic effects can increase the efficiency of light harvesting and boost the photocatalytic activity under visible light.

vi) **Photogenerated Charge Carrier Management:** Efficient charge separation and migration are crucial for photocatalysis. Nanotechnology allows for the design of materials with well-defined structures that help the separation and carriage of photogenerated charge carriers, preventing recombination and improving overall efficiency.

vii) **Quantum Dots and Nanostructures:** Quantum dots, nanowires, and other nanostructures exhibit unique electronic and optical properties. These materials can be tailored for specific photocatalytic applications, offering advantages in terms of efficiency and selectivity.

viii) **Environmental Applications:** VLPC using nanomaterials has applications in environmental remediation, such as the degradation of pollutants in water and air. Nanocatalysts can be designed to target specific pollutants and enhance the overall degradation process.

20.2 Mechanism of Photocatalysis

Photocatalysis is a fusion of photochemistry and catalysis, with the term itself consisting of two elements: "photo" denoting light and "catalysis" indicating the modification of the chemical transformation rate in the presence of a catalyst. It is crucial to note that the catalyst remains unchanged and unaffected throughout the process, experiencing no alterations or depletion. In the process of photocatalysis, a photocatalyst is utilized, and the rate of the reaction is predominantly influenced by the crystal structure of the catalyst and the energy carried by incoming photons, whether in the visible or UV light spectrum. Acting as a sensitizer, the materials employed as catalysts facilitate light-induced redox processes based on their electronic structure, characterized by the occupied valence band (VB) and vacant conduction band (CB) [3].

The efficiency of the catalyst is closely tied to its band gap. The electrons in the VB absorb photons, if band gap is less than or equal to the energy of the incident light, exciting them to the CB and leaving holes in the VB (Figure 20.1) [4]. These holes act as oxidants for donor molecules, generating hydroxyl (a potent oxidizer) when interacting with water. Simultaneously, the electrons in the CB combine with water to produce superoxide ions, acting as reducing agents. Essentially, these liberated electrons drive redox reactions, and the resulting pairs of free electrons and holes can engage in oxidation-reduction reactions with any substance in contact with the catalyst, transforming it into the desired end product.

20.2.1 Oxidation Mechanism

When the catalyst is illuminated, it generates electron-hole pairs by elevating a valence electron to the CB. The resulting hole in the catalyst seizes a water molecule, initiating its oxidation into a hydroxyl radical, known for its potent oxidizing

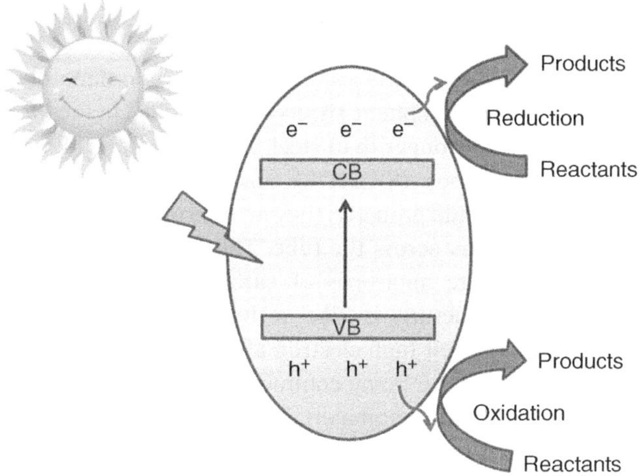

Figure 20.1 Mechanism of photocatalysis. *Source:* barbaliss/Adobe Stock Photos.

capabilities. When these hydroxyl radicals interact with any organic pollutant present, they facilitate its decomposition. In the presence of oxygen, a chain reaction ensues between the intermediate radicals of the organic molecules and oxygen molecules. The outcome of this process, in the case of organic pollutants, is the production of CO_2 and water (H_2O).

20.2.2 Reductive Mechanism

The reduction of atmospheric oxygen involves a pairing reaction. Oxygen, being readily reducible, provides an alternative avenue for hydrogen production. Electrons from the CB engage in a reaction with oxygen, leading to the formation of superoxide ions. These anions can further participate in oxidation reactions by attaching themselves to intermediates, resulting in the production of peroxides, which ultimately transform into water.

The reduction process is more favorable in organic matter than in water. As a result, a high concentration of organic matter boosts photocatalytic activity by increasing the likelihood of generating more holes. This, in turn, decreases the recombination rate of carriers, leading to a more efficient photocatalytic process [5].

20.3 Classification of Nanomaterials

Nanomaterials are categorized based on their morphology, size, and chemical and physical properties. These categories include carbon-based nanomaterials, metal nanoparticles, ceramic nanoparticles, semiconductor nanoparticles, and polymeric nanoparticles [6].

20.4 Types of Carbon-Based Nanoparticles

The main sub-types of carbon-based nanoparticles (CNPs) are carbon nanotubes (CNTs) and carbon fullerenes. CNTs are graphene sheets that are rolled in the form of tubes, which are a hundred times stronger than steel. Furthermore, CNTs are sub-divided into single-wall nanotubes (SWCNTs) and multi-walled CNTs (MWCNTs). These tubes possess a unique property: they are thermally conductive along their length but act as insulators across the tube. The second category of CNPs is carbon fullerenes, which are allotropes of carbon, contain sixty or more carbon atoms, and their structure is like a hollow football called Buckminsterfullerene (C-60). Due to their high electron affinity, strength, electrical conductivity, and structure, these have many commercial applications. Here, our focus is on the role of carbon-based nanomaterials in green energy applications. Hence, their use for hydrogen production, treatment of organic and inorganic pollutants, environmental pollutants, and photocatalytic disinfection using VLPC becomes an important area of investigation. C60/g-C_3N_4, TiO_2@MWCNTs, and C60/$Bi_2TiO_4F_2$ reported against degradation of organic pollutants, such as methylene blue (MB) and Rhodamine B (RhB). Similarly, C60-modified MWCNT and CNF-TiNT find use in degradation of inorganic pollutants [6].

Another type of CNPs is carbon dots, which are characterized by their small size typically less than 10 nm. They display excellent photoluminescence properties and are used in imaging, sensing, and biological applications. Carbon dots can be derived from various carbon sources, including CNTs and graphene. CQDs are small CNPs with quantum confinement effects, leading to unique electronic properties [7].

20.5 Application of CNPs as Photocatalysts

CNPs, including materials like CNTs and graphene, have garnered significant interest as photocatalysts due to their unique properties [8, 9]. Here are some applications of carbon nanoparticles as photocatalysts:

i) **Water Splitting for Hydrogen Production:** CNPs such as modified graphene and carbon nitride materials, can serve as efficient photocatalysts for water splitting, harnessing solar energy to drive the reaction, and produce hydrogen as a clean and renewable fuel.
ii) **Pollutant Degradation:** CNPs can be employed in the photocatalytic degradation of organic pollutants in water and air. Their high surface area and chemical stability make them effective in breaking down various pollutants under the influence of light.

iii) **Carbon Dioxide Reduction:** Photocatalytic reduction of CO_2 to valuable chemicals or fuels is a promising area of research. Carbon nanoparticles, particularly those doped or functionalized with specific groups, can serve as photocatalysts for converting CO_2 into useful products.
iv) **Photocatalytic Antibacterial Applications:** Carbon nanoparticles can be functionalized to enhance their photocatalytic properties and used in antibacterial applications. When exposed to light, these materials generate reactive oxygen species that can kill bacteria, making them useful in water purification and medical applications.
v) **Solar Cells and Photovoltaics:** Carbon nanoparticles, especially graphene, can be incorporated into solar cells to enhance their efficiency. Graphene's excellent electrical conductivity and transparency make it a suitable material for improving charge transport and collection in photovoltaic devices.
vi) **Organic Synthesis:** Carbon nanoparticles can be utilized as photocatalysts in organic synthesis reactions. Photocatalysis can enable mild and selective synthesis routes for various organic compounds, and carbon nanoparticles can play a role in facilitating these reactions under light irradiation.
vii) **Environmental Remediation:** Photocatalysis with carbon nanoparticles can be applied for environmental remediation, such as the removal of persistent organic pollutants, dyes, and other contaminants from air and water sources.
viii) **Energy Storage:** Carbon nanoparticles can be used as photocatalysts for energy storage applications. For example, they can be involved in light-driven processes that store solar energy in chemical forms, contributing to the development of efficient energy storage systems.
ix) **Hydrogen Peroxide Production:** Certain carbon-based photocatalysts can be employed for the light-induced production of hydrogen peroxide. This has applications in both environmental remediation and as a green oxidant in chemical processes.

The use of carbon nanoparticles as photocatalysts is a rapidly evolving field, and ongoing research aims to optimize their properties for specific applications, enhance their efficiency, and explore new avenues for sustainable energy and environmental technologies.

20.5.1 Photocatalytic Splitting of Water for Hydrogen Production

Photocatalysis can be employed to generate hydrogen gas, serving as a promising source of energy. Consequently, the energy thus obtained is cost-effective and devoid of harmful byproducts that could potentially harm the environment. As a result, numerous researchers are actively engaged in exploring this avenue for obtaining clean energy. The initial discovery of water splitting occurred in 1972, utilizing TiO_2 as a catalyst. However, it received limited attention at that time due to its low efficiency. The subsequent oil crises in 1973 and global energy crises in 1979 brought increased focus to the water-splitting photocatalytic reaction. In the context of twenty-first-century crises, energy scarcity stands out as a significant concern. The contemporary world necessitates an energy resource that can seamlessly replace existing natural sources, given the challenges associated with their limited availability. The photocatalytic process offers advantages such as reduced energy consumption, prolonged lifespan, and environmental friendliness for hydrogen production. It operates under natural solar light and relies on a photocatalyst [10–13].

To date, a diverse array of photocatalysts has been developed for water splitting to produce hydrogen, including metal-free photocatalysts, metal oxides, and metal chalcogenides [14–18]. Among semi-conductor catalysts, carbon-based materials have been extensively investigated due to their exceptional catalytic performance. Graphitic carbon nitride (g-C_3N_4), in particular, has been widely studied; however, its photocatalytic efficiency is hampered by inadequate light-harvesting capability and low charge mobilities. Fortunately, the alleviation of these issues can be achieved through the construction of heterojunctions or modifications.

The effective separation and storage of generated hydrogen molecules during photocatalytic water-splitting processes are critical. To prevent the reverse reaction and potentially dangerous explosions, it is essential to isolate hydrogen from oxygen products during water oxidation. Recent work by Yang et al. demonstrated that a two-dimensional multi-layer system, with a carbon nitride monolayer sandwiched between two graphene monolayers, facilitates both photocatalytic hydrogen generation and hydrogen storage [9]. The outer graphene layers selectively allow protons to pass through, ensuring the separation of hydrogen and oxygen production. However, the storage rate of hydrogen is constrained by the interlayer distance, with an effective interlayer charge transfer distance between 2D materials being approximately 7 Å.

CNTs have garnered extensive attention for hydrogen storage due to their abundance of carbon elements, light mass density, controllable porous structure, and favorable interaction with hydrogen molecules. One challenge in storing hydrogen within CNTs is the method of introducing hydrogen into them. Notably, various nanowires, including carbon chains, metal nanowires, nitride nanowires, and carbide nanowires, have been successfully encapsulated in CNTs. If a photosensitive material is placed inside a CNT, it can undergo a photocatalytic reaction, generating hydrogen molecules within the CNT. The selective permeability of the CNT wall allows only protons to pass through, ensuring the safe and stable storage of hydrogen inside CNTs. This goal can also be achieved with double-walled nanotubes.

Recent reports through first-principle calculations, have proposed a strategy involving a CNT containing a carbon nitride nanowire (CNNW). The investigation delved into the electronic structures, separation of photogenerated carriers, and subsequent chemical reactions within this photocatalytic system. The results indicated that the oxygen evolution reaction (OER) and hydrogen evolution reaction (HER) occur on the outer wall of the CNTs and the inner CNNWs, respectively. These reactions are propelled by the photogenerated carriers, facilitating the effective separation and storage of hydrogen within the CNTs. The designed core-shell structure based on CNTs emerges as a promising candidate system for both photocatalytic water-splitting and secure hydrogen storage [9].

20.5.1.1 Factors Affecting Photocatalytic Splitting of Water

As per previous report studies [19–21], there are several factors affecting hydrogen production through splitting of water:

i) **Band Gap Energy:** The distinction in energy levels between the conduction and VBs in a photocatalyst is denoted by the band gap (Eg). For the successful evolution of hydrogen, the thermodynamic potential of the CB must exceed the potential level of the acceptor. The band gap plays a crucial role in facilitating charge separation, thereby diminishing recombination rates and promoting activity under specific wavelengths. The overarching goal is to minimize the band gap, enabling the activation of the photocatalyst in the visible spectrum. This adjustment aims to amplify the efficiency of hydrogen production. Notably, it has been observed that the introduction of dopants, whether metallic or non-metallic, serves to decrease the band gap and shift it into the visible region.
ii) **Surface Area/Structure:** Modifying the surface texture of a catalyst stands out as a technique for reducing the band gap. The photocatalytic activity undergoes influence based on diverse crystal structures and surface area. A greater surface area harbors an increase in the number of active sites, elevating the chances of reactions occurring. Furthermore, smaller-sized particles facilitate the photoelectron and hole in reaching the surface more easily, covering shorter distances and thereby diminishing the recombination rate. This augmentation contributes to an increased activity of a photocatalyst.
iii) **Light Intensity:** The intensity of incident light plays a pivotal role in the efficiency of photocatalysis, operating within two distinct regimes. Generally, enhancing the activity involves elevating the intensity of incident light, provided the light's energy exceeds or equals the material's band gap or threshold energy. In the first mode, this results in a higher rate of the forward reaction, where photoelectron and hole consumption lead to the production of hydrogen and oxygen, outweighing the backward reaction. The second mode,

characterized by a half-order reaction, shows that further increasing intensity diminishes activity. Here, the rate of the backward reaction exceeds that of the forward reaction, leading to a decreased hydrogen yield.

iv) **Temperature:** Increased temperatures facilitate the enhancement of photocatalytic activity by promoting the generation of electron-hole pairs. Moreover, elevated temperatures promote greater desorption of products from the catalyst surface, thereby further boosting activity. This heightened desorption provides more opportunities for the generation of electron-hole pairs, which are subsequently employed in oxidation and reduction reactions. Conversely, lower temperatures lead to a decrease in the desorption rate of products, resulting in a reduction of photocatalytic activity.

v) **pH:** The production of hydrogen via water splitting depends on the concentration of protons, which correlates with the pH of the solution. Here, photogenerated electrons are involved in reduction of protons to produce hydrogen. In the literature, studies generally indicate a higher rate of H_2 production in a basic environment compared to an acidic one. Specifically, significant H_2 production is observed under strongly basic conditions ($pH > 10$). However, there are contrasting findings from certain studies suggesting that H_2 production in an acidic environment can exceed that in a basic environment, particularly under strongly acidic conditions. This is attributed to the enhanced absorption of H^+ ions by the photocatalyst in a strongly acidic medium, which increases the likelihood of their reduction to H_2. In their experiments using a TiO_2/RuO_2-MV^{2+} photocatalyst in a methanol-water mixture, researchers reported these observations. In summary, the prevailing trend suggests that the highest levels of H_2 production occur in strongly basic conditions.

20.5.2 Photodegradation of Organic Pollutants

Organic pollutants pose a significant threat to both human health and ecosystems, exhibiting persistence in soil and water, and proving resistant to natural degradation. As a result, there is a pressing need for innovative treatment technologies, given the challenge of breaking them down through traditional water treatment methods like coagulation, sedimentation, filtration, disinfection, ozone oxidation separation, and membrane separation technology. Photocatalytic technology has emerged as a promising solution, offering the ability to convert solar energy directly into chemical energy for effective organic pollutant degradation, garnering considerable attention in recent times [22]. Notably, the cost-effectiveness and environmental friendliness of photocatalysis, coupled with its ability to achieve complete pollutant degradation (resulting in CO_2 and H_2O through complete mineralization) make it a compelling option for achieving greener surroundings. Through extensive research, various semiconductor materials such as TiO_2, ZnO, WO_x, graphitic carbon nitride (g-C_3N_4), and graphene have demonstrated photocatalytic capabilities. Among these, g-C_3N_4 stands out for its adjustable band gap, strong chemical resistance, ease of modification, and composition of abundant elements C and N found in nature [23, 24].

g-C_3N_4, a novel material derived from s-triazine (C_3N_3) or s-heptazine (C_6N_7) units connected through secondary or tertiary amino groups is depicted in Figure 20.2. With a VB position of 1.6 eV, a CB position of −1.1 eV, and a band gap of 2.7 eV, it qualifies as a layered two-dimensional material capable of photoinduction in visible light, forming reactive groups ($O_2^{\bullet}-$, $^{\bullet}OH$) with O_2 and H_2O on its surface.

Additionally, g-C_3N_4 possesses properties such as stability at high temperatures (up to 600 °C), resistance to mild acids and bases, and insolubility in both water and organic solvents, meeting environmental requirements for organic pollutant degradation. However, bulk g-C_3N_4 has inherent drawbacks such as limited visible-light absorption, a high rate of photogenerated carrier complexation, structural disorder, and a small specific surface area. Consequently, modification is necessary to enhance its effectiveness in photocatalytically treating organic pollutants.

Several approaches are available to improve the photocatalytic performance of g-C_3N_4, including morphological control, modification with carbonaceous nanomaterials, and doping with metals, non-metals, and molecules.

Carbon quantum dots (CQDs), a novel zero-dimensional nanomaterial, exhibit favorable optical and electronic properties, such as aqueous solubility, size-dependent effects, photoinduced conversion effects, photochemical stability, non-toxicity, and cost-effectiveness. Being a nanoscale conjugated π-bonded structured non-metallic material, CQDs can form heterojunctions with g-C_3N_4 through π–π

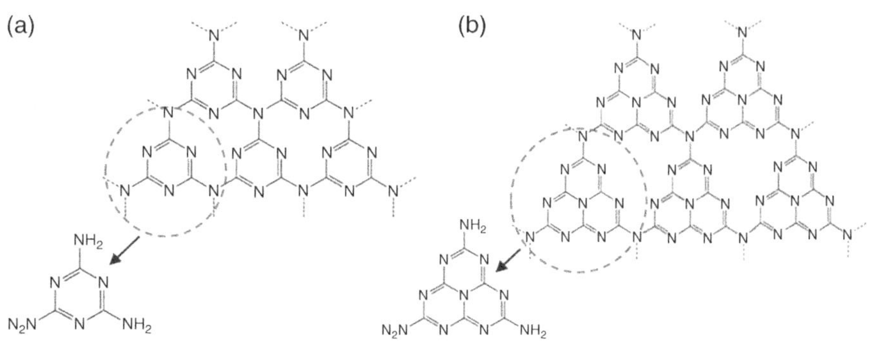

Figure 20.2 Structure of g-C_3N_4: (a) the s-triazine; and (b) s-heptazine units.

interactions, showcasing efficient conversion properties and functioning as electron-trapping hydrazines.

Moreover, CQDs contribute to broadening the visible light absorption range of g-C_3N_4, thereby enhancing its responsiveness to visible light. Notably, the eco-friendly nature of CQD doping serves as a non-metallic modification for g-C_3N_4, eliminating concerns related to metal leakage pollution. Consequently, CQDs can be employed to modify g-C_3N_4, specifically through the modification by carbonaceous nanomaterials, leading to a substantial improvement in its photocatalytic performance. This enhancement aligns with the requirements for efficiently removing organic pollutants, emphasizing the potential of CQDs as a valuable tool in the modification of g-C_3N_4 for enhanced photocatalysis.

Nevertheless, an increasing number of catalysts have been modified using g-C_3N_4/CQDs, driven by the necessity to address the breakdown of various organic pollutants, as g-C_3N_4/CQDs alone often prove ineffective in treating actual organic wastewater. Liu et al. devised N-doped CQDs modified defect-rich g-C_3N_4 (NCDs/DCN) to address the photocatalytic degradation challenges posed by complex pollutants such as bisphenol-A/hexavalent chromium, ofloxacin/hexavalent chromium, and ciprofloxacin/hexavalent chromium. Additionally, beyond effectively degrading common organic pollutants like aldehydes and phenol, g-C_3N_4/CQDs and their related modification have demonstrated capability in meeting the degradation requirements of emerging organic pollutants such as estrogen and diclofenac [25]. Furthermore, the photocatalytic degradation by g-C_3N_4/CQDs doesn't entail partial disintegration of pollutants; instead, it achieves complete mineralization, producing CO_2, H_2O, and inorganic salt as byproducts.

20.5.2.1 Mechanism of Photocatalytic Degradation

As a photocatalyst, g-C_3N_4 demonstrates the ability to degrade real organic wastewater through specific properties detailed as follows: (1) under visible light, it generates active substances like electrons (e^-) and hydrogen ions (H^+); (2) it features active functional groups on its surface; and (3) its distinctive material structure includes a porous composition, multilayer stacking, and substantial specific surface area. CQDs play a crucial role as spectral converters, facilitating the overall photocatalytic process and optimizing electron arrangement in type-II heterojunctions or Z-type photocatalytic systems as electron conduction mediators. Electron spin resonance (ESR) spectra demonstrate that CQDs-doped catalysts generate more radicals under both dark and light conditions, highlighting CQDs' remarkable conversion capability. Furthermore, CQDs collaboratively enhance g-C_3N_4's photocatalytic activity, elucidating their fundamental mechanism.

In general, photogenerated electrons exhibit potent reducing capabilities, effectively reducing water and heavy metals, while photogenerated holes demonstrate potent oxidizing power, oxidizing organic pollutants and water. Both entities can impact bacteria and viruses. Photogenerated electrons additionally play a role in activating and generating free radicals, which are central to photocatalysis due to their significant redox power, thereby enabling pollutant mineralization. The catalysis and activation of the photocatalyst are facilitated by functional groups, such as C—N, C—C, —NH—, and —NH_2 groups, on the catalyst's surface, with defective sites serving as reaction sites. Moreover, these functional groups possess adsorption properties that promote photocatalysis by adsorbing heavy metal ions such as Cu^{2+}, Zn^{2+}, and Pb^{2+}.

Furthermore, the material's inherent structure significantly influences its catalytic effectiveness. For example, the layered and porous structure of g-C_3N_4 exhibits a notable adsorption effect that greatly aids in pollutant degradation. These collective features enable g-C_3N_4/CQDs to effectively degrade authentic organic wastewater (Figure 20.3) [26]. The fundamental pathway of photocatalytic degradation of organic pollutants by g-C_3N_4/CQDs in aquatic environments and similar conventional conditions is represented by the following equations:

$$\text{g-C}_3\text{N}_4 + \text{visible light} \rightarrow e^- + h^+ \tag{20.1}$$

$$e^- + O_2 \rightarrow O_{2\cdot}^- \tag{20.2}$$

$$e^- + O_2 + h+ \rightarrow H_2O_2 \tag{20.3}$$

$$e^- + H_2O_2 \rightarrow \cdot OH + OH^- \tag{20.4}$$

$$e^- + O_2 \rightarrow O_{2\cdot}^- + H+ \rightarrow HO_2 \tag{20.5}$$

$$2HO_{2\cdot} \rightarrow O_2 + H_2O \tag{20.6}$$

$$H_2O_2 + O_{2\cdot}^- \rightarrow \cdot OH + OH^- + O_2 \tag{20.7}$$

$$\text{Organic pollutants} + h + O_{2\cdot}^-, \text{and} \cdot OH \rightarrow CO_2 + H_2O + \text{other inorganic salts} \tag{20.8}$$

20.5.2.2 Factors Influencing Degradation of Organic Pollutants by g-C_3N_4/CQDs

There are numerous factors influencing the photocatalytic deterioration of organic pollutants [23]. Some of these are discussed as follows:

i) **Temperature:** Temperature is critical in both the preparation and modification of g-C_3N_4/CQDs, exerting significant influence on structural characteristics like pore size, specific surface area, and active sites. These temperature-induced alterations lead to pronounced changes in the

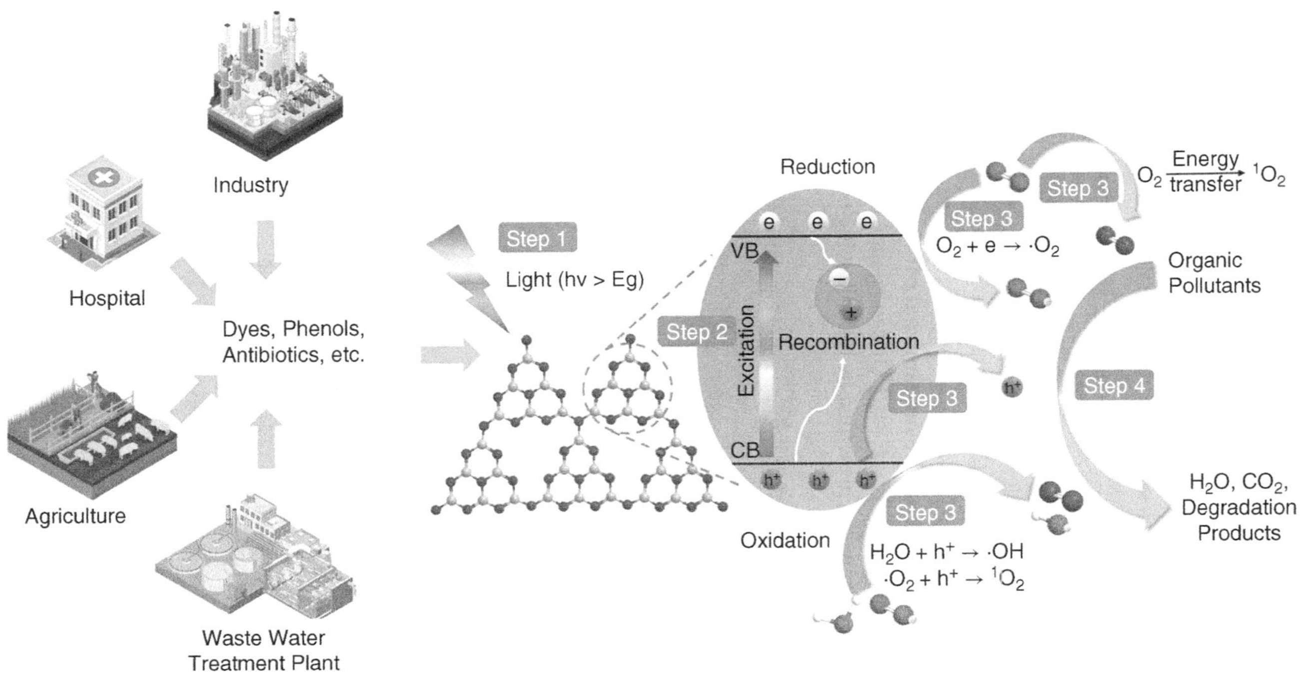

Figure 20.3 Fundamental mechanism of g-C_3N_4/CQDs (1) sunlight; (2) charge excitation; (3) charge separation and transfer; (4) charge recombination.

catalyst's performance. The meticulous preparation of g-C_3N_4 nanosheet films adorned with CQDs imposes stringent temperature requirements. For instance, precise temperature control is essential when synthesizing g-C3N4 nanosheets with CQDs, where heating melamine to 550 °C and subsequent ambient temperature conditions are crucial steps. Temperature also affects molecular mobility during catalyst preparation and organic pollutant treatment, influencing radical generation and stabilization in degradation processes.

ii) **pH:** The pH of the environment has a key role in the generation of radicals (·OH) for g-C_3N_4, with optimal conditions observed in weakly basic environments. pH stability affects reactive group integrity, impacting pollutant degradation efficiency. Extreme acidity can damage surface hydroxyl groups, reducing catalytic efficacy. Studies on CQDs/CNC catalysts for diclofenac degradation show pH's substantial influence on degradation kinetics, emphasizing the need for optimal pH determination.

iii) **Catalyst Dosage:** Increasing catalyst quantities theoretically enhance pollutant degradation by providing more active sites, but excessive amounts can cloud the system and hinder light absorption. Therefore, finding the optimal dosage, as demonstrated in studies using N-CQDs/g-C_3N_4/PDS for tetracycline hydrochloride degradation, is crucial for maximizing efficiency while minimizing costs. Various catalyst dosages were examined, revealing that 0.5 $mg\,l^{-1}$ of catalyst dose yielded the most effective deterioration.

iv) **Coexisting Ions:** Various ions, such as NaCl, $NaNO_3$ and, Na_2SO_4 often coexist in g-C_3N_4/CQDs systems and can significantly influence photocatalytic processes. Understanding their effects, as seen in studies of sulfamethazine degradation under LED irradiation, is essential for accurately assessing catalyst performance [27]. The experiments evaluated the effect of wavelength and found the significant ions among several co-existing ions present in the system that had an impact on reaction kinetics. The results suggest that the concentrations of bicarbonate and bromide ion had a significant impact on kinetics, while the influence of other co-existing ions such as Cl^-, NO^{3-}, and $SO_4{}^{2-}$ was less pronounced.

v) **Doping Ratio:** Adjusting doping ratios in g-C_3N_4/CQDs materials is challenging yet critical for optimizing catalytic activity. Research has shown that specific doping ratios enhance degradation rates of pollutants like RhB [28] and methyl blue [29], underscoring the importance of precise ratio determination for achieving desired photocatalytic outcomes. It is important to note that increasing the dopant may not always increase the efficacy, hence doping ratio should be determined carefully.

20.6 Conclusions

The photocatalytic technique emerges as a promising strategy, providing solutions to environmental challenges by offering a renewable, cost-effective, and pollution-free approach for both energy production using green methods and offering a greener environment by pollution control. In navigating these critical issues, innovative approaches like photocatalysis become indispensable for forging a sustainable and healthier future for all. In this aspect, nanotechnology provides a versatile toolbox for designing and tailoring materials with enhanced properties for VLPC. This has significant implications for a wide range of applications, including environmental remediation, energy conversion, and sustainable development such as H_2 evolution through splitting of water and degradation of organic pollutants by using various forms of CNPs. In summary, the application of VLPC using CNPs offers an environment-friendly and green route for solving the global challenges related to energy and pollution.

20.7 Future Scope

The future scope of CNPs as visible-light photocatalysts is promising and continues to be a subject of active research. Further research may focus on development of CNPs in context such as:

i) Optimizing their photocatalytic efficiency for large-scale environmental cleanup,
ii) Tailoring CNPs for efficient and sustainable hydrogen production under visible light,
iii) Advancing carbon dots as visible-light-responsive agents for imaging and therapeutic purposes in medicine.
iv) Utilizing CNT in energy storage devices, such as supercapacitors and batteries, driven by visible light.
v) Exploring synergistic effects that improve charge separation and overall efficiency.
vi) Developing efficient visible-light-responsive materials for disinfection and water purification.
vii) Visible-light-driven electrocatalysis, particularly in the reduction of CO_2 to produce valuable chemicals and fuels. Thus, addressing carbon emissions and promoting sustainable energy conversion. Precise functionalization of CNPs to enhance their photocatalytic properties for specific applications.
viii) Engineering of CNPs for controlled release of chemicals or drugs in response to visible light, hence advancing drug delivery systems and smart materials.

ix) Scaling up the production of carbon-based photocatalysts for practical and large-scale industrial applications, thus, optimizing synthesis methods, ensuring cost-effectiveness, and addressing challenges associated with stability and recyclability.

References

1 Tvaronavičienė, M., Baublys, J., Raudeliūnienė, J., and Jatautaitė, D. (2020). Global energy consumption peculiarities and energy sources: role of renewables. In: *Energy Transformation Towards Sustainability* (ed. M. Tvaronavičienė and B. Ślusarczyk), 1–49. Elsevier https://doi.org/10.1016/B978-0-12-817688-7.00001-X.

2 Tripathi, S.L. and Padmanaban, S. (ed.) (2020). *Green Energy: Solar Energy, Photovoltaics, and Smart Cities*. Wiley https://doi.org/10.1002/9781119760801.ch8.

3 Yang, X. and Wang, D. (2018). Photocatalysis: from fundamental principles to materials and applications. *ACS Applied Energy Materials* 1 (12): 6657–6693. https://doi.org/10.1021/acsaem.8b01345.

4 He, X. and Zhang, C. (2019). Recent advances in structure design for enhancing photocatalysis. *Journal of Materials Science* 54 (12): 8831–8851. https://doi.org/10.1007/s10853-019-03417-8.

5 Tahir, M.B., Sohaib, M., Sagir, M., and Rafique, M. (2022). Role of nanotechnology in photocatalysis. *Encyclopedia of Smart Materials* 578. https://doi.org/10.1016%2FB978-0-12-815732-9.00006-1.

6 Baig, N., Kammakakam, I., and Falath, W. (2021). Nanomaterials: a review of synthesis methods, properties, recent progress, and challenges. *Materials Advances* 2 (6): 1821–1871. https://doi.org/10.1039/d0ma00807a.

7 Liu, J., Li, R., and Yang, B. (2020). Carbon dots: a new type of carbon-based nanomaterial with wide applications. *ACS Central Science* 6 (12): 2179–2195. https://doi.org/10.1021/acscentsci.0c01306.

8 Song, X., Bu, H., Fan, Y. et al. (2022). Photocatalytic hydrogen production and storage in carbon nanotubes: a first-principles study. *RSC Advances* 12 (27): 17029–17035. https://doi.org/10.1039/D2RA02349K.

9 Yang, L., Li, X., Zhang, G. et al. (2017). Combining photocatalytic hydrogen generation and capsule storage in graphene based sandwich structures. *Nature Communications* 8 (1): 16049. https://doi.org/10.1038/ncomms16049.

10 Guo, S., Li, X., Li, J., and Wei, B. (2021). Boosting photocatalytic hydrogen production from water by photothermally induced biphase systems. *Nature Communications* 12 (1): 1343. https://doi.org/10.1038/s41467-021-21616-4.

11 Villa, K., Galán-Mascarós, J.R., López, N., and Palomares, E. (2021). Photocatalytic water splitting: advantages and challenges. *Sustainable Energy & Fuels* 5 (18): 4560–4569. https://doi.org/10.1039/D1SE00001A.

12 Xu, F. and Weng, B. (2023). Photocatalytic hydrogen production: an overview of newly advances on structural tuning strategies. *Journal of Materials Chemistry A* https://doi.org/10.1039/D3TA00252A.

13 Kumar, S., Kumar, A., Bahuguna, A. et al. (2017). Two-dimensional carbon-based nanocomposites for photocatalytic energy generation and environmental remediation applications. *Beilstein Journal of Nanotechnology* 8 (1): 1571–1600. https://doi.org/10.3762/bjnano.8.159.

14 Liao, G., Tao, X., and Fang, B. (2022). An innovative synthesis strategy for high-efficiency and defects-switchable-hydrogenated TiO2 photocatalysts. *Matter* 5 (2): 377–379. https://doi.org/10.1016/j.matt.2021.11.022.

15 Shandilya, P., Sambyal, S., Sharma, R. et al. (2022). Properties, optimized morphologies, and advanced strategies for photocatalytic applications of WO3 based photocatalysts. *Journal of Hazardous Materials* 428: 128218. https://doi.org/10.1016/j.jhazmat.2021.128218.

16 Li, C., Liao, G., and Fang, B. (2022). Magnetic-field-promoted photocatalytic overall water-splitting systems. *Chem Catalysis* 2 (2): 238–241. https://doi.org/10.1016/j.checat.2022.02.001.

17 Liao, G., Li, C., Liu, S.Y. et al. (2022). Emerging frontiers of Z-scheme photocatalytic systems. *Trends in Chemistry* 4 (2): 111–127. https://doi.org/10.1016/j.trechm.2021.12.005.

18 Ida, S., Kim, N., Ertekin, E. et al. (2015). Photocatalytic reaction centers in two-dimensional titanium oxide crystals. *Journal of the American Chemical Society* 137 (1): 239–244. https://doi.org/10.1021/ja510128k.

19 Kumara, B.N., Kalimuthu, P., and Prasad, K.S. (2023). Synthesis, properties and potential applications of photoluminescent carbon nanoparticles: a review. *Analytica Chimica Acta* 341430. https://doi.org/10.1016/j.aca.2023.341430.

20 Chen, X., Shen, S., Guo, L., and Mao, S.S. (2010). Semiconductor-based photocatalytic hydrogen generation. *Chemical Reviews* 110 (11): 6503–6570. https://doi.org/10.1021/cr1001645.

21 Guo, Y., Zhou, Q., Zhu, B. et al. (2023). Advances in organic semiconductors for photocatalytic hydrogen evolution reaction. *EES Catalysis* https://doi.org/10.1039/D3CY00263E.

22 Long, Z., Li, Q., Wei, T. et al. (2020). Historical development and prospects of photocatalysts for pollutant removal in water. *Journal of Hazardous Materials* 395: 122599. https://doi.org/10.1016/j.jhazmat.2020.122599.

23 Wen, J., Zhou, L., Tang, Q. et al. (2023). Photocatalytic degradation of organic pollutants by carbon quantum dots functionalized g-C3N4: a review.

Ecotoxicology and Environmental Safety 262: 115133. https://doi.org/10.1016/j.ecoenv.2023.115133.

24 Wang, T., Nie, C., Ao, Z. et al. (2020). Recent progress in g-C_3N_4 quantum dots: synthesis, properties and applications in photocatalytic degradation of organic pollutants. *Journal of Materials Chemistry A* 8 (2): 485–502. https://doi.org/10.1039/C9TA11368A.

25 Liu, H., Liang, J., Fu, S. et al. (2020). N doped carbon quantum dots modified defect-rich g-C_3N_4 for enhanced photocatalytic combined pollutions degradation and hydrogen evolution. *Colloids and Surfaces A: Physicochemical and Engineering Aspects* 591: 124552. https://doi.org/10.1016/j.colsurfa.2020.124552.

26 Luo, Y., Zhu, Y., Han, Y. et al. (2023). g-C_3N_4-based photocatalysts for organic pollutant removal: a critical review. *Carbon Research* 2 (1): 14. https://doi.org/10.1007/s44246-023-00045-5.

27 Di, G., Zhu, Z., Dai, Q. et al. (2020). Wavelength-dependent effects of carbon quantum dots on the photocatalytic activity of g-C_3N_4 enabled by LEDs. *Chemical Engineering Journal* 379: 122296. https://doi.org/10.1016/j.cej.2019.122296.

28 Hong, Y., Meng, Y., Zhang, G. et al. (2016). Facile fabrication of stable metal-free CQDs/g-C_3N_4 heterojunctions with efficiently enhanced visible-light photocatalytic activity. *Separation and Purification Technology* 171: 229–237. https://doi.org/10.1016/j.seppur.2016.07.025.

29 Zhang, Z., Lin, S., Li, X. et al. (2017). Metal free and efficient photoelectrocatalytic removal of organic contaminants over g-C_3N_4 nanosheet films decorated with carbon quantum dots. *RSC Advances* 7 (89): 56335–56343. https://doi.org/10.1039/C7RA11205J.

21

Carbon-Based Nanomaterials in Day-to-Day Human Life

Manju Choudhary[1], Pooja Nain[2], Shivanshu Garg[3], and Himanshu Punetha[3]

[1] *Department of Bioscience and Biotechnology, Banasthali Vidyapeeth, NewaniTonk, Rajasthan, India*
[2] *Department of Soil Science, College of Agriculture, G. B. Pant University of Agriculture and Technology, Pantnagar, Uttarakhand, India*
[3] *Department of Biochemistry, College of Basic Sciences & Humanities, G. B. Pant University of Agriculture and Technology, Pantnagar, Uttarakhand, India*

21.1 Introduction

Like water carries, the exceptional property of being a universal solvent, the carbon nanomaterials withhold the credit of being the ones with the highest strength-to-weight ratio adding millions of values to their stretching power without losing their characteristics. The bacterial byproducts constantly seep into rhizosphere, the removal of which becomes necessary when they turn out to be pathogenic both to plants and water. Likewise, other harmful agents to be classified as pollutants in air, water, and soil can be single-handedly eradicated or minimized to a negligible level by utilization of carbon-based nanomaterials. The transition state of catalysis is the stage where events like bond breakage, charge development, and bond formation occur. The control of these technological insights in context of carbon nanomaterials provides us with a slow release of fertilizers and herbicides to crops and into the soil. The bioinformatic work categorized as insilico approach may be termed incarbo with the advent of faster processing speed in microprocessors achieved by replacing silicon with carbon nanomaterials, architecting the field of biosensors also. The outer surface of earth upon which the plants dwell, nanomaterials' presence can be exemplified on the basis of minerals and clays. However, it is carbon that can be highly polymerized at atomic level by virtue of four electrons present in its outer shell, opening the gateways of its

Carbon-based Nanomaterials for Green Applications, First Edition. Edited by Upendra Kumar, Piyush Kumar Sonkar, and Suman Lata Tripathi.

Published 2025 by John Wiley & Sons, Inc.

potential, which when manipulated by human mind leads to excessive bonding patterns and strategies, resulting in a useful product.

Carbon nanomaterials are the sole structures of carbon-enriched multi-composites of various forms of carbon. The animal–plant and other biological systems are composites of biomolecules (amino acids, nucleotides, nucleosides, fatty ester compounds, and polyhydroxy aldehydes and ketones) having the carbon fixed into structures of these biomolecules in one or another way. Carbon cannot be neglected as identity of living organisms and most of the non-biological entities are also withheld to one or other form of carbon. Multi-dimensional biology of universe coincides with fundamental property of entropy, that is the randomness or disorder of the system. The point of meeting the boundaries of biological and non-biological systems is attributed to the presence of carbon. Structural elucidations have shown the potential benefit of these carbon particles at nanoscale marking the beginning of these in day-to-day life. Numerous examples of this structure-to-function relationship exist that govern the use of carbon nanoparticles (CNPs). The methodology of carbon nanomaterials' design is presented in Figure 21.1.

Daily life encompasses the need for medical attention to survive long and to meet the ultimate diagnosis of disease is a hallmark of pioneering advancement.

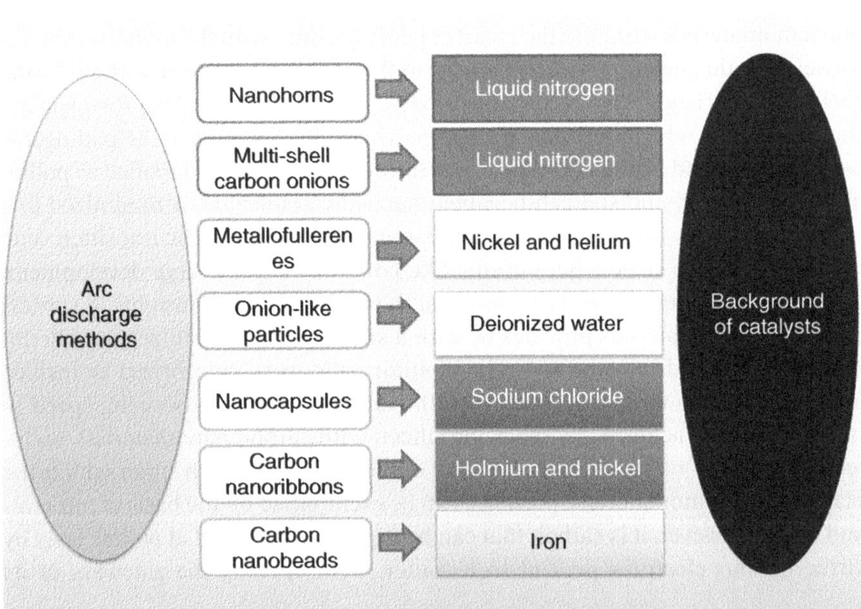

Figure 21.1 Carbon nanomaterial synthesis approaches.

To add to this, malignant tumor diagnosis and the pathways of its metabolic build-up, kidney filtration rate accompanied by presence/absence of key metabolites, the vaccinating scenario of secondary immune response, utilization of solar and non-solar radiations to get overall interims of body scans, implanting the foreign tissues to body without undermining tissue rejection and, finally, identifying the drug and drug delivery targets, the CNPs have deterred major lacking of field of bioimaging, detection, and diagnosis of various pathologies. Supply chains are the resource personnel of human needs and electricity has been part and parcel of human life without which one cannot withstand in today's world. CNPs cover the role of harnessing it in the form of nanobatteries, which have crucial technological advantages, meanwhile detecting harmful gases in the form of biosensors and also working for other aspects like blood sugar detection imprints. The large screens displaying colorful and animated improvisations correlated with televisions, computers, mobiles, and theaters are all due to virtue of the carbon's presence in these CNPs. Pollution control by CNPs is accounted for by their use in harnessing solar and wind energy. Following the law of conservation of energy, the very first of thermodynamics, the CNPs have been used to develop lithium-ion batteries (LiBs) for a longstanding range of household equipment as well as fore(electric)-vehicles used for transport and commerce purposes. Apart from the utilization of CNPs in build-up of human life, these have been cited for use in warfare as the development of radars inflicted their use. To hold the watt power for a long run, supercapacitors have proven useful by virtue of CNPs. Food generation is one aspect of meeting needs of global food security and sustaining the harvested product is another aspect of keeping it for the long run. CNPs site their potential in this aspect also, in the form of safe moisture-free packaging, nanoencapsulation, and emulsification potential.

Most of the CNPs used for the above applications are derived majorly from three types of carbon forms depicted in Figure 21.2. Apart from these, the role of CNPs and carbon-made composites has been deciphered out by mankind in natural environment (Table 21.1), cosmetic applications (Table 21.2), and wearable fashion (Table 21.3).

21.2 Utilization of CNPs in Medical Services

21.2.1 Pathological Condition Detections

To the point of infection spread, the earlier and specific diagnoses are crucial for outcomes of the treatment, including the detection of damaged tissues or cancerous tumors. CNPs have the ability to significantly improve medical imaging quality by emitting fluorescence upon photoexcitation. Common CNPs include

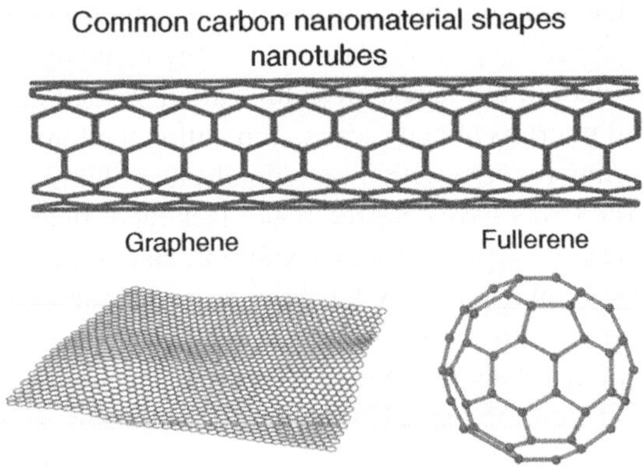

Figure 21.2 Common carbon nanomaterial shapes.

Table 21.1 Natural abundance of nanoparticles in environment boosting physiological processes.

S. No.	Environment type	Presence/indulgence of nanoparticle type	Scale of control over environment
1.	Air (stratosphere)	Airborne carbon clusters and aerosols	Cloud gathering and insulation of electrical charge present
2.	Water (ground water)	Airborne carbon clusters	Soil Cation/Anion Exchange Capacity
		Nanoparticles formed during geological evolution of aluminum, silicon, and carbon type	Weathering and mineralization of sedimentary rocks
3.	Rhizosphere	Present as silica-and carbon-fused forms	Hardening of plants and leaching of heavy elements
4.	Soil	Present as silicon-, carbon-, calcium-, aluminum-, and magnesium-fused forms	Colloid-facilitated transport of minerals, Formation of soil clusters and Carbon and phosphate sequestration

carbon-imprinted dots, graphene-imprinted quantum dots, and carbon composites of nanotubes [1]. These nanostructures can target individual organelles and cells, enhancing the fluorescence effect and enhancing the visualization dynamics and detection optima of imaging techniques chiefly mentioned as computed topography (CT) and magnetic resonance imaging (MRI) [2]. CNPs can also be used for analysis of tissue histopathology imaging through optics, named near-infrared (NIR) optical imaging, thereby offering higher resolution and no auto-fluorescence [3]. The practical intra-operative use of CNPs is highly noticeable in research literature, with

Table 21.2 Nanoparticles designed by mankind for cosmetic and health-care functions.

S. No.	Company	Product; use	Intended function
1.	Pharmanex	LifePaknano; face gelling agent	• Lowers activation energy of coenzyme Q_{10} • Anti-ageing
2.	L'Oreal	RevitaLift anti-wrinkle; skin anti-ageing agent	• Supplemented with pro-retinol A (a type of vitamin A) made as nanosome • Anti-Wrinkle agent
3.	Kara Vita	Eye tender; fibroblast stimulator and collagen builder	• Contains nanospheres with bioactive peptides • Inflammation reducer and skin tone builder
4.	Natural Korea	Cosil nano beauty soap; body cleanser	• Silver nanoparticle enrichment • Disinfectant and stain remover
5.	Zelens	Zelens Fullerene C-60 Moisture-based cream; anti-ageing agent	• Carbon nanoparticles with anti-oxidant properties • Free radical remover
6.	Derma Swiss	Nanosphereplus; anti-ageing agent	• Nanospheres enriched with viable stem cells • Heals skin and prevails longevity to dermal cells

Table 21.3 Nanoparticles designed by mankind for wearable fashion.

S. No.	Company	Product; use	Intended function
1.	DevarajCotton Mills	Nanofinishfabric; textile industry	Self-cleaning, water and flame retardant
2.	GlobalPhotonic	FlexPower™; fabrics and humidifiers	Capacity to generate electricity from mechanical and solar movements
3.	Devik Tradernanofabrics and nanogarments	Water repellent fabrics; agriculture, architecture and health care boosters	Super quick dry, Stain and odor resistant, anti-bacterial and cool temperature builder
4.	Asian Haat	Shirt; wearable scale	Smoothly textured, stain and wrinkle-free cloth
5.	SchoellerTechnologies	Nanosphere; fabric scale	Water, odor and stain resistant

studies showing negligible effects on human health after intravenous application. CNPs have also been used for pre-operative and post-operative detection of lymph nodes in patients with malignant tissues related to colon, breast, and thyroid gland origin, allowing accurate resection and reduced surgical trauma.

Nanodiamond nanoparticles with silicon-vacancy centers can be used for color-specific imaging with noticeable changes in live cells and detection within cellular boundaries by virtue of properties like stability under photochromic effect and narrow region of spectrum at the peripheries of NIR region intracellular tracking due to their photostability and narrow emissions in the NIR region [4]. Carbon isotopic segments of graphene impregnated into nanoscale sheets are able to detect the neural interface, with studies showing their potential in neuroscience research. Due to the good conduction capacity, the graphene electrodes can be applied without the need for conductive gel as a medium, making the electroencephalograph (EEG) examinations painless for patients. CNP-based materials have high electrical conductivity and are used for recording cardiac electrophysiological signals. Graphene or multi-walled CNTs (MWCNTs) are used for short-term and long-term monitoring in wearable devices [5].

21.2.1.1 Detection Relying on Adsorption of Metabolites

Radiology and nuclear medicine involve the use of radiative nuclei that emit high- and low-density particles aimed at tissues of the body, leaving people exposed to radiation, raising a concern for their health. CNP-based materials can help protect them effectively. Acting as an X-ray shield, graphene oxide (GO) coated with polyaniline has shown superior results in doing blockage of beams compared to materials not coated with GO. [6]. Nanodiamonds were found to be much more efficient in absorption of inflammation-causing biomolecules, which otherwise have resulted in causing inflammation and then tissue necrosis, which could now be avoided. The prepared CNPs with origin of carbon-chitin nanotube fabrication spheres along with lysine immobilization at the micro level were shown to have removed the bilirubin from blood, saving a person from bilirubin encephalopathy (a condition known by inflammation caused due to bilirubin). This overall improves blood flow and leads to better health in terms of tracking down absorption of metabolite [7].

CNTs were used to create a hemodialysis membrane, reducing fouling by 30% and increasing rate of urea and creatinine removal. The membrane also effectively absorbs cholesterol, which can be suitably used as a marker for build-up of cardiovascular-associated and liver pathologic disorders, also linked to cancer. CNT-bound cholesterol at both (inner and outer) surfaces of a membrane can move freely within the membrane boundaries, showing an overall lowering in cardiovascular and other disease risks [8].

21.2.1.2 Bioimaging by Photoacoustics

Photoacoustic (PA) imaging is a non-invasive biomedical technique that uses light/laser pulses to absorb agents having high contrast and convert them into heat radiations and vibrations having thermal inputs, which are then recorded as signals with ultrasound intensity [9]. PA has a resolution of angles covering spatial and temporal scales making it highly resolute and also comes with the ability to gain insights into tissue pathology with marked alterations, allowing for the diagnosis and cure for various types of melanomas. CNTs serve as good contrast agents, targeting specific tissues. Cyclic compounds of triaminopeptides configured as Arg-Gly-Asp destined onto top of single-walled CNTs have been shown to markedly affect the lowering in potential of tumors. A study done by Han et al. proved at both in vivo and in vitro scales has described a few nanometer-scaled carbon dots (CDs) possessing the capacity to penetrate and intervene in the cancerous tissue delivering the site-specific drug and also perceiving signals pertained to PA nature at NIR region [10].

21.2.1.3 Protection from Penetration Power of X-Rays

Radiology- and nuclear medicine-impregnated use leads to exposure of patients toward tissue and cell machinery damaging radiations, leading to a risk to their health. To protect them effectively, CNT-based materials have been suggested as potential solutions. A 20-kV X-ray beam was attenuated by upto 9.14% through poly (vinylidene) fluoride (PVDF) homopolymer/barium sulfate. The response in attenuation was elevated to a factor of 24.65% with incorporation of GO-coated nanosheet as 4% of the total weight [11]. In comparison to fullerenes and pyrolytic graphite, the CNTs have better attenuation power. Like, in a general textile clothing thickness of 25 mm when coated with CNTs, there was an increase in value of X-ray attenuation by a factor of more than 70%. Similarly, 0.1 mm-thick GO film was able to attenuate the high and low energy beams of X-rays at factors of 83 and 49% with an overall expense of energy of 7 and 8 keV [12].

Using polyaniline-impregnated hybrid GO nanoflakes with a trio of tungsten-bismuth-tin, an X-ray shield was made by researchers with confirmed experiments about their effectiveness in clothes protecting from solar radiation. These were found to be fit for tissue-preventive damage like in the form of thyroid and apron shields [13]. Other collaborations include the use of lightweight, flexible material, combining fabric and GO, to create a wearable shielding material having advantages of being lightweight with flexibility and proper aeration through nanopores fitted within the clothing making permeation of water and air possible. These CNPs ought to benefit by providing added X-ray protection [14].

21.2.1.4 Changing Dynamic State by Photodynamic Therapy

Photodynamic therapy (PDT) includes use of high-energy photons to target and destroy specific cells, particularly in cancer treatment. It uses reactive oxygen species (ROS) generated by CNTs to destroy cancer cells. Different types of ROS are generated by CNPs, such as superoxide anions ($^{\cdot}O_2-$) and singlet oxygen [15]. PDT is non-invasive, effective, and has minimal systemic toxicity. It can be combined with light, anticancer drugs, or photothermal therapy to enhance its effectiveness. The activation threshold levels of CNPs are concerned with NIR light, having the potential of deeper penetration into tissues and cell sections as compared to visible or ultraviolet light. PDT has demonstrated utilization in treating bacteria-, virus-, and fungi-led pathogenic infections. Studies have shown that GO can effectively eliminate bacteria, while GO loaded with dyes named indocyanine green is effective in eliminating the R (resistant) type of plasmid presented by *Staphylococcus aureus* methicillin-resistant strain [16].

21.2.1.5 Temperature-Induced Transformations Through Photothermal Treatment

In animal trials, photothermal treatment (PTT) is a twenty-first-century method for successfully curing cancer. The creation of effective and secure photosensitive agents is essential to the PTT's success. A relatively new PTT agent is activated carbon (AC), a substance that is used widely and safe to be utilized in both normal and emergency medical services. For the first time, the manufacturing and PTT application reports of in vitro compiled CNPs (with less than 9–10 nm diameter) from AC have come into limelight. In this approach, non-specific adsorption is minimized and biocompatibility is enhanced by transferring CNPs, which are produced by chemical oxidation, to PEGylated CNPs (PCNPs). PCNPs are coupled with fluorescein isothiocyanate to investigate the time-dependent absorption by MCF-7 human breast cancer cells. The experimentation perceiving the photothermal effect is an indication that PCNPs are far more capable of converting light energy into heat than CDs. While PCNPs can successfully deteriorate and damage the cancer-inflicted cells upon exposure to NIR light, they offer very little risk to the integrity of membranes and the viability of cells while they are in the dark. The utilization potential of the PCNPs can be attributed to a concise, secure, and effective PTT agent [17].

The treatment is selective, spatiotemporally regulated, minimally invasive, and compatible with various anticancer therapies, particularly systemic and local chemotherapy. The therapy (either loaded on nanoparticles or systemic chemotherapy) is made substantially more effective by loading the anticancer agent onto the particle and accelerating its release using NIR irradiation. Research findings at both in vitro and in vivo scales have demonstrated the efficacy of photothermal treatment in cells identified with breast bladder and lung cancer [18]. This treatment triggers an anticancer immune response in addition to killing cancer cells. Alarmins, other damage-associated molecules, and tumor-associated antigens were released more frequently by apoplastic and necroptotic cells. Death-associated molecular patterns (DAMPS) are molecular frameworks involved in promotion of both cell-mediated and humoral immunity [19]. Multi-functional CNPs were deciphered by Cheng et al. to be used as tracing agents in lymphoid system (only for lymph nodes) in the cases of metastases observed at node ends. The docetaxel-loaded CNPs were utilized for these treatments. The drug's release was enhanced by the laser's radiation, and the photothermal effect induced necrosis of cancer cell-inflicted damage [20]. The effectiveness of the multi-functional platform based on nanodiamonds in treating colon cancer was realized where orally administered CNPs were mostly localized in the colon and tumor, had extended intestinal retention periods, and inhibited tumor growth in vivo [21]. Mesoporous carbon nanoparticles (MCNs) are essential. For NIR-I and NIR-II, the MCNs have a larger absorption coefficient than graphene and SWCNTs enabling their dynamic use in exploring the absorption of high-energy radiations. Moreover, they can produce photothermal and PA signals with larger frequencies, which can improve the effectiveness of diagnostics-impregnated therapy [22].

21.2.1.6 Raising Immune Responses by Vaccination

In order to stop the disease from spreading or, at the very least, from progressing severely, the immune response has to be triggered by administration of antibody-specific vaccine, against the antigen to be targeted for lowering the disease severity. Vaccines are now increasingly commonly utilized to treat autoimmune disorders and cancer in addition to protecting against infections [23].

CNPs can act as adjuvants and antigen carriers because they can be encrypted by immunogenic structures and target antigens (of microbial or any other source) that lead to an overall development in disease-fighting efficiency of the improved immune system [24].

CNT-binding antigens can enhance count of cytolytic T-cells in response to metastases invasion or may cause the generation of certain IgGs that neutralize a pathogen in animal models [25]. The creation of the influenza vaccine nanoplatform embedded with polyethyleneimine-functionalized GO (having the capacity to bind hemagglutinin molecule inflicted with recombinant influenza) has shown positive signs of protection against the influenza viral attack. Vaccines given intraperitoneally to mice elicited remarkable humoral and cell-mediated immunity, which subsequently provided protection against influenza attack [26]. Additionally, nanodiamonds may act as a carrier of influenza viral antigen. They can triple the synthesis of neutralizing antibodies in mice, thus multiplying the immune response to 300-fold [27].

21.2.1.7 Generation of Artificial Tissue Implants

Because of their alluring chemical and physical characteristics, nanomaterials have potential and promise to be allografts (regenerative therapy) [28]. Implants that cause discomfort after administration are typically rejected; however, attaching proteins and amino acids to nanotubes has shown promise in preventing this rejection. Single- and multi-walled nanotubes (S-, M-W-) are utilized as implants in terms of prosthetic joints and other grafts that don't cause host rejection. Additionally, due to special qualities like comparatively higher tensile strength, the new-age CNTs function as implants and substitutes for bone formation, with calcium impregnation molded in the structure of the bone [29].

SWCNT and MWCNT composites have a noticeable effect on improvement of cell–cell attachment (cellular adhesion) and cell–cell multiplication (cellular proliferation). As a result of these metabolic advantages, these nanotubes have profound utilization in creation of nanocomposites (made up of both natural and synthetic material) [30].

21.2.1.8 Engineering Tissues and Tissue Metabolites

Engineering the tissues means to do the replacement, creating a fresh tissue or helping heal a tissue from ill or damaging agents with sustained biological alternatives, supported by materials science and engineering advancements. CNTs have profound use in chemo-sensing, tracking, and labeling, and in the creation of surface and sub-surface tissue matrices [31]. There are a few non-invasive methods to look into aspects of tissue environment, of which cell tracking and labeling are prominent ones that are utilized to dive deep into in vivo aspects of growth, development, and pathological implications. Helping in evaluating the viability of engineered tissue and understanding the biodistributions, migration, and movement pathways among cell peripheries, thus deciphering the major pathways of movement, CNTs have wide implications in tissue engineering aspects. CNTs are being explored as imaging contrast agents for magnetic resonance, optical, and radiotracer modalities. CNTs also have potential for biodistribution measurements and gamma scintigraphy using radiotracers can be synchronized with advancement of increased tissue counts [32].

Nanotubes have been tested in BALB/c mice to assess their biodistribution. The design of better-engineered tissues can improve monitoring of biophysiological aspects of cells, including metalloenzyme interactions, secretion of proteins in Golgi apparatus, cellular behavior, and ion transport. CNTs, with their larger surface-to-area ratio, ability to immobilize proteins, and electrical properties, are ideal for nanosensors. This tracks down the components responsible for causing abnormalities and also those which are responsible for maintaining overall cell health including charged groups of proteins participating in redox reactions along with the active site participation of amino acids concerned [33].Cell matrix is crucial in tissue engineering, but synthetic polymers like polylactic co-galactic acid (PLGA) and polylactic acid (PLA) are devoid of the kind of mechanical strength that is required to support a hard skeleton and thus are un-operational for functionalized systems. CNTs are suitable for functionalized systems offering potential to be used as tissue scaffolds. However, they are not biodegradable. Combining CNTs with polymers can enhance mechanical strength, as demonstrated by MWCNTs combined with chitosan, compared to chitosan alone [34].

21.2.1.9 Drug Targeting and Drug Delivery Approaches

Drug therapy is a systemic stream, meaning it affects the entire body. Targeting therapy has the potential to significantly improve efficacy and is able to avoid dangerous consequences of systemic inflammation. CNPs provide the perfect platform for delivering drugs to different tissues. Large loading capabilities are possessed by them for various pharmacological compounds, including mixtures of the medicine formulations. The abilities to target certain tissues or molecules, improve penetration and retention effects, and regulate drug release are all possible by the functionalization of CNPs. Temperature, pH, enzymatic activity, and light are agents of releasing the stimulus [35].

Graphene, fullerene, and CNTs are able to bind proteins (monoclonal and polyclonal antibodies), antimicrobials, and anticancer agents. With targeted therapy, pharmacological systemic toxicity is decreased and fewer dosages are required to produce the same therapeutic effect [36]. Moreover, CNPs can cure gliomas by delivering anticancer medications to the central nervous system. When they become functional, they are able to cross the blood–brain barrier in both directions of flow [37]. CNPs offer a vast array of medicinal applications. By employing smaller dosages of medications, they can improve the effectiveness and targeting of therapies, and lowering treatment costs and patient risks.

21.3 Applications Pertaining to Electrical and Electronics Sectors

21.3.1 Radar Waves Absorption by CNPs

Reducing the radar cross-section (RCS) of targets that are being monitored by radars is the most popular technique for hiding from detection and moving stealthily. The use of materials that absorb radar is the most significant and straightforward method of reducing the RCS. In 1930, research on electromagnetic waves was initiated and continues to this day [38].

Actually, the primary function of materials designed to absorb radar waves and signals is to behave as a tiny electrical resistor, absorbing the waves and messages and preventing their reflection. Radar-absorbing materials are broadly classified into two types based on their component parts: magnetic absorbent materials and dielectric absorbent materials. Conductive polymers, carbon-based materials, and ferromagnetic materials are examples of absorbent materials based on their dielectric and magnetic characteristics. Due to the heavyweight nature of magnetic adsorbents and the limited absorption breadth of dielectric adsorbents, hybrid magnetic and dielectric compounds are used to create radar absorbent nanostructures that improve absorption. Both kinds of adsorbents are employed in tandem to address these two issues with adsorbent manufacturing [39].

Compared to pure metals and their respective metal oxides, the absorption qualities of metal-based alloys of nanoscale in absorbing electromagnetic compounds are superior. Among the different types of nanoalloys, iron-nickel nanoalloy finds extensive use in absorbers concerning waves of electromagnetic origin, magnet-based sensors, antennas for receiving signals, catalysts enhancing reaction rates, and magnetic stability systems due to their stable structure, exceptional electromagnetic characteristics, and high magnetic permeability. Despite their low weight and excellent dielectric loss, CNTs' poor magnetic characteristics limit their application as radar-absorbent materials. In order to enhance the magnetic properties of CNTs and add filler, magnetic materials, viz. ferromagnetic or superparamagnetic compounds are frequently coated on nanotubes [40].

21.3.2 Nanoparticles Implanted on Chip

A chip is a tiny silicon piece with electrical circuits running all over it that may store enormous amounts of data or conduct mathematical or logical processes [41]. Recent advancements have substantially increased the sensitive levels of sensors based on optical fibers made up of metal nanoparticle arrays and single nanoparticles [42]. Surface plasmon resonance sensing devices based on nanochips have advantages of being highly sensitive in real time monitoring of molecular interactions. Such approaches have not yet been patented [43]. Like agarose is meant to be a support chemical of gel electrophoresis analysis of DNA, RNA, and proteins, a likewise programmable bionanochip (PBNC) technology relies on agarose units consisting of 300 nm nanonets to hold micro-fluidic elements through micro-electrochemical support [44].

21.3.3 Developing Nanosensors

The word "sensor" originates from the Latin sensire, which meaning "to sense" or "to detect." It can be applied to any object which may receive a signal or stimulus and certainly specifically to it, including photoelectric cells. Today a wide array of materials are there, which may be utilized in the production of sensors and these include nanocomposites, nanowires, nanofibers, nanorods, nanotubes, and nanopolymers [45].

21.3.3.1 Gas Phase Nanosensors

The fundamental requirements for efficient gas-sensing systems include high sensitivity and selectivity, quick reaction and recovery times, low analyst consumption, low operating temperatures, and temperature independence. The working principle of gas sensors is based on "extremely high surface-to-volume ratio and hollow structure of nanomaterials, which are ideal for adsorption of gas molecules" [46].

The agent of gas-sensing application beholds several keynote units like: (1) Gas sensors made up of carbon nanofiber (CNF) having higher vapor-sensing ability of 100 times magnitude from that of the actual one;(2) Gas sensors with improved electronic properties have a better chemical sense of the concerned environment;(3) Addition of the polymer material to the carbon black nanoparticle to obey the resistivity versus percolation theoretical aspect;(4) Gas sensors with reduced graphene having enhanced carrier scattering/scintillation of gas concerned and thereof the reduced time lag for absorbent layer that in turn concentrates intrinsic gas response by plotting graphene in H_2/Ar [46].

21.3.3.2 Optical Detection Nanosensors

The plasmonic materials that include optical sensing properties are able to expedite the photoabsorption process. One such example of a plasmonic material is a 3-D hybrid nanomaterial that is built on silver nanowires [47]. Due to their superior optical and chemical properties, Au, Ag, and Cu plasmonic metallic particles display localized surface plasmon resonance (LSPR) phenomenon, enabling their use as building blocks in the production of biosensors [48].

21.3.4 Light-Emitting Diode Screens Implanted With Nanomaterials

OLED stands for organic light-emitting diode defined as a cell that turns to emitting light while supplied with a certain voltage. Because it is faster, thinner, brighter, and less expensive than liquid crystal display (LCD) technology, OLED technology is being used as a display device in portable electronics like mobile phones. Theoretical findings suggest that altering defensive energy and particle diameter, as well as using different metal NPs like Ag (silver) and Pd (palladium) doped in LC, could affect nanoparticle liquid-crystal-mediated interactions [49]. According to Collins Dictionary, a touch screen is a type of visual display unit screen that enables users to interact with computers without a keyboard by touching specific areas of the screen [41, 50].

21.3.5 Conserving Charge via Nanobatteries

According to Collins Dictionary [41], a battery is any tiny device that supplies electricity for electrical equipment like radios and electronics. Now with added dual advantage of having a long life with higher energy production and lower self-discharging potential, LIBs are utilized in a number of electric-based communications ranging from their large-scale use in devices operating on electricity with portable tendency as well as devices with longer-time operations like either hybrid or fully electric vehicles (EVs) [51].

21.4 Applications Pertaining to Wind and Solar Energies

21.4.1 The Charge Held by Nanomaterials as Supercapacitors

Creating supercapacitors with high-efficiency rates has emerged as most promising fields of academic and industrial study in this new era. Currently, a lot of research is focused on carbon nanomaterial applications as the electrodes (anode and cathode) made up of supercapacitors. Supercapacitors, like ordinary capacitors, are made up of two electrodes (may be solid or porous) that are adjoined to a separator and an electrically charged substance. The maximum power output of any energy storage device reaches a maximum level from where it cannot be uplifted, only when the internal resistance of the cell equals the load resistance. The volumetric capacitance, the cell's energy, and power densities, and the ESR and the device's cell voltage are key performance indicators for energy storage devices, of particularly batteries and supercapacitors. The electrolyte is responsible for creating the mass transfer resistance in contrast to the ion movement it can hold via the electrode matrix's internal resistance, the pores in the matrix, and numerous additional resistances that are present in the cell [52].

The self-discharge rate and leakage current of a supercapacitor determine its performance primarily, while the composition and chemistry of materials for the electrodes determine the surface area, conductivity and stability for a long period of time. A supercapacitor that is completely charged has more free energy than one that is discharged according to thermodynamics. Consequently, it is the supercapacitor's inclination toward the self-discharge generally to equilibrate with the surroundings. There are different kinds of materials available for construction of the electrodes. These are alternating current (AC) materials with a large surface area and oxides of electroactive polymers with transition metals.

Scientific studies have led to an increase in the energy storage capacity of supercapacitors by examining the porous architectures of graphene and CNTs. H_2SO_4 electrolyte-filled freestanding mats of entangled MWCNTs have demonstrated notable enhancement of the electrodes' specific capacitance of supercapacitors [53]. There have been attempts to investigate the impact of random and vertically oriented CNTs focusing on enhancing the ability of electrodes to undergo specific capacitance concerned with their supercapacitors[54]. Mesoporous carbon spheres with CNTs implanted have been created as electrodes merged in an aqueous electrolyte to increase porosity and enhance electrical conductivity spheres of carbon [55]. Using chemical vapor deposition (CVD)-grown CNTs on ceramic foam, polypyrrole (PPy)/CNT composite was created. Chemical polymerization was then used to cover the PPy and was shown to serve as a binder as well as an assistant in linking the clusters of CNTs, which aids in maintaining the capacitance of the electrode following 5000 cycles of redox. PAni, or polyaniline, has carbon pores with high surface-to-area ratio-type material displayed with high specific capacitance and an enhanced life cycle. Nanocomposites of PAni, such as those involving graphene and CNFs, the combination of PAni/CNT generated particular capacitance values that were more than PAni individually [56, 57].

21.4.2 Energy Conservation by Fuel Cells

Fuel cells are energy-storing devices that are safe for the environment to use. They have a high power density and energy conversion rate, and they pose little to no risk of emissions. The primary applications of the polymer electrolyte membrane (PEM) fuel cell include applications for mobile devices at low temperatures [58]. Due to their sluggish electrode kinetics, a substantial amount of expensive and scarce catalyst materials are being exploited, particularly platinum (Pt) metal [59]. Fuel cells are similar to batteries in that they transform chemical energy into electrical energy as a byproduct, generating water and heat [60].

Carbon nanomaterials, majorly graphene and CNTs, have been investigated for their potential to improve fuel cell performance [61]. Electrochemical fuel cells, or microbial fuel cell gadgets transform organic waste's stored energy into functional electrical energy, made up of a cathode and an anode divided by a membrane specific to a certain cation [62]. The micro-organisms oxidize the organic fuel at the anode to produce protons that flow across a membrane to the cathode, and to create current-generated charge, the electrons travel from the anode to the extent of external circuit [63]. According to reports [64], the process of coating sponge with CNTs forms a macroporous structure in sponge/CNT composites, which are employed for treating wastewater effluents, removing salt components from water and generating hydrogen for power supplies. Direct methanol fuel cells (DMFCs), which don't even need a separate hydrogen-generating system, are being researched for their many uses in the automotive industry. An electrochemical device called a DMFC transforms stored energy inside the fuel's (methanol's) molecular bonds to produce usable electrical energy. The broad spectrum compound methanol's accessibility, affordability, and ease of processing make it a valuable fuel making the fuel cell technology known as DMFC appealing [65]. Pt has often been utilized as an electrode for removing the electrons from methanol (oxidation is the process of removal of electrons). Still, the Pt catalyst is poisoned by the intermediate carbon monoxide(CO) species that is produced during the intermittent stages [66]. To address this severe problem, catalyst systems with binary, ternary, and quaternary treatment operations have been created [67]. A further strategy to reduce this toxicity effect is to choose a suitable substance to support the catalyst. The main objectives of catalyst support are to: (1) provide stabilization to the catalyst particles;(2) help reveal the deposited catalyst's participating reaction active surface area;(3) participation in anode to cathode electron flow;(4) reduce the conditions destined for possible poisoning at the catalyst surface; and (5) enhance mass transfer. Potential uses for conductive polymer as catalyst-supporting matrices have been identified in this area [68]. Nitrogen-containing CNTs were created by impregnating silicon wafer with iron-tetramethoxyphenylporphyrin chloride (including a heat treatment step) after toluene and acrylonitrile, which were formed by the CVD technique, reacted to a fresh cathode in PEM fuel cells for the oxygen reduction reaction (ORR) [69]. According to reports [70], the production of sulfonated electroactivity of poly(ether ether ketone) was improved by adding iron tetramethoxyphenylporphyrin chloride and using a modified cast-drop process of the catalyst and improved fuel cell utilization. Incorporated utilization of ball-milled CNTs resulted in improved magnesium(Mg) absorption kinetics and also resulted in faster hybridizing kinetics of up to 6.6 wt%, which was achieved only in one minute at 150 °C at optimum time and temperature, respectively. Bipolar plates consisting of graphite and polypropylene composite have been made utilizing CNTs as a conducting interlayer and 304 stainless steel as supported by another finding in the year 2011 [71]. In comparison to the graphite bipolar plate, the efficiency of the composite has increased to 91%. The due credit is aforementioned for the addition of CNTs, which has also improved the mechanical strength and the electrical conductance of the material for up to 91% in which they were incorporated. CNTs generated using simple catalyst templates made of hexagonal mesoporous silica containing iron have been used to create Pt/CNT anodic electrocatalysts with applications concerned to creation of DMFC and the oxidation of methanol [72].

It has been discovered that the diameter of CNTs significantly affects both the electrocatalyst's performance and the dispersion of Pt nanoparticles. For instance, Pt/MWCNTs electrocatalyst, where the diameter of the CNTs is more than 8 nm, demonstrated improved electrocatalytic activity compared to the traditional Pt/Vulcan, Pt/SWCNT, and carbon catalysts XC-72. PtRu/CNTs and PtRuMo/CNTs are catalysts produced using a microwave-assisted method and utilized as a DMFC anode [73]. According to the report, the addition of Mo assisted in lowering the energy of every Pt site. As the temperature rose, it was discovered that the DMFC's performance improved. This finding was related to a similar rise in the membrane's ion conductivity electrode assembly (MEA), which improves the catalytic efficiency. Cathode catalysts PtCo/PPy-MWCNTs, produced via the one-pot synthetic method Pt and CoNx, which have two active sites, have been utilized for ORR in fuel cells [74]. The performance of the second transition metal was enhanced by the catalyst by way of a cooperative result.

21.4.3 Lithium-Ion Batteries: A Run for Electric Vehicles

The fundamental specifications for an electrode for being advanced LiBs are: (1) high electrical conductivity as the low electrical conductivity will tend to reduce required optimum voltage, along with fluctuations in power and energy density requirements; (2) a larger surface area because the influence of greater surface area would result in generation of more active sites), increasing the reaction rates at the surface interface);and (3) the electrode's microstructure (the electrode's porous structure affects efficiency and performance as well between the electrodes). The involvement of a mesoporous structure actually shortens the transport distance for lithium ions but it has been of scientific interest that a large portion of it remains inaccessible to the electrolyte [75]. Irrespective of this, the macroporous structure is responsible for efficient Li-ion diffusion and mass transfer among energy-storing and energy-extracting devices. Researchers looked into how carbon nanomaterials, such as graphene and CNTs, can improve LiB performance [76]. It is commonly known that CNTs can be found in electrodes in three different configurations: binder-free CNT networks, arrays of CNTs and discontinuous/disrupted/non-uniform CNTs. Generally speaking, though, discontinuous CNTs are the better option when it comes to improving electrode performance. The synthesized electrodes using LFP/C/CNT composites allowed the CNT network to be effectively utilized to reduce internal resistance and improve electrode performance [77]. Even after 10,000 cycles, the $ZnFe_2O_4$-C/LFP-CNT electrode continued to function well. In metal-air batteries, graphene is used as a catalyst as well as an active material. It can serve as the best to fit all-time active material for double-layer capacitors undergoing electrochemistry by storing electrostatic charges or by hosting ions. In comparison to single-layer graphene, reduced GO (RGO) is thought to be the most popular material for Li-ion storage because of its high capacity values. However, because the process is irreversible, this charge is not entirely released after delithiation. The degree of flaws in the layers that cause the low Coulombic efficiency in the first cycle also affects the efficiency of the RGO electrode. Graphene is produced using nitric acid and ultrasonication, the GO has been etched with acid, and then thermally exfoliated or reduced claimed to improve LiB's high-rate performance [78]. For LiBs, solid hybrid polymer electrolytes based on lithium TFSI, PEO, and triethylsulfonium-bis (trifluorosulfonyl) imide (S_2TFSI) have demonstrated long-term cycle stability against metallic lithium in addition to more than 4.5 V, noticed as reversible cathode stability [79]. These days, easy-to-fit LiBs in automobiles of various sizes are available with excellent mechanical qualities and great charge mobility that are made possible only by graphene-based materials, a type of carbon nanomaterials [80].

Graphene sheets are used as both a current collector and a conducting agent. The anode materials have the properties of higher active or specific surface area, higher conductivity potential, good mechanistic qualities, and a high capacity for storing lithium (744 mAhg^{-1}) in graphene composites as anode materials [81]. Sol–gel-prepared $LiFePO_4$/graphene nanocomposites (used as a cathode) have been found to have enhanced rate capability, cyclic stability, and better electrical conductivity than pure $LiFePO_4$ (as high as six times) [82]. Sn/graphene nanocomposites have considerable use as anodes for LiBs. They were finalized by combinatorial sequence step of microwave hydrothermal and reducing the hydrogen gas reduction [83].

Anode consisting of Sn particles having 60–120 nm size range dynamics has demonstrated lower cycle stability and capacity than anode consisting of Sn particles of smaller magnitude (10–20 nm). There have also been reports of magnesium-ion batteries based on graphene recently. Here, graphene is used for the first time as the cathode material in magnesium-ion batteries, despite the battery's subpar electrochemical characteristics [84]. Additionally, according to reports [85], graphene was observed to increase the efficiency of vanadium redox-flow batteries (VRFBs). Furthermore, previous research has shown the exceptional electrochemical capabilities of TiO_2nanoneedle/graphene composites, which combine 1-D and 2-D features, which arise due to their distinct microstructure. While the 2-D structure of graphene nanosheet provides the electrode with superior electrical conductivity, the 1-D TiO_2 structure is responsible for shortening the Li-ion transit pathway, thereby enhancing the diffusion rate [86].

In order to meet the diverse demands of society, renewable energy sources shall have the ability to supplement non-renewable energy resources. The commercialization of renewable energy storage systems is hampered, nevertheless. Financial strategies need to be flexible and innovative in order to get beyond the obstacles that stand in the way of non-renewable energy projects becoming commercially successful. The following factors impact the commercialization of renewable energy technologies: high project development-to-investment cost ratio, high capital required for operation and maintenance, small total investment requirement, erroneous risk assessment, challenge in ensuring project cash flow, inadequate foundation for non-resource financing, and inexperienced project developers. Because fossil fuels are abundant and inexpensive, they currently provide 80% of the world's energy needs. Nonetheless, the cost advantage of renewable energy sources over fossil fuels has grown recently. In 2014, it is anticipated that as time goes on, the cost of renewable energy sources will decline. As a result, financial strategies will likely cause research trends to shift in favor of renewable energy sources. In order to examine the economy of the batteries, researchers have looked into the design of experiments and the process-based cost model (PBCM) [87]. The cost and process yield of MWCNT significantly impact cost per unit of satellite batteries, while cycle time and process yield determine unit cost of computers [88].

21.5 Application Pertaining to Food Industry Sector

21.5.1 Nanoencapsulation: An Approach to Increasing Shelf-Life

The required technological input to pack items in tiny destined for nanoscale also enhances food rehydration and is known as nanoencapsulation in food processing. The non-toxic edible coatings are applied to food commodities of daily use which act as barriers against gas, lipids, and moisture [89]. Nanoencapsulation is utilized in the production of confections, pharmaceuticals, and food systems with nanocarrier technology and dietary supplements [90].

There are three methods for achieving nanoencapsulation: (1) physical, which involves spray drying, coating, and extrusion; (2) physiochemical processes involving use of liposomes for entrapment; and (3) chemical processes involving polymerization at interfacial levels with inclusion boundaries at molecular level. [91].

21.5.2 Nanoemulsification: An Aid to Food Digestion

The process of creating nanoemulsions by methods like composition decision at phase conversion levels, homogenization at levels of high pressure, using sound waves to mix (ultrasonication) and microfluidization is known as nanoemulsification in the food processing industry [92]. For a given application in the food industry, nanoscale implementing systems such as liposomes, biopolymer nanoparticles, solid lipid nanoparticles, nanoemulsions, multiple emulsions, multi-layer emulsions, and microgels use need to be decided on the basis of: (1) safe use;(2) viability concerns with commercial and economic value;(3) compatibility of being a support system for food-based items;(4) long shelf-life with robustness; and (5) labeling and performance requirements [93].

21.5.3 Nanomaterial Levels in Food Packaging

As of now, a number of techniques have been employed to preserve food, including irradiation, salting, curing, canning, freezing, refrigeration, freezing, and using preservatives [94]. Labuza coined the phrase "active packaging" in 1987 to increase nutritional quality and shelf-life of food with a longer shelf-life and freshness [95]. Additionally, food packing methods should be created to preserve food flavor and quality as well as guard against microbial infestation, which causes food to decay [94]. Active, intelligent, and smart packaging kinds have lately been documented.

21.5.3.1 Type 1: Active Packaging

In addition to acting as an inert barrier between the unfinished boundaries of environment and food product, the active packaging interacts with the food product to prolong its shelf-life and maintain optimal conditions. Food packaging that is considered active consists of the following components: (1) nanocomposites (made up of metal ions such as Ag, Cu (Copper), Au (Gold), and metallic oxides of Ti(Titanium) and Mg(Magnesium); (2) film coatings especially of sodium benzoate, ethanol, zinc, and various metal ions like S (Sulfur), Al(Aluminum), Fe (Iron), and Ca (Calcium). Also incorporated is the use of natural anti-microbial deterring agents from plant-based origin, viz. pepper, coffee, and cinnamon; and (3) gas scavengers (titanium oxide, iron core powder, silicon particles) [96]. The alcohol oxidase enzyme and the alcohols produced in the food product interact chemically to provide enzymatic oxygen scavenging [97].

Carton board, polyethylene, aluminum sheet, repeatable polyethylene, and carton layers make up the materials and construction of Tetra Pak™ packaging, which has been used in Switzerland since 1951 for the storage and extension of the shelf-life of milk products. Nanoclay–polymer composites were used to improve the gas barrier properties of food packages. The most commonly used is montmorillonite (also known as bentonite), which has a natural nanoscalar layered structure that can restrict the permeation of gases when incorporated into a polymer [98].

21.5.3.2 Type 2: Intelligent and Responsive Packaging

In order to increase shelf-life, improve safety, improve quality, offer information, and alert users to potential issues, intelligent packaging can perform detecting, sensing, recording, tracing, communicating, and applying scientific logic [99], the indicators (methylene blue, acid, antioxidants, mineral oil and sugar, elements [Na, K, Ca, Si, Al, and Mg], and antioxidants) are the components of these kinds of intelligent packages [100]. The foundation of responsive packaging technologies is the incorporation of a sensor or sensing interface into the packaging film for continuous and real-time quality monitoring. This is achieved by putting three crucial design elements into practice: preventing cross-contamination, choosing the target analytes, and selecting the transduction system. The stimuli-sensitive materials, hydrogels, surfaces, particles, and supramolecules are the parts of responsive packages [101].

21.5.3.3 Type 3: Smart Packaging

Smart packaging is any small, low-cost label or tag that is affixed to primary packagings, like trays, bottles, and pouches, or, more frequently, to secondary packaging, like large containers used to ship in order to set communication among chains of supply. The use of absorbers/generators of carbon dioxide can be optimized in label or sachet form so as to bring down the respiration rate (eliminating the oxygen content), thereby improving the quality of vegetables and fruits. Modified atmospheric packaging (MAP) is usefully optimized so as to extend the shelf-life of edible major food items by preventing the occurrence of free radicals and thereby the oxidation reaction. But what come with major concern are the existing valid reports linking adverse health effects by intoxication upon usage of emulsified form CNTs [97, 102].

21.6 Nanoparticles Operating Within Soil

The formation of nanoparticles in soil is attributed to abiotic (non-living system) and biotic (living system) existences. The clay minerals are formed by abiotic routes and humic substance (HS) formation is attributed to biotic routes. Although the decomposition is the same scenario, it is the rate, medium, and effects of decomposition that classify the biotic and abiotic routes into distinct pathways, while there are nanoparticles of iron and manganese, which may be formed by both routes as both of these can be found in biotic and abiotic origins [103]. The high mobility potential of colloids makes the colloids-facilitated transport possible carrying along with them the contaminants [104]. The physical laws of nature like zig-zag random Brownian movement, substantial transport (or adjective flow) and the gravitational pull all contribute toward the formation of soil nanoparticles [105]. The revealing of particle attachment in process of colloid-facilitated transport was elucidated using a homogeneous saturated soil medium [106].

21.6.1 Clay Particles

An impure kaolinite is composed of different varying fractions of ferric oxide, magnesia, and alkali, whereas pure clay is a silicate of aluminum [$Al_2Si_2O_5{\cdot}OH)_4$] known by the name of kaolin or kaolinite. A white soil mineral known as kaolinite can be created when rocks weather. The kaolinite nano- and microcrystals form when phyllosilicates weather by growing over primary minerals by a "periodical 2D nucleation mechanism at the initial stage of its formation, and by a helical mechanism at the final stage." [107].

21.6.2 Humic Substances: Humidifying Agents in Soil

Humins are the components of soil that are insoluble in water and are found in alkali (high pH) and acid (low pH) media. HS can be further classified into three major fractions: (1) humin, (2) humic acids (HAs), and (3) fulvic acids (FAs) [108]. Humic materials influence nutrient uptake, enhance soil fertility, and improve soil structure, all of which have a positive impact on plant physiology. HAs are significant constituents of the organic matter found naturally in soil humus, sediments, and aquatic media. They have the ability to interact with the surface of mineral particles to form clay-humic complexes, which can have an impact on soil structure, soil erosion, and soil sequestration of carbon [109].

21.6.3 Fulvic Acids: A Response Toward Soil Salinity

Humic matter, organic phosphorus compounds, nucleic acid, other nitrogen compounds, hemicellulose, and organic acids are the components that make up the FA fraction in soil humus [110]. FAs are chemically reactive and have twice the oxygen content of HAs due to their smaller molecular weight (1000–10,000) and abundance of carboxyl (—COOH) and hydroxyl (—COH) groups [111].

21.7 Agricultural Aspects of Nanomaterials

Through their incorporation into fertilizer compounds and provision of essential nutrients for growing plants, nanomaterials are playing an increasingly important role in agriculture. Three categories—nanoscalefertilizer input, nanoscale additives, and nanoscale coatings or host materials for fertilizers—are utilized to describe nanomaterials with nanotechnology used in fertilizers and plant protection.

Broad advancements in agricultural research are made possible by nanotechnology, including advances in reproductive science and technology, the production of energy from food and agricultural wastes, as well as other valuable byproducts through enzymatic nanobioprocessing, disease prevention, and plant treatment with a variety of nanoscale devices with unique properties. To circumvent biological barriers for successful targeting, "smart delivery systems" in agriculture should combine highly controlled, precisely targeted, time-controlled, remotely managed, preprogrammed, self-regulated, and multi-functional properties [112]. Biotechnological uses for metal nanoparticles with size-dependent characteristics, such as photocatalytic degradation (metal oxide NPs), fluorescence, or magnetism (magnetic NPs), include sensor creation, soil remediation, and agrochemical degradation. Properties like stiffness, permeability, crystallinity, thermal stability, solubility, and biodegradability are essential for creating and manufacturing nanopesticides, which can improve agricultural formulations' wettability and dispersion while lowering runoff of organic solvents. These properties are possessed by nanomaterials and biocomposites [111].

21.8 Nanomaterials Bringing Out Latest Revolutions

21.8.1 Fuel of Nucleotide Triphosphates (NTPs)

A new type of motor, called a nanomotor, uses nucleotide triphosphates as fuel to attach letters to sentences. The motor requires energy from the "alphabet soup" produced by polymerase, which removes two phosphate groups to attach a new letter. The motor can only continue running when a sufficient number of nucleotide triphosphates are available. Researchers from Michigan and Arizona have demonstrated that the motor can carry out expected movement and can be combined with other structures to create a unique style. They are also planning to create a clutch that can only use the motor's power at certain times and leave it idle. In the long term, the motor could become the heart of a complex nanomachine [113].

21.8.2 Hybridization of Nanomaterials to Achieve Sustainability

Researchers have reviewed the progress and challenges in the development of Polyoxometalate-based nanohybrids, which could revolutionize sustainability across various industries. Researchers have developed a detailed overview of polyoxometalate-based nanohybrids, a new class of nanoscale hybrid materials with potential for sustainability in energy systems, transport, biosensors, water purification, and 3-D printing.

Nanomaterials, which have dimensions of 1–10 nm, exhibit unique physical, chemical, and biological properties. These materials can be made from various substances, including metals, semiconductors, ceramics, and polymers. Polyoxometalate (POM)-based nanohybrids are particularly interesting for their unique catalytic properties in photoelectrochemical reactions. POM nanohybrids have potential applications in clean energy conversion, storage, sensors, and electronics that don't rely on dirty energy sources. POMs can act as electron acceptors and donors, facilitating the transfer of electric charge and improving reaction efficiency. They also have the potential to combine the unique properties of three different materials, resulting in greater functionality and versatility. However, research on ternary POM-based nanohybrids is still in its infancy, and challenges such as solubility, non-uniform dispersion, and control of particle size and shape remain [114].

21.8.3 Membranous Carbon Nanotubes Role in Removing the Micro-Pollutants

Researchers from Donghua University and Harbin Institute of Technology have developed a flow-through $KMnO_4$/CNT system to enhance micro-pollutant degradation. The system uses CNTs as an electron transfer mediator, facilitating the oxidative decomposition of organic contaminants. The system outperforms conventional batch reactors, achieving over 70% removal of 80 μmol/L sulfamethoxazole solution at single-pass mode. The system's reusability and long-lasting reactivity make it a green solution for micropollutant remediation [115].

21.8.4 Drug Delivery Targeting to Malignant Neurons

A research team has developed a treatment for medulloblastoma, a deadly brain cancer, using nanoparticles administered directly into the cerebrospinal fluid (CSF). Professors Mark Saltzman and Ranjit Bindra led the team and administered the treatment to mice identified with medulloblastoma, which is prone to spread along the leptomeninges in the central nervous system. The CSF has no anatomic barriers to prevent further growth, making it difficult to target tumors. To overcome this obstacle, research experts of the team utilized nanoparticles with fabrication having the added ability to adhere to tumor cells with release of a chemical called talazoparib (known to inhibit DNA repair enzymes), widely used to treat cancer cells and also approved by Food and Drug Administration (FDA). The nanoparticle treatment was injected intrathecally between the leptomeninges protecting the CSF, allowing researchers to use the drug talazoparib (proven effective in peripheral brain boundary solid tumors). The drug was delivered intrathecally, avoiding convection-enhanced delivery, a challenging procedure. The mice receiving the nanoparticle-based treatment lived significantly longer than those without the nanoparticles and had less spreading of the cancer. The future advancements, which are in the need of the hour, are to do the validation in larger animals especially, experimental model organisms and clinical research trials on humans. The drug delivery to malignant tumors of brain needs to be studied at a wider yet intriguing scale [116].

21.8.5 DNA Origami Nanoturbine and Nanomotor Revolutions

Researchers at Delft University of Technology have developed a ground breaking nanomotor, the DNA origami nanoturbine. This device uses ion gradients or electrical potential to drive mechanical rotations in a solid-state nanopore. The design allows researchers to control rotation direction [116].

21.8.5.1 Nanoturbines and Generation of Mini-Multi-Powers

Flow-driven turbines are crucial for various machines, including windmills and airplanes. A 25-nm diameter rotor made from DNA material controls rotation direction. The nanoturbine, docked in a strong water flow, can drive a rigid rod up to 10 rev/s, according to lead author Xin Shi. The DNA origami nanoturbine's rotation is influenced by ion concentration, allowing it to spin clockwise or anticlockwise, a unique feature exclusive to nanoscale research. The study, supported by molecular dynamics simulations and theoretical modeling, suggests that DNA-origami could be used to create nanomachines capable of delivering drugs to specific cells in the human body. At Technical University of Munich, researchers have developed a turbine using DNA origami techniques with full control over its design and operation. The "DNA origami" technique uses specific interactions between complementary DNA base pairs to build dynamic 3D nanoobjects, allowing the direction of rotation in nanopores to be controlled through the blade's handedness. This breakthrough marks the next phase of the DNA active nanorotor, which can harness energy from salt gradients [117].

21.8.5.2 Induction of Power Performance of Potassium-Ion Batteries

Researchers at Xi'an Jiaotong University have developed a bimetallic selenide, $NiCo_2Se_4$, for potassium-ion batteries. $NiCo_2Se_4$ is a hydrothermal process devised under two steps that synthesize a structure of nanotube embedded with clusters of flower-like appearance, creating and optimizing the channels required for potassium cation and anion transfer. The researchers found that $NiCo_2Se_4$ has more active sites, uniformly distributed elements, and outperformed other electrode materials, comprising cobalt selenide (Co_3Se_4) and nickel selenide (Ni_3Se_4). The technical achievements in terms of better electrochemical performance were outcome of cyclic and rate stability offered by unique structure of nanotube and synergy combinations of two different metals. The bimetallic $NiCo_2Se_4$ nanotube electrode also had a higher capacity, which is beneficial for maintaining cyclic stability and high-rate performance. To achieve this are new insights gained for designing the micro-/nanostructured binary metal selenides for optimized production of potassium-ion batteries [118].

21.9 Conclusion and Future Scope

This chapter describes the importance and incorporation of CNPs in day-to-day human life along with important contributions of nanomaterials in human life. The individuality of each nanomaterial has been described here ranging from social, self, and research fields. The utilization potential of these nanomaterials and thus their efficiency needs to be tested at certain levels where toxicity is a major concern nonetheless, undetected so far in majority of applications involving nanomaterials. Moreover, the thinking of human mind has left with us a huge improvement in fields of health, personal care, electronics and communication, agricultural and several other industrial sectors. The presence of nanomaterials in natural environments ranging from soil to water to air has boosted scientific knowledge and deep insight has been provided into principles of their action.

Recent advances in last five years have laid down the knowledge poles deep into the futuristic mindsets of new generation, whereby a door of new innovations of carbon nanomaterials and their potential use in day-to-day human life is being sighted to improve lifestyle of future generations. Green energy utilization through the advent of nanomaterials has opened the gateways of resource utilization for human needs. To even make recycling efficient and successful, the dependency on non-renewable resources needs to be switched to carbon nanomaterials for longevity and sustainability for the upcoming future generations. Green energy applications need to be widened for the scope of a better life style along with minimum detrimental effects on environment, thus making utilization of green energy an essential tool to uncover for future generations.

References

1 Liu, J., Li, R., and Yang, B. (2020). Carbon dots: a new type of carbon-based nanomaterial with wide applications. *ACS Central Science* 6 (12): 2179–2195.

2 Cherukula, K., Manickavasagam Lekshmi, K., Uthaman, S. et al. (2016). Multifunctional inorganic nanoparticles: recent progress in thermal therapy and imaging. *Nanomaterials* 6 (4): 76.

3 Cherukula, Z., Fang, C., Li, B. et al. (2020). First-in-human liver-tumour surgery guided by multispectral fluorescence imaging in the visible and near-infrared-I/II windows. *Nature Biomedical Engineering* 4 (3): 259–271.

4 Liu, W., Alam, M.N.A., Liu, Y. et al. (2022). Silicon-vacancy nanodiamonds as high performance near-infrared emitters for live-cell dual-color imaging and thermometry. *Nano Letters* 22 (7): 2881–2888.

5 Cui, T.R., Li, D., Huang, X.R. et al. (2022). Graphene-based flexible electrode for electrocardiogram signal monitoring. *Applied Sciences* 12 (9): 4526.

6 Hashemi, S.A., Mousavi, S.M., Faghihi, R. et al. (2020). Superior X-ray radiation shielding effectiveness of biocompatible polyaniline reinforced with hybrid graphene oxide-iron tungsten nitride flakes. *Polymers* 12 (6): 1407.

7 Wu, S., Duan, B., Zeng, X. et al. (2017). Construction of blood compatible lysine-immobilized chitin/carbon nanotube microspheres and potential applications for blood purified therapy. *Journal of Materials Chemistry B* 5: 2952–2963.

8 Holmannova, D., Borsky, P., Svadlakova, T. et al. (2022). Carbon nanoparticles and their biomedical applications. *Applied Sciences* 12 (15): 7865.

9 Hamblin, M.R. and Avci, P. (ed.) (2015). *Applications of Nanoscience in Photomedicine*. Elsevier.

10 Han, Y., Liu, H., Fan, M. et al. (2022). Near-infrared-II photothermal ultra-small carbon dots promoting anticancer efficiency by enhancing tumor penetration. *Journal of Colloid and Interface Science* 616: 595–604.

11 Fujimori, T., Tsuruoka, S., Fugetsu, B. et al. (2011). Enhanced X-ray shielding effects of carbon nanotubes. *Materials Express* 1 (4): 273–278.

12 Viegas, J., Silva, L.A., Batista, A.M. et al. (2017). Increased X-ray attenuation efficiency of graphene-based nanocomposite. *Industrial & Engineering Chemistry Research* 56 (41): 11782–11790.

13 Zarei, M., Sina, S., and Hashemi, S.A. (2021). Superior X-ray radiation shielding of biocompatible platform based on reinforced polyaniline by decorated graphene oxide with interconnected tungsten–bismuth–tin complex. *Radiation Physics and Chemistry* 188: 109588.

14 Türkaslan, S.S., Ugur, Ş.S., Türkaslan, B.E., and Fantuzzi, N. (2022). Evaluating the X-ray-shielding performance of graphene-oxide-coated nanocomposite fabric. *Materials* 15 (4): 1441.

15 Hsieh, H.S., Wu, R., and Jafvert, C.T. (2014). Light-independent reactive oxygen species (ROS) formation through electron transfer from carboxylated single-walled carbon nanotubes in water. *Environmental Science & Technology* 48 (19): 11330–11336.

16 Chen, Y., Huang, W., Dong, Y. et al. (2022). Enhanced antibacterial activity of indocyanine green-loaded graphene oxide via synergistic contact killing, photothermal and photodynamic therapy. *Journal of Biomedical Nanotechnology* 18 (1): 185–192.

17 Tu, X., Ma, Y., Cao, Y. et al. (2014). PEGylated carbon nanoparticles for efficient in vitro photothermal cancer therapy. *Journal of Materials Chemistry B* 2 (15): 2184–2192.

18 McKernan, P., Virani, N.A., Faria, G.N. et al. (2021). Targeted single-walled carbon nanotubes for photothermal therapy combined with immune checkpoint inhibition for the treatment of metastatic breast cancer. *Nanoscale Research Letters* 16: 1–9.

19 Xu, P. and Liang, F. (2020). Nanomaterial-based tumor photothermal immunotherapy. *International Journal of Nanomedicine* 15: 9159–9180.

20 Cheng, J., Liu, Y., He, L. et al. (2020). Novel multifunctional nanoagent for visual chemo/photothermal therapy of metastatic lymph nodes via lymphatic delivery. *ACS Omega* 5 (7): 3194–3206.

21 Li, Y., Su, Y., Pan, H. et al. (2022). Nanodiamond-based multifunctional platform for oral chemo-photothermal combinational therapy of orthotopic colon cancer. *Pharmacological Research* 176: 106080.

22 Zhou, L., Jing, Y., Liu, Y. et al. (2018). Mesoporous carbon nanospheres as a multifunctional carrier for cancer theranostics. *Theranostics* 8 (3): 663.

23 van Eden, W. (2020). Vaccination against autoimmune diseases moves closer to the clinic. *Human Vaccines & Immunotherapeutics* 16 (2): 228–232.

24 Gottardi, R. and Douradinha, B. (2013). Carbon nanotubes as a novel tool for vaccination against infectious diseases and cancer. *Journal of Nanobiotechnology* 11 (1): 1–7.

25 Scheinberg, D.A., McDevitt, M.R., Dao, T. et al. (2013). Carbon nanotubes as vaccine scaffolds. *Advanced Drug Delivery Reviews* 65 (15): 2016–2022.

26 Dong, C., Wang, Y., Gonzalez, G.X. et al. (2021). Intranasal vaccination with influenza HA/GO-PEI nanoparticles provides immune protection against homo-and heterologous strains. *Proceedings of the National Academy of Sciences of the United States of America* 118 (19): e2024998118.

27 Ho, T.T., Pham, V.T., Nguyen, T.T. et al. (2021). Effects of size and surface properties of nanodiamonds on the immunogenicity of plant-based H5 protein of A/H5N1 virus in mice. *Nanomaterials* 11 (6): 1597.

28 Calvert, P. (1999). A recipe for strength. *Nature* 399 (6733): 210–211.

29 Bian, Z., Wang, R.J., Wang, W.H. et al. (2004). Carbon-nanotube-reinforced Zr-based bulk metallic glass composites and their properties. *Advanced Functional Materials* 14 (1): 55–63.

30 Peigney, A. and Laurent, C.H. (2006). Carbon nanotubes-ceramic composites. In: *Ceramic-Matrix Composites* (ed. I.M. Low), 309–333. Woodhead Publishing.

31 Wang, Y., Da, S., Kim, M.J. et al. (2004). Ultrathin "bed-of-nails" membranes of single-wall carbon nanotubes. *Journal of the American Chemical Society* 126 (31): 9502–9503.

32 Chen, S., Yuan, R., Chai, Y. et al. (2009). Electrochemical sensing platform based on tris (2,2′-bipyridyl) cobalt (III) and multiwall carbon nanotubes–Nafion composite for immunoassay of carcinoma antigen-125. *ElectrochimicaActa* 54 (28): 7242–7247.

33 Zhang, L. and Webster, T.J. (2009). Nanotechnology and nanomaterials: promises for improved tissue regeneration. *Nano Today* 4 (1): 66–80.

34 Kam, N.W.S., O'Connell, M., Wisdom, J.A., and Dai, H. (2005). Carbon nanotubes as multifunctional biological transporters and near-infrared agents for selective cancer cell destruction. *Proceedings of the National Academy of Sciences of the United States of America* 102 (33): 11600–11605.

35 Zare, H., Ahmadi, S., Ghasemi, A. et al. (2021). Carbon nanotubes: smart drug/gene delivery carriers. *International Journal of Nanomedicine* 16: 1681–1706.

36 Daniyal, M., Liu, B., and Wang, W. (2020). Comprehensive review on graphene oxide for use in drug delivery system. *Current Medicinal Chemistry* 27 (22): 3665–3685.

37 Wang, B., Guo, H., Xu, H. et al. (2022). The role of graphene oxide nanocarriers in treating gliomas. *Frontiers in Oncology* 12: 736177.

38 Roco, M.C., Williams, R.S., and Alivisatos, P. (ed.) (2000). *Nanotechnology Research Directions: IWGN Workshop Report: Vision for Nanotechnology in the Next Decade*. Springer Science & Business Media.

39 Liang, C., Liu, C., Wang, H. et al. (2014). SiC–Fe_3O_4 dielectric–magnetic hybrid nanowires: controllable fabrication, characterization and electromagnetic wave absorption. *Journal of Materials Chemistry A* 2 (39): 16397–16402.

40 Bystrzejewski, M., Karoly, Z., Szepvolgyi, J. et al. (2009). Continuous synthesis of carbon-encapsulated magnetic nanoparticles with a minimum production of amorphous carbon. *Carbon* 47 (8): 2040–2048.

41 Collins Dictionary Online.(2016). Copyright © HarperCollins Publishers, https://www.collinsdictionary.com/dictionary/english/silicon-chip, https://www.collinsdictionary.com/dictionary/english/biosensor, https://www.collinsdictionary.com/dictionary/english/battery, https://www.collinsdictionary.com/dictionary/english/organic-light-emitting-diode, https://www.collinsdictionary.com/dictionary/english/touch-screen, https://www.collinsdictionary.com/dictionary/english/solar-cell.

42 Anker, J.N., Hall, W.P., Lyandres, O. et al. (2008). Biosensing with plasmonic nanosensors. *Nature Materials* 7 (6): 442–453.

43 Pang, L.andFainman, Y. (2016). U.S. Patent No. 9,513,226. Washington, DC: U.S. Patent and Trademark Office.

44 Jokerst, J.V., Chou, J., Camp, J.P. et al. (2011). Location of biomarkers and reagents within agarose beads of a programmable bio-nano-chip. *Small* 7 (5): 613–624.

45 Márquez, F. and Morant, C. (2015). Nanomaterials for sensor applications. *Soft Nanoscience Letters* 5 (01): 1–2.

46 Llobet, E. (2013). Gas sensors using carbon nanomaterials: a review. *Sensors and Actuators B: Chemical* 179: 32–45.

47 Park, S.G., Mun, C., Lee, M. et al. (2015). 3D hybrid plasmonic nanomaterials for highly efficient optical absorbers and sensors. *Advanced Materials* 27 (29): 4290–4295.

48 Shi, L., Jing, C., Gu, Z., and Long, Y.T. (2015). Brightening gold nanoparticles: new sensing approach based on plasmon resonance energy transfer. *Scientific Reports* 5 (1): 10142.

49 Nayek, P. and Li, G. (2015). Superior electro-optic response in multiferroic bismuth ferrite nanoparticle doped nematic liquid crystal device. *Scientific Reports* 5 (1): 10845.

50 Kim, D.J., Shin, H.I., Ko, E.H. et al. (2016). Roll-to-roll slot-die coating of 400 mm wide, flexible, transparent Ag nanowire films for flexible touch screen panels. *Scientific Reports* 6 (1): 34322.

51 Liu, Y. and Yang, Y. (2016). Recent progress of TiO_2-based anodes for Li ion batteries. *Journal of Nanomaterials* 2016: 2.

52 Arico, A.S., Bruce, P., Scrosati, B. et al. (2005). Nanostructured materials for advanced energy conversion and storage devices. *Nature Materials* 4 (5): 366–377.

53 Niu, C., Sichel, E.K., Hoch, R. et al. (1997). High power electrochemical capacitors based on carbon nanotube electrodes. *Applied Physics Letters* 70 (11): 1480–1482.

54 Futaba, D.N., Hata, K., Yamada, T. et al. (2006). Shape-engineerable and highly densely packed single-walled carbon nanotubes and their application as super-capacitor electrodes. *Nature Materials* 5 (12): 987–994.

55 Yi, B., Chen, X., Guo, K. et al. (2011). High-performance carbon nanotube-implanted mesoporous carbon spheres for supercapacitors with low series resistance. *Materials Research Bulletin* 46 (11): 2168–2172.

56 Lee, H., Kim, H., Cho, M.S. et al. (2011). Fabrication of polypyrrole (PPy)/carbon nanotube (CNT) composite electrode on ceramic fabric for supercapacitor applications. *ElectrochimicaActa* 56 (22): 7460–7466.

57 Luo, Y., Kong, D., Jia, Y. et al. (2013). Self-assembled graphene@ PANI nanoworm composites with enhanced supercapacitor performance. *RSC Advances* 3 (17): 5851–5859.

58 Dutta, K., Kumar, P., Das, S., and Kundu, P.P. (2014). Utilization of conducting polymers in fabricating polymer electrolyte membranes for application in direct methanol fuel cells. *Polymer Reviews* 54 (1): 1–32.

59 Kumar, P., Dutta, K., Das, S., and Kundu, P.P. (2014). An overview of unsolved deficiencies of direct methanol fuel cell technology: factors and parameters affecting its widespread use. *International Journal of Energy Research* 38 (11): 1367–1390.

60 Dutta, K., Das, S., and Kundu, P.P. (2016). Effect of the presence of partially sulfonated polyaniline on the proton and methanol transport behavior of partially sulfonated PVdF membrane. *Polymer Journal* 48 (3): 301–309.

61 Gharibi, H., Golmohammadi, F., and Kheirmand, M. (2013). Palladium/cobalt coated on multi-walled carbon nanotubes as an electro-catalyst for oxygen reduction reaction in passive direct methanol fuel cells. *Fuel Cells* 13 (6): 987–1004.

62 Nandy, A., Kumar, V., Mondal, S. et al. (2015). Performance evaluation of microbial fuel cells: effect of varying electrode configuration and presence of a membrane electrode assembly. *New Biotechnology* 32 (2): 272–281.

63 Das, S., Dutta, K., and Rana, D. (2018). Polymer electrolyte membranes for microbial fuel cells: a review. *Polymer Reviews* 58 (4): 610–629.

64 Kundu, P.P. and Dutta, K. (2018). *Progress and Recent Trends in Microbial Fuel Cells*. Elsevier.

65 Liu, H., Song, C., Zhang, L. et al. (2006). A review of anode catalysis in the direct methanol fuel cell. *Journal of Power Sources* 155 (2): 95–110.

66 Kumar, P., Dutta, K., and Kundu, P.P. (2014). Enhanced performance of direct methanol fuel cells: a study on the combined effect of various supporting electrolytes, flow channel designs and operating temperatures. *International Journal of Energy Research* 38 (1): 41–50.

67 Das, S., Dutta, K., Kundu, P.P., and Bhattacharya, S.K. (2018). Nanostructured polyaniline: an efficient support matrix for platinum-ruthenium anode catalyst in direct methanol fuel cell. *Fuel Cells* 18 (4): 369–378.

68 Kundu, P.P. and Dutta, K. (2016). Hydrogen fuel cells for portable applications. In: *Compendium of Hydrogen Energy* (ed. M. Ball, A. Basile, and T.N. Veziroğlu), 111–131. Woodhead Publishing.

69 Prehn, K., Warburg, A., Schilling, T. et al. (2009). Towards nitrogen-containing CNTs for fuel cell electrodes. *Composites Science and Technology* 69 (10): 1570–1579.

70 Chen, B.H., Kuo, C.H., Ku, J.R. et al. (2013). Highly improved with hydrogen storage capacity and fast kinetics in Mg-based nanocomposites by CNTs. *Journal of Alloys and Compounds* 568: 78–83.

71 Lee, Y.B., Lee, C.H., Kim, K.M., and Lim, D.S. (2011). Preparation and properties on the graphite/polypropylene composite bipolar plates with a 304 stainless steel by compression molding for PEM fuel cell. *International Journal of Hydrogen Energy* 36 (13): 7621–7627.

72 Lo, A.Y., Yu, N., Huang, S.J. et al. (2011). Fabrication of CNTs with controlled diameters and their applications as electrocatalyst supports for DMFC. *Diamond and Related Materials* 20 (3): 343–350.

73 Chen, S., Ye, F., and Lin, W. (2010). Effect of operating conditions on the performance of a direct methanol fuel cell with PtRuMo/CNTs as anode catalyst. *International Journal of Hydrogen Energy* 35 (15): 8225–8233.

74 Zhao, H., Pan, L., Jin, J. et al. (2012). PtCo/polypyrrole-multiwalled carbon nanotube complex cathode catalyst containing two types of oxygen reduction active sites used in direct methanol fuel cells. *Fuel Cells* 12: 876–882.

75 Yuan, W., Zhang, Y., Cheng, L. et al. (2016). The applications of carbon nanotubes and graphene in advanced rechargeable lithium batteries. *Journal of Materials Chemistry A* 4 (23): 8932–8951.

76 Tao, L., Zai, J., Wang, K. et al. (2012). Co_3O_4 nanorods/graphene nanosheets nanocomposites for lithium ion batteries with improved reversible capacity and cycle stability. *Journal of Power Sources* 202: 230–235.

77 Wu, X.L., Guo, Y.G., Su, J. et al. (2013). Carbon-nanotube-decorated nano-$LiFePO_4$@ C cathode material with superior high-rate and low-temperature performances for lithium-ion batteries. *Advanced Energy Materials* 3 (9): 1155–1160.

78 Varzi, A., Bresser, D., von Zamory, J. et al. (2014). $ZnFe_2O_4$-C/$LiFePO_4$-CNT: A novel high-power lithium-ion battery with excellent cycling performance. *Advanced Energy Materials* 4 (10): 1400054.

79 Zhao, X., Hayner, C.M., Kung, M.C., and Kung, H.H. (2011). Flexible holey graphene paper electrodes with enhanced rate capability for energy storage applications. *ACS Nano* 5 (11): 8739–8749.

80 Fisher, A.S., Khalid, M.B., Widstrom, M., and Kofinas, P. (2011). Solid polymer electrolytes with sulfur based ionic liquid for lithium batteries. *Journal of Power Sources* 196 (22): 9767–9773.

81 Mao, M., Hu, J., and Liu, H. (2015). Graphene-based materials for flexible electrochemical energy storage. *International Journal of Energy Research* 39 (6): 727–740.

82 Rao, C.N.R., Maitra, U., and Matte, H.R. (2012). Synthesis, characterization, and selected properties of graphene. In: *Graphene: Synthesis, Properties, and Phenomena* (ed. C.N.R. Rao and A.K. Sood), 1–47. Wiley.

83 Dhindsa, K.S., Mandal, B.P., Bazzi, K. et al. (2013). Enhanced electrochemical performance of graphene modified $LiFePO_4$ cathode material for lithium ion batteries. *Solid State Ionics* 253: 94–100.

84 Chen, S., Wang, Y., Ahn, H., and Wang, G. (2012). Microwave hydrothermal synthesis of high performance tin–graphene nanocomposites for lithium ion batteries. *Journal of Power Sources* 216: 22–27.

85 Wang, Y., Zhamu, A., andJang, B.Z.(2016). U.S. Patent No. 9,455,469. Washington, DC: U.S. Patent and Trademark Office.

86 Weber, A.Z., Mench, M.M., Meyers, J.P. et al. (2011). Redox flow batteries: a review. *Journal of Applied Electrochemistry* 41: 1137–1164.

87 Gan, Y., Zhu, L., Qin, H. et al. (2015). Hybrid nanoarchitecture of rutile TiO_2 nanoneedle/graphene for advanced lithium-ion batteries. *Solid State Ionics* 269: 44–50.

88 Timmons, D., Harris, J.M., and Roach, B. (2014). *The Economics of Renewable Energy*. Global Development and Environment Institute, Tufts University, Medford.

89 Hakimian, A., Kamarthi, S., Erbis, S. et al. (2015). Economic analysis of CNT lithium-ion battery manufacturing. *Environmental Science: Nano* 2 (5): 463–476.

90 Couch, L.M., Wien, M., Brown, J.L., and Davidson, P. (2016). Food nanotechnology: proposed uses, safety concerns and regulations. *Agro Food Industry Hi Tech* 27: 36–39.

91 Ezhilarasi, P.N., Karthik, P., Chhanwal, N., and Anandharamakrishnan, C. (2013). Nanoencapsulation techniques for food bioactive components: a review. *Food and Bioprocess Technology* 6: 628–647.

92 Quintanilla-Carvajal, M.X., Camacho-Díaz, B.H., Meraz-Torres, L.S. et al. (2010). Nanoencapsulation: a new trend in food engineering processing. *Food Engineering Reviews* 2: 39–50.

93 McClements, D.J. (2015). Nanoscale nutrient delivery systems for food applications: improving bioactive dispersibility, stability, and bioavailability. *Journal of Food Science* 80 (7): N1602–N1611.

94 Labuza, T.P. and Breene, W.M. (1989). Applications of "active packaging" for improvement of shelf-life and nutritional quality of fresh and extended shelf-life foods 1. *Journal of Food Processing and Preservation* 13 (1): 1–69.

95 Pitt, J.I. and Hocking, A.D. (2009). *Fungi and Food Spoilage*, 519. Springer.

96 Pradhan, N., Singh, S., Ojha, N. et al. (2015). Facets of nanotechnology as seen in food processing, packaging, and preservation industry. *BioMed Research International* 99: 1–18.

97 Brockgreitens, J. and Abbas, A. (2016). Responsive food packaging: recent progress and technological prospects. *Comprehensive Reviews in Food Science and Food Safety* 15 (1): 3–15.

98 Lee, S.Y., Lee, S.J., Choi, D.S., and Hur, S.J. (2015). Current topics in active and intelligent food packaging for preservation of fresh foods. *Journal of the Science of Food and Agriculture* 95 (14): 2799–2810.

99 Wyser, Y., Adams, M., Avella, M. et al. (2016). Outlook and challenges of nanotechnologies for food packaging. *Packaging Technology and Science* 29 (12): 615–648.

100 Yam, K.L. (ed.) (2010). *The Wiley Encyclopedia of Packaging Technology*. John Wiley & Sons.

101 Ahvenainen, R. (ed.) (2003). *Novel Food Packaging Techniques*. Elsevier.

102 Yam, K.L., Takhistov, P.T., and Miltz, J. (2005). Intelligent packaging: concepts and applications. *Journal of Food Science* 70 (1): R1–R10.

103 Theng, B.K.G. and Yuan, G. (2007). Nanoparticles in the soil environment. *Elements* 4: 395–399.

104 Birdi, K.S. (2016). *Handbook of Surface and Colloid Chemistry*, 4e. CRC Press.

105 Huang, P.M., Li, Y., and Sumner, M.E. *Handbook of Soil Sciences: Properties and Processes*, 2e. CRC Press.

106 Morales, V.L.(2011). Facilitated transport of groundwater contaminants in the vadose zone: colloids and preferential flow paths. PhD thesis. Cornell University.

107 Samotoin, N.D. and Bortnikov, N.S. (2011). Formation of kaolinite nano-and microcrystals by weathering of phyllosilicates. *Geology of Ore Deposits* 53: 340.

108 Kloster, N. and Avena, M. (2014). Interaction of humic acids with soil minerals: adsorption and surface aggregation induced by Ca^{2+}. *Environment and Chemistry* 12 (6): 731–738.

109 Kobo, K. and Tatsukawa, R. (1959). On the colored material of fulvic acid. *Zeitschriftfür Pflanzenernährung und Bodenkunde* 84: 137–147.

110 Pettit, R.E.(2013).Organicmatter, humus, humate, humic acid, fulvic acid and humin: Their Importance in Soil Fertility and Plant Health. https://humates.com/wp-content/uploads/2020/04/ORGANICMATTERPettit.pdf. Regelink, I.C. PhD thesis (accessed 4 December, 2023).

111 Rai, M., Ribeiro, C., Mattoso, L., and Duran, N. (2015). *Nanotechnologies in Food and Agriculture*. Springer https://doi.org/10.1007/978-3-319-14024-7.

112 Nair, R., Varghese, S.H., Nair, B.G. et al. (2010). Nanoparticulate material delivery to plants. *Plant Science* 3: 154–163.

113 Centola, M., Poppleton, E., Ray, S. et al. (2023). A rhythmically pulsing leaf-spring DNA-origami nanoengine that drives a passive follower. *Nature Nanotechnology* 1–11.

114 Zhang, S., Liu, R., Streb, C., and Zhang, G. (2023). Design and synthesis of novel polyoxometalate-based binary and ternary nanohybrids for energy conversion and storage. *Polyoxometalates* 2 (3): 9140037.

115 Wang, X., Guo, D., Zhang, J. et al. (2023). Insights into the electron transfer mechanisms of permanganate activation by carbon nanotube membrane for enhanced micropollutants degradation. *Frontiers of Environmental Science & Engineering* 17 (9): 106.

116 Khang, M., Lee, J.H., Lee, T. et al. (2023). Intrathecal delivery of nanoparticle PARP inhibitor to the cerebrospinal fluid for the treatment of metastatic medulloblastoma. *Science Translational Medicine* 15 (720): eadi1617.

117 Shi, X., Pumm, A.K., Maffeo, C. et al. (2023). A DNA turbine powered by a transmembrane potential across a nanopore. *Nature Nanotechnology* 19: 338.

118 Wang, M., Li, Y., Yao, S. et al. (2023). Conversion mechanism of $NiCo_2Se_4$ nanotube sphere anodes for potassium-ion batteries. *Energy Materials and Devices* 1 (1): 9370001.

Index

Carbon-based Nanomaterials for Green Applications, First Edition. Edited by Upendra Kumar, Piyush Kumar Sonkar, and Suman Lata Tripathi.

Published 2025 by John Wiley & Sons, Inc.

d

e

h

i

l

m

Printed and bound by CPI Group (UK) Ltd, Croydon, CR0 4YY

24/08/2026

14947829-0008